From Radiobiology to Radiation Oncology

Farzad Taghizadeh-Hesary
Sumel Ashique • Neeraj Mishra • Babak Behnam
Editors

From Radiobiology to Radiation Oncology

Springer

Editors
Farzad Taghizadeh-Hesary
ENT and Head and Neck Research Center and Department
The Five Senses Health Institute
School of Medicine, Iran University of Medical Sciences
Tehran, Iran

Neeraj Mishra
Amity Institute of Pharmacy
Amity University Madhya Pradesh
Gwalior, Madhya Pradesh, India

Sumel Ashique
Department of Pharmaceutical Technology
Bharat Technology
Uluberia, West Bengal, India

Babak Behnam
Department of Biology
College of Arts and Sciences
American University
Washington, DC, USA

ISBN 978-3-031-95723-9 ISBN 978-3-031-95724-6 (eBook)
https://doi.org/10.1007/978-3-031-95724-6

This Springer imprint is published by the registered company Springer Nature Switzerland AG
The registered company address is: Gewerbestrasse 11, 6330 Cham, Switzerland

Preface

Radiation oncology stands at the crossroads of innovation and tradition, where decades of foundational science meet transformative clinical applications. This book, *From Radiobiology to Radiation Oncology*, aims to bridge the gap between the underlying biological principles of radiation and their translation into modern therapeutic strategies. Beginning with the hallmarks of cancer and the foundational "Seven R's" of radiobiology, the text progresses through key determinants of radiosensitivity, and mechanisms of radiation-induced cell fate.

The chapters in this volume reflect the evolving landscape of radiotherapy, encompassing both conventional and cutting-edge approaches. Readers will find discussions on novel methods to enhance radiosensitivity, advances in site-specific treatments, and innovative techniques aimed at minimizing toxicity. The integration of emerging themes such as FLASH radiotherapy, artificial intelligence, radiogenomics, and the influence of the microbiome highlights the multidisciplinary nature of contemporary radiation research and its promise for personalized care.

Designed for radiation oncologists, researchers, and students alike, this book offers a rich synthesis of radiobiological concepts and clinical advances, fostering a deeper understanding of how science continues to shape and elevate the art of radiation therapy.

Tehran, Iran — Farzad Taghizadeh-Hesary
Uluberia, West Bengal, India — Sumel Ashique
Gwalior, Madhya Pradesh, India — Neeraj Mishra
Washington, DC, USA — Babak Behnam

Contents

About the Editors

Farzad Taghizadeh-Hesary, MD is an assistant professor of radiation oncology at the Iran University of Medical Sciences. His research and clinical expertise lie at the intersection of radiation oncology, radiobiology, and cancer metabolism. Dr. Taghizadeh-Hesary's work focuses on integrating the tumor microenvironment and mitochondrial metabolism into cancer research and personalized radiotherapy. He has published over 90 peer-reviewed articles in high-impact journals and serves on the editorial boards of leading oncology journals. Dr. Taghizadeh-Hesary has received numerous awards, including the prestigious Razi Medical Science Research Prize. With a deep understanding of the complex interplay between radiation, tumor biology, and metabolic pathways, Dr. Taghizadeh-Hesary's work aims to develop innovative strategies for improving cancer treatment outcomes and enhancing the patients' quality of life.

Sumel Ashique is an assistant professor at the Department of Pharmaceutical Technology, Bharat Technology, West Bengal 711316, India. He has 5 years of teaching experience and has published 80+ papers in internationally and nationally accredited journals, including research, reviews, column articles, and an editorial in Scopus and Web of Science (WOS). He has knowledge in drug delivery, nanotechnology, and targeted treatment strategy. He holds five patents from IP and Australia and has authored 20 book chapters published by Bentham, Springer, Elsevier, and Taylor & Francis. Currently, he is editing 14 books under CRC Press, Elsevier, Taylor and Francis, Wiley, Nova Science, and Springer. Also, he is now serving as a guest editor in the JoVE journal (IF 1.2, Scopus, and WOS), Current Probiotics (Bentham Science), and Current Indian Science Journal. He is a potential reviewer in Elsevier, Springer, Taylor & Francis, Dove Press, Bentham, and Wiley journals due to his expertise in the field of targeted drug delivery, nanotechnology, gut-microbiome, and postbiotics. Currently, he is pursuing his PhD degree from Lovely Professional University, Punjab, India (NAAC A++) in the pharmaceutical science department.

Neeraj Mishra is working as Professor and Head, Department of Pharmaceutics in Amity Institute of Pharmacy, Gwalior since July 2019. He has around twenty years of teaching and research experience. He has more than 140 publications of International and National repute in recent concepts of novel drug delivery systems, localized drug delivery, targeted and controlled drug delivery of nanocarriers/ microparticles for the treatment of breast, colon cancer, and neurodegenerative disorders. Dr. Mishra has supervised 30 M Pharm and 2 PhD students. Dr. Mishra more than 4200 citations with h- index 36. The total cumulative Impact Factor of his published papers is more than 500 (as per SCOPUS). He has also written 25 Books and 45 book chapters in Elsevier, Bentham, Springer and Wiley publications, etc. He has also been granted 3 international patents and 2 Indian patents. Dr. Mishra is recipient of "Distinguished Professor Award" 2019 from DST-NSTMIS, SPAICS, Indore, and M.P. in September 2019. Dr. Mishra has also received the "Outstanding Scientist Award" 2020, 6th International Scientist Awards on Engineering, Science, and Medicine, 20–21 June 2020, Chennai, India. Dr. Mishra has received the "Distinguished Professor Award" 2021 from Indian Pharmaceutical Association, MP State Branch, Indore. Dr Mishra has received the Best Academician Award in the Indian Pharmacy Graduate Association, M.P. State International Conference held on 06th May 2023 at DAVV Auditorium, Indore, India. Recently, honored to be in World Top 2% Scientists by prestigious Stanford University, CA, USA and Elsevier in 2025. Honored to be in the Stanford University Scientists Rankings conducted by AD Scientific Index - World Scientists rankings in 2024, 2022 and 2025.

Babak Behnam, MD, PhD is an academic researcher in the field of medical molecular biology and translational medicine at the American University and professor at the Floret Genomic Center. He earned his M.D. from Iran University of Medical Sciences (IUMS) in 1997 before specializing in genomic medicine, completing an M.Sc. in molecular medicine at the University of Sheffield and a Ph.D. in human genetics at the University College London (UCL) in 2005. After his Ph.D., Dr. Behnam moved to the USA, where he was awarded an NIH-based Ruth L. Kirschstein NRSA fellowship and completed four years of postdoctoral training at the University of Michigan and the University of Central Florida. His research focuses on molecular metastatic pathways, cancer biology, and mitochondrial genetics. At IUMS, he served as associate professor of human genetics, established and directed the medical genetics laboratory at OB/GYN and Children's University Hospitals for seven years, and taught and mentored numerous students. Upon returning to the USA, Dr. Behnam joined the NIH Undiagnosed Disease Program while completing an ABMGG-accredited clinical biochemical genetics fellowship at NHGRI, NIH. He later transitioned to Regulatory Affairs at NSF International, overseeing FDA submissions and compliance. As Co-Founder and Chief Scientific Officer (CSO) of Avicenna Biotech Research, he leads IP strategy and serves as a Medical Science Liaison (MSL). A prolific academic, Dr. Behnam has authored over 80 peer-reviewed publications, contributed to several chapters of five books, and holds editorial roles at MDPI Cancers and Frontiers. He also holds multiple patents in addiction therapies and cancer therapeutics.

Chapter 1
Overview on Cancer Biology and Cancer Hallmarks

Sanzia Mehjabin and Md. Khokon Miah Akanda

Abstract Cancer is a complicated and multifaceted disease that is caused by changes in genes and epigenetics. These problems are what cause cells to multiply and stay alive without control. This chapter gives a complete overview of cancer biology by focusing on the most important molecular and cellular processes that cause tumors to grow from the inside out. Among the key characteristics of cancer that we investigate are persistent proliferative signals, evasion of growth suppressors, resistance to cell death, the capacity to create replicative immortality, the activation of invasion and metastases, and the generation of angiogenesis. All of these traits are necessary for the development of cancer. Additionally, discuss new markers and enabling traits, including genomic instability, inflammation, unregulated cellular energetics, and immunological evasion. Understanding these markers helps explain cancer and design targeted medicines. It discusses how cancer cells and their microenvironment interact and how the tumor microenvironment affects cancer growth and medication resistance. This chapter closes the distance between basic cancer research and clinical practice by including molecular biology, genetics, and clinical oncology. This is crucial to develop cancer treatment from radiology to radiation oncology, which demands a strategy involving several disciplines using radiology as the basis.

Keywords Cancer · Cellular pathways · Cell death · Evasion · Growth · Radiation

S. Mehjabin
Department of Pharmacy, Varendra University, Rajshahi, Bangladesh

M. K. M. Akanda (✉)
Department of Pharmacy, University of Asia Pacific, Dhaka, Bangladesh
https://scholar.google.com/citations?user=q7lgBRsAAAAJ&hl=en

F. Taghizadeh-Hesary et al. (eds.), *From Radiobiology to Radiation Oncology*,
https://doi.org/10.1007/978-3-031-95724-6_1

Abbreviation

ALDH	Aldehyde dehydrogenase
ALT	Alternative lengthening of telomeres
APC	Argon plasma coagulation
ATP	Adenosine triphosphate
BCR-ABL	Breakpoint cluster region (BCR) and Abelson (ABL)
BRCA	Breast cancer gene
CAR-T	Chimeric antigen receptor T-cell therapy
CDKN2A	Cyclin-dependent kinase inhibitor 2A
CML	Chronic myeloid leukemia
cMYC	Cellular myelocytomatosis oncogene
CpG	Clinical Practice Guidelines
CSC	Cancer stem cells
DNA	Deoxyribonucleic acid
ECM	Extracellular matrix
EGFR	Estimated glomerular filtration rate
EMT	Epithelial-mesenchymal transition
FGF	Fibroblast growth factor
HER2	Human epidermal growth factor receptor 2
HER2-positive	Human epidermal growth factor receptor 2-positive
HIF	Hypoxia-inducible factor
HRAS	Harvey Rat sarcoma virus
KRAS	Kirsten rat sarcoma viral oncogene homolog
MAPK/ERK	Mitogen-activated protein kinase/extracellular signal-regulated kinase
MDSC	Myeloid-derived suppressor cells
MDSC	Myeloid-derived suppressor cells
MHC	Major histocompatibility complex
MMP	Matrix metalloproteinase
mRNA	Messenger RNA
NGS	Next Generation Sequencing
NK	Natural killer
NRAS	Neuroblastoma ras viral oncogene homolog
PD-1	Programmed death-1
PET	Proton-emission tomography
PI3K/AKT/mTOR	Phosphatidylinositol-3-kinase, protein kinase B, and mammalian target of rapamycin signaling pathway
pRb	Retinoblastoma protein
PTEN	Phosphatase and tensin homolog deleted on chromosome 10
RAS/RAF/MEK/ERK	Ras/Raf/mitogen-activated protein kinase (MEK)/extracellular signal-regulated kinase (ERK) pathway
RNA	Ribonucleic acid

TAM	Tumor-associated macrophages
TCA	Tricyclic antidepressant
TME	Tumor microenvironment
TMS	Transmembrane space
TSG	Tumor suppressor gene
TSGs	Tumor suppressor genes
UV	Ultraviolet
VEGF	Vascular endothelial growth factor
WGS	Whole-Genome Sequencing

1.1 Introduction

Cancer, a challenging and formidable disease, is characterized by the uncontrolled growth and spread of abnormal cells and, despite significant advances in understanding and treatment, yet remains a leading cause of morbidity and mortality worldwide [101]. Each type of cancer is unique, involving different cell types and exhibiting distinct behaviors. However, common threads run through these variations, defined by the biological hallmarks that unify the diverse manifestations of cancer [70]. Cancer fundamentally arises from unregulated cellular proliferation. Cellular proliferation is typically regulated by a network of signals that maintains equilibrium among cell growth, division, and apoptosis. In cancer, these regulatory processes are impaired, resulting in unchecked proliferation. Typically, genetic anomalies that either activate oncogenes or deactivate tumor suppressor genes, which regulate cell division and promote apoptosis, precipitate this disturbance. These genetic modifications may be inherited, triggered by environmental influences, or arise spontaneously [8]. In conclusion, the chapter comprehensively intends to discuss an overview of cancer biology and its defining hallmarks, setting the stage for further exploration of its complexities and the ongoing efforts to combat it through various medical disciplines, including radiology and radiation oncology.

1.1.1 Definition and Scope of Cancer Biology

Cancer biology is the study of the molecular, genetic, and cellular mechanisms that underlie the development and progression of cancer. It encompasses a broad range of topics, including the genetic mutations that drive cancer, the signaling pathways that regulate cell proliferation and death, and the interactions between cancer cells and their microenvironment [3]. Cancer biology examines metastasis and treatment resistance. Cancer biology fundamentally investigates the transformation of normal cells into malignant ones, the growth and dissemination of tumors, and their evasion of the body's innate defenses. By elucidating these processes, researchers aim to identify new targets for therapy and develop more effective strategies for prevention, diagnosis, and treatment [74]. Over 100 types of cancer affect humans, with tobacco use accounting for 22% of cancer deaths. Other factors include obesity, poor diet, lack of physical activity, and excessive alcohol consumption [48]. Environmental factors, including viruses, radiation, and pollution, cause 16–18% of cancers worldwide. Helicobacter pylori, hepatitis B and C, human papillomavirus, Epstein-Barr virus, HSV-1, Kaposi's sarcoma-associated herpesvirus, and Merkel cell polyomavirus are infectious. HIV increases cancer risk from other infections. The vaccine can nearly reduce cancer risk from these viruses [65]. In 2019, cancer cases increased by 23.6 million people and deaths by ten million, representing a 26% and 21% increase over the previous decade [90]. Lung, prostate, colorectal, and stomach cancers affect men, while breast, colorectal, lung, and cervical cancers affect women. Skin cancer, including melanoma, accounts for 40% of new cases [15]. Child acute lymphoblastic leukemia and brain tumors are prevalent, with 165,000 diagnosed in 2012. Risk increases with age, with developed countries having higher rates. Global economic costs of cancer are estimated at $1.62 trillion annually [49].

1.1.2 Historical Perspectives

Cancer's history dates back thousands of years, with the earliest descriptions found in ancient Egyptian manuscripts like the Edwin Smith Papyrus around 1600 BCE, highlighting its recognition as a distinct disease [44]. In ancient Greece, Hippocrates (460–370 BCE) proposed the humoral theory, which posited cancer as a result of excess black bile, and it was a dominant medical concept for centuries [71]. In the eighteenth century, Giovanni Battista Morgagni, known as the "Father of modern pathology," revolutionized cancer understanding by correlating clinical symptoms with post-mortem findings. His work, De Sedibus et Causis Morborum per Anatomen Indagatis (1761), demonstrated cancer as an organ-specific disease caused by abnormal tissue growth. His detailed autopsies laid the foundation for evidence-based medicine and shifted focus from humoral theories to anatomical and pathological analyses, laying the groundwork for modern oncology [36]. In the nineteenth century, Rudolf Virchow, the father of cellular pathology, revolutionized cancer research by proving that diseases, including cancer, stem from cellular abnormalities. In his work, Die Cellularpathologie, Virchow debunked earlier humoral theories and demonstrated that cancer develops from uncontrolled cell proliferation. His emphasis on cellular changes advanced cancer's understanding and laid the foundation for modern histopathology and cellular biology, which remain crucial in cancer diagnosis and treatment [89]. In the early twentieth century, the somatic mutation theory suggested cancer is caused by genetic mutations in somatic cells, supported by the discovery of carcinogens like tobacco smoke [92]. The 1970s and 1980s saw the discovery of oncogenes and tumor suppressor genes, which drive cell proliferation or fail to regulate cell division and repair [19]. In 2000, Douglas

Hanahan and Robert Weinberg's paper, "Hallmarks of Cancer", outlines six cell physiology changes that drive tumor growth and metastasis. The framework was updated in 2011 to include deregulating cellular energetics, avoiding immune destruction, and enabling characteristics like genome instability and mutation, as well as tumor-promoting inflammation [55]. Targeting tumor growth and progression molecules and pathways using tyrosine kinase inhibitors and monoclonal antibodies has transformed cancer treatment. These therapies, like imatinib for chronic myeloid leukemia and trastuzumab for HER2-positive breast cancer, have significantly improved treatment outcomes and ushered in an era of precision oncology [57]. Immunotherapy boosts the immune system's cell-detection and killing abilities to fight cancer. CAR-T cell therapy, cancer vaccines, cytokine treatments, and immune checkpoint inhibitors like pembrolizumab are used. Immunotherapy has proven successful in treating melanoma and lung cancer, offering durable responses and transforming cancer care [53]. Genomic medicine uses genetic information to personalize disease prevention, diagnosis, and treatment. It identifies specific mutations in cancer, enabling targeted therapies and improved patient outcomes. Integrating next-generation sequencing allows precise tumor profiling, advancing precision oncology and fostering healthcare innovations [23].

1.2 Fundamentals of Cancer Biology

Cancer biology is a multidisciplinary field that investigates the development, progression, and treatment of cancer. It focuses on the uncontrolled proliferation of cells due to genetic and epigenetic alterations. Key processes include mutations in proto-oncogenes and tumor suppressor genes, which activate oncogenes and inactivate tumor-suppressing pathways. Understanding cancer's complexity is crucial, and molecular biology advances have revealed key cancer signaling pathways like "PI3K/AKT/mTO" and "RAS/RAF/MEK/ERK". These insights have led to the development of targeted therapies and immunotherapies, revolutionizing cancer treatment strategies [45].

1.2.1 Cellular and Molecular Basis of Cancer

Disruption of normal cell regulatory systems causes cancerous cell proliferation and invasion. Cancer cells bypass these mechanisms due to genetic mutations and alterations in signaling pathways, leading to tumor formation [87]. Cancer is a cellular transformation where normal cells acquire malignant properties, often triggered by mutations in proto-oncogenes, tumor suppressor genes, and DNA repair mechanisms. Mutations in these genes promote uncontrolled cell division and inhibit cell growth and proliferation [50]. Our bodies consist of trillions of cells that follow genetic instructions to grow, work, divide, and die. However, when DNA is damaged or mutated, these instructions can cause cells to grow out of control, leading to cancer. Normal cells replicate, but cancer cells have gene alterations. Mutations can be inherited, develop over time, or be produced by cigarette smoke, alcohol, or UV radiation. Cancer cells act differently than normal cells, growing and dividing out of control instead of dying and staying immature. Despite various types of cancer, they all start due to abnormal and out-of-control cell growth and can start in any cell in the body [99]. Dysregulation of the "PI3K/AKT/mTOR, RAS/RAF/MEK/ERK, and p53" pathways, which involve growth factors, receptors, intracellular signaling molecules, and transcription factors, promotes cancer. Stromal cells, immune cells, and extracellular matrix components in the tumor microenvironment also help cancer grow and metastasize [78].

1.2.2 Genetic and Epigenetic Changes in Cancer

Oncogenes, tumor suppressor genes, and DNA repair genes can be inherited or acquired, causing DNA sequence mutations and cancer. Common mutations include KRAS, MYC, TP53, RB1, BRCA1, and BRCA2, respectively [1]. Genetic and epigenetic alterations in cancer-related genes and molecules reactivate telomerase, inactivate p53, overexpress cyclin E, and degrade p27 KIP1, causing human gastric carcinomas. Well-differentiated gastric cancer is predominantly related to CpG hypermethylation-induced hMLH1 mismatch repair gene inactivation, microsatellite instability, c-erb B2 oncogene amplification, APC tumor suppressor gene inactivation, and K-ras mutations. Poorly differentiated or scirrhous stomach carcinomas require mutation, CpG hypermethylation, and K-sam and c-met oncogene amplification to reduce or lose E-cadherin and catenin. Gastric cancer morphogenesis involves c-met-expressing cancer cells and stromal hepatocyte growth factor. Epigenetic alterations affect gene expression but do not modify DNA sequence. They include DNA methylation, histone modifications, and non-coding RNA expression. Epigenetic alterations can activate or silence tumor suppressor genes. Hypermethylation of the tumor suppressor gene CDKN2A's promoter region can silence it and cause uncontrolled cell proliferation [2, 103].

1.2.3 Mechanisms of Carcinogenesis

Normal cells become cancer cells through initiation, promotion, and advancement. Genetic mutations in normal cells begin with initiation. These mutations can be produced by chemical carcinogens, radiation, and viruses. The started cell has permanent genetic alterations that predispose it to future transformation [19]. Multistage cancer development requires tumor promotion to enable preneoplastic and premalignant cell clonal growth and genetic instability. Continuous cellular signal transduction disturbances overstimulate metabolic pathways that produce cell growth, inflammation, and genotoxic byproducts. This stage promotes cell growth in started cells. Hormones, persistent inflammation, and chemicals promote clonal proliferation of started cells. If the stimulating factors are eliminated, this stage of selective development of altered cells can be reversed [62]. Enlargement of the tumor, new metastasis to lymph nodes or other organ sites, or disease recurrence were signs of progression. Tumor cells acquire malignant genetic and epigenetic alterations in this final stage. Genetic instability, apoptosis resistance, angiogenesis, and invasion and metastasis are among these modifications. This stage is more aggressive and difficult to treat due to tumor cell heterogeneity. Throughout these stages, cancer cells and their surroundings must interact. Tumor growth and therapeutic resistance are supported by the tumor microenvironment, which includes immune cells, fibroblasts, and extracellular matrix [18].

1.3 The Hallmarks of Cancer

Malignant development and survival are enabled by cancer hallmarks. These include persistent proliferative signals and growth suppressor evasion, allowing unrestrained cell division. Repeatable immortality and cell death resistance help cancer cells survive and develop indefinitely. Angiogenesis, invasion, and metastasis promote tumor growth and spread while escaping immune destruction and changing energy metabolism to maintain malignancy and hallmarks of cancer, as summarized in Fig. 1.1 [59].

1.3.1 Sustaining Proliferative Signaling

Cancer cells may survive persistent proliferation, unlike normal cells. In normal conditions, growth signals carefully govern cell proliferation for growth, division, and survival. However, cancer cells can make their own growth signals (autocrine stimulation), adjust their sensitivity, or change their surroundings to promote continuous division. By repeatedly activating pathways like the mitogen-activated protein kinase (MAPK)/ERK pathway, oncogenes like the Ras family of genes promote uncontrollable cellular proliferation. Additionally, cancer cells can impair negative feedback mechanisms that limit excessive growth signal propagation [5, 30].

1.3.2 Evading Growth Suppressors

Normal cells' proliferative signals are balanced by tumor suppressor genes (TSGs). The most important TSGs regulate cell growth and division, including pRb and many kinase-inhibitor proteins. An intracellular monitoring mechanism, the p53

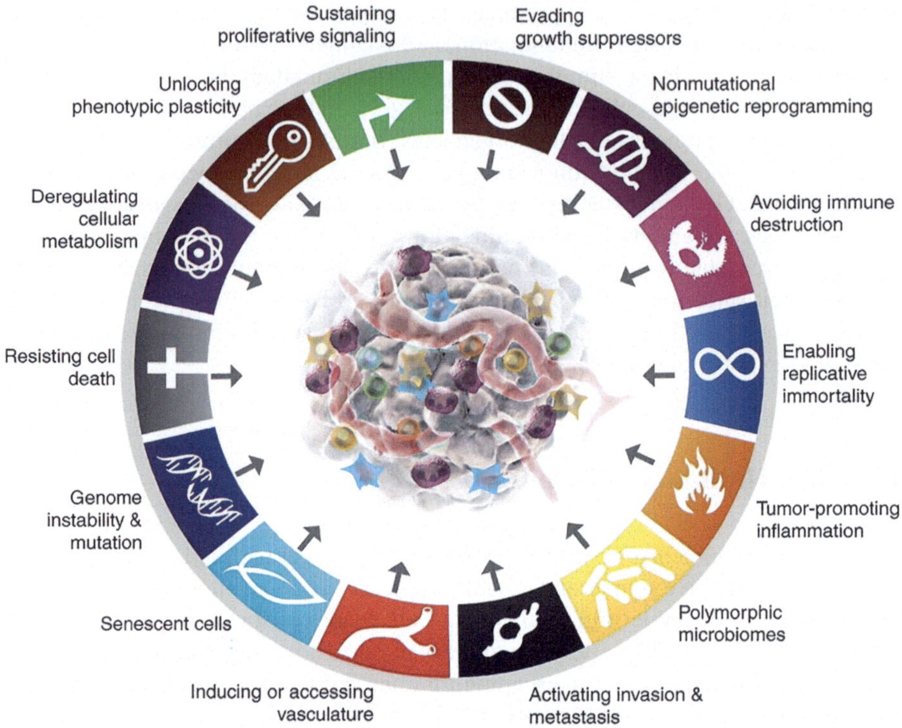

Fig. 1.1 Hallmarks of Cancer. ("Reprinted from Lythgoe et al. [59], with permission via the Creative Commons Attribution License, https://creativecommons.org/licenses/by/4.0/")

protein, ensures cells grow and divide at the right time. About 40% of human malignancies have p53 gene mutations, and many others have genetic lesions or epigenetic changes that impair p53 signaling. Genetic profiling shows that most human cancers have TSG abnormalities. Evacuating growth suppressors is a defining capability to ensure cancer cell proliferation and growth is not interrupted by cell-division braking mechanisms [72].

1.3.3 Resisting Cell Death

The intrinsic mechanisms of cell growth can produce programmed cell death. The most common kind of planned cell death is apoptosis, caused by cell abnormalities. Chromosomes and organelles are degraded, and phagocytes or nearby cells devour them. Certain cellular regulators can activate necroptosis, the passive breakdown of dead cells. Oxygen and energy deficit, viral infection, and inflammation cause necroptosis. Cell death can be caused by autophagy, an organelle recycling pathway. However, excessive food shortage or acute stress can hyperactivate autophagic recycling, causing "autophagy-associated" cell death. Cancer cells must bypass or inhibit these systems to continue proliferating and evolving to higher malignancy [22, 66].

1.3.4 Enabling Replicative Immortality

Telomeres shorten with each cell division, limiting normal somatic cells' replicative capacity, known as the Hayflick limit. Maintaining telomere length allows cancer cells to replicate immortally. This is done by upregulating telomerase, an enzyme that adds telomeric repeats to chromosomal ends to prevent shortening and allow ongoing division. Some cancer cells retain telomere length using alternative lengthening of telomeres (ALT). Cancer cells can grow and advance endlessly due to this ability [10].

1.3.5 Inducing Angiogenesis

Tumor cells need oxygen, glucose, and nutrition to survive and grow. The tumor-associated vasculature does this. Normal tissue ischemia kills cells and damages organs. Cancer cells stop growing when they can't get blood-borne nutrients. Many tumor forms benefit from angiogenesis, the growth of new blood vessels. Cancer cells without oxygen and glucose perish by necrosis, apoptosis, or autophagy. Tortuous, dilated, and leaky blood arteries characterize tumor-associated neovasculature. Some cancers use their invasion and metastatic abilities to coexist with normal tissue vasculature. Most cancers grow through persistent angiogenesis, but some tolerate quasi-hypoxic conditions [32].

1.3.6 Activating Invasion and Metastasis

Five hallmarks of cancer cell chronic proliferation include invasion and migration. Cancer cells use tissue, blood, and lymphatic arteries as pipelines to spread to local and distant anatomical areas through invasive growth programs. These arteries produce lymph node metastases that enter the circulation. Cell-intrinsic programming and tissue microenvironment accessory cells regulate these capabilities in a complicated manner. Normal organogenesis cell migrations and tissue invasions are regulated by the epithelial-mesenchymal transition (EMT), a primary cancer cell-intrinsic process. The hypoxia response system activates hypoxia-inducible factors (HIFs), HIF1α, and HIF2α, affecting gene expression. These hallmark skills can evolve throughout tumor growth and progression due to mutational or epigenetic evolution or global regulatory network activation [28].

1.4 Emerging Hallmarks and Enabling Characteristics

New hallmarks and enabling properties of cancer have expanded our tumor biology knowledge. These include cellular energetic reprogramming, immune evasion, genome instability, mutation, and tumor-promoting inflammation. These concepts highlight the complexity of cancer and emphasize the need for targeted therapeutic intervention [43].

1.4.1 Deregulating Cellular Energetics

Otto Warburg proposed over 90 years ago that cancer cells use glucose to proliferate. This "aerobic glycolysis" provides macromolecules needed for cell growth and division, and cancer cells' metabolism resembles actively dividing normal cells. Cancer cells use oxidative phosphorylation and glycolysis at different rates. PET can indirectly measure aerobic glycolysis. Glutamate is becoming a major blood energy source and lipid and amino acid precursor. Lactate, once harmful waste, promotes tumors in many ways. KRAS, cMYC, and p53 loss can reprogram cancer cell energy metabolism, making cellular energetics and metabolism reprogramming an emerging hallmark [98].

1.4.2 Avoiding Immune Destruction

Cancer cells evade the immune system, making treatment difficult. The immune system generally eliminates abnormal cells. Cancer cells avoid immune attack. They express immunological checkpoint proteins such as PD-L1, which bind to PD-1 receptors to inhibit T-cell activation. Cancer cells can recruit immune-lowering Tregs and MDSCs to create an immunosuppressive tumor microenvironment. Cancer cells downregulate tumor antigens or MHC molecules, making them less visible to immune cells. Immunotherapy, including immune checkpoint inhibitors and CAR-T cells, aims to bypass these evasion mechanisms and combat cancer [33].

1.4.3 Genome Instability and Mutation

Genome instability drives cancer origins and development. It refers to a heightened genome mutation, chromosomal rearrangement, and aneuploidy risk. Damage to mismatch repair, homologous recombination, and non-homologous end joining causes instability. Genetic changes give cancer cells growth advantages, apoptosis resistance, and other malignant features. Mutations can influence oncogenes, tumor suppressors, and cell cycle genes. Enhancing DNA repair pathways and targeting genomic instability may slow cancer growth and enhance therapy outcomes [75].

1.4.4 Tumor-Promoting Inflammation

Cancers establish hallmark capacities through tumor-promoting immune cell infiltration (IIC). IIC promotes proliferation, survival, angiogenesis, local invasion, and blood-borne metastasis. Some IIC can actively block immune system-produced cytotoxic T-lymphocytes that kill cancer cells. Different tumor types and multistep carcinogenesis phases recruit IIC in different ways. In early lesions, IIC can assist cancer cells to grow, survive, evade antigrowth regulators, and initiate angiogenesis. Later on, IIC at tumor margins can aid invasiveness. Chronic inflammation and wound healing induce angiogenesis and stimulate cell survival, proliferation, and migration or invasion [45, 79].

1.5 Tumor Microenvironment

A tumor's microenvironment is its internal ecosystem. Immune cells, extracellular matrix, blood arteries, and fibroblasts are included. A tumor and its microenvironment constantly affect each other, either positively or adversely [24].

1.5.1 Interaction Between Cancer Cells and Their Microenvironment

Cancer cells and their environment interact bidirectionally, affecting growth and metastasis. These interactions involve cell-to-cell contact and soluble cytokines, chemokines, and growth factors. CAFs strongly alter the tumor microenvironment (TME) and contribute to tumor formation by secreting growth factors like TGF-β and VEGF, changing the ECM, and encouraging angiogenesis. They can also increase cancer cell invasiveness by producing matrix metalloproteinases (MMPs), enzymes that tear down extracellular matrix components and help malignancies infiltrate [104]. Tregs, TAMs, and MDSCs in the TME reduce anti-tumor immune responses, enabling tumor-immune evasion. Despite being suppressed in the TME, cytotoxic T-lymphocytes (CTLs) and natural killer (NK) cells can kill cancer cells [76]. Cancer and stromal cells release cytokines and chemokines that can stimulate or inhibit tumor growth. IL-6 and IL-8 increase cancer cell proliferation and survival, promoting tumor growth. Conversely, IFN-γ exhibits anti-tumor properties [67, 82]. VEGF promotes angiogenesis, which gives the tumor the blood it needs to develop and disseminate metastases. In its early stages, TGF-β suppresses tumors. Due to its propensity to induce EMT, it promotes tumor growth later on [20]. Cancer cells' metabolism is disrupted, and the Warburg effect's enhanced glycolysis is common. Metabolic reprogramming can affect the TME by changing nutrition and pH levels, which affects stromal and immune cell activity [56].

1.5.2 Role of the Extracellular Matrix

Transmembrane space (TMS) requires extracellular matrix (ECM). Biochemical and mechanical components support and govern cellular processes. Collagen, fibronectin, laminin, and hyaluronic acid make up the ECM, a network of proteins, glycoproteins, and proteoglycans [93]. MMPs break down ECM, making tumor invasion and spread easier. Cancer and stromal cells can upregulate MMP expression, which can remodel the ECM and promote cancer [14]. Density and stiffness of the ECM can alter cancer cell activity. A rigid extracellular matrix increases tumor aggressiveness and poor prognosis. Cancer cells get mechanical cues from the ECM via cell surface receptors such as integrins, which affect their motility, survival, and proliferation [26]. Integrins are transmembrane receptors that bind cells to the extracellular matrix. They transmit signals from the ECM to the cell interior to regulate migration, survival, and differentiation. Cancer often has aberrant integrin signaling, which stimulates tissue development [51]. ECM can sequester growth factors, creating a reservoir that alters their availability and activity. For instance, heparan sulfate proteoglycans can bind to growth hormones like FGF and VEGF, influencing their signaling pathways. When tumor growth outpaces blood vessel growth, the TME can become hypoxic. When maintained under low oxygen, HIFs can induce angiogenesis, metabolism, and extracellular matrix remodeling genes. Hypoxia can also increase ECM stiffness and cancer cell invasion [61].

1.6 Cancer Stem Cells

1.6.1 Concept and Characteristics

CSCs, or cancer stem cells, can self-renew and specialize into many tumor cell types. CSCs are based on the hierarchical cancer concept. According to this hypothesis, tumors are constructed like normal tissues, with a tiny population of stem cells sustaining the malignancy [35]. Continuous CSCs can divide several times without getting differentiated. These traits are necessary for long-term tumor development and survival [77]. Cancer stem cells can grow into tumor-forming cell lineages like normal stem cells. This allows cancer cells in the same tumor to have such diverse phenotypes [85]. CSCs often resist chemotherapy and radiation. This resistance comes from DNA repair, a quiescent cell cycle, and drug efflux pumps. Traditional chemotherapy and radiation often fail to treat CSCs. Drug efflux pump development, DNA repair efficiency, and quiescent cell cycle phase contribute to this resistance [73]. Surface indicators can be used to identify and isolate CSCs; however, they vary by cancer type. Common markers include CD133, CD44, and ALDH1 [38].

1.6.2 Role in Tumor Initiation and Progression

CSCs play a crucial role in tumor initiation and progression, influencing various stages of cancer development. CSCs originate from normal stem cells or progenitor cells that acquire oncogenic mutations, enhancing their self-renewal abilities and tumor development. The formation of CSCs can also be traced back to differentiated cancer cells that undergo dedifferentiation and acquire stem-like characteristics [37]. CSCs are key to the expansion of primary tumors due to their self-renewal capacity, which ensures a continuous supply of tumorigenic cells, contributing to the tumor's growth through its expansion [7]. CSCs play a crucial role in the metastatic spread of cancer, causing detachment from the main tumor, invasion of surrounding tissues, and entry into the bloodstream or lymphatic system. They can stimulate the creation of additional tumors once they reach a distant location. CSCs are often linked to the EMT process, which enhances their migratory and invasive capabilities [13].

1.7 Molecular Pathways in Cancer

Molecular processes in cancer change normal cells into malignant ones through intricate genetic and metabolic interactions. The PI3K/AKT/mTOR pathway, RAS/RAF/MEK/ERK system, and Wnt/β-catenin pathway are frequently dysregulated in cancer, affecting cell growth, survival, proliferation, differentiation, destiny, and migration. DNA repair and apoptosis depend on the p53 tumor suppressor pathway, which tumors inactivate. Mutations, amplifications, and deletions in these pathways cause uncontrolled cell proliferation, apoptosis resistance, and metastatic potential, promoting tumor growth. Targeted medicines and personalized anticancer treatment techniques need to understand these biological mechanisms [6, 12].

1.7.1 Key Signaling Pathways

The PI3K/AKT/mTOR pathway is essential for cell growth and survival. PI3K triggers AKT, which activates mTOR. PI3K mutations, PTEN loss, and AKT amplification cause uncontrolled cell proliferation and apoptosis resistance. Breast, ovarian, and endometrial malignancies often dysregulate this pathway [52]. RAS-RAF-MEK-ERK, or the MAPK pathway, regulates cell division, differentiation, and migration. Mutations in RAS (KRAS, NRAS, HRAS) and RAF (BRAF) genes activate this system constitutively, causing oncogenesis. This route is important in melanoma, colorectal, and thyroid cancers [47]. Extracellular cytokines regulate gene expression, promoting cell proliferation and survival, while mutations in JAK or STAT genes cause leukemia and lymphoma by abnormally activating JAK/STAT signaling [83].

1.7.2 Targeted Therapies and Their Mechanisms

Targeted therapies target specific molecules involved in cancer growth and progression, unlike traditional chemotherapy, which targets all rapidly dividing cells [57]. Signal transduction pathways require tyrosine kinases to transfer phosphate groups to protein tyrosine residues. Inhibiting these kinases disrupts cancer cell signaling. In chronic myeloid leukemia, imatinib targets the BCR-ABL fusion protein. Erlotinib and Gefitinib target epidermal growth factor receptor (EGFR) in NSCLC [9]. Monoclonal antibodies identify cancer cells for trastuzumab, which targets the HER2/neu receptor in HER2-positive breast cancer. Rituxan targets CD20 on non-Hodgkin lymphoma B-cells [91]. Pembrolizumab and nivolumab are immune checkpoint inhibitors that target T-cell PD-1 receptors, disrupting proteins that prevent the immune system from attacking cancer cells [58]. PARP inhibitors use synthetic lethality in BRCA-mutated malignancies. PARP is needed to repair DNA in cancer cells due to BRCA1 and BRCA2 mutations. PARP inhibition damages cancer cells' DNA and kills them. Olaparib and rucaparib treat BRCA-mutated ovarian and breast malignancies [21].

1.8 Cancer Metabolism

Cancer metabolism is the altered metabolic pathways cancer cells use to develop and survive in the tumor microenvironment. The Warburg effect occurs when cancer cells use aerobic glycolysis instead of oxidative phosphorylation. Cancer cells can synthesize nucleotides, amino acids, and lipids even with oxygen thanks to this metabolic change. Cancer cells increase glucose intake and lactate generation, which supports energy needs and creates an acidic milieu that promotes spread and immune evasion. Cancer metabolism is also linked to signaling pathways involving oncogenes like MYC and tumor suppressors like p53, which govern energy metabolism. Understanding these metabolic changes has led to new therapeutic treatments targeting cancer cell metabolism enzymes and pathways, which may improve cancer therapy outcomes [25].

1.8.1 Metabolic Reprogramming in Cancer Cells

Metabolism of Glutamine Along with glucose, glutamine is essential for cancer cells. The TCA cycle and anabolic activities are fueled by glutamine conversion to glutamate and α-ketoglutarate. Cancer cells commonly upregulate glutaminase, which converts glutamine to glutamate, making it a therapeutic target [3, 68]. Metabolizing Lipids: For fast-dividing cells, cancer cells synthesize more lipids de novo. In malignancies, fatty acid synthase (FASN) is overexpressed. Altered lipid metabolism also contributes to the production of signaling molecules and energy storage, further promoting cancer cell survival and proliferation [69]. Cancer cells modify amino acid metabolism for growth and survival. Upregulated serine and glycine metabolism provides one-carbon units for nucleotide and methyl group production. Other amino acid metabolism changes, such as proline and arginine, contribute to cancer cell metabolic flexibility [100].

1.8.2 Therapeutic Implications

Utilizing mitochondrial dysfunction, metformin, a mitochondrial targeting drug, specifically affects cancer cells with altered mitochondrial function by disrupting mitochondrial respiration and ATP generation. In cancer cells with redox imbalance, elesclomol can produce mitochondrial oxidative stress and apoptosis [54]. Drug combinations, such as chemotherapy and radiotherapy, can be more effective when used with metabolic inhibitors to sensitize cancer cells [80].

1.9 Cancer Genomics

Cancer genomics is a fast-evolving discipline that studies genetic and molecular alterations that cause cancer. Cancer genomics analyzes the full genome of cancer cells to find mutations, chromosomal abnormalities, and other genetic changes that cause cancer. Cancer genomics is an important research topic for targeted therapeutics, tailored treatments, and early diagnostics due to genomic technology breakthroughs [95].

1.9.1 Advances in Genomic Technologies

next-generation sequencing (NGS) is one of the most revolutionary genomics technologies, sequencing complete genomes at record speeds and cheaper costs. NGS can evaluate genomes, exomes, and transcriptomes, unlike Sanger sequencing, which only analyzed discrete gene areas. It detects point mutations, insertions, deletions, copy number changes, and structural variants in cancer cells [40]. Whole-genome sequencing (WGS) uncovers cancer cells' whole genome, including coding and non-coding sections. WGS can find novel mutations and uncommon genomic changes that focused sequencing may miss. It helps identify cancer-causing genetic rearrangements and structural alterations [11]. RNA-Seq is a valuable technique for investigating cancer gene expression profiles. It analyzes mRNA levels, splicing variants, and gene fusions to reveal cancer's molecular pathways. Researchers can find cancer-specific biomarkers and therapeutic targets by comparing normal and malignant transcriptomes [64]. Cancer genomics has been transformed by CRISPR-Cas9 genome-editing technology, which allows precise gene manipulation. Researchers can now insert specific mutations or fix genetic abnormalities into cancer cells to better understand genetic alterations and carcinogenesis [39]. A non-invasive alternative to tissue biopsy, liquid biopsy analyzes blood samples for cell-free DNA (cfDNA), circulating tumor DNA(ctDNA), and other molecular markers. Liquid biopsy can detect genetic changes in real time for early identification, treatment response tracking, and minimal residual illness assessment [81].

1.9.2 Genomic Alterations and Their Clinical Relevance

Tumors grow uncontrollably due to genomic changes in cancer cells. These changes can influence oncogenes, TSGs, DNA repair pathways, and chromatin regulators. Identifying these changes affects diagnosis, prognosis, and treatment [16]. Cancer cells often have chromosomal rearrangements, amplifications, and deletions. A translocation between chromosomes 9 and 22 causes the Philadelphia chromosome, which is indicative of chronic myelogenous leukemia. Chromosome anomalies can assist in identifying cancer and choosing treatment, such as CML tyrosine kinase inhibitors [86]. Tumor heterogeneity is when genetically heterogeneous subpopulations of cells in a tumor make up cancer. Mutations, epigenetic changes, and clonal evolution cause tumor heterogeneity. This diversity makes cancer cell treatment difficult since various subpopulations respond differently to medicines. Genomic technologies like single-cell RNA sequencing are mapping heterogeneity and identifying therapeutic targets [63, 88]. Targeted Therapies and Precision Medicine. The detection of genetic abnormalities has led to tailored medicines that inhibit cancer-related genes or proteins. EGFR inhibitors may help EGFR-mutated NSCLC patients, while trastuzumab, a monoclonal antibody targeting HER2, may help HER2-positive breast cancer patients. Clinicians can find the best medicines by examining tumor genetic profiles, lowering side effects and ineffectiveness [41].

1.10 Cancer Immunology

Cancer immunology studies tumor-immune interactions. It investigates how the immune system detects and responds to tumor cells and how cancer cells avoid detection. Cancer immunotherapy is a rising field of research because cancers typically try to suppress or evade the immune system, which protects the body from malignant alterations [31].

1.10.1 Immune Surveillance and Evasion

Immune surveillance is the immune system's ability to detect and kill cancer cells. Specified immune cells, such as CTLs, NK cells, macrophages, and dendritic cells, patrol the body for aberrant or altered cells. TAAs, aberrant antigens found in cancer cells, are recognized by immune cells as "non-self" and destroyed [84]. Antigen presentation is when the dendritic cells and other APCs acquire and process tumor antigens, presenting them as MHC-bound peptides on their surfaces. T-cells are activated by this immune response cascade [96]. Cytotoxic T-lymphocytes Once activated, CTLs can recognize tumor cells with the same tumor antigens and release perforins and granzymes to kill them [29]. NK cells Cancer cells are identified and destroyed by NK cells without antigen sensitization. They can identify and kill tumor cells that have downregulated MHC class I molecules, a frequent means of immune evasion [97]. Inflammatory Cytokines and Macrophages: Macrophages are recruited to tumor sites and can improve immune surveillance. Depending on the circumstances, they release pro-inflammatory cytokines that promote or inhibit tumor growth and the immune response [34].

1.11 Clinical Implications and Future Directions

Clinical implications and future healthcare orientations stress integrating emerging research discoveries into clinical practice to improve patient outcomes. Clinicians must adapt to new treatments like personalized medicine, gene therapies, and precision diagnostics as medical research improves. These advances could revolutionize therapy, but accessibility, cost, and ethics remain issues. Improve interdisciplinary collaboration, increase clinical trials, and use AI and ML in diagnoses and treatment planning. Healthcare inequities and global health initiatives must be addressed to ensure developments benefit all populations. By adopting these innovations, healthcare can improve efficiency, effectiveness, and equity in patient care [60].

1.11.1 Translating Cancer Biology into Clinical Practice

Over the past few decades, molecular and genetic breakthroughs have transformed cancer diagnosis, therapy, and prognosis. Identifying tumor-causing molecular targets has enabled customized therapy and tailored therapeutics. Biomarker identification and NGS have helped physicians classify tumors by their genetic and epigenetic profiles. This move from one-size-fits-all to personalized therapy offers tumor-specific treatment strategies [4]. Therapy targeting the EGFR in NSCLC or HER2 in breast cancer has shown therapeutic advantages. Due to tumor heterogeneity, patient-specific variables, and the requirement for rigorous clinical studies to validate innovative treatments, transferring laboratory findings into ordinary clinical practice is difficult [17].

1.11.2 Emerging Therapies and Innovations

Rapid advances in immunotherapy, targeted therapy, gene therapy, and combination medicines are changing cancer treatment. Emerging treatments could enhance patient outcomes and change cancer therapy. Immunotherapy, particularly immune checkpoint inhibitors like pembrolizumab and nivolumab, has shown promise in melanoma, non-small cell lung cancer, and bladder cancer. These treatments activate the immune system to fight cancer. Chimeric antigen receptor T-cell (CAR-T) immunotherapy has redefined treatment for blood cancers, including leukemia and lymphoma, by modifying a patient's own T-cells to target and kill cancer cells [94]. Gene therapy could repair or replace cancer-causing genetic alterations. This method can fix mutations in cancer cells or introduce new genes to prevent cancer growth or boost immunity [102].

1.11.3 Challenges and Future Prospects

Despite cancer therapy advances, patient outcomes remain difficult to improve. From cancer's molecular intricacy to side effects and long-term outcomes, these hurdles exist. The genetic makeup of a tumor can vary greatly within and between metastases, making tumor heterogeneity a substantial challenge. Due to this variety, medicines targeting a certain mutation or pathway may not work for all tumor cells, causing relapse or progression. Drug resistance is another major issue. Initial targeted treatments may work, but tumor cells can evolve and bypass the medications. Resistance mechanisms and long-term efficacy are being studied [46]. Novel medicines' toxicity and side effects are another issue. Immune checkpoint inhibitors and targeted therapy may generate immune-related side effects. Clinicians struggle to balance treatment efficacy with acceptable side effects [27]. Personalized medicine will evolve as AI and machine learning enable the analysis of massive patient data to predict treatment results and optimize treatments. Combining molecular diagnostics, real-time monitoring, and patient-specific data can help clinicians create more precise and successful treatment regimens [42].

1.12 Conclusions

Cancer biology, with its complex molecular and cellular pathways, is one of the most difficult but fascinating disciplines of medical research. Proliferative signaling, cell death resistance, angiogenesis, invasion, and metastasis show cancer's complexity, influenced by metabolism, immunological evasion, genetic instability, and inflammation. Understanding these basic traits helps us understand cancer development and develop targeted therapies. Genetics, proteomics, and immunotherapy have revolutionized cancer treatment, giving tailored medicine hope. Despite these advances, tumor heterogeneity, medicine resistance, and cancer's systemic effects remain challenges. Future research must combine knowledge from many domains to create holistic strategies that solve multiple hallmarks. Early diagnosis, therapeutic measures, and patient care innovations must all emphasize lowering the global cancer burden, particularly in resource-limited areas. Finally, a thorough understanding of cancer biology and its hallmarks will continue to impact the future of oncology, fostering advances that improve patient outcomes and quality of life.

References

1. Aaltonen LA, Peltomäki P, Leach FS, Sistonen P, Pylkkänen L, Mecklin JP, et al. Clues to the pathogenesis of familial colorectal cancer. Science. 1993;260(5109):812–6. https://doi.org/10.1126/science.8484121.
2. Akanda MKM, Hasan AHMN. Characterization of pharmacological properties of methanolic seed and stem bark extracts of *Ziziphus mauritiana* (BAU Kul) using in-vitro and in-vivo animal (Swiss albino male mice) model. Clin Phytoscience. 2021;7(1):1–13. https://doi.org/10.1186/s40816-020-00246-0.
3. Akanda MKM, Mehjabin S. Potential role of probiotics, prebiotics and synbiotics in management of metabolic syndrome and chronic metabolic disease. In: Applications of synbiotics in metabolic disorder management. CRC Press/Taylor & Francis; 2024. p. 39–54. https://doi.org/10.1007/978-981-97-9056-2_10.
4. Akanda MKM, Mehjabin S. Genomic biomarkers: unraveling the DNA of cancer. In: Prabhakar PK, editor. The potential of cancer biomarkers: from discovery to clinical application. Academic; 2025. https://doi.org/10.1016/B978-0-443-29279-8.00004-1.
5. Akanda MKM, Mehjabin S, Mosaddik A (2025a) Quinones in the treatment of cardiovascular diseases. In: Dar UA, Shahnawaz Mohd, Rehman Hakeem K (eds) Quinone-based compounds in drug discovery. Academic, pp. 103–117. https://doi.org/10.1016/B978-0-443-24126-0.00006-8

6. Akanda MKM, Mehjabin S, Hasan AHMN, Parvez GMM. Inflammatory bowel disease and gut microbiota: current strategies and future prospects of nanomedicine. In: Mishra N, Ashique S, Garg A, editors. Inflammatory bowel disease and gut microbiota: current strategies and future prospects of nanomedicine. Apple Academic Press, Inc. Co-published with CRC Press (Taylor & Francis); 2025b. p. 349–73.
7. Alieva M, van Rheenen J, Broekman MLD. Potential impact of invasive surgical procedures on primary tumor growth and metastasis. Clin Exp Metastasis. 2018;35(4):319–31. https://doi.org/10.1007/s10585-018-9896-8.
8. Anand P, Kunnumakara AB, Sundaram C, Harikumar KB, Tharakan ST, Lai OS, Sung B, Aggarwal BB. Cancer is a preventable disease that requires major lifestyle changes. Pharm Res. 2008;25(9):2097–116. https://doi.org/10.1007/s11095-008-9661-9.
9. Arteaga CL, Johnson DH. Tyrosine kinase inhibitors—ZD1839 (Iressa). Curr Opin Oncol. 2001;13(6):491.
10. Bailey SM. Editorial: Hallmark of cancer: replicative immortality. Front Oncol. 2023;13 https://doi.org/10.3389/fonc.2023.1204094.
11. Balloux F, Brynildsrud OB, van Dorp L, Shaw LP, Chen H, Harris KA, Wang H, Eldholm V. From theory to practice: translating whole-genome sequencing (WGS) into the clinic. Trends Microbiol. 2018;26(12):1035–48. https://doi.org/10.1016/j.tim.2018.08.004.
12. Berindan-Neagoe I, Calin GA. Molecular pathways: microRNAs, cancer cells, and microenvironment. Clin Cancer Res. 2014;20(24):6247–53. https://doi.org/10.1158/1078-0432.CCR-13-2500.
13. Brabletz T, Hlubek F, Spaderna S, Schmalhofer O, Hiendlmeyer E, Jung A, Kirchner T. Invasion and metastasis in colorectal cancer: epithelial-mesenchymal transition, mesenchymal-epithelial transition, stem cells and β-catenin. Cells Tissues Organs. 2005;179(1–2):56–65. https://doi.org/10.1159/000084509.
14. Brown GT, Murray GI. Current mechanistic insights into the roles of matrix metalloproteinases in tumour invasion and metastasis. J Pathol. 2015;237(3):273–81. https://doi.org/10.1002/path.4586.
15. Cakir BÖ, Adamson P, Cingi C. Epidemiology and economic burden of nonmelanoma skin cancer. Facial Plast Surg Clin North Am. 2012;20(4):419–22. https://doi.org/10.1016/j.fsc.2012.07.004.
16. Carotenuto M, Sacco A, Forgione L, Normanno N. Genomic alterations in cholangiocarcinoma: clinical significance and relevance to therapy. Explor Target Antitumor Ther. 2022;3(2):200–23. https://doi.org/10.37349/etat.2022.00079.
17. Chin L, Gray JW. Translating insights from the cancer genome into clinical practice. Nature. 2008;452(7187):553–63. https://doi.org/10.1038/nature06914.
18. Conti CJ. Mechanisms of tumor progression. In: McQueen CA, editor. Comprehensive toxicology. 2nd ed. Oxford: Elsevier; 2010. p. 335–47.
19. Cooper GM. The development and causes of cancer. In: The cell: a molecular approach. 2nd ed. Sinauer Associates; 2000.
20. Courau T, Nehar-Belaid D, Florez L, Levacher B, Vazquez T, Brimaud F, Bellier B, Klatzmann D. TGF-β and VEGF cooperatively control the immunotolerant tumor environment and the efficacy of cancer immunotherapies. JCI Insight. 2016;1(9):e85974. https://doi.org/10.1172/jci.insight.85974.
21. Curtin NJ. PARP inhibitors for cancer therapy. Expert Rev Mol Med. 2005;7(4):1–20. https://doi.org/10.1017/S146239940500904X.
22. D'Amico M, De Amicis F. Challenges of regulated cell death: implications for therapy resistance in cancer. Cells. 2024;13(13):1083. https://doi.org/10.3390/cells13131083.
23. Das SK, Menezes ME, Bhatia S, Wang X-Y, Emdad L, Sarkar D, Fisher PB. Gene therapies for cancer: strategies, challenges and successes. J Cell Physiol. 2015;230(2):259–71. https://doi.org/10.1002/jcp.24791.
24. de Visser KE, Joyce JA. The evolving tumor microenvironment: from cancer initiation to metastatic outgrowth. Cancer Cell. 2023;41(3):374–403. https://doi.org/10.1016/j.ccell.2023.02.016.

25. DeBerardinis RJ, Chandel NS. Fundamentals of cancer metabolism. Sci Adv. 2016;2(5):e1600200. https://doi.org/10.1126/sciadv.1600200.
26. Deng B, Zhao Z, Kong W, Han C, Shen X, Zhou C. Biological role of matrix stiffness in tumor growth and treatment. J Transl Med. 2022;20(1):540. https://doi.org/10.1186/s12967-022-03768-y.
27. Denison TA, Bae YH. Tumor heterogeneity and its implication for drug delivery. J Control Release. 2012;164(2):187–91. https://doi.org/10.1016/j.jconrel.2012.04.014.
28. Duffy M, McGowan P, Gallagher W. Cancer invasion and metastasis: changing views. J Pathol. 2008;214(3):283–93. https://doi.org/10.1002/path.2282.
29. Farhood B, Najafi M, Mortezaee K. CD8+ cytotoxic T lymphocytes in cancer immunotherapy: a review. J Cell Physiol. 2019;234(6):8509–21. https://doi.org/10.1002/jcp.27782.
30. Feitelson MA, Arzumanyan A, Kulathinal RJ, Blain SW, Holcombe RF, Mahajna J, et al. Sustained proliferation in cancer: mechanisms and novel therapeutic targets. Semin Cancer Biol. 2015;35:S25–54. https://doi.org/10.1016/j.semcancer.2015.02.006.
31. Finn OJ. Cancer immunology. N Engl J Med. 2008;358(25):2704–15. https://doi.org/10.1056/NEJMra072739.
32. Folkman J. The role of angiogenesis in tumor growth. Semin Cancer Biol. 1992;3(2):65–71.
33. Fridman WH, Pagès F, Sautès-Fridman C, Galon J. The immune contexture in human tumours: impact on clinical outcome. Nat Rev Cancer. 2012;12(4):298–306. https://doi.org/10.1038/nrc3245.
34. Fujiwara N, Kobayashi K. Macrophages in inflammation. Curr Drug Targ Inflamm Allergy. 2005;4(3):281–6. https://doi.org/10.2174/1568010054022024.
35. Fulawka L, Donizy P, Halon A. Cancer stem cells – the current status of an old concept: literature review and clinical approaches. Biol Res. 2014;47(1):66. https://doi.org/10.1186/0717-6287-47-66.
36. Ghosh SK. Giovanni Battista Morgagni (1682–1771): father of pathologic anatomy and pioneer of modern medicine. Anat Sci Int. 2017;92(3):305–12. https://doi.org/10.1007/s12565-016-0373-7.
37. Gonzalez H, Hagerling C, Werb Z. Roles of the immune system in cancer: from tumor initiation to metastatic progression. Genes Dev. 2018;32(19–20):1267–84. https://doi.org/10.1101/gad.314617.118.
38. Gopalan V, Islam F, Lam AK. Surface markers for the identification of cancer stem cells. In: Papaccio G, Desiderio V, editors. Cancer stem cells: methods and protocols. New York: Springer; 2018. p. 17–29.
39. Gori JL, Hsu PD, Maeder ML, Shen S, Welstead GG, Bumcrot D. Delivery and specificity of CRISPR/Cas9 genome editing technologies for human gene therapy. Human Gene Ther. 2015;26(7):443–51. https://doi.org/10.1089/hum.2015.074.
40. Grada A, Weinbrecht K. Next-generation sequencing: methodology and application. J Invest Dermatol. 2013;133(8):1–4. https://doi.org/10.1038/jid.2013.248.
41. Greenwalt I, Zaza N, Das S, Li BD. Precision medicine and targeted therapies in breast cancer. Surg Oncol Clin N Am. 2020;29(1):51–62. https://doi.org/10.1016/j.soc.2019.08.004.
42. Hamamoto R, Komatsu M, Yamada M, Kobayashi K, Takahashi M, Miyake M, et al. Current status and future direction of cancer research using artificial intelligence for clinical application. Cancer Sci. 2025;116(2):297–307. https://doi.org/10.1111/cas.16395.
43. Hanahan D. Hallmarks of cancer: new dimensions. Cancer Discov. 2022;12(1):31–46. https://doi.org/10.1158/2159-8290.CD-21-1059.
44. Hanahan D, Weinberg RA. The hallmarks of cancer. Cell. 2000;100(1):57–70. https://doi.org/10.1016/s0092-8674(00)81683-9.
45. Hanahan D, Weinberg RA. Hallmarks of cancer: the next generation. Cell. 2011;144(5):646–74. https://doi.org/10.1016/j.cell.2011.02.013.
46. Heppner GH, Shapiro WR, Rankin JK. Tumor heterogeneity. In: Humphrey GB, Dehner LP, Grindey GB, Acton RT, editors. Pediatric oncology 1: with a special section on rare primitive neuroectodermal tumors. Dordrecht: Springer; 1981. p. 99–116.

47. Hilger RA, Scheulen ME, Strumberg D. The Ras-Raf-MEK-ERK pathway in the treatment of cancer. Onkologie. 2003;25(6):511–8. https://doi.org/10.1159/000068621.
48. Jayasekara H, MacInnis RJ, Room R, English DR. Long-term alcohol consumption and breast, upper aero-digestive tract and colorectal cancer risk: a systematic review and meta-analysis. Alcohol Alcohol. 2016;51(3):315–30. https://doi.org/10.1093/alcalc/agv110.
49. Jemal A, Bray F, Center MM, Ferlay J, Ward E, Forman D. Global cancer statistics. CA Cancer J Clin. 2011;61(2):69–90. https://doi.org/10.3322/caac.20107.
50. Jiang WG, Puntis MCA, Hallett MB. Molecular and cellular basis of cancer invasion and metastasis: implications for treatment. Br J Surg. 1994;81(11):1576–90. https://doi.org/10.1002/bjs.1800811107.
51. Kanchanawong P, Calderwood DA. Organization, dynamics and mechanoregulation of integrin-mediated cell–ECM adhesions. Nat Rev Mol Cell Biol. 2023;24(2):142–61. https://doi.org/10.1038/s41580-022-00531-5.
52. Karar J, Maity A. PI3K/AKT/mTOR pathway in angiogenesis. Front Mol Neurosci. 2011;4 https://doi.org/10.3389/fnmol.2011.00051.
53. Kirkwood JM, Butterfield LH, Tarhini AA, Zarour H, Kalinski P, Ferrone S. Immunotherapy of cancer in 2012. CA Cancer J Clin. 2012;62(5):309–35. https://doi.org/10.3322/caac.20132.
54. Kluza J, Jendoubi M, Ballot C, Dammak A, Jonneaux A, Idziorek T, et al. Exploiting mitochondrial dysfunction for effective elimination of Imatinib-resistant leukemic cells. PLoS One. 2011;6(7):e21924. https://doi.org/10.1371/journal.pone.0021924.
55. Kocarnik JM, Compton K, Dean FE, Fu W, Gaw BL, Harvey JD, et al. Cancer incidence, mortality, years of life lost, years lived with disability, and disability-adjusted life years for 29 cancer groups from 2010 to 2019: a systematic analysis for the global burden of disease study 2019. JAMA Oncol. 2022;8(3):420–44. https://doi.org/10.1001/jamaoncol.2021.6987.
56. Lee M, Yoon J-H. Metabolic interplay between glycolysis and mitochondrial oxidation: the reverse Warburg effect and its therapeutic implication. World J Biol Chem. 2015;6(3):148–61. https://doi.org/10.4331/wjbc.v6.i3.148.
57. Lee YT, Tan YJ, Oon CE. Molecular targeted therapy: treating cancer with specificity. Eur J Pharmacol. 2018;834:188–96. https://doi.org/10.1016/j.ejphar.2018.07.034.
58. Li B, Chan HL, Chen P. Immune checkpoint inhibitors: basics and challenges. Curr Med Chem. 2019;26(17):3009–25. https://doi.org/10.2174/0929867324666170804143706.
59. Lythgoe MP, Mullish BH, Frampton AE, Krell J. Polymorphic microbes: a new emerging hallmark of cancer. Trends Microbiol. 2022;30(12):1131–4. https://doi.org/10.1016/j.tim.2022.08.004.
60. MacConaill LE, Garraway LA. Clinical implications of the cancer genome. JCO. 2010;28(35):5219–28. https://doi.org/10.1200/JCO.2009.27.4944.
61. Macri L, Silverstein D, Clark RAF. Growth factor binding to the pericellular matrix and its importance in tissue engineering. Adv Drug Deliv Rev. 2007;59(13):1366–81. https://doi.org/10.1016/j.addr.2007.08.015.
62. Marks F, Fürstenberger G, Müller-Decker K. Tumor promotion as a target of cancer prevention. Recent Results Cancer Res. 2007;174:37–47. https://doi.org/10.1007/978-3-540-37696-5_3.
63. Marusyk A, Polyak K. Tumor heterogeneity: causes and consequences. Biochim Biophys Acta Rev Cancer. 2010;1805(1):105–17. https://doi.org/10.1016/j.bbcan.2009.11.002.
64. McGettigan PA. Transcriptomics in the RNA-seq era. Curr Opin Chem Biol. 2013;17(1):4–11. https://doi.org/10.1016/j.cbpa.2012.12.008.
65. McGuire S. World Cancer Report 2014. Geneva: World Health Organization, International Agency for Research on Cancer, WHO Press; 2016, 2015. Advances in Nutrition 7(2):418–419. https://doi.org/10.3945/an.116.012211
66. Mehjabin S, Akanda MKM, Hoque N, Hasan AHMN, Parvez GMM, Mosaddik A. Flavonoid intake and risk of toxicity in chronic metabolic diseases. In: Role of flavonoids in chronic metabolic diseases: from bench to clinic. Wiley; 2024. p. 511–34.
67. Mehjabin S, Akanda MKM, Hasan AHMN, Parvez GMM. Role of inulin in the management of metabolic disorders. In: Akram W, Mishra N, Haider T, editors. Inulin for pharmaceutical applications: a versatile biopolymer. Singapore: Springer Nature; 2025. p. 189–214.

68. Meister A. Metabolism of glutamine. Physiol Rev. 1956;36(1):103–27. https://doi.org/10.1152/physrev.1956.36.1.103.
69. Milazzotto MP, Goissis MD, Chitwood JL, Annes K, Soares CA, Ispada J, et al. Early cleavages influence the molecular and the metabolic pattern of individually cultured bovine blastocysts. Mol Reprod Dev. 2016;83(4):324–36. https://doi.org/10.1002/mrd.22619.
70. Mosaddik A, Akanda MKM, Pal TR, Elaissari A. Advances in nano-biomaterials for effective antimicrobial therapy. In: Pradhan M, Yadav K, Singh Chauhan N, editors. Biomaterial-inspired nanomedicines for targeted therapies. Singapore: Springer Nature; 2024. p. 425–50. https://doi.org/10.1007/978-981-97-3925-7_16.
71. Nachef L, Bouchet A, Bourguignon M, Foray N. When DNA mutations interplay with cellular proliferation: a narrative history of theories of carcinogenesis. Cancers (Basel). 2024;16(11):2104. https://doi.org/10.3390/cancers16112104.
72. Nahta R, Al-Mulla F, Al-Temaimi R, Amedei A, Andrade-Vieira R, Bay SN, et al. Mechanisms of environmental chemicals that enable the cancer hallmark of evasion of growth suppression. Carcinogenesis. 2015;36(Suppl_1):S2–S18. https://doi.org/10.1093/carcin/bgv028.
73. Najafi M, Mortezaee K, Majidpoor J. Cancer stem cell (CSC) resistance drivers. Life Sci. 2019;234:116781. https://doi.org/10.1016/j.lfs.2019.116781.
74. National Cancer Institute. What is cancer? 2007. https://www.cancer.gov/about-cancer/understanding/what-is-cancer. Accessed 2 Feb 2025.
75. Negrini S, Gorgoulis VG, Halazonetis TD. Genomic instability—an evolving hallmark of cancer. Nat Rev Mol Cell Biol. 2010;11(3):220–8. https://doi.org/10.1038/nrm2858.
76. Noy R, Pollard JW. Tumor-associated macrophages: from mechanisms to therapy. Immunity. 2014;41(1):49–61. https://doi.org/10.1016/j.immuni.2014.06.010.
77. O'Brien CA, Kreso A, Jamieson CHM. Cancer stem cells and self-renewal. Clin Cancer Res. 2010;16(12):3113–20. https://doi.org/10.1158/1078-0432.CCR-09-2824.
78. Oliff A. Farnesyltransferase inhibitors: targeting the molecular basis of cancer. Biochim Biophys Acta Rev Cancer. 1999;1423(3):C19–30. https://doi.org/10.1016/S0304-419X(99)00007-4.
79. Parvez GMM, Akanda KM. Foods and arthritis: an overview. In: Bioactive food as dietary interventions for arthritis and related inflammatory diseases. Elsevier; 2019. p. 3–22.
80. Postiglione I, Chiaviello A, Palumbo G. Enhancing photodynamyc therapy efficacy by combination therapy: dated. Curr Oncoming Strateg Cancers. 2011;3(2):2597–629. https://doi.org/10.3390/cancers3022597.
81. Poulet G, Massias J, Taly V. Liquid biopsy: general concepts. Acta Cytol. 2019;63(6):449–55. https://doi.org/10.1159/000499337.
82. Raman D, Baugher PJ, Thu YM, Richmond A. Role of chemokines in tumor growth. Cancer Lett. 2007;256(2):137–65. https://doi.org/10.1016/j.canlet.2007.05.013.
83. Rane SG, Reddy EP. Janus kinases: components of multiple signaling pathways. Oncogene. 2000;19(49):5662–79. https://doi.org/10.1038/sj.onc.1203925.
84. Ribatti D. The concept of immune surveillance against tumors: the first theories. Oncotarget. 2016;8(4):7175–80. https://doi.org/10.18632/oncotarget.12739.
85. Rich JN. Cancer stem cells: understanding tumor hierarchy and heterogeneity. Medicine (Baltimore). 2016;95(Suppl 1):S2–7. https://doi.org/10.1097/MD.0000000000004764.
86. Rieseberg LH. Chromosomal rearrangements and speciation. Trends Ecol Evol. 2001;16(7):351–8. https://doi.org/10.1016/S0169-5347(01)02187-5.
87. Russo J, Hu YF, Yang X, Russo IH. Developmental, cellular, and molecular basis of human breast cancer. J Natl Cancer Inst Monogr. 2000;27:17–37. https://doi.org/10.1093/oxfordjournals.jncimonographs.a024241.
88. Saha SK, Hasan AHMN, Anjum R, Rimun SN, Rahman NZ, Akanda MKM, Buhiyan MA, Islam SA. From prodrug to multimatrix: recent advancement of colon specific drug delivery system. Sci Pharm. 2024;3(3):1–15. https://doi.org/10.58920/sciphar0303233.
89. Schultz M. Rudolf Virchow. Emerg Infect Dis. 2008;14(9):1480–1. https://doi.org/10.3201/eid1409.086672.

90. Sciacovelli M, Schmidt C, Maher ER, Frezza C. Metabolic drivers in hereditary cancer syndromes. Annu Rev Cancer Biol. 2020;4(4, 2020):77–97. https://doi.org/10.1146/annurev-cancerbio-030419-033612.
91. Singh S, Tank NK, Dwiwedi P, Charan J, Kaur R, Sidhu P, Chugh VK. Monoclonal antibodies: a review. Curr Clin Pharmacol. 2018;13(2):85–99. https://doi.org/10.2174/1574884712666170809124728.
92. Soto AM, Sonnenschein C. The somatic mutation theory of cancer: growing problems with the paradigm? BioEssays. 2004;26(10):1097–107. https://doi.org/10.1002/bies.20087.
93. Stetler-Stevenson WG, Liotta LA, Kleiner DE Jr. Extracellular matrix 6: role of matrix metalloproteinases in tumor invasion and metastasis. FASEB J. 1993;7(15):1434–41. https://doi.org/10.1096/fasebj.7.15.8262328.
94. Till SJ, Francis JN, Nouri-Aria K, Durham SR. Mechanisms of immunotherapy. J Allergy Clin Immunol. 2004;113(6):1025–34. https://doi.org/10.1016/j.jaci.2004.03.024.
95. Tran B, Dancey JE, Kamel-Reid S, McPherson JD, Bedard PL, Brown AMK, et al. Cancer genomics: technology, discovery, and translation. JCO. 2012;30(6):647–60. https://doi.org/10.1200/JCO.2011.39.2316.
96. Unanue ER, Cerottini J-C. Antigen presentation. FASEB J. 1989;3(13):2496–502. https://doi.org/10.1096/fasebj.3.13.2572499.
97. Vivier E, Tomasello E, Baratin M, Walzer T, Ugolini S. Functions of natural killer cells. Nat Immunol. 2008;9(5):503–10. https://doi.org/10.1038/ni1582.
98. Ward PS, Thompson CB. Metabolic reprogramming: a cancer hallmark even Warburg did not anticipate. Cancer Cell. 2012;21(3):297–308. https://doi.org/10.1016/j.ccr.2012.02.014.
99. Watanabe H, Carmi P, Hogan V, Raz T, Silletti S, Nabi IR. Purification of human tumor cell autocrine motility factor and molecular cloning of its receptor. J Biol Chem. 1991;266(20):13442–8. https://doi.org/10.1016/S0021-9258(18)98859-9.
100. Wei Z, Liu X, Cheng C, Yu W, Yi P. Metabolism of amino acids in cancer. Front Cell Dev Biol. 2021;8 https://doi.org/10.3389/fcell.2020.603837.
101. WHO Cancer. 2018. https://www.who.int/health-topics/cancer. Accessed 2 Feb 2025.
102. Wirth T, Parker N, Ylä-Herttuala S. History of gene therapy. Gene. 2013;525(2):162–9. https://doi.org/10.1016/j.gene.2013.03.137.
103. Yokozaki H, Yasui W, Tahara E. Genetic and epigenetic changes in stomach cancer. In: International review of cytology. Academic; 2001. p. 49–95.
104. Zhang H, Yue X, Chen Z, Liu C, Wu W, Zhang N, et al. Define cancer-associated fibroblasts (CAFs) in the tumor microenvironment: new opportunities in cancer immunotherapy and advances in clinical trials. Mol Cancer. 2023;22(1):159. https://doi.org/10.1186/s12943-023-01860-5.

Chapter 2
Introduction to the Seven "Rs" of Radiobiology

Robab Anbiaee, Mohammad Houshyari, and Farzad Taghizadeh-Hesary

Abstract Radiation therapy (RT) is a cornerstone in treating cancer, aiming to eliminate tumor cells while saving surrounding normal tissues. Advances in personalized RT underscore the need for tailored treatment approaches that account for individual patient and tumor-specific factors, maximizing therapeutic efficacy and minimizing side effects. Radiobiology, the basic science of RT, presents crucial insights into cancer and normal tissue responses to irradiation. Central to radiobiology are the seven "Rs," including DNA Repair, Reoxygenation, Regeneration, Redistribution, Radiosensitivity, immune Reactivation, and Reinforcement by tumor microenvironment, which delineate key determinants of RT outcomes. Understanding these principles not only explains the variability in RT responses but also provides a framework for addressing clinical challenges such as radioresistance. Integrating the "Rs" into personalized RT offers immense potential to refine treatment strategies. The application of these radiobiological principles in combination with advanced imaging, biomarkers, and AI-driven treatment planning is pivotal for optimizing RT outcomes. This chapter highlights the transformative potential of the "Rs" in personalizing RT, bridging the gap between biological understanding and clinical innovation.

Keywords Cancer · Radiotherapy · "Rs" of Radiobiology · Radioresistance

R. Anbiaee · M. Houshyari
Radio oncology Department, Shahid Beheshti University of Medical Sciences, Tehran, Iran

F. Taghizadeh-Hesary (✉)
ENT and Head and Neck Research Center and Department, The Five Senses Health Institute, School of Medicine, Iran University of Medical Sciences, Tehran, Iran
e-mail: taghizadeh_hesary.f@iums.ac.ir

F. Taghizadeh-Hesary et al. (eds.), *From Radiobiology to Radiation Oncology*,
https://doi.org/10.1007/978-3-031-95724-6_2

2.1 Introduction

Radiation therapy (RT) is a cornerstone of treating malignancies, used in over 50 percent of patients with cancer during their treatment journey [1]. Radiobiology, the fundamental science underpinning RT, mainly focuses on enhancing the impact of radiation on cancer cells while protecting nearby organs at risk (OAR) from its harmful effects. The "Rs" of radiobiology are English terms starting with the letter R that constitutes of the framework of the radiobiology science. These terms outline a hierarchy of scientific concepts that are structured to be easily understood. In fact, these terms are the determinants of living tissues, either tumoral or normal tissues, to RT.

The first four "Rs" were introduced by Rodney H. Withers from the United States in 1975. In his novel article, he posited that "Regeneration," "Repair," "Reoxygenation," and "Radiosensitivity" are four determinants of living tissue response to fractionated RT [2]. In 1989, Gordon Steel et al. from England introduced "Radiosensitivity" as the fifth "R" by observing different survival curves of different cell lines to a 2-Gy radiation fraction [3]. "R" composition was not improved until recently when the advances in cell biology revealed the interactions between RT and the immune system. With this scope, Jihane Boustani et al. from France introduced "Reactivation" denoting immune reactivation as the 6th "R" [4]. Recently, Farzad Taghizadeh-Hesary from Iran highlighted the intricate interplay among cancerous cells and the surrounding tumor microenvironment (TME) by positing "Reinforcement" as the seventh "R" of radiobiology [5]. Figure 2.1

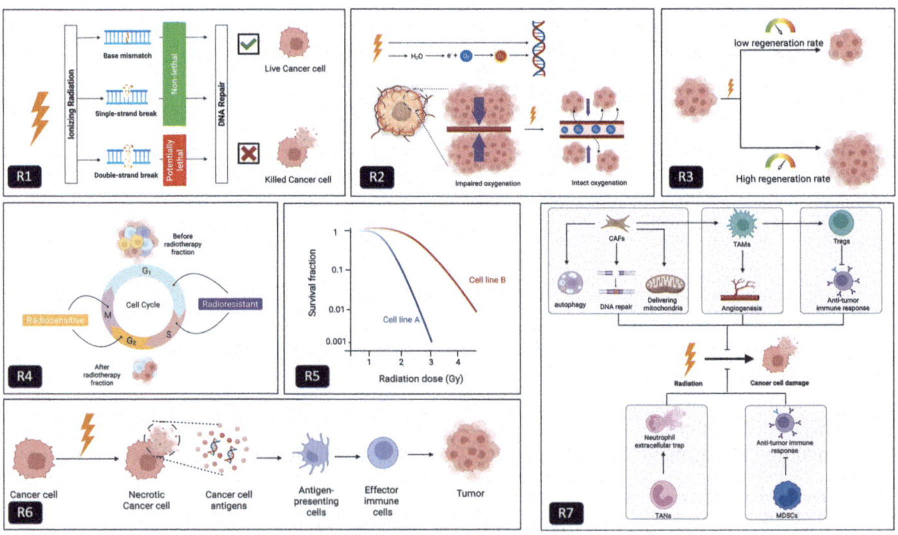

Fig. 2.1 The seven "Rs" of Radiobiology. The first R, DNA repair; second R, Reoxygenation; third R, Regeneration (Repopulation); fourth R, Redistribution (Reassortment); fifth R, intrinsic Radiosensitivity; sixth R, immune Reactivation; seventh R, Reinforcement of cancer cells by tumor microenvironment. [The source of R7 image is [6]; CC BY-NC-ND)

provides an overview picture of the seven defined "Rs" of radiobiology. This chapter is dedicated to providing a comprehensive, practical explanation of the "Rs" of radiobiology.

2.2 First R: DNA Repair

DNA (deoxyribonucleic acid) is the primary aim of RT. Ionizing radiation can directly disrupt DNA molecules or indirectly through releasing ROS (reactive-oxygen species) [7]. The former interaction becomes more dominant in high-LET (linear energy transfer) radiation [8]. In response to radiation, DNA molecules can sustain SSBs (single-strand breaks) or DSBs (double-strand breaks), each leading to vastly different outcomes. SSBs are generally nonlethal due to the efficient response of DNA repair complexes. In contrast, DSBs can induce cell death if not properly fixed [9] (Fig. 2.1-R1).

Repairing responses to DNA damage can be classified based on the damage type. SSBs can be fixed by BER (base excision repair), NER (nucleotide excision repair), or MMR (mismatch repair), while DSBs are repaired by HR (homologous recombination) or NHEJ (nonhomologous end joining) [10]. Each DNA repair mechanism involves a hierarchical network of proteins that work together to efficiently repair damaged DNA. These proteins include those which detect the damaged DNA (sentinel proteins), those which facilitate the recruiting of DNA repairing enzymes (signal transmitters), and the proteins responsible for DNA repair (effectors) (Table 2.1).

Notably, the importance of DNA repair mechanisms in guaranteeing the living of the living units is different. An experimental study on lung fibroblast cell lines revealed that cells deficient in NHEJ proteins, such as DNA-PKcs and XRCC4, exhibited greater sensitivity to proton irradiation compared to those lacking HR proteins, including XRCC2 and XRCC3. In contrast, this study demonstrated that the major mechanism in DNA repair in response to carbon RT is HR [11]. Considering this, the impact of radiation on living tissues is influenced by several factors, including radiation type, the extent of radiation-induced DNA damage, defined by the CDD (cluster of DNA damage), and to which extent target cells can fix the damages.

The crucial role of DNA repair in cellular responses to radiation was leveraged within various clinical studies evaluating DNA repair inhibitors. For instance, PARP inhibitors have been studied in patients with breast cancer [12], glioblastoma [13], and pancreatic cancer [14]. DNA-PK inhibitors have been investigated in HNSCC (squamous cell carcinoma of head and neck region) [15] and glioblastoma [16]. In addition, a topoisomerase I Inhibitor (irinotecan) was examined in a phase I clinical study on upper gastrointestinal cancer with acceptable toxicity profiles [17]. These studies have consistently shown the clinical safety of targeted therapies, paving the way for further clinical research.

Table 2.1 The proteins involved in different DNA repair mechanisms

Repair mechanisms	Sentinel proteins	Signal transmitters	Effectors
Base excision repair	OGG1, UNG, NTHL1	APE1, PARP1, PNKP	POLβ, XRCC1, LIG3α, LIG1
Nucleotide excision repair	XPC-RAD23B, DDB1, DDB2, RNA Polymerase II	TFIIH, XPA, RPA	XPF-ERCC1, XPG, POLδ, POLε, POLκ, LIG3α, LIG1
Mismatch repair	MSH2, MSH3, MSH6	MLH1, PMS2, EXO1, PCNA	POLδ, POLε, RFC, LIG1
Homologous recombination	MRE11, RAD50, NBS1, ATM, γH2AX	CtIP, BRCA1, BRCA2, RPA, PALB2	RAD51, RAD54, POLδ, MUS81, EME1, LIG1, XRCC2, XRCC3
Nonhomologous end joining	Ku70/Ku80, DNA-PKcs,	Artemis, XRCC4, PAXX, XLF	Pol μ, Pol λ, LIG4, XRCC4

Abbreviations: APE1 AP Endonuclease 1, *PARP1* Poly ADP-Ribose Polymerase 1, *PNKP* Polynucleotide Kinase Phosphatase, *ATM* Ataxia-Telangiectasia Mutated, *BRCA1/BRCA2* Breast Cancer Susceptibility Proteins 1 & 2, *PALB2* Partner and Localizer of BRCA2, *LIG1, LIG3α, LIG4* DNA Ligases, *POL* DNA Polymerases: Beta, Delta, Epsilon, Kappa, Lambda, Mu, *MLH1* MutL Homolog 1, *MSH2, MSH3, MSH6* MutS Homologs, *XPA* Xeroderma Pigmentosum Group A, *XPC* Xeroderma Pigmentosum Complementation Group C, *XPF-ERCC1* Xeroderma Pigmentosum Group F & Excision Repair Cross-Complementing Group 1, *XPG* Xeroderma Pigmentosum Group G, *MRE11* MRE11 Homolog, *RAD50, RAD51, RAD54* RAD Homologs, *Ku70/Ku80* Ku Heterodimer, *XRCC1, XRCC4* X-Ray Repair Cross-Complementing Proteins, *XLF* XRCC4-Like Factor, *PAXX* Paralog of XRCC4 and XLF, *TFIIH* Transcription Factor II H, *DDB1, DDB2* Damage-Specific DNA-Binding Proteins, *PCNA* Proliferating Cell Nuclear Antigen, *RPA* Replication Protein A, *RFC* Replication Factor C, *CtIP* C-Terminal Binding Protein-Interacting Protein, *γH2AX* Gamma-Histone 2AX, *MUS81* MUS81 Endonuclease, *UNG* Uracil DNA Glycosylase, *RAD23B* RAD23 Homolog B, *NBS1* Nibrin, *OGG1* OGG1 DNA Glycosylase, *NTHL1* Endonuclease III-Like Protein 1, *EXO1* Exonuclease 1

2.3 Second R: Reoxygenation

Reoxygenation reflects to increasing the oxygen pressure of intratumoral tissue. Generally speaking, hypoxia is a characteristic of TME [18]. Hypoxia favors cancerous cell proliferation, invasion, and spread by reducing the degradation of HIF-1 (hypoxia-inducible factor-1), the main regulator of cancer progression [19]. Meanwhile, cancer cells lower the surrounding oxygen levels by releasing angiogenic factors. It was previously believed that these factors could increase the oxygen pressure within the TME. However, recent studies have shown that angiogenesis can ironically reduce intratumoral oxygen pressure. This may be due to the leaky and immature nature of the newly formed microvessels with discontinuous basement membranes, which are prone to collapse under the high physical pressure within the TME [18]. Reoxygenation may also occur during fractionated RT, where the reduction in tumor cell numbers and the decrease in interstitial pressure on the fragile microvessels contribute to improved oxygenation (Fig. 2.1-R2). This phenomenon has been confirmed through 18F-Misonidazole PET scans during fractionated RT for prostate cancer [20]. This process may help explain the rationale behind the fractionated RT schedule.

Furthermore, experimental studies revealed that exposure to high oxygen levels can directly prevent tumor growth [21]. Given that oxygen is the substrate for ROS generation, hypoxia can diminish the effects of ionizing radiation on cancer cells by reducing the radiation's indirect impact [22]. Therefore, tissue reoxygenation can enhance the effects of radiation by increasing intratumoral ROS levels, particularly in low-LET radiation modalities. In radiobiology, the OER (oxygen enhancement ratio) measures how much more effective radiation is at eliminating cells in the presence of oxygen in comparison to low-oxygen conditions. Typically, for low-LET radiation (e.g., X-rays, electron), the OER is 2.5–3.0, while regarding high-LET radiation (e.g., alpha particles), it is lower, around 1.0–1.5, due to its direct damage mechanism [23]. Gray et al. showed that tumor cells with severe hypoxia needed a radiation dose three times higher than well-oxygenated tissues to produce a comparable extent of damage [24].

Overall, tumor reoxygenation has the potential to enhance the response to RT by degrading HIF-1, inhibiting tumor growth, and increasing the generation of ROS, thereby improving the effects of ionizing radiation. In this context, several early-stage clinical studies evaluated the adverse effects and clinical efficacy of intratumoral reoxygenation in enhancing radiation response. For example, a phase I trial showed that injection of an H_2O_2 solution (KORTUC) into the tumor tissue is a safe procedure [25, 26] Additionally, a phase II clinical trial on patients with glioblastoma demonstrated that combining bevacizumab (an anti-angiogenic agent) with re-irradiation is a safe procedure and well-tolerated by patients [27].

2.4 Third R: Regeneration

Regeneration, or repopulation, refers to the ability of cells to divide and replace lost or damaged cells (Fig. 2.1-R3). Experimental studies consistently show that cancer types vary widely in their proliferation rates. At one end of the spectrum, HNSCC has high proliferative rates, while prostate adenocarcinoma is characterized by low proliferation rates. This knowledge has guided researchers to explore altered fractionation regimens, aligning with the proliferative rate of a target tissue, and comparing it with conventional RT (1.8–2.0 Gy/fraction). For instance, hypofractionated regimens (2.5–3.3 Gy/fraction) are increasingly used worldwide in prostate cancer due to their promising results in tumor control with comparable toxicity profiles [28, 29].

Notably, extending the treatment duration for tumors with high proliferative rates can compromise tumor control by allowing cancer cells more time to regenerate. To address this, altered RT schedules for HNSCC have adopted smaller per-fraction doses (1.1–1.2 Gy) delivered twice daily to minimize overall treatment time. The MARCH meta-analysis demonstrated a survival benefit of altered fractionated RT over conventional fractionation in HNSCC patients [30]. Nevertheless, patient-specific factors such as comorbidities, age, and treatment tolerance are also crucial in selecting the optimal fractionation schedule [31].

Understanding the regenerative capacity of a tumor is crucial for designing personalized RT schedules that optimize treatment outcomes. Advanced PET imaging modalities and specific biological markers have been shown to aid in identifying the regeneration potential of irradiated tumors [32, 33]. Further research is needed to uncover additional biological markers that can provide deeper insights into tumor regeneration capacity and improve RT strategies.

2.5 Fourth R: Redistribution

The sensitivity to radiation varies across different phases of the cell cycle. Studies have indicated that cells in the G_2/M phases are more sensitive to radiation than those in the G_1/S phases, likely because of greater exposure of DNA molecules during the G_2/M phases (Fig. 2.1-R4) [34]. Furthermore, cancer cells can arrest in the G1 phase to fix the damaged DNA. This behavior, mediated by P53, can further enhance their resistance to RT [35]. Notably, CSCs (cancer stem cells) benefit from a resting G_0 phase with slower progression along the cell cycle making them more radioresistant [36].

Redistribution, also known as reassortment, refers to changes in the cell cycle phases of viable cancer cells following a radiation fraction. After radiation, cell cycle progression leads to an accumulation of more cells in the G_2/M phases, where they are more vulnerable to subsequent radiation damage. In addition, fractionated RT can increase *Notch-1* expression, propelling CSCs toward the cell cycle [37]. These benefits are the reasons why fractionated RT offers significant therapeutic benefits.

2.6 Fifth R: Radiosensitivity

In the late 1980s, Gordon Steel et al. introduced the 5th "R" of Radiobiology: intrinsic Radiosensitivity. This term represents the inherent radiosensitivity of cancer cells, a defining characteristic of each cancer type that determines its response to radiation. In their seminal article, Steel et al. conceptualized radiosensitivity as the steepness of the survival curve following RT.

Advancements in radiobiological understanding during the 1980s played a key role in shaping this concept. As illustrated in Fig. 2.1-R5, the comparison of two cell lines demonstrates this principle: Cell line A exhibits greater radiosensitivity than Cell line B, as reflected by the steeper slope of its survival curve following radiation exposure. This insight highlights the importance of intrinsic radiosensitivity in predicting and tailoring RT outcomes. Chapter 4 presents more discussion on this topic.

2.7 Sixth R: Reactivation

In mid-2019, Boustani et al. from the Unicancer-Georges-François Leclerc Cancer Center (Dijon, France) introduced the sixth "R" of Radiobiology: Reactivation [4]. This term refers to immune reactivation following RT, a concept rooted in radiobiological studies showing that radiation can change the tumor immune microenvironment (TIME) [38]. Compelling experimental evidence has demonstrated that radiation-induced release of tumor-specific antigens is captured by resident dendritic cells. These activated dendritic cells function as antigen-presenting cells, recruiting and priming lymphocytes ($CD8^+$ and $CD4^+$) as well as NK (natural killer) cells within the TME (Fig. 2.1-R6). The resulting immune response contributes to additional damage to cancer cells, enhancing the therapeutic effects of RT [39].

This concept has been largely applied in recent clinical studies to trial the combination of immunomodulatory agents in combination to enhance the RT effects in different cancers with promising results, for example, colorectal cancers [reviewed in [40]], lung cancer [41], bladder cancer [42], renal cell carcinoma [43], and esophago-gastric adenocarcinoma [44]. However, pembrolizumab (an immunotherapeutic agent) could not enhance the RT clinical outcomes in patients with HNSCC [45].

Emerging evidence has shown that different RT schedules can have varying effects on immune reactivation. For instance, Vanpouille-Box et al. demonstrated that radiation doses exceeding 12–18 Gy per-fraction can upregulate TREX1 expression [46]. TREX1 (three-prime repair exonuclease 1) has a pivotal function in the response to radiation by regulating the accumulation of cytosolic DNA and preventing excessive immune activation. TREX1 degrades cytosolic DNA fragments, thereby limiting their accumulation. This, in turn, suppresses the activation of the cGAS (cyclic GMP-AMP synthase)-stimulator of interferon genes (STING) pathway, a central regulator of the innate immune response. When TREX1 is overexpressed, cytosolic DNA is effectively depleted, preventing activation of the cGAS-STING pathway. Consequently, the production of pro-inflammatory cytokines and type I interferons is inhibited, which impairs the recruitment of immune cells such as T cells, dendritic cells, and NK cells [47]. A clinical study showed that hyperfractionated RT could effectively reduce the PD-1 (an immune checkpoint) expression in patients with refractory/relapsed DLBCL (diffuse large B-cell lymphoma) [48]. These findings highlight the importance of understanding how different RT regimens influence immune responses.

2.8 Seventh R: Reinforcement

The seventh "R" of Radiobiology is about TME and how its constituents *reinforce* intermingled cancer cells against RT effects. Advances in cell biology have demonstrated that cancer cells can reshape their surrounding microenvironment, known as *tumor microenvironment*, into a particular state to support their survival in the hypoxic, acidic TME milieu, enhance their progression into more advanced features, and improve their resistance against different treatment modalities, including chemotherapy, immunotherapy, and RT. TME is composed of two general components: cellular and noncellular. The cellular part is composed of several cell types with different impacts on cancerous cells, including TAMs (tumor-associated macrophages), CAFs (cancer-associated fibroblasts), TANs (tumor-associated neutrophils), Tregs (regulatory T cells), MDSCs (myeloid-derived suppressor cells), and TiNs (tumor-infiltrating nerves). The noncellular section is composed of a lymphovascular network and an extracellular matrix. The TME reinforces cancer cells' resistance to RT through several mechanisms (Fig. 2.2):

Immune evasion: TME elements suppress anti-tumor immune responses in several ways. Tregs release adenosine, TGF-β, and interleukins that block the activity of NK cells and cytotoxic T cells [49]. MDSCs and TAMs can also have immune suppressive effects by recruiting more Tregs to TME [5]. Additionally, tumor hypoxia creates conditions that limit the invasion and function of immune cells in the TME, which require physiologic oxygen levels for proper action [50]; (b) metabolic support: CAFs supply cancer cells with essential metabolic resources, including glucose, lactate, and fatty acids, enabling them to meet energy demands [51]. These resources also support mitochondrial activity, which facilitates DNA repair and remodeling of chromatin following radiation-induced damage. Furthermore, CAFs can deliver mitochondria to the adjacent cancerous cells through nanotubes [52]; (c) DNA damage repair: The TME promotes DNA dam-

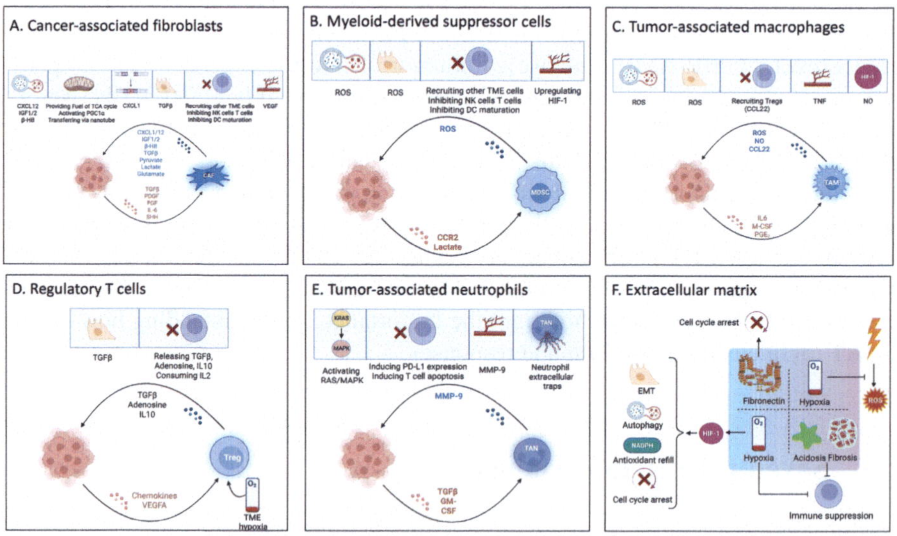

Fig. 2.2 The interplay between cancerous cells and their tumor microenvironment and the mechanisms by which different tumor microenvironment components enhance the radioresistance of cancer cells

age repair pathways in cancer cells. Cytokines like CXCL1 enhance the DNA damage response, and ATP-dependent repair processes supported by mitochondria ensure efficient recovery from radiation-induced lesions [53]; (d) epithelial-mesenchymal transition (EMT): Different TME components can induce EMT by releasing ROS and TGF β, leading to increased tumor cell invasiveness and migration while reducing radiosensitivity [5]. EMT also contributes to the formation of CSCs, which exhibit enhanced resistance to RT [54].

Moreover, the harsh and nonphysiological conditions of the TME contribute to tumor progression and hinder effective anti-tumor responses. For instance, tumor angiogenesis, leads to the generation of immature and leaky blood vessels, susceptible to collapse under physical stress. This process exacerbates hypoxia, which, in turn, promotes radioresistance by stabilizing HIFs and reducing ROS generation (see above) [18]. Various TME components, including CAFs, MDSCs, TAMs, and TANs, actively contribute to neoangiogenesis by releasing angiogenic factors like TNF-α and VEGF [5]. Additionally, both cancer cells and TME-associated cells adapt to hypoxic conditions by shifting their metabolism from oxidative phosphorylation to glycolysis. The resulting glycolysis-driven lactate is exported into the TME, increasing its acidity [55]. This acidic environment further impairs the recruitment of immune cells and facilitates cancer cell progression [56].

As noted above TiNs are also crucial components of TME. Emerging evidence has put forward several mechanisms by which TiNs can affect cancer progression and treatment resistance. An animal study by Uccello et al. indicated that beta-adrenergic receptor activation using a sympathomimetic (isoproterenol) reduced the radiation efficacy on rectal cancer. This detrimental effect was reversed by propranolol to block the beta-adrenergic receptor [57]. A study on patients with breast cancer showed that short-term beta-blockade using propranolol improved the number of anti-tumor $CD8^+$ cells [58]. In support, another study found that adrenergic stimulation using norepinephrine could significantly decrease $CD8^+$ cells in the lung adenocarcinoma [59]. Therefore, beta-adrenergic blockade can potentially improve the radiation effects by improving immune response (6th R of Radiobiology).

In addition to these, tumor-infiltrating microbiomes (TiMs) are also involved in the RT resistance of cancer cells. For example, TiMs in nasopharyngeal cancer (NPC) can impact RT response by exacerbating tumor hypoxia and immune dysfunction [60]. The latter effect pertains to the 6th "R" of Radiobiology (see above). Recently, Guo et al. showed that fusobacterium nucleatum can improve NPC radioresistance by protecting mitochondria from radiation effects by promoting mitochondrial fusion and by inhibiting PANoptosis through the SLC7A5/leucine-mTORC1 axis [61]. *PANoptosis* is a newly identified type of cellular death having the features of pyroptosis, apoptosis, and necroptosis. Another recent study demonstrated that TiMs composition can affect the hypoxic gene expression in colorectal cancer thereby response to radiation [62]. In addition, TiMs can affect radiation response by releasing metabolites. Colbert et al. showed that intratumoral lactobacillus iners can improve radioresistance by releasing lactate into TME [63]. The protective impact of lactate on radiation effects might be due to mitochondrial activation. Cai et al. demonstrated that lactate can directly activate the ETC (electron transport chain) of mitochondria [64]. This action has a pair of benefits for cells favoring their radioresistance. ETC has two main end products: ATP and NADPH (Reduced nicotinamide adenine dinucleotide phosphate). Enhanced cellular ATP content can improve radioresistance by enabling ATP-dependent enzymes to repair the radiation-induced DNA damage [65] (see above); while NADPH can improve the redox capability of cancer cells by replenishing their glutathione contents [66].

Given the critical role of the TME in driving radioresistance, there has been growing global attention on developing targeted therapies against various TME components. For instance, Huang et al. identified that CAFs with activated STING/IFN-γ signaling exhibit enhanced radioresistance [67]. Similarly, an experimental study by Jang and Kim (2022) demonstrated that the recruitment of M_2 macrophages in glioblastoma tissue is associated with poor responses to RT. [68]. These and similar findings have prompted researchers to design clinical trials aimed at targeting TME components to improve the RT clinical efficacy.

However, despite the promising results seen in experimental studies, clinical trials have so far failed to deliver significant benefits. For example, pexidartinib, a drug designed to convert M_2 macrophages to the more tumor-suppressive M1 phenotype, was found to be ineffective in improving PFS (progression-free survival) or OS (overall survival) in patients with glioblastoma [69].

This lack of clinical success may be due to the complex interplay between cancer cells and the TME, as well as interactions among TME components themselves. Targeting a single TME component could be offset or compensated for by the activities of other TME cells. Therefore, future studies should consider strategies that target multiple TME components simultaneously, which may provide a more effective approach for overcoming radioresistance.

2.9 Conclusions

The evolution of radiobiology has been marked by the identification of seven distinct “Rs,” each representing a critical determinant of tumor and normal tissue responses to RT. From basic concepts like DNA repair and radiosensitivity to more recent insights into immune reactivation and the reinforcing role of the TME, these principles have greatly enhanced our understanding of RT mechanisms and its optimization.

As radiobiology continues to evolve, several promising avenues of research emerge. Advanced imaging and biomarker studies are expected to provide real-time insights into tumor oxygenation, DNA repair status, and immune activity, enabling personalized treatment strategies. Moreover, combining RT with immunotherapies, DNA repair inhibitors, or TME-modulating agents offers significant potential to overcome treatment resistance. In addition, leveraging AI-driven tools to optimize treatment planning based on the “R” framework may revolutionize personalized RT delivery.

In conclusion, the seven "Rs" of radiobiology provide a robust framework for understanding and enhancing RT outcomes. Continued interdisciplinary research is essential to improve these principles into practice, ultimately improving the patients' survival and quality of life.

References

1. Delaney G, Jacob S, Featherstone C, Barton M. The role of radiotherapy in cancer treatment: estimating optimal utilization from a review of evidence-based clinical guidelines. Cancer. 2005;104(6):1129–37.
2. Withers HR. The four R's of radiotherapy. In: Advances in radiation biology. Elsevier; 1975. p. 241–71.
3. Steel GG, McMillan TJ, Peacock JH. The 5Rs of radiobiology. Int J Radiat Biol. 1989;56(6):1045–8.
4. Boustani J, Grapin M, Laurent PA, Apetoh L, Mirjolet C. The 6th R of radiobiology: reactivation of anti-tumor immune response. Cancers (Basel). 2019;11(6)
5. Taghizadeh-Hesary F. Reinforcement by tumor microenvironment: the seventh "R" of radiobiology. Int J Radiat Oncol Biol Phys. 2023;
6. Houshyari M. A critical review on exploring tumor microenvironment's impacts on radioresistance. J Radiat Res Appl Sci. 2024;17(2):100937.
7. Holley AK, Miao L, St Clair DK, St Clair WH. Redox-modulated phenomena and radiation therapy: the central role of superoxide dismutases. Antioxid Redox Signal. 2014;20(10):1567–89.
8. Melia E, Parsons JL. DNA damage and repair dependencies of ionising radiation modalities. Biosci Rep. 2023;43(10):BSR20222586.
9. Khanna KK, Jackson SP. DNA double-strand breaks: signaling, repair and the cancer connection. Nat Genet. 2001;27(3):247–54.
10. Rahimian E, Amini A, Alikarami F, Pezeshki SMS, Saki N, Safa M. DNA repair pathways as guardians of the genome: therapeutic potential and possible prognostic role in hematologic neoplasms. DNA Repair. 2020;96:102951.
11. Gerelchuluun A, Manabe E, Ishikawa T, Sun L, Itoh K, Sakae T, et al. The major DNA repair pathway after both proton and carbon-ion radiation is NHEJ, but the HR pathway is more relevant in carbon ions. Radiat Res. 2015;183(3):345–56.
12. Loap P, Loirat D, Berger F, Rodrigues M, Bazire L, Pierga JY, et al. Concurrent Olaparib and radiotherapy in patients with triple-negative breast cancer: the phase 1 Olaparib and radiation therapy for triple-negative breast cancer trial. JAMA Oncol. 2022;8(12):1802–8.
13. Derby S, Jackson MR, Williams K, Stobo J, Kelly C, Sweeting L, et al. Concurrent Olaparib and radiation therapy in older patients with newly diagnosed glioblastoma: the phase 1 dose-escalation PARADIGM trial. Int J Radiat Oncol Biol Phys. 2024;118(5):1371–8.
14. Tuli R, Shiao SL, Nissen N, Tighiouart M, Kim S, Osipov A, et al. A phase 1 study of veliparib, a PARP-1/2 inhibitor, with gemcitabine and radiotherapy in locally advanced pancreatic cancer. EBioMedicine. 2019;40:375–81.
15. Samuels M, Falkenius J, Bar-Ad V, Dunst J, van Triest B, Yachnin J, et al. A phase 1 study of the DNA-PK inhibitor Peposertib in combination with radiation therapy with or without cisplatin in patients with advanced head and neck tumors. Int J Radiat Oncol Biol Phys. 2024;118(3):743–56.
16. Majd N, Yeboa D, Wang S, Weathers SPS, Yuan Y, Al Anssari H, et al. Phase I clinical trial of peposertib plus radiation in adults with newly diagnosed MGMT-unmethylated glioblastoma. JCO. 2024;42(16_suppl):2076.

17. Komaki R, Janjan NA, Ajani JA, Lynch PM, Fairweather JS, Raijman I, et al. Phase I study of irinotecan and concurrent radiation therapy for upper GI tumors. Oncology (Williston Park). 2000;14(12 Suppl 14):34–7.
18. Chen Z, Han F, Du Y, Shi H, Zhou W. Hypoxic microenvironment in cancer: molecular mechanisms and therapeutic interventions. Sig Transduct Target Ther. 2023;8(1):70.
19. Jun JC, Rathore A, Younas H, Gilkes D, Polotsky VY. Hypoxia-inducible factors and cancer. Curr Sleep Med Rep. 2017;3:1–10.
20. Supiot S, Rousseau C, Dore M, Chèze-Le-Rest C, Kandel-Aznar C, Potiron V, et al. Reoxygenation during radiotherapy in intermediate-risk prostate cancer. Radiother Oncol. 2019;133:16–9.
21. Mast JM, Kuppusamy P. Hyperoxygenation as a therapeutic supplement for treatment of triple negative breast cancer. Front Oncol. 2018;8:527.
22. Wang H, Jiang H, Van De Gucht M, De Ridder M. Hypoxic radioresistance: can ROS be the key to overcome it? Cancers (Basel). 2019;11(1):112.
23. Brahme A. Accurate analytical description of the cell survival and dose-response relationships at low and high doses and LETs. In: Comprehensive biomedical physics [Internet]. Elsevier; 2014 [cited 2025 Jan 19]. pp. 121–42. Available from: https://linkinghub.elsevier.com/retrieve/pii/B9780444536327008078.
24. Gray LH, Conger AD, Ebert M, Hornsey S, Scott OC. The concentration of oxygen dissolved in tissues at the time of irradiation as a factor in radiotherapy. Br J Radiol. 1953;26(312):638–48.
25. Nimalasena S, Anbalagan S, Box C, Yu S, Boult JKR, Bush N, et al. Tumour reoxygenation after intratumoural hydrogen peroxide (KORTUC) injection: a novel approach to enhance radiosensitivity. BJC Rep. 2024;2(1):78.
26. Nimalasena S, Gothard L, Anbalagan S, Allen S, Sinnett V, Mohammed K, et al. Intratumoral hydrogen peroxide with radiation therapy in locally advanced breast cancer: results from a phase 1 clinical trial. Int J Radiat Oncol Biol Phys. 2020;108(4):1019–29.
27. Tsien CI, Pugh SL, Dicker AP, Raizer JJ, Matuszak MM, Lallana EC, et al. NRG oncology/RTOG1205: a randomized phase II trial of concurrent bevacizumab and reirradiation versus bevacizumab alone as treatment for recurrent glioblastoma. JCO. 2023;41(6):1285–95.
28. Ingrosso G, Ponti E, Francolini G, Caini S, Fondelli S, Santini R, et al. Image-guided moderately hypofractionated radiotherapy for localized prostate cancer: a multicentric retrospective study (IPOPROMISE). Radiol Med. 2024;129(4):643–52.
29. Pryor DI, Martin JM, Millar JL, Day H, Ong WL, Skala M, et al. Evaluation of hypofractionated radiation therapy use and patient-reported outcomes in men with nonmetastatic prostate cancer in Australia and New Zealand. JAMA Netw Open. 2021;4(11):e2129647.
30. Lacas B, Bourhis J, Overgaard J, Zhang Q, Grégoire V, Nankivell M, et al. Role of radiotherapy fractionation in head and neck cancers (MARCH): an updated meta-analysis. Lancet Oncol. 2017;18(9):1221–37.
31. Rakhsha A, Azghandi S, Taghizadeh-Hesary F. COVID-19 pandemic and patients with cancer: the protocol of a clinical oncology center in Tehran. Iran Rep Pract Oncol Radiother. 2020;25(5):765–7.
32. Koike I, Ohmura M, Hata M, Takahashi N, Oka T, Ogino I, et al. FDG-PET scanning after radiation can predict tumor regrowth three months later. Int J Radiat Oncol Biol Phys. 2003;57(5):1231–8.
33. Krause M, Dubrovska A, Linge A, Baumann M. Cancer stem cells: radioresistance, prediction of radiotherapy outcome and specific targets for combined treatments. Adv Drug Deliv Rev. 2017;109:63–73.
34. Pajonk F, Vlashi E, McBride WH. Radiation resistance of cancer stem cells: the 4 R's of radiobiology revisited. Stem Cells. 2010;28(4):639–48.
35. Pellegata NS, Antoniono RJ, Redpath JL, Stanbridge EJ. DNA damage and p53-mediated cell cycle arrest: a reevaluation. Proc Natl Acad Sci USA. 1996;93(26):15209–14.
36. Schulz A, Meyer F, Dubrovska A, Borgmann K. Cancer stem cells and radioresistance: DNA repair and beyond. Cancers (Basel). 2019;11(6):862.

37. Phillips TM, McBride WH, Pajonk F. The response of CD24 –/low /CD44 + breast cancer–initiating cells to radiation. JNCI J Natl Cancer Inst. 2006;98(24):1777–85.
38. Hu W, Pei Y, Ning R, Li P, Zhang Z, Hong Z, et al. Immunomodulatory effects of carbon ion radiotherapy in patients with localized prostate cancer. J Cancer Res Clin Oncol. 2023;149(8):4533–45.
39. McMahon RA, D'Souza C, Neeson PJ, Siva S. Innate immunity: looking beyond T-cells in radiation and immunotherapy combinations. Neoplasia. 2023;46:100940.
40. Jin Y, Jiang J, Mao W, Bai M, Chen Q, Zhu J. Treatment strategies and molecular mechanism of radiotherapy combined with immunotherapy in colorectal cancer. Cancer Lett. 2024;591:216858.
41. Chang JY, Lin SH, Dong W, Liao Z, Gandhi SJ, Gay CM, et al. Stereotactic ablative radiotherapy with or without immunotherapy for early-stage or isolated lung parenchymal recurrent node-negative non-small-cell lung cancer: an open-label, randomised, phase 2 trial. Lancet. 2023;402(10405):871–81.
42. Gupta NK, Reichard CA, Large M, Leagre CA, Ney K, Liebross RH, et al. Pembrolizumab and concurrent radiation is an effective regimen for muscle invasive bladder cancer. JCO. 2020;38(15_suppl):e17020.
43. Siva S, Bressel M, Wood ST, Shaw MG, Loi S, Sandhu SK, et al. Stereotactic radiotherapy and short-course pembrolizumab for oligometastatic renal cell carcinoma—the RAPPORT trial. Eur Urol. 2022;81(4):364–72.
44. Karukonda P, Czito BG, Duffy E, Uronis HE, D'Amico TA, Strickler JH, et al. Pembrolizumab, radiotherapy, and chemotherapy in neoadjuvant treatment of malignant esophago-gastric diseases (PROCEED): assessment of pathologic response and toxicity in a prospective, phase II single-arm trial. JCO. 2023;41(16_suppl):4062.
45. Machiels JP, Tao Y, Licitra L, Burtness B, Tahara M, Rischin D, et al. Pembrolizumab plus concurrent chemoradiotherapy versus placebo plus concurrent chemoradiotherapy in patients with locally advanced squamous cell carcinoma of the head and neck (KEYNOTE-412): a randomised, double-blind, phase 3 trial. Lancet Oncol. 2024;25(5):572–87.
46. Vanpouille-Box C, Alard A, Aryankalayil MJ, Sarfraz Y, Diamond JM, Schneider RJ, et al. DNA exonuclease Trex1 regulates radiotherapy-induced tumour immunogenicity. Nat Commun. 2017;8(1):15618.
47. Cai X, Chiu YH, Chen ZJ. The cGAS-cGAMP-STING pathway of cytosolic DNA sensing and signaling. Mol Cell. 2014;54(2):289–96.
48. Ruan J, Zhou D, Zhang Y, Zhao D, Wei C, Hu K, et al. Hyper-fractionated radiotherapy as a bridging strategy to enhance CAR-T efficacy by regulating T-cell co-stimulatory molecules in relapsed/refractory diffuse large B-cell lymphoma. Front Immunol. 2024;15:1481080.
49. Koyama S, Nishikawa H. Mechanisms of regulatory T cell infiltration in tumors: implications for innovative immune precision therapies. J Immunother Cancer. 2021;9(7):e002591.
50. Liu YN, Yang JF, Huang DJ, Ni HH, Zhang CX, Zhang L, et al. Hypoxia induces mitochondrial defect that promotes T cell exhaustion in tumor microenvironment through MYC-regulated pathways. Front Immunol. 2020;11:1906.
51. Nocquet L, Juin PP, Souazé F. Mitochondria at center of exchanges between cancer cells and cancer-associated fibroblasts during tumor progression. Cancers (Basel). 2020:12.
52. Ippolito L, Morandi A, Taddei ML, Parri M, Comito G, Iscaro A, et al. Cancer-associated fibroblasts promote prostate cancer malignancy via metabolic rewiring and mitochondrial transfer. Oncogene. 2019;38:5339–55.
53. Zhang H, Yue J, Jiang Z, Zhou R, Xie R, Xu Y, et al. CAF-secreted CXCL1 conferred radioresistance by regulating DNA damage response in a ROS-dependent manner in esophageal squamous cell carcinoma. Cell Death Dis. 2017;8(5):e2790.
54. Behnam B, Fazilaty H, Ghadyani M, Fadavi P, Taghizadeh-Hesary F. Ciliated, mitochondria-rich postmitotic cells are immune-privileged, and mimic immunosuppressive microenvironment of tumor-initiating stem cells: from molecular anatomy to molecular pathway. Front Biosci (Landmark Ed). 2023;28(10):261.

55. Pérez-Tomás R, Pérez-Guillén I. Lactate in the tumor microenvironment: an essential molecule in cancer progression and treatment. Cancers (Basel). 2020;12(11):3244.
56. Bellone M, Calcinotto A, Filipazzi P, De Milito A, Fais S, Rivoltini L. The acidity of the tumor microenvironment is a mechanism of immune escape that can be overcome by proton pump inhibitors. Onco Targets Ther. 2013;2(1):e22058.
57. Uccello TP, Lesch ML, Ullman NA, Kintzel SA, Gradzewicz LB, Velagaleti T, et al. Radiation therapy exacerbates tumor-promoting innervation and nerve signaling in rectal cancer. Int J Radiat Oncol Biol Phys. 2023;115(3):733–45.
58. Hiller JG, Cole SW, Crone EM, Byrne DJ, Shackleford DM, Pang JMB, et al. Preoperative β-blockade with propranolol reduces biomarkers of metastasis in breast cancer: a phase II randomized trial. Clin Cancer Res. 2020;26(8):1803–11.
59. Geng Q, Li L, Shen Z, Zheng Y, Wang L, Xue R, et al. Norepinephrine inhibits CD8+ T-cell infiltration and function, inducing anti-PD-1 mAb resistance in lung adenocarcinoma. Br J Cancer. 2023;128(7):1223–35.
60. Ashique S, Houshyari M, Islam A, Pal R, Ghazanfar S, Taghizadeh-Hesary F. The role of microbiota in nasopharyngeal cancer: where do we stand? Oral Oncol. 2024;158:106982.
61. Guo S, Xing S, Wu Z, Chen F, Pan X, Li Q, et al. Leucine restriction ameliorates Fusobacterium nucleatum-driven malignant progression and radioresistance in nasopharyngeal carcinoma. Cell Rep Med. 2024;5(10):101753.
62. Benej M, Hoyd R, Kreamer M, Wheeler CE, Grencewicz DJ, Choueiry F, et al. The tumor microbiome reacts to hypoxia and can influence response to radiation treatment in colorectal cancer. Cancer Res Commun. 2024;4(7):1690–701.
63. Colbert LE, El Alam MB, Wang R, Karpinets T, Lo D, Lynn EJ, et al. Tumor-resident Lactobacillus iners confer chemoradiation resistance through lactate-induced metabolic rewiring. Cancer Cell. 2023;41(11):1945–1962.e11.
64. Cai X, Ng CP, Jones O, Fung TS, Ryu KW, Li D, et al. Lactate activates the mitochondrial electron transport chain independently of its metabolism. Mol Cell. 2023;83(21):3904–3920.e7.
65. Akbari H, Taghizadeh-Hesary F, Heike Y, Bahadori M. Cell energy: a new hypothesis in decoding cancer evolution. Arch Iran Med. 2019;22:733–5.
66. Ju HQ, Lin JF, Tian T, Xie D, Xu RH. NADPH homeostasis in cancer: functions, mechanisms and therapeutic implications. Sig Transduct Target Ther. 2020;5(1):231.
67. Huang L, Lu W, Wu R, Xu R, Zhang Z. Interferon-dependent cancer-associated fibroblasts render the responsiveness of neoadjuvant radiotherapy in colorectal cancer. Int J Radiat Oncol Biol Phys. 2024;120(2):e368.
68. Jang BS, Kim IA. Relationship between macrophage and radiosensitivity in human primary and recurrent glioblastoma: in silico analysis with publicly available datasets. Biomedicines. 2022;10(2):292.
69. Mendez JS, Cohen AL, Eckenstein M, Jensen RL, Burt LM, Salzman KL, et al. Phase 1b/2 study of orally administered pexidartinib in combination with radiation therapy and temozolomide in patients with newly diagnosed glioblastoma. Neurooncol Adv. 2024;6(1):vdae202.

Chapter 3
Radiosensitivity and Its Contributing Biologic Factors

Pedram Fadavi, Mohammad Houshyari, and Farzad Taghizadeh-Hesary

Abstract Radiosensitivity (RS) is a key determinant of radiotherapy (RT) efficacy, influencing how cancer cells respond to ionizing radiation. This response is governed by a complex interplay of biological mechanisms that regulate oxidative stress mitigation, repair mechanisms, and cell death evasion. This chapter explores the key factors contributing to intrinsic RS, the fifth "R" of radiobiology. To enhance their survival following radiation exposure, cancer cells upregulate antioxidant systems, which neutralize radiation-induced reactive oxygen species. Additionally, mitochondrial adaptations contribute to maintain redox homeostasis and promote radioresistance. Cancer cells also exploit DNA repair pathways, enabling efficient recovery from radiation-induced DNA damage. Beyond various repair mechanisms, cancerous cells prevent the apoptosis process through upregulating anti-apoptotic mediators and suppressing death-inducing signaling complex formation. Furthermore, resistance to other regulated cellular death pathways, such as pyroptosis, necroptosis, and ferroptosis, enhances tumor survival after RT. In addition, senescence induction can impair radiation response by creating an immunosuppressive tumor microenvironment that hinders antitumor immune activity. A deeper understanding of these resistance mechanisms is required for designing targeted therapeutics to enhance RT efficacy and clinical outcomes in radioresistant cancers.

Keywords Cancer · Radiotherapy · Radiosensitivity · Mitochondria · DNA repair

P. Fadavi
Radio Oncology Department, Iran University of Medical Sciences, Tehran, Iran

Breast health & cancer research center, Iran University of Medical Sciences, Tehran, Iran

M. Houshyari
Radio Oncology Department, Shahid Beheshti University of Medical Sciences, Tehran, Iran

F. Taghizadeh-Hesary (✉)
ENT and Head and Neck Research Center and Department, The Five Senses Health Institute, School of Medicine, Iran University of Medical Sciences, Tehran, Iran
e-mail: taghizadeh_hesary.f@iums.ac.ir

F. Taghizadeh-Hesary et al. (eds.), *From Radiobiology to Radiation Oncology*,
https://doi.org/10.1007/978-3-031-95724-6_3

3.1 Introduction

Radiotherapy (RT) stands as one of the mainstays of conventional cancer treatment [1]. Its success relies heavily on the intricate balance between maximizing damage to cancerous cells while saving healthy tissues. While the broad efficacy of RT is well-established, its clinical application reveals an intriguing puzzle: why do patients with identical cancer categories respond differently to the same RT regimen [2]? Compelling evidence suggests that the biological features of tumors, collectively named as *intrinsic radiosensitivity (RS)*, have a crucial role in shaping their reply to radiation.

The concept of inherent (intrinsic) RS was first proposed in the early 1980s through a novel study by Fertil and Malaise. This study demonstrated distinct survival curve patterns across different cell lines—variations that could not be explained by previously established radiation response-determining factors [3]. In the following years, several other studies confirmed these results [4–6]. These findings motivated Steel et al. to coin inherent RS as the 5th "R" of radiobiology [7]. This concept may partly explain why patients with the same type of cancer often exhibit varying clinical responses to a standard RT schedule. Consequently, researchers worldwide have sought to identify the factors that determine the inherent RS of cancers. This chapter explores the current understanding of the key factors influencing inherent RS—factors that are not addressed by the previous four "Rs" (see Chap. 2). Chapter 4 provides a detailed discussion on the current understanding about how radiosensitizers can improve the tumor response to radiation.

This section provides current knowledge on how cancerous cells can improve their radioresistance at three levels: (1) preventing cellular damage, (2) repairing cellular damage, and (3) preventing cellular death and senescence due to damage.

3.2 Preventing Cellular Damage: Counteracting Radiation-Induced Oxidative Stress

3.2.1 What Is Oxidative Stress?

Reactive oxygen species (ROS), including hydrogen peroxide (H_2O_2), superoxide anions (O_2^-), and hydroxyl radicals (OH•), are natural by-products of cellular metabolism, particularly through mitochondrial oxidative phosphorylation. Excessive accumulation of ROS, named oxidative stress (OxSt), is toxic and can lead to damage to cell components, ultimately triggering cell death [8]. The OxSt can arise in conditions in which the generated ROS is more than the cell's antioxidant defense capacity, such as RT [9].

Low levels of ROS have an important role in preserving cellular functions, including signaling and metabolism. They contribute to essential pathways such as the Nrf2/Notch pathway, which regulates the physiological antioxidant response [10]; the adenosine monophosphate-activated protein kinase (AMPK)/PGC-1α signaling pathway, which aids in restoring adenosine triphosphate (ATP) levels to safeguard DNA integrity [11, 12]; and various pathways involved in immune regulation [13]. Regarding cancer cells, a moderate level of ROS is beneficial, leading to genomic instability and facilitating oncogenic transformation [14]. It has been demonstrated that cancer cells benefit from sophisticated redox systems to keep their ROS level at an ideal level. These mechanisms include enhanced antioxidant systems, redox signaling alterations, and mitochondrial adaptation, all of which contribute to cancer progression and therapy resistance [15].

3.2.2 How Cancer Cells Counteract Radiation-Induced Oxidative Stress?

Antioxidants

To counteract ROS accumulation, cancer cells upregulate antioxidant defense mechanisms that neutralize OxSt and maintain redox homeostasis. These defenses are broadly classified into enzymatic and non-enzymatic antioxidants, each playing a distinct role in detoxifying ROS and protecting cellular components.

The enzymatic antioxidants include several antioxidant enzymes that are overexpressed in cancer cells to maintain a controlled ROS balance, including (1) superoxide dismutases (SODs), which catalyze the conversion of O_2^- into H_2O_2, a less reactive molecule [16]; (2) catalase that further detoxifies H_2O_2 by converting it into water and oxygen, thus preventing its accumulation and subsequent conversion into more harmful hydroxyl radicals [17].

In addition, cancer cells benefit from several non-enzymatic antioxidants, further improving their resistance against OxSt, including (1) glutathione (GSH), which is the most abundant intracellular antioxidant. GSH is essential for neutralizing ROS. Increased GSH levels in cancer cells contribute to resistance against oxidative damage induced by RT [18]. GSH overexpression has been associated with radioresistance in several malignancies, including breast cancer [19, 20], oral cancer [21], and melanoma [22], among others; (2) thioredoxin (Trx) system, which includes thioredoxin, thioredoxin reductase, and nicotinamide adenine dinucleotide phosphate (NADPH), helps to regulate redox signaling and supports cancer cell survival under OxSt [23]. Trx overexpression has been associated with radioresistance in several malignancies, including lung cancer [24], breast cancer [25], and glioma [26], among others.

Furthermore, cancer cells can utilize exogenous antioxidants such as vitamin E [27], which scavenge ROS and protect cellular components from OxSt-induced damage. Notably, a recent clinical study on vitamin C did not show its radioprotective effects [28].

Taken together, various antioxidant systems enable cancer cells to thrive in oxidative environments while avoiding apoptosis. Understanding these mechanisms provides opportunities for therapeutic interventions aimed at disrupting redox homeostasis in cancer cells.

Adapted Mitochondrial Metabolism

Historically, it was believed that mitochondria in cancer cells are dysfunctional. This assumption led researchers to propose a key concept in cancer metabolism, referred to as "aerobic glycolysis" or the "Warburg effect" [29]. According to this theory, cancerous cells mainly depend on glycolysis for energy production, even when oxygen is readily available. However, further studies have shown that a cancer cell's predominant metabolic pathway depends on its stage of progression. Specifically, oxic epithelial cells (i.e., living with enough environmental oxygen levels) rely on oxidative phosphorylation (OXPHOS), hypoxic epithelial cells depend on glycolysis, mesenchymal cells are OXPHOS-dependent, and cancer stem cells utilize both glycolysis and OXPHOS [30].

Compelling evidence shows that cancerous cells depend on functional mitochondria to meet their extensive metabolic demands. They enhance mitochondrial quality by upregulating metabolic pathways such as the AMPK/peroxisome proliferator-activated receptor-gamma coactivator (PGC-1) signaling pathway [31]. Additionally, cancerous cells can increase their mitochondrial quantity by stealing mitochondria from immune cells through nanoscale tunneling nanotubes [32]. This ability of cancer cells further diminishes the effects of radiation by interfering with immune reactivation, often referred to as the 6th "R" of radiobiology (see Chap. 2). Ikeda et al. showed that cancer cells can transfer their dysfunctional mitochondria to nearby immune cells, impairing their function and enabling cancerous cells to evade immune detection [33]. Beyond these mechanisms, components of the tumor microenvironment (TME) also enhance the mitochondrial capacity of cancer cells [34]. Notably, cancer-associated fibroblasts (CAFs) have been shown to transfer functional mitochondria to cancer cells [35], thereby increasing their radioresistance.

How can mitochondria prevent OxSt in cancerous cells? Mitochondria are essential in regulating OxSt in cancer cells by maintaining redox balance and preventing excessive damage from ROS. To counteract the OxSt, cancer cells rely on several mitochondrial defense mechanisms, including antioxidant production, redox buffering, and metabolic adaptation. One key antioxidant defense mechanism is the generation of NADPH, a crucial byproduct of the mitochondrial electron transport (ETC) chain. NADPH is essential for regenerating glutathione, the most abundant intracellular antioxidant [36]. The glutathione system plays an important function in detoxifying ROS and preserving cellular redox homeostasis. Additionally, mitochondria contain other antioxidant enzymes like SOD and catalase (see above). Cancerous cells can upregulate mitochondrial metabolic pathways like the pentose phosphate pathway (PPP) and the malic enzyme pathway to boost NADPH production, further enhancing antioxidant capacity [37, 38]. These adaptations allow cancer cells to survive in RT-induced OxSt. By leveraging these mitochondrial antioxidant systems, cancer cells can maintain redox homeostasis, prevent oxidative damage, and enhance their survival, ultimately contributing to radioresistance.

3.3 Repairing Cellular Damage

Cancer cells can counteract the detrimental effects of RT by enhancing their ability to repair damaged organelles and molecules or by synthesizing new ones. This section outlines the key mechanisms that cancer cells use in this process.

3.3.1 Improved DNA Damage Response

DNA molecules are the primary targets of RT, meaning that RT schedules or modalities that induce greater DNA damage can result in more effective tumor destruction. Radiation-induced DNA damage can occur either directly by destructing DNA or indirectly through the release of ROS that results in oxidative damage to DNA components. Tumor cells, however, often possess an enhanced ability to repair radiation-induced DNA damages [39]. Disabling these mechanisms can enhance cancer cells' RS [40]. A study using a panel of cell lines with DNA repair defects generated by CRISPR-Cas9 showed that while the extent of induced and residual double-strand breaks (DSBs) are key determinants of RS, these factors alone do not fully explain the variability in RS. The authors suggest that other signaling pathways, particularly those involved in metabolism and proliferation, may also play a crucial role in determining RS [41]. For further details, please refer to Chap. 2. In addition to DNA damage response, cancer cells benefit from dedicated mechanisms to recover their different components from radiation damage, as follows.

3.3.2 Autophagy

Autophagy, a term originating from the Greek words "auto" (self) and "phagein" (to consume), describes the cellular process where cells break down and reuse their own organelles. This process enables cancer cells to survive by cleaning up damaged organelles and providing essential building blocks for the synthesis of new organelles [42]. Kim et al. demonstrated that upon radiation, pro-survival autophagy is activated following endoplasmic reticulum stress [43]. Moreover, autophagy plays a critical role in mitigating radiation-induced damage. For instance, during mitophagy (the autophagic degradation of mitochondria), cancer cells attempt to replenish their mitochondrial content. This process enhances their resistance to radiation by improving their ability to scavenge OxSt generated by radiation (see above). Enhanced mitochondrial content can therefore serve as a key index of the intrinsic RS of cancerous cells. In support of this, cancer stem cells (CSCs), which exhibit more radioresistant characteristics, have been shown to possess higher mitochondrial content compared to surrounding cancer cells [44].

Additionally, a recent review by our team emphasized that mitochondrial rehabilitation contributes to radioresistance through multiple mechanisms. It facilitates DNA repair (1st "R" of radiobiology) by ensuring sufficient ATP supply, supports cell regeneration (3rd "R" of radiobiology) by providing essential macromolecules, promotes cell cycle progression (4th "R" of radiobiology) through the regulation of cyclin E function, and mitigates immune response activation (6th "R" of radiobiology) [45]. The latter effect is mediated by increasing PD-L1 (programmed cell death protein ligand 1) expression, an immune checkpoint, on cancer cells [46]. Consequently, targeting mitochondrial metabolism presents a promising therapeutic approach to improve the intrinsic RS of cancerous cells.

3.3.3 *Mitochondrial Biogenesis*

As noted in the previous section, mitochondria can serve as safeguards for cancer cells to protect them from radiation effects. In addition to mitophagy, cancer cells can improve their mitochondrial content in terms of quantity by improving their mitochondrial biogenesis pathways. One of the well-known up-regulators of mitochondrial biogenesis is the AMPK/PGC-1 α signaling pathway. Activation of this pathway can promote the replication of mitochondrial DNA and the production of mitochondrial proteins [47]. Furthermore, mitochondrial homeostasis is maintained by quality control processes such as fusion, fission, and mitophagy, which are regulated by the transcriptional coactivator PGC-1α [48]. RT can stimulate the overexpression of AMPK in cancerous cells [49]. As noted in the section 3.2.2 under the subheading of "Adapted Mitochondrial Metabolism," cancerous cells can increase their mitochondrial number by stealing mitochondria from immune cells [32] and receiving mitochondria from surrounding CAFs [35]. These processes would enhance the cancerous cells' resistance against RT effects.

3.3.4 *Endoplasmic Reticulum Stress*

Endoplasmic reticulum (ER) is a key biosynthetic hub for synthesizing proteins and lipids. It is divided into rough ER, which aids in protein synthesis, and smooth ER, which focuses on lipid production. Under normal conditions, the ER ensures proper protein folding and quality control. However, when exposed to stressors like radiation, disruptions in protein folding, Ca^{2+} homeostasis, or other cellular disturbances can overwhelm the ER's capacity—a phenomenon known as *ER stress*. This triggers a response known as the ER stress response, which initiates various adaptive mechanisms to restore cellular balance, such as the unfolded protein response (UPR) and pathways involved in protein degradation [50]. ER stress can improve cancer cells' radioresistance. Qiao et al. found that the protein kinase RNA-like endoplasmic reticulum kinase (PERK) / eukaryotic initiation factor-2 (eIF2α) signaling pathway of ER stress promotes radioresistance in oropharyngeal carcinoma through the activation of nuclear factor kappa-light-chain-enhancer of activated B cells (NF-κB). This study demonstrated that PERK-mediated NF-κB activation improves radioresistance by inhibiting cell apoptosis and G_2/M arrest as well as improving DNA damage response [51].

3.3.5 Protein Homeostasis

Proteome homeostasis, or proteostasis, is another mechanism by which cancer cells enhance their radioresistance. Through this process, cancer cells detect and degrade damaged proteins, thereby increasing their survival against the effects of RT. In human cells, the protein quality control system primarily consists of three key pathways: the ubiquitin-proteasome system, molecular chaperones, and the autophagy-lysosomal system [52]. Emerging evidence highlights the crucial role of the protein quality control systems in cancer cell radioresistance. For instance, deubiquitinases, a component of the ubiquitin-proteasome system, contribute to radioresistance by regulating DNA damage response, arresting the cell cycle progression through stabilizing cyclins and CDKs, inhibiting cellular death, and promoting the stabilization of HIF-1 (hypoxia-inducible factor-1) [53]. Similarly, chaperone proteins can improve cancer cell radioresistance. Schneider et al. demonstrated that the chaperone protein GRP78 enhances the radioresistance of head and neck squamous carcinoma (HNSCC) cells. Their study found that irradiated cancer cells can enhance the radioresistance of neighboring cancer cells by releasing GRP78-containing extracellular vesicles [54]. These findings highlight the importance of proteostasis in cancer cell radioresistance. Therefore, targeting these regulatory mechanisms presents a promising strategy for enhancing RS in cancer treatment.

3.4 Preventing Cellular Death and Inducing Senescence

3.4.1 Cells Fates Following Radiation

In response to radiation exposure, cells may undergo various forms of cell death, each characterized by distinct intracellular signaling pathways. The different kinds of programmed cell death include apoptosis, ferroptosis, cuproptosis, pyroptosis, necroptosis, and senescence (Fig. 3.1). Interestingly, some cancer cells may survive radiation exposure by entering cellular senescence, which is a viable but non-proliferative state in which cells cease dividing but remain metabolically active. This state of "dormancy" can allow cancer cells to evade cell death and potentially contribute to disease recurrence or treatment resistance [55].

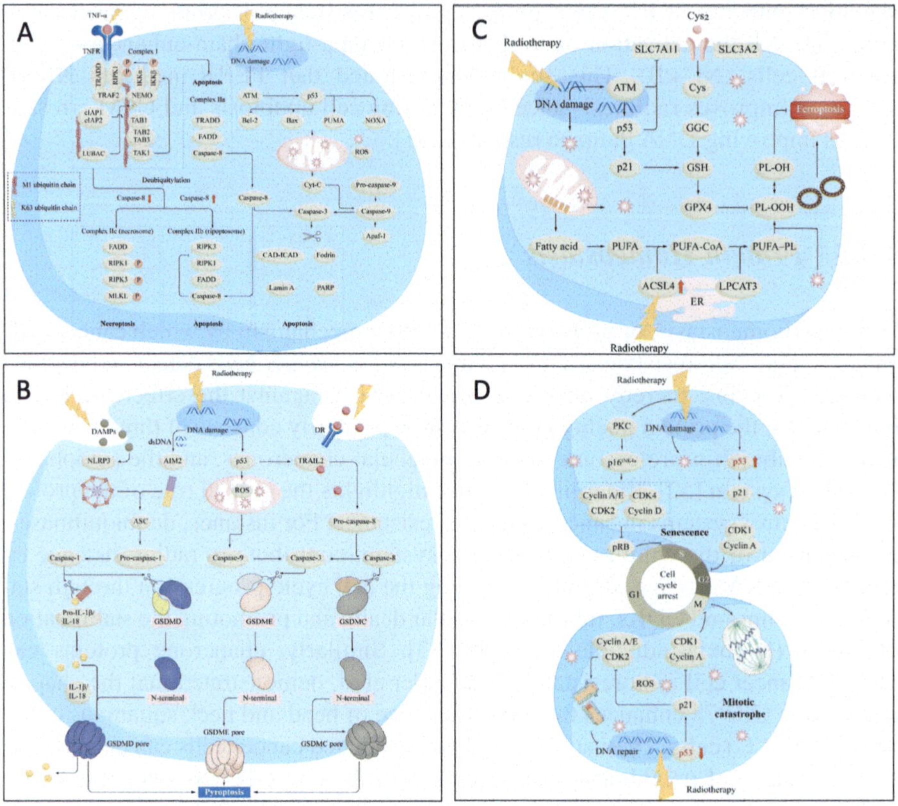

Fig. 3.1 Different cell fates following irradiation. (**a**) apoptosis and necroptosis; (**b**) pyroptosis; (**c**) ferroptosis; and (**d**) senescence. (The source of images is [55] under the CC BY 4.0 license)

In the following, the mechanisms by which cancer cells prevent various cellular death or induce senescence pathways are outlined.

3.4.2 Inhibition of Apoptotic Pathways

Inhibiting programmed cell death, or apoptosis, is a strategy that cancerous cells use to evade destruction, especially in response to DNA damage or therapeutic interventions like radiation. Apoptosis, which can occur through both intrinsic and extrinsic pathways, is a delicately regulated process that serves as an important defense mechanism. By eliminating cells with irreparable damage, apoptosis helps prevent further genetic instability or malignant transformation. However, cancer cells often manipulate the apoptotic pathways to ensure their survival, even in the face of significant cellular damage [56]. The key molecular mechanisms involved in apoptosis evasion are as follows:

Alterations in the p53 Pathway

P53, commonly known as the "guardian of the genome," is crucial in controlling cell fate decisions following DNA damage. Under normal conditions, p53 acts to halt the cell cycle, promote DNA repair, and trigger apoptosis when DNA damage is extensive [57]. However, in many cancers, p53 is either mutated or inactivated through genetic or epigenetic alterations, enabling cancer cells to escape apoptosis and continue to proliferate despite damage [58]. Meanwhile, several p53 mutations result in gain-of-function features in p53 supporting cancer cells' survival and proliferation [59]. A recent study revealed that the loss-of-function mutations in *TP53*, rather than gain-of-function mutations, are essential for promoting cancer survival, growth, metastasis, and resistance [60]. The p53 loss of function in cancerous cells allows them to bypass apoptosis, even under conditions of significant genomic instability, improving their radioresistance [58].

Regulation of the Bcl-2 Family of Proteins

The Bcl-2 family of proteins is crucial for controlling the intrinsic (mitochondrial) apoptotic pathway, which plays a vital role in initiating cell death in response to stress signals like DNA damage and OxSt. To avoid apoptosis driven by mitochondria, cancer cells can either reduce the expression of pro-apoptotic proteins (such as Bak and Bax) or enhance the levels of anti-apoptotic proteins (such as Bcl-xl, Bcl-2, and Mcl-1). These modifications can either take place following genetic or epigenetic alterations. An experimental study demonstrated that reversing these pathways can promote apoptosis in cancer cells [61]. Different experiments have demonstrated the association between downregulation of Bax [62] or Bak [63] and upregulation of Bcl-2 [64], Mcl-1 [65], or Bcl-xl [66] with radioresistance in different cancer cell types.

Inactivation of Death Receptor Signaling

The extrinsic apoptosis pathway is initiated when FasL or tumor necrosis factor α (TNF-α) attaches to their death receptors, Fas and TNFR1. This interaction triggers the assembly of the death-inducing signaling complex (DISC) and activates caspase-8, ultimately leading to apoptosis [67]. However, many cancer cells evade this pathway by disrupting DISC formation. They achieve this by downregulating the expression of death receptors [68], upregulating decoy receptors (binding death ligands without triggering apoptosis) [69], and increasing the expression of inhibitors that suppress death receptor signaling [70]. These mechanisms allow cancer cells to avoid radiation-induced apoptosis.

Upregulation of Inhibitors of Apoptosis Proteins (IAPs)

Cancer cells can directly suppress apoptosis by increasing the levels of inhibitors of apoptosis proteins (IAPs). This family of proteins blocks cell death by preventing the activation of caspases, which are the key enzymes involved in executing apoptosis. Examples of these include survivin, X-linked IAP (XIAP), and c-IAP1/2 (cellular IAPs) [71]. The importance of different IAPs in improving the radioresistance of cancer cells has been demonstrated in different experimental [72] and clinical studies [73].

3.4.3 Resistance to Necroptosis

Necroptosis is a controlled type of cell death resembling necrosis, which relies on the activation of receptor-interacting protein kinases (RIPK1,3) along with mixed lineage kinase domain-like protein (MLKL). Unlike apoptosis, which is caspase-dependent, necroptosis occurs when caspase-8 is inhibited or inactivated (Fig. 3.1a) [55].

Necroptosis can potentially contribute to radiation effects by aggravating antitumor immunity (6th "R" of radiobiology). Yang et al. found that ZBP1-MLKL activation in irradiated cancer cells can aggravate inflammatory cascades, thereby the antitumor immune response [74]. In response, cancerous cells can exploit RIP1 to prevent necroptosis and consequently reduce immune reactivation following radiation [75]. On the other hand, necroptosis can improve tumor resistance to radiation by inducing repopulation. Wang et al. demonstrated that radiation-induced necroptosis can propel cancer cells toward repopulation through the RIP1/RIP3/MLKL/JNK/IL-8 pathway [76]. These findings reflect the dual function of necroptosis in response to radiation. Further studies are invited to reveal the uncharted dimensions of radiation-induced necroptosis.

3.4.4 Resistance to Pyroptosis

Pyroptosis is a form of programmed cell death characterized by intense inflammation, driven by gasdermin proteins and executed through inflammasomes such as NLRP3, AIM2, and NLRC4. This mechanism triggers the formation of pores in the cell's outer membrane, leading to the extracellular release of pro-inflammatory signaling molecules (Fig. 3.1b) [77]. Similar to necroptosis, pyroptosis can stimulate an inflammatory response by releasing these pro-inflammatory cytokines from cancer cells. This characteristic of pyroptosis could potentially be exploited to enhance the immunogenic response to cancer cells during RT [78]. A recent study on colorectal cancer cells showed that cancerous cells can enhance their radioresistance by downregulating gasdermin E. The study found that upregulating gasdermin E aggravated the recruitment of natural killer (NK) cells, proposing that this protein can activate the immune system [79]. Therefore, understanding the mechanisms by which cancer cells inhibit pyroptosis and targeting these mechanisms could offer a promising strategy to enhance the effectiveness of RT in future studies.

3.4.5 Resistance to Ferroptosis

This is an iron-dependent cellular death featured by lipid peroxidation and oxidative damage. It is regulated by glutathione peroxidase 4 (GPX4), System Xc– (cystine/glutamate antiporter), and ferroptosis suppressor protein 1 (FSP1) (Fig. 3.1c) [55]. Cancer cells can enhance their resistance against ferroptosis in several ways, including FSP1 overexpression [80], GPX4 overexpression [81], and SLC7A11 (System Xc–) upregulation [82], among others. Experiments have demonstrated that ferroptosis-promoting agents can enhance cancer cells' RS. For example, Pan et al. realized that radioresistant non-small cell lung cancers have overexpression of GPX4. They found that erastin could improve RS of cancer cells by targeting GPX4 [83].

3.4.6 Resistance to Cuproptosis

Cuproptosis is a copper-dependent type of cellular death induced by the aggregation of mitochondrial lipoylated proteins and disruption of ferredoxin (FDX1). This kind of cellular death was introduced in 2022 [84]. To resist cuproptosis, tumors adjust their lipoylation pathways and downregulate FDX1, a key regulator of cuproptosis, in order to avoid cell death [85].

Because cuproptosis has been recently discovered, the information about its role in RS is scarce. Residual tumors post-RT can increase sensitivity to cuproptosis by upregulating FDX1 and LIAS. This study also suggested a radiosensitization strategy using a copper-containing nanocapsule to enhance cuproptosis upon radiation exposure, overcoming acquired radiation resistance and achieving a 40% cure rate in resistant tumor models [86]. Thus, targeting cuproptosis offers a promising approach to improve RT outcomes.

3.4.7 Senescence

Senescence is a cellular state in which cancer cells permanently exit the cell cycle while remaining metabolically active [87]. This unique characteristic initiates a cascade of processes that ultimately contribute to radioresistance. Senescent cancer cells exhibit a SASP (senescence-associated secretory phenotype), releasing a variety of factors, including chemokines. Earlier studies suggested that SASP could transform an immunologically "cold" tumor into a "hot" tumor, thereby enhancing radiation response through immune reactivation (the sixth "R" of radiobiology) [88]. However, several other studies present a contrasting perspective.

Experimental studies have shown that senescent cancer cells acidify the TME by increasing lactate production and secretion [89]. This acidic shift inhibits an effective antitumor immune response, as immune cells require a physiological pH for optimal function. Additionally, senescent cancer cells can recruit monocytes with immunosuppressive properties, further dampening immune-mediated tumor clearance [90]. These mechanisms collectively impair the effectiveness of RT by creating an environment that favors tumor persistence. Further studies are needed to explore targeted senolytic therapies (targeting senescence) that selectively eliminate senescent cancer cells, as well as strategies to neutralize their immunosuppressive effects, ultimately improving RT outcomes.

3.5 Conclusions

RS is a multifaceted trait governed by intricate biological mechanisms that dictate tumor response to RT. This chapter has explored the various factors contributing to radioresistance in cancer cells, including their ability to counteract radiation-induced OxSt, repair cellular damage, and evade cell death. Cancer cells employ robust antioxidant systems, mitochondrial adaptations, and proteostasis mechanisms to mitigate oxidative damage. Additionally, they exploit DNA repair pathways, autophagic processes, and metabolic reprogramming to sustain survival under radiotherapeutic stress. Furthermore, cancer cells evade different types of cell death while inducing senescence to enhance their persistence post-radiation exposure. Understanding these resistance mechanisms provides a foundation for future studies for developing novel therapeutic strategies aimed at improving RS.

References

1. Abdel-Wahab M, Gondhowiardjo SS, Rosa AA, Lievens Y, El-Haj N, Polo Rubio JA, et al. Global radiotherapy: current status and future directions—white paper. JCO Glob Oncol. 2021;7:827–42.
2. Baskar R, Dai J, Wenlong N, Yeo R, Yeoh KW. Biological response of cancer cells to radiation treatment. Front Mol Biosci. 2014;1:24.
3. Fertil B, Malaise EP. Inherent cellular radiosensitivity as a basic concept for human tumor radiotherapy. Int J Radiat Oncol Biol Phys. 1981;7(5):621–9.
4. Deacon J, Peckham MJ, Steel GG. The radioresponsiveness of human tumours and the initial slope ofthe cell survival curve. Radiother Oncol. 1984;2(4):317–23.
5. Gordon Steel G, Deacon JM, Duchesne GM, Horwich A, Kelland LR, Peacock JH. The dose-rate effect in human tumour cells. Radiother Oncol. 1987;9(4):299–310.
6. Kelland LR, Steel GG. Dose-rate effects in the radiation response of four human tumour xenografts. Radiother Oncol. 1986;7(3):259–68.

7. Steel GG, McMillan TJ, Peacock JH. The 5Rs of radiobiology. Int J Radiat Biol. 1989;56(6):1045–8.
8. An X, Yu W, Liu J, Tang D, Yang L, Chen X. Oxidative cell death in cancer: mechanisms and therapeutic opportunities. Cell Death Dis. 2024;15(8):556.
9. Park J, Choi J, Cho I, Sheen YY. Radiotherapy-induced oxidative stress and fibrosis in breast cancer are suppressed by vactosertib, a novel, orally bioavailable TGF-β/ALK5 inhibitor. Sci Rep. 2022;12(1):16104.
10. Paul MK, Bisht B, Darmawan DO, Chiou R, Ha VL, Wallace WD, et al. Dynamic changes in intracellular ROS levels regulate airway basal stem cell homeostasis through Nrf2-dependent notch signaling. Cell Stem Cell. 2014;15(2):199–214.
11. Rabinovitch RC, Samborska B, Faubert B, Ma EH, Gravel SP, Andrzejewski S, et al. AMPK maintains cellular metabolic homeostasis through regulation of mitochondrial reactive oxygen species. Cell Rep. 2017;21(1):1–9.
12. St-Pierre J, Drori S, Uldry M, Silvaggi JM, Rhee J, Jäger S, et al. Suppression of reactive oxygen species and neurodegeneration by the PGC-1 transcriptional coactivators. Cell. 2006;127(2):397–408.
13. Franchina DG, Dostert C, Brenner D. Reactive oxygen species: involvement in T cell signaling and metabolism. Trends Immunol. 2018;39(6):489–502.
14. Nakamura H, Takada K. Reactive oxygen species in cancer: current findings and future directions. Cancer Sci. 2021;112(10):3945–52.
15. Li P, Zhang D, Shen L, Dong K, Wu M, Ou Z, et al. Redox homeostasis protects mitochondria through accelerating ROS conversion to enhance hypoxia resistance in cancer cells. Sci Rep. 2016;6(1):22831.
16. Che M, Wang R, Li X, Wang HY, Zheng XFS. Expanding roles of superoxide dismutases in cell regulation and cancer. Drug Discov Today. 2016;21(1):143–9.
17. Glorieux C, Buc CP. Targeting catalase in cancer. Redox Biol. 2024;77:103404.
18. Chatterjee A. Reduced glutathione: a radioprotector or a modulator of DNA-repair activity? Nutrients. 2013;5(2):525–42.
19. Kim JC, Shin SO. Inductoin of Radioresistance by overexpression of glutathione S-transferase K1 (hGSTK1) in MCF-7 cells. Radiat Oncol J. 2001;19(4):381–8.
20. Nakashima R, Goto Y, Koyasu S, Kobayashi M, Morinibu A, Yoshimura M, et al. UCHL1-HIF-1 axis-mediated antioxidant property of cancer cells as a therapeutic target for radiosensitization. Sci Rep. 2017;7(1):6879.
21. Matsuoka Y, Yoshida R, Kawahara K, Sakata J, Arita H, Nkashima H, et al. The antioxidative stress regulator Nrf2 potentiates radioresistance of oral squamous cell carcinoma accompanied with metabolic modulation. Lab Investig. 2022;102(8):896–907.
22. Prezioso JA, Shields D, Wang N, Rosenstein M. Role of γ-glutamyltranspeptidase-mediated glutathione transport on the radiosensitivity of B16 melanoma variant cell lines. Int J Radiat Oncol Biol Phys. 1994;30(2):373–81.
23. Patwardhan RS, Sharma D, Sandur SK. Thioredoxin reductase: an emerging pharmacologic target for radiosensitization of cancer. Transl Oncol. 2022;17:101341.
24. Wang L, Fu JN, Wang JY, Jin CJ, Ren XY, Tan Q, et al. Selenium-containing thioredoxin reductase inhibitor ethaselen sensitizes non-small cell lung cancer to radiotherapy. Anti-Cancer Drugs. 2011;22(8):732–40.
25. Wang H, Bouzakoura S, de Mey S, Jiang H, Law K, Dufait I, et al. Auranofin radiosensitizes tumor cells through targeting thioredoxin reductase and resulting overproduction of reactive oxygen species. Oncotarget. 2017;8(22):35728–42.
26. Zhang H, Gu C, Yu J, Wang Z, Yuan X, Yang L, et al. Radiosensitization of glioma cells by TP53-induced glycolysis and apoptosis regulator knockdown is dependent on thioredoxin-1 nuclear translocation. Free Radic Biol Med. 2014;69:239–48.
27. Singh A, Bodas M, Wakabayashi N, Bunz F, Biswal S. Gain of Nrf2 function in non-small-cell lung cancer cells confers radioresistance. Antioxid Redox Signal. 2010;13(11):1627–37.
28. Khazaei S, Nilsson L, Adrian G, Tryggvadottir H, Konradsson E, Borgquist S, et al. Impact of combining vitamin C with radiation therapy in human breast cancer: does it matter? Oncotarget. 2022;13:439–53.

29. Warburg O, Posener K, Negelein E. The metabolism of cancer cells. Biochem Z. 1924;152:319–44.
30. Taghizadeh-Hesary F, Ghadyani M, Kashanchi F, Behnam B. Exploring TSGA10 function: a crosstalk or controlling mechanism in the signaling pathway of carcinogenesis? Cancers. 2024;16(17):3044.
31. Chaube B, Malvi P, Singh SV, Mohammad N, Viollet B, Bhat MK. AMPK maintains energy homeostasis and survival in cancer cells via regulating p38/PGC-1α-mediated mitochondrial biogenesis. Cell Death Discov. 2015;1:15063.
32. Saha T, Dash C, Jayabalan R, Khiste S, Kulkarni A, Kurmi K, et al. Intercellular nanotubes mediate mitochondrial trafficking between cancer and immune cells. Nat Nanotechnol. 2022;17(1):98–106.
33. Ikeda H, Kawase K, Nishi T, Watanabe T, Takenaga K, Inozume T, et al. Immune evasion through mitochondrial transfer in the tumour microenvironment. Nature [Internet]. 2025 Jan 22 [cited 2025 Jan 29]; Available from: https://www.nature.com/articles/s41586-024-08439-0.
34. Taghizadeh-Hesary F. Reinforcement by tumor microenvironment: the seventh "R" of radiobiology. Int J Radiat Oncol Biol Phys. 2023;
35. Ippolito L, Morandi A, Taddei ML, Parri M, Comito G, Iscaro A, et al. Cancer-associated fibroblasts promote prostate cancer malignancy via metabolic rewiring and mitochondrial transfer. Oncogene. 2019;38:5339–55.
36. Gansemer ER, McCommis KS, Martino M, King-McAlpin AQ, Potthoff MJ, Finck BN, et al. NADPH and glutathione redox link TCA cycle activity to endoplasmic reticulum homeostasis. iScience. 2020;23(5):101116.
37. Murai S, Ando A, Ebara S, Hirayama M, Satomi Y, Hara T. Inhibition of malic enzyme 1 disrupts cellular metabolism and leads to vulnerability in cancer cells in glucose-restricted conditions. Oncogenesis. 2017;6(5):e329.
38. Patra KC, Hay N. The pentose phosphate pathway and cancer. Trends Biochem Sci. 2014;39(8):347–54.
39. Lomax ME, Folkes LK, O'Neill P. Biological consequences of radiation-induced DNA damage: relevance to radiotherapy. Clin Oncol. 2013;25(10):578–85.
40. Hu L, Wu QQ, Wang WB, Jiang HG, Yang L, Liu Y, et al. Suppression of Ku80 correlates with radiosensitivity and telomere shortening in the U2OS telomerase-negative osteosarcoma cell line. Asian Pac J Cancer Prev. 2013;14(2):795–9.
41. Liberal FDCG, McMahon SJ. Characterization of intrinsic radiation sensitivity in a diverse panel of normal, cancerous and CRISPR-modified cell lines. Int J Mol Sci. 2023;24(9):7861.
42. Liu J, Wu Y, Meng S, Xu P, Li S, Li Y, et al. Selective autophagy in cancer: mechanisms, therapeutic implications, and future perspectives. Mol Cancer. 2024;23(1):22.
43. Kim KW, Moretti L, Mitchell LR, Jung DK, Lu B. Endoplasmic reticulum stress mediates radiation-induced autophagy by perk-eIF2alpha in caspase-3/7-deficient cells. Oncogene. 2010;29(22):3241–51.
44. Mauro-Lizcano M, Di Pisa F, Larrea Murillo L, Sugden CJ, Sotgia F, Lisanti MP. High mitochondrial DNA content is a key determinant of stemness, proliferation, cell migration, and cancer metastasis in vivo. Cell Death Dis. 2024;15(10):745.
45. Taghizadeh-Hesary F, Houshyari M, Farhadi M. Mitochondrial metabolism: a predictive biomarker of radiotherapy efficacy and toxicity. J Cancer Res Clin Oncol. 2023;149(9):6719–41.
46. Akbari H, Taghizadeh-Hesary F, Bahadori M. Mitochondria determine response to anti-programmed cell death protein-1 (anti-PD-1) immunotherapy: an evidence-based hypothesis. Mitochondrion. 2022;62:151–8.
47. Wu J, Li J, Feng B, Bi Z, Zhu G, Zhang Y, et al. Activation of AMPK-PGC-1α pathway ameliorates peritoneal dialysis related peritoneal fibrosis in mice by enhancing mitochondrial biogenesis. Ren Fail. 2022;44(1):1546–58.
48. Halling JF, Pilegaard H. PGC-1α-mediated regulation of mitochondrial function and physiological implications. Appl Physiol Nutr Metab. 2020;45(9):927–36.
49. Zannella VE, Cojocari D, Hilgendorf S, Vellanki RN, Chung S, Wouters BG, et al. AMPK regulates metabolism and survival in response to ionizing radiation. Radiother Oncol. 2011;99(3):293–9.

50. Zhang W, Shi Y, Oyang L, Cui S, Li S, Li J, et al. Endoplasmic reticulum stress—a key guardian in cancer. Cell Death Discov. 2024;10(1):343.
51. Qiao Q, Sun C, Han C, Han N, Zhang M, Li G. Endoplasmic reticulum stress pathway PERK-eIF2α confers radioresistance in oropharyngeal carcinoma by activating NF-κB. Cancer Sci. 2017;108(7):1421–31.
52. Chen XQ, Shen T, Fang SJ, Sun XM, Li GY, Li YF. Protein homeostasis in aging and cancer. Front Cell Dev Biol. 2023;11:1143532.
53. Cao X, Yan Z, Chen Z, Ge Y, Hu X, Peng F, et al. The emerging role of deubiquitinases in radiosensitivity. Int J Radiat Oncol Biol Phys. 2024;118(5):1347–70.
54. Schneider M, Winkler K, Kell R, Pfaffl MW, Atkinson MJ, Moertl S. The chaperone protein GRP78 promotes survival and migration of head and neck cancer after direct radiation exposure and extracellular vesicle-transfer. Front Oncol. 2022;12:842418.
55. Chen H, Han Z, Luo Q, Wang Y, Li Q, Zhou L, et al. Radiotherapy modulates tumor cell fate decisions: a review. Radiat Oncol. 2022;17(1):196.
56. Vucic D, Fairbrother WJ. The inhibitor of apoptosis proteins as therapeutic targets in cancer. Clin Cancer Res. 2007;13(20):5995–6000.
57. Feroz W, Sheikh AMA. Exploring the multiple roles of guardian of the genome: P53. Egypt J Med Hum Genet. 2020;21(1):49.
58. Marusyk A, Porter CC, Zaberezhnyy V, DeGregori J. Irradiation selects for p53-deficient hematopoietic progenitors. PLoS Biol. 2010;8(3):e1000324.
59. Alvarado-Ortiz E, de la Cruz-López KG, Becerril-Rico J, Sarabia-Sánchez MA, Ortiz-Sánchez E, García-Carrancá A. Mutant p53 gain-of-function: role in cancer development, progression, and therapeutic approaches. Front Cell Dev Biol. 2020;8:607670.
60. Wang Z, Burigotto M, Ghetti S, Vaillant F, Tan T, Capaldo BD, et al. Loss-of-function but not gain-of-function properties of mutant TP53 are critical for the proliferation, survival, and metastasis of a broad range of cancer cells. Cancer Discov. 2024;14(2):362–79.
61. Naseri MH, Mahdavi M, Davoodi J, Tackallou SH, Goudarzvand M, Neishabouri SH. Up regulation of Bax and down regulation of Bcl2 during 3-NC mediated apoptosis in human cancer cells. Cancer Cell Int. 2015;15:55.
62. Kim R, Inoue H, Toge T. Bax is an important determinant for radiation sensitivity in esophageal carcinoma cells. Int J Mol Med. 2004;14(4):697–706.
63. Yamana K, Inoue J, Yoshida R, Sakata J, Nakashima H, Arita H, et al. Extracellular vesicles derived from radioresistant oral squamous cell carcinoma cells contribute to the acquisition of radioresistance via the miR-503-3p-BAK axis. J Extracell Vesicle. 2021;10(14):e12169.
64. Hara T, Omura-Minamisawa M, Chao C, Nakagami Y, Ito M, Inoue T. Bcl-2 inhibitors potentiate the cytotoxic effects of radiation in Bcl-2 overexpressing radioresistant tumor cells. Int J Radiat Oncol Biol Phys. 2005;61(2):517–28.
65. Liang YY, Niu FY, Xu AA, Jiang LL, Liu CS, Liang HP, et al. Increased MCL-1 synthesis promotes irradiation-induced nasopharyngeal carcinoma radioresistance via regulation of the ROS/AKT loop. Cell Death Dis. 2022;13(2):131.
66. Pesch AM, Chandler BC, Michmerhuizen AR, Carter HM, Hirsh NH, Wilder-Romans K, et al. Bcl-xL inhibition Radiosensitizes *PIK3CA/PTEN* wild-type triple-negative breast cancers with low mcl-1 expression. Cancer Res Commun. 2022;2(7):679–93.
67. Mahdizadeh SJ, Thomas M, Eriksson LA. Reconstruction of the Fas-based death-inducing signaling complex (DISC) using a protein–protein docking meta-approach. J Chem Inf Model. 2021;61(7):3543–58.
68. Eichholtz-Wirth H, Sagan D. Altered signaling of TNFalpha-TNFR1 and SODD/BAG4 is responsible for radioresistance in human HT-R15 cells. Anticancer Res. 2002;22(1A):235–40.
69. Tang J, Cao Y, Zhang H, Wang R. Oxymatrine inhibits the development of radioresistance in NSCLC cells by reversing EMT through the DcR3/AKT/GSK-3β pathway. Arch Med Sci. 2024;20(5):1631–54.
70. Luo A, Wang W, Sima N, Lu Y, Zhou J, Xu G, et al. Short hairpin RNA targeting c-FLIP sensitizes human cervical adenocarcinoma Hela cells to chemotherapy and radiotherapy. Cancer Lett. 2008;271(2):323–32.

71. Altieri DC. Survivin and IAP proteins in cell-death mechanisms. Biochem J. 2010;430(2):199–205.
72. Asanuma K, Kobayashi D, Furuya D, Tsuji N, Yagihashi A, Watanabe N. A role for survivin in radioresistance of pancreatic cancer cells. Jpn J Cancer Res. 2002;93(9):1057–62.
73. Rödel F, Hoffmann J, Distel L, Herrmann M, Noisternig T, Papadopoulos T, et al. Survivin as a radioresistance factor, and prognostic and therapeutic target for radiotherapy in rectal cancer. Cancer Res. 2005;65(11):4881–7.
74. Yang Y, Wu M, Cao D, Yang C, Jin J, Wu L, et al. ZBP1-MLKL necroptotic signaling potentiates radiation-induced antitumor immunity via intratumoral STING pathway activation. Sci Adv. 2021;7(41):eabf6290.
75. Mannion J, Gifford V, Bellenie B, Fernando W, Ramos Garcia L, Wilson R, et al. A RIPK1-specific PROTAC degrader achieves potent antitumor activity by enhancing immunogenic cell death. Immunity. 2024;57(7):1514–32.e15.
76. Wang Y, Zhao M, He S, Luo Y, Zhao Y, Cheng J, et al. Necroptosis regulates tumor repopulation after radiotherapy via RIP1/RIP3/MLKL/JNK/IL8 pathway. J Exp Clin Cancer Res. 2019;38(1):461.
77. Wei X, Xie F, Zhou X, Wu Y, Yan H, Liu T, et al. Role of pyroptosis in inflammation and cancer. Cell Mol Immunol. 2022;19(9):971–92.
78. Li H, Yang T, Zhang J, Xue K, Ma X, Yu B, et al. Pyroptotic cell death: an emerging therapeutic opportunity for radiotherapy. Cell Death Discov. 2024;10(1):32.
79. Tan G, Lin C, Huang C, Chen B, Chen J, Shi Y, et al. Radiosensitivity of colorectal cancer and radiation-induced gut damages are regulated by gasdermin E. Cancer Lett. 2022;529:1–10.
80. Bersuker K, Hendricks JM, Li Z, Magtanong L, Ford B, Tang PH, et al. The CoQ oxidoreductase FSP1 acts parallel to GPX4 to inhibit ferroptosis. Nature. 2019;575(7784):688–92.
81. Wang Y, Zheng L, Shang W, Yang Z, Li T, Liu F, et al. Wnt/beta-catenin signaling confers ferroptosis resistance by targeting GPX4 in gastric cancer. Cell Death Differ. 2022;29(11):2190–202.
82. Wang X, Chen Y, Wang X, Tian H, Wang Y, Jin J, et al. Stem cell factor SOX2 confers ferroptosis resistance in lung cancer via upregulation of SLC7A11. Cancer Res. 2021;81(20):5217–29.
83. Pan X, Lin Z, Jiang D, Yu Y, Yang D, Zhou H, et al. Erastin decreases radioresistance of NSCLC cells partially by inducing GPX4-mediated ferroptosis. Oncol Lett. 2019;17(3):3001–8.
84. Tsvetkov P, Coy S, Petrova B, Dreishpoon M, Verma A, Abdusamad M, et al. Copper induces cell death by targeting lipoylated TCA cycle proteins. Science. 2022;375(6586):1254–61.
85. Yang W, Wang Y, Huang Y, Yu J, Wang T, Li C, et al. 4-Octyl itaconate inhibits aerobic glycolysis by targeting GAPDH to promote cuproptosis in colorectal cancer. Biomed Pharmacother. 2023;159:114301.
86. Liao Y, Wang D, Gu C, Wang X, Zhu S, Zheng Z, et al. A cuproptosis nanocapsule for cancer radiotherapy. Nat Nanotechnol. 2024;19(12):1892–902.
87. Wyld L, Bellantuono I, Tchkonia T, Morgan J, Turner O, Foss F, et al. Senescence and cancer: a review of clinical implications of senescence and Senotherapies. Cancers (Basel). 2020;12(8):2134.
88. Ruscetti M, Morris JP, Mezzadra R, Russell J, Leibold J, Romesser PB, et al. Senescence-induced vascular remodeling creates therapeutic vulnerabilities in pancreas cancer. Cell. 2020;181(2):424–41.e21.
89. Liao EC, Hsu YT, Chuah QY, Lee YJ, Hu JY, Huang TC, et al. Radiation induces senescence and a bystander effect through metabolic alterations. Cell Death Dis. 2014;5(5):–e1255.
90. Maggiorani D, Le O, Lisi V, Landais S, Moquin-Beaudry G, Lavallée VP, et al. Senescence drives immunotherapy resistance by inducing an immunosuppressive tumor microenvironment. Nat Commun. 2024;15(1):2435.

Chapter 4
Innovative Strategies to Enhance Radiosensitivity: Hyperthermia, Nanoparticles, and DNA Repair Inhibition

Reza Taleei, Dario Rodrigues, Arlene Oei, Sunil Krishnan, Khadijeh Koushki, Ramin Abolfath, and Michael Weinfeld

Abstract Radiation therapy (RT) is a key component of modern cancer care. It uses high-intensity ionizing radiation to target and control tumor cells. Despite advancements in RT techniques, challenges such as tumor heterogeneity and radio-resistant cells limit its clinical efficacy. This chapter explores innovative strategies to enhance radiosensitivity, focusing on hyperthermia, nanoparticle-based radiosensitizers, and DNA repair inhibition. Hyperthermia involves elevating tumor temperatures to 40–44 °C, enhancing blood flow, reoxygenating hypoxic regions, and disrupting DNA repair mechanisms, thereby increasing radiosensitivity. Clinical applications have demonstrated its efficacy across various cancer types, including cervical, breast, and head-and-neck cancers. Nanoparticle-based radiosensitizers utilize high atomic number elements like gold, gadolinium, and hafnium to amplify local radiation doses within tumors. These nanoparticles enhance the photoelectric effect, generating secondary electrons that increase oxidative stress and DNA dam-

R. Taleei (✉)
Medical Physics Division, Department of Radiation Oncology, Sidney Kimmel Medical College at Thomas Jefferson University, Philadelphia, PA, USA
e-mail: reza.taleei@jefferson.edu

D. Rodrigues
Department of Radiation Oncology, University of Maryland School of Medicine, Baltimore, MD, USA

A. Oei
Department of Radiation Oncology, and Laboratory for Experimental Oncology and Radiobiology (LEXOR), Cancer Center Amsterdam, Amsterdam UMC, University of Amsterdam, Amsterdam, The Netherlands

S. Krishnan · K. Koushki
Department of Neurosurgery, The University of Texas Health Science Center at Houston (UTHealth), Houston, TX, USA

R. Abolfath
Department of Physics and Astronomy, Howard University, Washington, DC, USA

M. Weinfeld
Department of Oncology, University of Alberta, Edmonton, AB, Canada

F. Taghizadeh-Hesary et al. (eds.), *From Radiobiology to Radiation Oncology*,
https://doi.org/10.1007/978-3-031-95724-6_4

age in cancer cells. The chapter discusses the potential of various nanoparticles, including gold, gadolinium, silver, and iron, highlighting their mechanisms, therapeutic benefits, and challenges in clinical translation. DNA repair inhibition targets specific pathways involved in repairing radiation-induced DNA damage. By inhibiting key enzymes in these pathways, such as PARP, ATR, and DNA-PK, radiosensitizers can enhance the effectiveness of RT. The chapter reviews the mechanisms of DNA repair inhibitors, their preclinical and clinical data, and the potential for combination therapies to improve treatment outcomes. In conclusion, integrating hyperthermia, nanoparticle-based radiosensitizers, and DNA repair inhibition represents a promising frontier in enhancing RT efficacy. These approaches offer multifaceted strategies to overcome tumor radioresistance, potentially revolutionizing cancer therapy and improving patient survival rates.

Keywords Hyperthermia · Nanoparticles · DNA repair · Radiosensitization

4.1 Introduction

Radiation therapy (RT) is a cornerstone of modern cancer treatment that uses high-intensity ionizing radiation, such as photons (X-rays, gamma rays) and particles (electrons, protons, carbon ions), to target and destroy tumor cells while minimizing damage to surrounding healthy tissues. RT induces cancer cell death directly by damaging DNA and indirectly by generating reactive oxygen species (ROS) through water radiolysis. However, its non-selective nature can damage normal tissues, causing dose-dependent side effects like skin burns, fibrosis, and organ dysfunction, requiring significant dose titration to limit collateral toxicity [1, 2]. Additionally, RT effectiveness depends on tissue oxygenation, as hypoxic tumors are more radioresistant and require higher radiation doses for effectiveness [3, 4]. Despite advancements in RT techniques, such as brachytherapy, intensity-modulated radiation therapy (IMRT), image-guided radiation therapy (IGRT), and particle therapy (protons and carbon ions), tumor heterogeneity and radioresistant cells limit the clinical efficacy of RT [1, 5, 6]. To overcome these challenges, various strategies have been explored to enhance RT efficacy while minimizing damage to healthy tissues. Radiosensitizers improve tumor cell susceptibility to radiation by maximizing cancer cell kill and overcoming radioresistance. Hyperthermia therapy, which elevates tumor temperature (40–44 °C), enhances blood flow, reoxygenates hypoxic regions, and disrupts DNA repair mechanisms, thereby increasing radiosensitivity. Additionally, nanoparticle-based radiosensitizers, composed of high atomic number elements like gold and gadolinium, amplify local radiation dose within tumors, improving treatment outcomes. Synthetic lethality, which targets specific DNA repair pathways, presents another promising approach to sensitize cancer cells to radiation selectively. This chapter explores these innovative approaches' mechanisms, clinical applications, and future directions, highlighting their potential to enhance RT efficacy in cancer therapy.

4.2 Hyperthermia Therapy

A key factor contributing to the resistance of tumors to radiotherapy and chemotherapy is the presence of hypoxic and nutrient-deprived regions within the tumor volume [7, 8]. Hyperthermia therapy, which involves local heating of the tumor to 40–44 °C for 1 hour, specifically targets these challenging areas. The efficacy of hyperthermia as an adjuvant treatment in combination with radiotherapy and/or chemotherapy has been well established and successfully applied across various cancer types, including locally advanced cervical cancer [9], recurrent breast cancer [10], glioblastoma [11], melanoma [12], sarcomas [13], and head-and-neck cancer [14], among many other tumor locations [15]. In this section, we will provide an overview of hyperthermia and discuss its radiosensitizing and chemosensitization effects.

The effectiveness of hyperthermia relies on its thermal effect on tumors that is dependent on both temperature and heating duration, as demonstrated in extensive preclinical [16–18] and clinical studies [19–23]. The range of therapeutic goals in tumors differs in literature, with treatment protocols varying by a few degrees centigrade, spanning within 39–45 °C [24]. This variation is attributed to evolving knowledge of hyperthermia effects in tissues and cells but also depends on the multimodality approach, as well as the anatomical location of the tumor. An expert panel at the 32nd Annual Meeting of the European Society for Hyperthermic Oncology (Germany, 2018) defined the goal temperatures to be 40–44 °C. This temperature range is based on robust clinical evidence supporting the combination of hyperthermia with radiotherapy. For chemosensitization, the same expert panel defined a slightly lower temperature goal of 39–43 °C to achieve a therapeutic effect. It should be emphasized that the benefit of hyperthermia has been clinically demonstrated only when applied in combination with radiotherapy and/or chemotherapy. Other thermal therapy modalities, like ablation therapy, aim at lethal thermal doses in localized tumors, i.e., acting as a standalone therapy, with temperatures typically in excess of 48 °C up to 100 °C for a few minutes [25].

In current practice, hyperthermia is typically administered 1–3 times per week before/after radiotherapy within a 0.5–4 h window, while with chemotherapy, hyperthermia is applied simultaneously or shortly before the chemotherapeutic regimen [26]. As heat is applied, an increase in blood flow alters the heating dynamics, requiring precise control of the heat delivery. If temperatures reach levels above 43–44 °C, patients can experience pain, formation of edema, and thermal blistering, but continuous temperature monitoring, surface water cooling, and patient feedback contribute to a favorable treatment safety profile [27, 28]. Most commonly used hyperthermia devices are based on electromagnetic techniques, such as radiofrequency (RF) and microwave (MW) applicators. These devices can be invasive or non-invasive and are classified as local, regional, or whole-body hyperthermia systems [29]. Local hyperthermia can be applied to both superficial and deep-seated tumors using external, intraluminal, or interstitial MW methods. Regional hyperthermia, typically used for treating deep pelvic or abdominal tumors, is delivered via RF-based applicators. The key distinction between these modalities lies in the

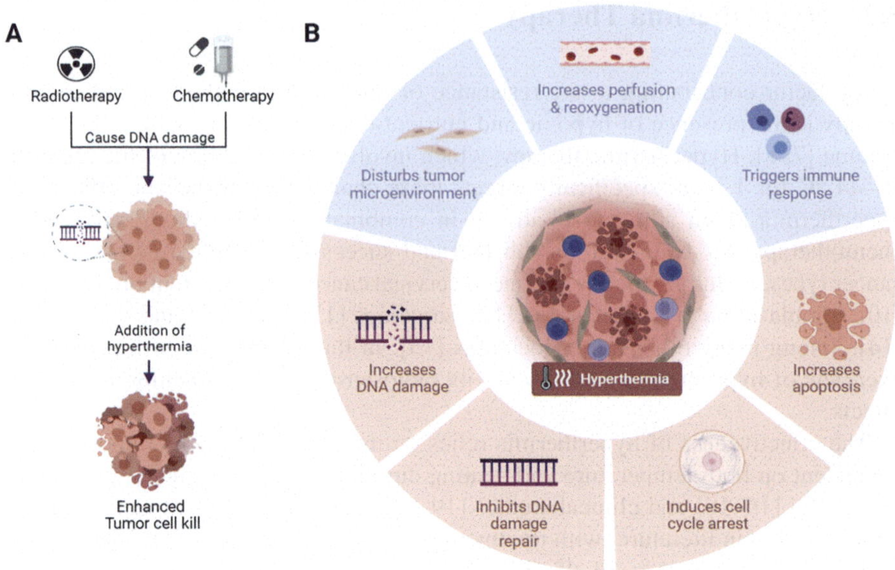

Fig. 4.1 Mechanisms of hyperthermia in cancer treatment. (**a**) Hyperthermia enhances the effectiveness of radiotherapy and chemotherapy, resulting in enhanced tumor cell death. (**b**) Biological mechanisms of hyperthermia: Its effects on the tumor microenvironment (outlined in purple) and tumor cells (highlighted in red) contribute to the sensitization of radiotherapy and chemotherapy. (This figure was made using BioRender)

operating frequency: MW (300 MHz - 300 GHz) is used for localized heating, whereas RF (3 kHz - 300 MHz) is utilized for deep regional heating. Whole-body hyperthermia uses infrared radiant heat devices and is most often used in advanced or metastasized cancers of the lung, ovary, colon, pancreas, and sarcomas [30].

Hyperthermia enhances the effect of both radiotherapy and chemotherapy (Fig. 4.1a). Local heating of the tumor induces radiosensitization by affecting tumor cells and their microenvironment (Fig. 4.1b). On a macroscopic level, HT enhances tumor blood perfusion, leading to reoxygenation of hypoxic regions and thereby changes in the tumor microenvironment [18]. This improved oxygenation increases the radiation-induced DNA damage and facilitates the fixation of DNA breaks, thereby enhancing the effectiveness of radiotherapy. In addition, improved blood perfusion attracts and promotes leakage of chemotherapeutic agents into the locally heated tumor tissue, further sensitizing the tumor to chemotherapy. On a microscopic level, hyperthermia impairs DNA damage repair mechanisms induced by radiotherapy and chemotherapy, resulting in a higher accumulation of unrepaired DNA breaks. DNA repair occurs primarily through two major mechanisms: homologous recombination (HR) and non-homologous end joining (NHEJ). HR operates mainly during the S-phase of the cell cycle, as it requires a sister chromatid to accurately repair the DNA break, using the sister chromatid. As a consequence, this pathway is considered error-free. In contrast, NHEJ is active throughout the entire cell cycle and is thus an error-prone process that repairs approximately 80% of DNA breaks within hours. Hyperthermia has been shown to temporarily inhibit critical repair proteins such as BRCA2 (in HR) and DNA-PKcs (in NHEJ), resulting in increased cell cycle arrest and elevated levels of apoptosis [31, 32]. As a result, the synergistic effect of radiotherapy and hyperthermia is most potent when these treatments are administered simultaneously or within a 4-h window [23, 33–35].

In addition to disrupting DNA repair, hyperthermia interferes with critical cell cycle regulators, especially in tumors originating from an infection with the human papillomavirus [36, 37]. Disruption of these viral tumor suppressor proteins contributes to increased cancer-related cell death. Recent studies have further demonstrated that hyperthermia can induce local and systemic immune responses, further enhancing its therapeutic potential [36, 38–44]. Despite robust clinical evidence supporting its efficacy, hyperthermia remains underutilized in radiation and medical oncology. Barriers to its widespread adoption include complex hardware requirements, limited treatment planning software, a shortage of trained practitioners, and prolonged treatment times. However, as technologies for the safe and effective delivery of hyperthermia continue to advance, its broader integration into multidisciplinary cancer care will become increasingly attainable, potentially revolutionizing cancer treatment care.

4.3 Enhancement of Radiation Effects by Nanoparticles

Nanoparticle (NP) formulations enable selective, preferential, and high-dose drug delivery to cancer cells [45]. Aside from delivering radiosensitizers to tumor cells, when NPs are made of high atomic number (Z) elements, they can focally amplify radiation dose within tumors, serving as radiosensitizers themselves without the need for a drug to synergize with RT.

4.3.1 Radiosensitization by Physical Dose Enhancement

High-Z metallic NPs can enhance the effectiveness of RT by augmenting the local radiation dose within tumors, resulting in a higher absorbed dose than the delivered dose. By enhancing photoelectric interactions between incident radiation and tumor tissues, these NPs increase secondary electron production, thereby amplifying tumor cell killing by both direct and indirect mechanisms.

Such high-Z metals like gadolinium (Z = 64) and gold (Z = 79) absorb X-ray radiation strongly compared to normal tissues owing to their higher mass-energy absorption coefficients [46, 47]. Enhanced photoelectric interactions between incident radiation and inner shell electrons increase the yield of secondary electrons per delivered dose, scaling as a function of Z^{3-4}. For gold, this translates to the equivalent of 10^3–10^4 assuming, that hydrogen, carbon, nitrogen, and oxygen have an

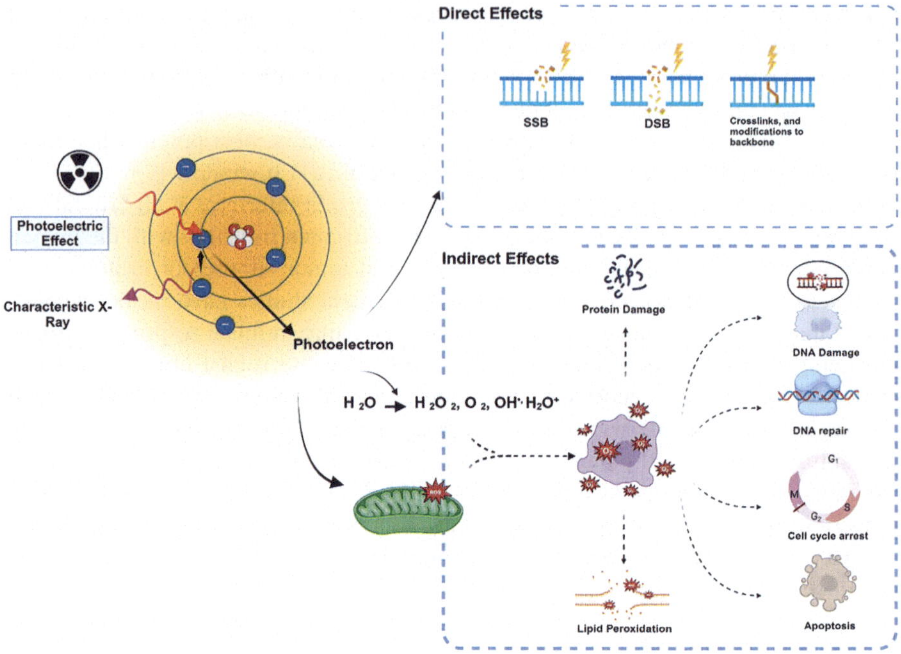

Fig. 4.2 Schematic illustration of interactions of ionizing radiation with atomic structures and consequent downstream biological consequences. *SSB* single-strand breaks, *DSB* double-strand breaks

average Z of roughly 8 and gold has a Z of roughly 80 (i.e., a tenfold increase of Z in tumor tissues can increase dose deposited by 1000–10,000-fold). In turn, these secondary electrons generate more free radicals to effectively amplify oxidative stress from RT and enhance DNA damage following RT (see Fig. 4.2). As an extension of these downstream biological effects, the increased 'absorbed dose' beyond the 'delivered dose' could transform radioresistant tumors into radiosensitive ones. Depending on the nature and location of the high-Z metallic NP in the cell, the resulting complex DNA damage may also overwhelm DNA repair pathways and mechanisms, another limitation of conventional RT [48–50].

4.3.2 *Gold Nanoparticles*

Gold nanoparticles (AuNPs) present desirable characteristics such as adjustable size, low toxicity, biocompatibility, high mass-attenuation coefficient, and strong photoelectric absorption, which make them ideal radiosensitizers. Indeed, Monte Carlo simulation algorithms suggest that the dose enhancement ratio (DER) using gold as in nanoparticle-enhanced RT is among the highest for a metallic NP seeking to enhance RT outcomes. [51].

The radiosensitization mechanism of AuNPs is complex and still under investigation. The most widely accepted explanation is that the high Z of AuNPs enhances photoelectron production, increasing radiation dose delivery to tumor cells [52]. Upon irradiation, AuNPs generate ROS and damage double-stranded DNA [53]. Additionally, AuNPs may modulate the cell cycle by accelerating the G1/S-phase and arresting the G2/M phase to radiosensitize cancer cells [54, 55]. The cell cycle changes are accompanied by p53 and cyclin A downregulation and cyclin B1 and E upregulation [54]. AuNPs have also been found to bind to DNA via electrostatic interactions, promoting strand breakage. Their proximity to nuclear DNA further amplifies radiation-induced cytotoxicity, maximizing therapeutic efficacy [1, 56].

AuNPs can be decorated with biomolecules such as polyethylene glycol (PEG) to evade reticuloendothelial capture and increase circulation time and/or with peptides and antibodies to afford selective anchoring to cell-surface receptors on cancer cells for uptake and retention within cancers, thereby increasing tumor-to-normal tissue accumulation ratios [57]. As needed, AuNPs can also be further functionalized to penetrate the nucleus of cancer cells, where the close proximity to nuclear DNA could amplify complex DNA damage and result in more significant radiation-induced cytotoxicity [1, 56]. These properties, along with their affordability, ease of fabrication, and an established history of clinical use of gold salts as drugs, have established AuNPs at the forefront of biocompatible and efficient nano-radiosensitizers.

In the past two decades, numerous in vitro and in vivo studies have explored the efficacy and therapeutic potential of AuNPs as nano enhancers. Multiple research groups have employed AuNPs of varying sizes in diverse tumor types and models to significantly enhance radiotherapeutic efficacy [58–60], translating to drastically increased median tumor doubling time, one-year survival rate, and long-term survival even in classically 'radioresistant' tumors such as gliomas, sarcomas, and some head-and-neck squamous cell carcinomas [61–65]. The challenge has been to then translate this from the bench to the bedside, with many roadblocks related to the dependence of radiosensitization magnitude on size, shape, concentration, surface chemistry, and active targeting moiety; challenges with good manufacturing practice fabrication and good laboratory practice toxicological testing; and scale-up production [58–60].

4.3.3 Gadolinium Nanoparticles

Similar to AuNPs, another high-Z metal is gadolinium (Z = 64), which can significantly enhance radiosensitization [66]. Gadolinium (^{64}Gd)-based NPs exhibit a strong ability to capture X-ray photons and enhance photoelectric scattering [67]. These NPs have been widely utilized for radiosensitization [68] and neutron capture therapy [69]. Additionally, their inherent paramagnetic properties allow easy visualization in vivo by magnetic resonance imaging (MRI) [70].

Several studies have reported the radiosensitization potential of gadolinium-based NPs [71–73]. Hyaluronic acid-functionalized gadolinium oxide (Gd_2O_3) NPs exhibit low cytotoxicity, good biocompatibility, and effective tumor cell uptake via endocytosis, which enables tumor radiosensitization by increasing apoptosis and arresting the cell cycle [74]. Ultrasmall Gd-based NPs (AGuIX) have shown promise as both radiosensitizers and MRI contrast agents [75]. These are sub-5 nm NPs made of a polysiloxane core and an average of 10–12 dodecane tetraacetic acid (DOTA) pendants, each complexing a gadolinium atom within it. Preclinical studies have demonstrated their ability to enhance radiation efficiency, improve survival in various tumor models, and amplify radiotherapy-induced oxidative stress [76, 77]. Currently, AGuIXs are being investigated in clinical trials to treat brain metastases and cervical cancer, among other cancer types [78]. Motexafin gadolinium (gadolinium texaphyrin) is a synthetic, aromatic macrocycle compound that forms an irreversible, stable coordination complex with gadolinium (III) that has been evaluated for its ability to sensitize tumors to RT. Although not technically a nanoparticle, we include this here as it falls within the broad theme of a metallic construct with radiosensitizing potential. Early results suggested that motexafin gadolinium significantly prolonged time to neurological progression when added to whole-brain radiation therapy in lung cancer patients with brain metastases; however, an improvement in overall survival was more elusive [79, 80].

4.3.4 Silver Nanoparticles

Silver (^{47}Ag) NPs have demonstrated significant radiosensitization potential, as noted in studies on glioma [81, 82], breast cancer [83], hepatocellular cancer [84], lung cancer [85], and leukemia cells. While their radiosensitization efficacy is variable, one study suggested that smaller AgNPs (around 20 nm) outperform larger particles in radiosensitization capability, possibly via interaction with the cellular DNA repair system [82]. AgNPs have even demonstrated more prominent radiosensitization compared to AuNPs in glioma therapy, possibly by inducing higher levels of autophagy and apoptosis [86, 87]. However, AgNPs are less inert than AuNPs, and their mechanisms of radiation sensitization effects may be more complex. Findings have shown the potential of AgNPs to enhance radiotherapy outcomes, especially when combined with targeted modifications or hybrid formulations, in order to increase their intracellular uptake and improve tumor accumulation [82, 88, 89].

4.3.5 Hafnium Nanoparticles

Hafnium (^{72}Hf) and its oxide nanoparticles (HfO_2) are significant potential radiosensitizers. Their inert features and negligible toxicity, coupled with radiosensitization capabilities consequent to ROS generation, apoptosis induction, and reduced DNA repair, make them promising candidates for radiosensitization [90]. Functionalized HfO_2 NPs, such as NBTXR3, have demonstrated promising results in both preclinical and clinical trials [91]. Additionally, NBTXR3 NPs have been shown to activate anti-tumor immune responses, which could enhance anti-cancer therapy [92]. Phase II/III clinical trials have demonstrated promising results in patients with soft tissue sarcomas, indicating the potential of HfO_2 nanoparticles in enhancing radiotherapeutic efficacy [93].

4.3.6 Iron Nanoparticles

Superparamagnetic iron oxide nanoparticles (SPIONs), composed of magnetite (Fe_3O_4) or maghemite (γFe_2O_3), are promising agents in cancer therapies. They can be visualized by MRI and used for targeted delivery of drugs/genes to tumors driven by an external magnetic force or for thermal ablation of tumors exposed to alternating magnetic fields [94, 95]. SPIONs are effective radiosensitizers using multiple mechanisms, including generating ROS leading to DNA damage and apoptosis [96, 97], inducing ferroptosis [98–100], and cell cycle arrests [101, 102]. SPIONs have been used as carriers for chemotherapeutic agents like cisplatin [103] and genetic material [104] and in combination with other nanoparticles [105] to enhance radiosensitivity [67].

Other high-Z metal NPs have also been investigated, including platinum (^{78}Pt) NPs that have been reported to significantly enhance lethal DNA damage nearly twofold by amplifying local energy deposition following RT [106] and bismuth (^{83}Bi) NPs that could, in principle, exhibit higher radiation dose enhancement factors than gold and platinum NPs under similar conditions due to their slightly higher atomic number [107].

4.3.7 Non-metal-Based Radiosensitizers

Silica has been explored as a carrier or coating for heavy metal radiosensitizers like gold [108, 109] and FeO_4 [110]. Silica NPs alone show limited radiosensitization, while modifications such as the aminosilanized oxidized silicon NPs enhance ROS generation, mitochondrial stress, and therapeutic efficacy [111]. Polymeric NPs are also extensively utilized as effective delivery vehicles for radiosensitizers. Poly(lactic-co-glycolic acid) (PLGA)-functionalized nanographene oxide (NGO) and polycaprolactone (PCL)-PEG-coated iron oxide NPs loaded with IUdR (5-iodo2′deoxyuridine) demonstrated enhanced radiosensitization and hyperthermia effects, significantly reducing glioblastoma cell colony formation in combination with radiotherapy and near-infrared (NIR) irradiation [112, 113]. Paclitaxel PLGA NPs have also demonstrated the sensitization of tumor cell lines to radiation [114]. These findings revealed the potential of PNPs to improve treatment efficacy by effective delivery into cells.

4.3.8 Challenges Using NPs for Radiosensitization

Although NPs hold promise as radiosensitizers, numerous biological obstacles limit their clinical translatability. While in circulation, opsonization by serum proteins and the subsequent entrapment of NPs by the reticuloendothelial system (RES) lead to nonspecific accumulation in the liver and the spleen. Long-term exposure has reportedly resulted in direct tissue damage and altered gene expression [115, 116]. At the tissue level, ensuring intratumoral uptake of NPs relies on the enhanced permeability and retention (EPR) effect, which exploits the leaky tumor vasculature and impaired lymphatic drainage to effectively drain and retain NPs in tumors [117]. However, the efficiency of such passive targeting is limited by tumor vascular heterogeneity, more so in the necrotic regions. Further, the clinical manifestation of the EPR effect remains uncertain and rather limited [118].

Modulating the physicochemical characteristics of NPs, such as their size, shape, charge, and surface coating, can limit their capture by the RES while ensuring prolonged circulation time to then enable EPR-mediated tumor accumulation. While larger NPs exploit the EPR effect to increase accumulation, they are prone to RES capture [119]. Conversely, small NPs (10–30 nm) exhibit longer circulatory half-lives and higher penetration in tumors [120–123]. Ultrasmall particles (5–10 nm) are cleared by the kidneys and, therefore, avoid sequestration in the liver and spleen. Further, endogenous stimuli (low pH, upregulated enzymes) can be combined with exogenous light and ultrasound to impart size- and charge-modulation to trigger NP aggregation or disassembly [124–126]. Surface modification of NPs by passivating agents such as PEG [127] and zwitterionic groups [126, 128–130] can reduce nonspecific protein adsorption and opsonization. This aids in evading immune surveillance and RES uptake [130, 131]. Biomimetic cell membrane coatings such as red blood cells, leukocytes, platelet membranes, and exosomes are also viewed as biocompatible solutions to evade RES uptake and possibly offer some tumor targeting owing to surface proteins that recognize molecular markers on tumor cells [132–136]. Finally, active targeting strategies can augment NP accumulation by virtue of targeting moieties (antibodies, aptamers, peptides, small high-affinity ligands) on their surface against antigens in the tumor vasculature, the tumor milieu, or the tumor cells [137–139].

4.4 Radiosensitization by DNA Repair Inhibition

Since DNA is considered the primary cellular target for radiation-induced toxicity, it is not surprising that DNA repair is a major focus of research into the clinical response to radiation. Much of this was stimulated by the early finding linking the radiosensitivity of ataxia telangiectasia cells to defective DNA damage response [140, 141]. Since then, we have come to learn considerably more about the key repair pathways responsible for responding to and repairing IR-induced DNA damage. These include the base-excision repair (BER), single-strand break repair (SSBR), and double-strand break repair (DSBR) pathways, as well as regulatory pathways involving cell cycle checkpoints following the induction of DNA damage and stalled replication forks. Thus, targeting key genes and proteins within these pathways, primarily by small molecule inhibitors, has become an important approach to increasing radiation sensitivity.

BER and SSBR pathways handle relatively minor, dispersed base lesions, such as 8-oxoguanine and thymine glycol, and single-strand breaks that frequently contain termini that require processing in addition to rejoining [142, 143]. Key enzymes in these pathways include several DNA glycosylases, e.g., OGG1, poly(ADP-ribose) polymerases 1 and 2 (PARP1 and 2), AP endonuclease 1 (APE1), DNA polymerase β (Pol β), polynucleotide kinase/phosphatase (PNKP), FEN1, and DNA ligases I and III.

As introduced in Sect. 4.2, the two major DSBR pathways are NHEJ and HR, with NHEJ being regarded as error-prone and active throughout the cell cycle (except during mitosis) [144]. On the other hand, HR requires an intact template, i.e., a sister chromatid, and therefore is active only during the late S and G2 phases of the cell cycle, thus being considered to be error-free [145]. Key enzymes in these DSBR pathways include DNA-PKcs, RAD51 recombinase, MRE11 nuclease, BRCA1/2, and WRN DNA helicase. There are also backup DSBR pathways, such as "alternative end-joining" (AltEJ). A key player in this pathway is Pol θ, which is required to rejoin resected DNA strand break termini possessing a degree of microhomology [146].

The repair of stalled replication forks is regulated by the ataxia telangiectasia and Rad3-related protein (ATR), which senses single-stranded DNA and, following activation by TOPBP1, phosphorylates the downstream kinase CHK1 to trigger a G2/M cell cycle arrest checkpoint [147, 148]. This provides the opportunity for repair and resolution of the stalled replication forks. A similar role is played by the ataxia telangiectasia mutated protein (ATM) in the response to DSB induction. Following recruitment of ATM to DSB by the MRN complex, ATM phosphorylates and activates the CHK2 kinase. CHK2, in turn, phosphorylates CDC25A, leading to its degradation and resulting in the inactivation of the CDK1-cyclin B complex with concomitant cell cycle arrest [149]. An additional important player in regulating the cell cycle in response to DNA damage is WEE1, which can phosphorylate both CDK1 and CDK2, thereby activating G2/M and S cell cycle checkpoints, respectively [150, 151]. WEE1-mediated phosphorylation of CDK1 can also enable exit from mitosis [152, 153].

Small molecule inhibitors have been developed against a number of these enzymes and have been shown in preclinical cell and animal model systems to increase radiosensitivity, with several having undergone or undergoing clinical trials. Examples are provided in Table 4.1. Of these, PARP inhibitors have been the most studied both as single agents due to their use as synthetic lethal agents targeting BRCA1/2- and other HR-deficient tumors [166] and as radiosensitizers. They are also the only repair inhibitors to have entered regular clinical practice. Features of the PARP inhibitors can, therefore, serve as examples with regard to radiosensitization by DNA repair inhibitors. There is now a general consensus that the PARP inhibitors act by retaining PARP at DNA strand breaks due to catalytic inhibition and allosteric trapping [167, 168], thereby impeding the repair of SSBs and thus increasing the likelihood of collapsed replication forks and persistent DSBs [169].

Additional benefit can accrue from combining PARP inhibitors with inhibitors of other DNA damage response proteins, especially, although not exclusively, members of the HR pathway, due to the synthetic lethal relationship between PARP and HR. Examination of HR-deficient models has indicated a more potent radiosensitization by PARP inhibitors than in HR-proficient models [170, 171], although this is not always the case [172]. Nonetheless, combinations involving PARP inhibitors

Table 4.1 Examples of DNA repair inhibitors shown to enhance radiosensitivity

DNA repair inhibitor	Preclinical data showing radiosensitization	Clinical Trials	References
ATM inhibitor M3541	Enhanced radiosensitivity in head-and-neck cancer xenografts	Phase I trial in combination with palliative radiotherapy in patients with solid tumors	[154, 155]
ATR inhibitor AZD6738	Enhanced radiosensitivity in head-and-neck cancer xenografts	Phase I trial in combination with palliative radiotherapy in patients with solid tumors	[156, 157]
DNA-PK inhibitor Peposertib (formerly M3814)	Enhanced radiosensitivity in head-and-neck and lung cancer xenografts	Phase I trial in combination with palliative radiotherapy in patients with tumors or metastases located in the head-and-neck region or thorax	[158, 159]
PARP inhibitor Olaparib	Enhanced radiosensitivity in prostate cancer xenografts	Phase I trial in combination with palliative radiotherapy in patients with triple-negative breast cancer	[160, 161]
CHK1 inhibitor Prexasertib	Enhanced radiosensitivity in head-and-neck cancer xenografts	Phase Ib trial in combination with chemoradiation in patients with head-and-neck squamous cell carcinoma	[162, 163]
WEE1 inhibitor Adavosertib (formerly AZD1775, MK-1775)	Enhanced radiosensitivity in esophageal cancer xenografts	Phase I trial in combination with chemoradiation in patients with head-and-neck squamous cell carcinoma	[164, 165]

with inhibitors of other DNA repair and cell cycle checkpoint proteins, including ATR [173], RAD51 [174], WEE1 [175, 176], CHK1 [177], and DNA-PKcs [178], have shown enhanced efficacy in model systems.

The potential of PARP-inhibitor-mediated radiosensitization has spawned a number of clinical trials. To date, almost all the published reports are of Phase 1 clinical trials employing olaparib [160, 179–183], providing information primarily on dose-limiting toxicity and maximum tolerated doses, but not efficacy. One report from a Phase 2 randomized clinical trial, in which veliparib was combined with whole-brain radiation therapy for patients with brain metastases from non-small cell lung cancer, indicated a slightly longer, but not statistically significant, median overall survival of 209 days for patients treated with drug + radiotherapy vs. 185 days for patients treated with placebo + radiotherapy [184].

Ongoing research that may lead to improved outcomes for efficacy through radiosensitization includes the development of nanoformulations of PARP inhibitors [185, 186] and combining PARP inhibitors with radiation and immune checkpoint inhibitors, such as the one in reference [187].

4.5 Conclusion

In conclusion, the integration of hyperthermia, nanoparticle-based radiosensitizers, and DNA repair inhibition approaches represents a promising frontier in enhancing the efficacy of radiotherapy for cancer treatment. Hyperthermia‘s ability to improve tumor oxygenation and impair DNA repair mechanisms, combined with the targeted delivery and dose amplification capabilities of nanoparticles, offers a multifaceted strategy to overcome tumor radioresistance. Additionally, exploiting DNA repair inhibitors further sensitizes cancer cells to radiation, potentially transforming treatment outcomes. Continued advancements in these areas, along with overcoming current challenges, hold the potential to revolutionize cancer therapy and improve patient survival rates.

References

1. Hainfeld JF, Dilmanian FA, Slatkin DN, Smilowitz HM. Radiotherapy enhancement with gold nanoparticles. J Pharm Pharmacol. 2008;60(8):977–85.
2. Greish K. Enhanced permeability and retention of macromolecular drugs in solid tumors: a royal gate for targeted anticancer nanomedicines. J Drug Target. 2007;15(7–8):457–64.
3. Peitzsch C, Perrin R, Hill RP, Dubrovska A, Kurth I. Hypoxia as a biomarker for radioresistant cancer stem cells. Int J Radiat Biol. 2014;90(8):636–52.
4. Rey S, Schito L, Koritzinsky M, Wouters BG. Molecular targeting of hypoxia in radiotherapy. Adv Drug Deliv Rev. 2017;109:45–62.
5. Marie-Egyptienne DT, Lohse I, Hill RP. Cancer stem cells, the epithelial to mesenchymal transition (EMT) and radioresistance: potential role of hypoxia. Cancer Lett. 2013;341(1):63–72.

6. Lalla RV, Treister N, Sollecito T, Schmidt B, Patton LL, Mohammadi K, et al. Oral complications at 6 months after radiation therapy for head and neck cancer. Oral Dis. 2017;23(8):1134–43.
7. Elming PB, Sorensen BS, Oei AL, Franken NAP, Crezee J, Overgaard J, Horsman MR. Hyperthermia: the optimal treatment to overcome radiation resistant hypoxia. Cancers (Basel). 2019;11(1):60.
8. Oei AL, Kok HP, Oei SB, Horsman MR, Stalpers LJA, Franken NAP, Crezee J. Molecular and biological rationale of hyperthermia as radio- and chemosensitizer. Adv Drug Deliv Rev. 2020;163-164:84–97.
9. Datta NR, Rogers S, Klingbiel D, Gomez S, Puric E, Bodis S. Hyperthermia and radiotherapy with or without chemotherapy in locally advanced cervical cancer: a systematic review with conventional and network meta-analyses. Int J Hyperth. 2016;32(7):809–21.
10. Datta NR, Puric E, Klingbiel D, Gomez S, Bodis S. Hyperthermia and radiation therapy in locoregional recurrent breast cancers: a systematic review and meta-analysis. Int J Radiat Oncol Biol Phys. 2016;94(5):1073–87.
11. Sneed PK, Stauffer PR, McDermott MW, Diederich CJ, Lamborn KR, Prados MD, et al. Survival benefit of hyperthermia in a prospective randomized trial of brachytherapy boost +/− hyperthermia for glioblastoma multiforme. Int J Radiat Oncol Biol Phys. 1998;40(2):287–95.
12. Overgaard J, Gonzalez Gonzalez D, Hulshof MC, Arcangeli G, Dahl O, Mella O, Bentzen SM. Hyperthermia as an adjuvant to radiation therapy of recurrent or metastatic malignant melanoma. A multicentre randomized trial by the European Society for Hyperthermic Oncology. Int J Hyperth. 1996;12(1):3–20.
13. Issels RD, Lindner LH, Verweij J, Wessalowski R, Reichardt P, Wust P, et al. Effect of Neoadjuvant chemotherapy plus regional hyperthermia on long-term outcomes among patients with localized high-risk soft tissue sarcoma: the EORTC 62961-ESHO 95 randomized clinical trial. JAMA Oncol. 2018;4(4):483–92.
14. Datta NR, Rogers S, Ordonez SG, Puric E, Bodis S. Hyperthermia and radiotherapy in the management of head and neck cancers: a systematic review and meta-analysis. Int J Hyperth. 2016;32(1):31–40.
15. Chia BSH, Ho SZ, Tan HQ, Chua MLK, Tuan JKL. A review of the current clinical evidence for loco-regional moderate hyperthermia in the adjunct management of cancers. Cancers (Basel). 2023;15(2)
16. Horsman MR, Overgaard J. Hyperthermia: a potent enhancer of radiotherapy. Clin Oncol. 2007;19(6):418–26.
17. van den Tempel N, Horsman MR, Kanaar R. Improving efficacy of hyperthermia in oncology by exploiting biological mechanisms. Int J Hyperth. 2016;32(4):446–54.
18. Song CW, Park H, Griffin RJ. Improvement of tumor oxygenation by mild hyperthermia. Radiat Res. 2001;155(4):515–28.
19. Thrall DE, Rosner GL, Azuma C, Larue SM, Case BC, Samulski T, Dewhirst MW. Using units of CEM 43°C T90, local hyperthermia thermal dose can be delivered as prescribed. Int J Hyperth. 2000;16(5):415–28.
20. Jones EL, Oleson JR, Prosnitz LR, Samulski TV, Vujaskovic Z, Yu D, et al. Randomized trial of hyperthermia and radiation for superficial tumors. J Clin Oncol. 2005;23(13):3079–85.
21. Maguire PD, Samulski TV, Prosnitz LR, Jones EL, Rosner GL, Powers B, et al. A phase II trial testing the thermal dose parameter CEM43° T90 as a predictor of response in soft tissue sarcomas treated with pre-operative thermoradiotherapy. Int J Hyperth. 2001;17(4):283–90.
22. Franckena M, Fatehi D, de Bruijne M, Canters RA, van Norden Y, Mens JW, et al. Hyperthermia dose-effect relationship in 420 patients with cervical cancer treated with combined radiotherapy and hyperthermia. Eur J Cancer. 2009;45(11):1969–78.
23. Kroesen M, Mulder HT, van Holthe JML, Aangeenbrug AA, Mens JWM, van Doorn HC, et al. The effect of the time interval between radiation and hyperthermia on clinical outcome in 400 locally advanced cervical carcinoma patients. Front Oncol. 2019;9(MAR):Article 134.

24. Datta NR, Ordóñez SG, Gaipl US, Paulides MM, Crezee H, Gellermann J, et al. Local hyperthermia combined with radiotherapy and−/or chemotherapy: recent advances and promises for the future. Cancer Treat Rev. 2015;41(9):742–53.
25. Diederich CJ. Thermal ablation and high-temperature thermal therapy: overview of technology and clinical implementation. Int J Hyperth. 2005;21(8):745–53.
26. Paulides MM, Dobsicek Trefna H, Curto S, Rodrigues DB. Recent technological advancements in radiofrequency- andmicrowave-mediated hyperthermia for enhancing drug delivery. Adv Drug Deliv Rev. 2020;163–164:3–18.
27. Linthorst M, Baaijens M, Wiggenraad R, Creutzberg C, Ghidey W, van Rhoon GC, van der Zee J. Local control rate after the combination of re-irradiation and hyperthermia for irresectable recurrent breast cancer: results in 248 patients. Radiother Oncol. 2015;117(2):217–22.
28. Longo TA, Gopalakrishna A, Tsivian M, Van Noord M, Rasch CR, Inman BA, Geijsen ED. A systematic review of regional hyperthermia therapy in bladder cancer. Int J Hyperth. 2016;32(4):381–9.
29. Rodrigues DB, Dobsicek-Trefna H, Curto S, Winter L, Molitoris JK, Vrba J, et al. Radiofrequency and microwave hyperthermia in cancer treatment. In: Prakash P, Srimathveeravalli G, editors. Principles and technologies for electromagnetic energy based therapies. Academic Press; 2022. p. 281–311.
30. Peeters H, van Zwol EM, Brancato L, da Cunha MGMC, Bogers J. Systematic review of the registered clinical trials for oncological hyperthermia treatment. Int J Hyperth. 2022;39(1):806–12.
31. Krawczyk PM, Eppink B, Essers J, Stap J, Rodermond H, Odijk H, et al. Mild hyperthermia inhibits homologous recombination, induces BRCA2 degradation, and sensitizes cancer cells to poly (ADP-ribose) polymerase-1 inhibition. Proc Natl Acad Sci USA. 2011;108(24):9851–6.
32. Oei AL, Ahire VR, van Leeuwen CM, Ten Cate R, Stalpers LJA, Crezee J, et al. Enhancing radiosensitisation of BRCA2-proficient and BRCA2-deficient cell lines with hyperthermia and PARP1-i. Int J Hyperth. 2018;34(1):39–48.
33. van Leeuwen CM, Crezee J, Oei AL, Franken NAP, Stalpers LJA, Bel A, Kok HP. The effect of time interval between radiotherapy and hyperthermia on planned equivalent radiation dose. Int J Hyperth. 2018;34(7):901–9.
34. Mei X, Ten Cate R, van Leeuwen CM, Rodermond HM, de Leeuw L, Dimitrakopoulou D, et al. Radiosensitization by hyperthermia: the effects of temperature, sequence, and time interval in cervical cell lines. Cancers (Basel). 2020;12(3)
35. Mei X, Kok HP, Rodermond HM, van Bochove GGW, Snoek BC, van Leeuwen CM, et al. Radiosensitization by hyperthermia critically depends on the time interval. Int J Radiat Oncol Biol Phys. 2024;118(3):817–28.
36. Oei AL, Vriend LE, Crezee J, Franken NA, Krawczyk PM. Effects of hyperthermia on DNA repair pathways: one treatment to inhibit them all. Radiat Oncol. 2015;10:165.
37. Conley MJ, Epifano I, Kirk A, Stevenson A, Graham SV. Microwave hyperthermia represses human papillomavirus oncoprotein activity and induces cell death due to cell stress in 3D tissue models of anogenital precancers and cancers. EBioMedicine. 2023;91:104577.
38. Oei AL, Korangath P, Mulka K, Helenius M, Coulter JB, Stewart J, et al. Enhancing the abscopal effect of radiation and immune checkpoint inhibitor therapies with magnetic nanoparticle hyperthermia in a model of metastatic breast cancer. Int J Hyperth. 2019;36(sup1):47–63.
39. Evans SS, Repasky EA, Fisher DT. Fever and the thermal regulation of immunity: the immune system feels the heat. Nat Rev Immunol. 2015;15(6):335–49.
40. van der Zee J, Gonzalez Gonzalez D, van Rhoon GC, van Dijk JD, van Putten WL, Hart AA. Comparison of radiotherapy alone with radiotherapy plus hyperthermia in locally advanced pelvic tumours: a prospective, randomised, multicentre trial. Dutch Deep Hyperthermia Group Lancet. 2000;355(9210):1119–25.
41. Westermann AM, Jones EL, Schem BC, van der Steen-Banasik EM, Koper P, Mella O, et al. First results of triple-modality treatment combining radiotherapy, chemotherapy, and

hyperthermia for the treatment of patients with stage IIB, III, and IVA cervical carcinoma. Cancer. 2005;104(4):763–70.
42. Datta NR, Jain BM, Mathi Z, Datta S, Johari S, Singh AR, et al. Hyperthermia: a potential game-changer in the management of cancers in low-middle-income group countries. Cancers (Basel). 2022;14(2)
43. Frey B, Weiss E-M, Rubner Y, Wunderlich R, Ott OJ, Sauer R, et al. Old and new facts about hyperthermia-induced modulations of the immune system. Int J Hyperth. 2012;28(6):528–42.
44. Repasky EA, Evans SS, Dewhirst MW. Temperature matters! And why it should matter to tumor immunologists. Cancer Immunol Res. 2013;1(4):210–6.
45. Yao Y, Zhou Y, Liu L, Xu Y, Chen Q, Wang Y, et al. Nanoparticle-based drug delivery in cancer therapy and its role in overcoming drug resistance. Front Mol Biosci. 2020;7:193.
46. Liu Y, Zhang P, Li F, Jin X, Li J, Chen W, Li Q. Metal-based nanoenhancers for future radiotherapy: radiosensitizing and synergistic effects on tumor cells. Theranostics. 2018;8(7):1824.
47. Kolyvanova MA, Belousov AV, Krusanov GA, Isagulieva AK, Morozov KV, Kartseva ME, et al. Impact of the spectral composition of kilovoltage X-rays on high-Z nanoparticle-assisted dose enhancement. Int J Mol Sci. 2021;22(11):6030.
48. Sharma RA, Plummer R, Stock JK, Greenhalgh TA, Ataman O, Kelly S, et al. Clinical development of new drug–radiotherapy combinations. Nat Rev Clin Oncol. 2016;13(10):627–42.
49. Rosa S, Connolly C, Schettino G, Butterworth KT, Prise KM. Biological mechanisms of gold nanoparticle radiosensitization. Cancer Nanotechnol. 2017;8(1):2.
50. Butterworth KT, McMahon SJ, Taggart LE, Prise KM. Radiosensitization by gold nanoparticles: effective at megavoltage energies and potential role of oxidative stress. Transl Cancer Res. 2013;2(4):269–79.
51. Martelli S, Chow JCL. Dose enhancement for the flattening-filter-free and flattening-filter photon beams in nanoparticle-enhanced radiotherapy: a monte carlo phantom study. Nanomaterials (Basel). 2020;10:4.
52. Castillo MH, Button TM, Doerr R, Homs MI, Pruett CW, Pearce JI. Effects of radiotherapy on mandibular reconstruction plates. Am J Surg. 1988;156(4):261–3.
53. Misawa M, Takahashi J. Generation of reactive oxygen species induced by gold nanoparticles under x-ray and UV irradiations. Nanomedicine. 2011;7(5):604–14.
54. Roa W, Zhang X, Guo L, Shaw A, Hu X, Xiong Y, et al. Gold nanoparticle sensitize radiotherapy of prostate cancer cells by regulation of the cell cycle. Nanotechnology. 2009;20(37):375101.
55. Wang C, Jiang Y, Li X, Hu L. Thioglucose-bound gold nanoparticles increase the radiosensitivity of a triple-negative breast cancer cell line (MDA-MB-231). Breast Cancer. 2015;22:413–20.
56. Liu Y, Liu X, Jin X, He P, Zheng X, Dai Z, et al. The dependence of radiation enhancement effect on the concentration of gold nanoparticles exposed to low-and high-LET radiations. Phys Med. 2015;31(3):210–8.
57. Koushki K, Keshavarz Shahbaz S, Keshavarz M, Bezsonov EE, Sathyapalan T, Sahebkar A. Gold nanoparticles: multifaceted roles in the management of autoimmune disorders. Biomol Ther. 2021;11(9):1289.
58. Schuemann J, Berbeco R, Chithrani DB, Cho SH, Kumar R, McMahon SJ, et al. Roadmap to clinical use of gold nanoparticles for radiation sensitization. Int J Radiat Oncol Biol Phys. 2016;94(1):189–205.
59. Schuemann J, Bagley AF, Berbeco R, Bromma K, Butterworth KT, Byrne HL, et al. Roadmap for metal nanoparticles in radiation therapy: current status, translational challenges, and future directions. Phys Med Biol. 2020;65(21):21RM02.
60. Cui L, Her S, Borst GR, Bristow RG, Jaffray DA, Allen C. Radiosensitization by gold nanoparticles: will they ever make it to the clinic? Radiother Oncol. 2017;124(3):344–56.
61. Rahman WN, Bishara N, Ackerly T, He CF, Jackson P, Wong C, et al. Enhancement of radiation effects by gold nanoparticles for superficial radiation therapy. Nanomedicine. 2009;5(2):136–42.

62. Hainfeld JF, Smilowitz HM, O'Connor MJ, Dilmanian FA, Slatkin DN. Gold nanoparticle imaging and radiotherapy of brain tumors in mice. Nanomedicine. 2013;8(10):1601–9.
63. Hainfeld JF, Dilmanian FA, Zhong Z, Slatkin DN, Kalef-Ezra JA, Smilowitz HM. Gold nanoparticles enhance the radiation therapy of a murine squamous cell carcinoma. Phys Med Biol. 2010;55(11):3045.
64. Hainfeld JF, Slatkin DN, Smilowitz HM. The use of gold nanoparticles to enhance radiotherapy in mice. Phys Med Biol. 2004;49(18):N309.
65. Herold DM, Das IJ, Stobbe CC, Iyer RV, Chapman JD. Gold microspheres: a selective technique for producing biologically effective dose enhancement. Int J Radiat Biol. 2000;76(10):1357–64.
66. Sun W, Zhou Z, Pratx G, Chen X, Chen H. Nanoscintillator-mediated X-ray induced photodynamic therapy for deep-seated tumors: from concept to biomedical applications. Theranostics. 2020;10(3):1296–318.
67. Dufort S, Bianchi A, Henry M, Lux F, Le Duc G, Josserand V, et al. Nebulized gadolinium-based nanoparticles: a Theranostic approach for lung tumor imaging and radiosensitization. Small. 2015;11(2):215–21.
68. Dufort S, Le Duc G, Salomé M, Bentivegna V, Sancey L, Bräuer-Krisch E, et al. The high radiosensitizing efficiency of a trace of gadolinium-based nanoparticles in tumors. Sci Rep. 2016;6(1):29678.
69. Narmani A, Farhood B, Haghi-Aminjan H, Mortezazadeh T, Aliasgharzadeh A, Mohseni M, et al. Gadolinium nanoparticles as diagnostic and therapeutic agents: their delivery systems in magnetic resonance imaging and neutron capture therapy. J Drug Deliv Sci Technol. 2018;44:457–66.
70. Lux F, Sancey L, Bianchi A, Crémillieux Y, Roux S, Tillement O. Gadolinium-based nanoparticles for theranostic MRI-radiosensitization. Nanomedicine. 2015;10(11):1801–15.
71. Miladi I, Aloy M-T, Armandy E, Mowat P, Kryza D, Magné N, et al. Combining ultrasmall gadolinium-based nanoparticles with photon irradiation overcomes radioresistance of head and neck squamous cell carcinoma. Nanomedicine. 2015;11(1):247–57.
72. Wu C, Cai R, Zhao T, Wu L, Zhang L, Jin J, et al. Hyaluronic acid-functionalized gadolinium oxide nanoparticles for magnetic resonance imaging-guided radiotherapy of tumors. Nanoscale Res Lett. 2020;15(1):94.
73. Andoh T, Nakatani Y, Suzuki M, Sakurai Y, Fujimoto T, Ichikawa H. Influence of the particle size of gadolinium-loaded chitosan nanoparticles on their tumor-killing effect in neutron capture therapy in vitro. Appl Radiat Isot. 2020;164:109270.
74. Wu C, Cai R, Zhao T, Wu L, Zhang L, Jin J, et al. Hyaluronic acid-functionalized gadolinium oxide nanoparticles for magnetic resonance imaging-guided radiotherapy of tumors. Nanoscale Res Lett. 2020;15:1–12.
75. Dufort S, Appelboom G, Verry C, Barbier EL, Lux F, Bräuer-Krisch E, et al. Ultrasmall theranostic gadolinium-based nanoparticles improve high-grade rat glioma survival. J Clin Neurosci. 2019;67:215–9.
76. Verry C, Sancey L, Dufort S, Le Duc G, Mendoza C, Lux F, et al. Treatment of multiple brain metastases using gadolinium nanoparticles and radiotherapy: NANO-RAD, a phase I study protocol. BMJ Open. 2019;9(2):e023591.
77. Lux F, Tran VL, Thomas E, Dufort S, Rossetti F, Martini M, et al. AGuIX® from bench to bedside—transfer of an ultrasmall theranostic gadolinium-based nanoparticle to clinical medicine. Br J Radiol. 2019;92(1093):20180365.
78. Sancey L, Lux F, Kotb S, Roux S, Dufort S, Bianchi A, et al. The use of theranostic gadolinium-based nanoprobes to improve radiotherapy efficacy. Br J Radiol. 2014;87(1041):20140134.
79. Mehta MP, Shapiro WR, Phan SC, Gervais R, Carrie C, Chabot P, et al. Motexafin gadolinium combined with prompt whole brain radiotherapy prolongs time to neurologic progression in non–small-cell lung cancer patients with brain metastases: results of a phase III trial. Int J Radiat Oncol Biol Phys. 2009;73(4):1069–76.

80. Forouzannia A, Richards GM, Khuntia D, Mehta MP. Motexafin gadolinium: a novel radiosensitizer for brain tumors. Expert Rev Anticancer Ther. 2007;7(6):785–94.
81. Zhao J, Liu P, Ma J, Li D, Yang H, Chen W, Jiang Y. Enhancement of Radiosensitization by silver nanoparticles functionalized with polyethylene glycol and Aptamer As1411 for glioma irradiation therapy. Int J Nanomedicine. 2019;14:9483–96.
82. Xu R, Ma J, Sun X, Chen Z, Jiang X, Guo Z, et al. Ag nanoparticles sensitize IR-induced killing of cancer cells. Cell Res. 2009;19(8):1031–4.
83. Swanner J, Mims J, Carroll DL, Akman SA, Furdui CM, Torti SV, Singh RN. Differential cytotoxic and radiosensitizing effects of silver nanoparticles on triple-negative breast cancer and non-triple-negative breast cells. Int J Nanomedicine. 2015;10:3937–53.
84. Saratale RG, Shin HS, Kumar G, Benelli G, Kim DS, Saratale GD. Exploiting antidiabetic activity of silver nanoparticles synthesized using Punica granatum leaves and anticancer potential against human liver cancer cells (HepG2). Artif Cells Nanomed Biotechnol. 2018;46(1):211–22.
85. Sunil Gowda SN, Rajasowmiya S, Vadivel V, Banu Devi S, Celestin Jerald A, Marimuthu S, Devipriya N. Gallic acid-coated sliver nanoparticle alters the expression of radiation-induced epithelial-mesenchymal transition in non-small lung cancer cells. Toxicol In Vitro. 2018;52:170–7.
86. Liu P, Jin H, Guo Z, Ma J, Zhao J, Li D, et al. Silver nanoparticles outperform gold nanoparticles in radiosensitizing U251 cells in vitro and in an intracranial mouse model of glioma. Int J Nanomedicine. 2016;11:5003–14.
87. Boca SC, Potara M, Gabudean AM, Juhem A, Baldeck PL, Astilean S. Chitosan-coated triangular silver nanoparticles as a novel class of biocompatible, highly effective photothermal transducers for in vitro cancer cell therapy. Cancer Lett. 2011;311(2):131–40.
88. Habiba K, Aziz K, Sanders K, Santiago CM, Mahadevan LSK, Makarov V, et al. Enhancing colorectal cancer radiation therapy efficacy using silver Nanoprisms decorated with graphene as radiosensitizers. Sci Rep. 2019;9(1):17120.
89. Zhao J, Li D, Ma J, Yang H, Chen W, Cao Y, Liu P. Increasing the accumulation of aptamer AS1411 and verapamil conjugated silver nanoparticles in tumor cells to enhance the radiosensitivity of glioma. Nanotechnology. 2021;32(14):145102.
90. Gerken LRH, Keevend K, Zhang Y, Starsich FHL, Eberhardt C, Panzarasa G, et al. Lanthanide-doped hafnia nanoparticles for multimodal theranostics: tailoring the physicochemical properties and interactions with biological entities. ACS Appl Mater Interfaces. 2019;11(1):437–48.
91. Bonvalot S, Rutkowski PL, Thariat J, Carrère S, Ducassou A, Sunyach M-P, et al. NBTXR3, a first-in-class radioenhancer hafnium oxide nanoparticle, plus radiotherapy versus radiotherapy alone in patients with locally advanced soft-tissue sarcoma (Act. In. Sarc): a multicentre, phase 2–3, randomised, controlled trial. Lancet Oncol. 2019;20(8):1148–59.
92. Zhang P, Marill J, Darmon A, Mohamed Anesary N, Lu B, Paris S. NBTXR3 radiotherapy-activated functionalized hafnium oxide nanoparticles show efficient antitumor effects across a large panel of human cancer models. Int J Nanomedicine. 2021:2761–73.
93. Bonvalot S, Rutkowski P, Thariat J, Carrere S, Sunyach M-P, Saada E, et al. A phase II/III trial of hafnium oxide nanoparticles activated by radiotherapy in the treatment of locally advance soft tissue sarcoma of the extremity and trunk wall. Ann Oncol. 2018;29:viii753.
94. Wadajkar AS, Menon JU, Kadapure T, Tran RT, Yang J, Nguyen KT. Design and application of magnetic-based theranostic nanoparticle systems. Recent Patents Biomed Eng. 2013;6(1):47–57.
95. Zaloga J, Pöttler M, Leitinger G, Friedrich RP, Almer G, Lyer S, et al. Pharmaceutical formulation of HSA hybrid coated iron oxide nanoparticles for magnetic drug targeting. Eur J Pharm Biopharm. 2016;101:152–62.
96. Yu S, Zhang H, Zhang S, Zhong M, Fan H. Ferrite nanoparticles-based reactive oxygen species-mediated cancer therapy. Front Chem. 2021;9:651053.

97. Huang G, Chen H, Dong Y, Luo X, Yu H, Moore Z, et al. Superparamagnetic iron oxide nanoparticles: amplifying ROS stress to improve anticancer drug efficacy. Theranostics. 2013;3(2):116.
98. Fernández-Acosta R, Iriarte-Mesa C, Alvarez-Alminaque D, Hassannia B, Wiernicki B, Díaz-García AM, et al. Novel iron oxide nanoparticles induce ferroptosis in a panel of cancer cell lines. Molecules. 2022;27(13):3970.
99. Wen J, Chen H, Ren Z, Zhang P, Chen J, Jiang S. Ultrasmall iron oxide nanoparticles induced ferroptosis via Beclin1/ATG5-dependent autophagy pathway. Nano Converg. 2021;8(1):10.
100. Huang QT, Hu QQ, Wen ZF, Li YL. Iron oxide nanoparticles inhibit tumor growth by ferroptosis in diffuse large B-cell lymphoma. Am J Cancer Res. 2023;13(2):498–508.
101. Wu J, Sun J. Investigation on mechanism of growth arrest induced by iron oxide nanoparticles in PC12 cells. J Nanosci Nanotechnol. 2011;11(12):11079–83.
102. Ye P, Ye Y, Chen X, Zou H, Zhou Y, Zhao X, et al. Ultrasmall Fe3O4 nanoparticles induce S-phase arrest and inhibit cancer cells proliferation. Nanotechnol Rev. 2020;9(1):61–9.
103. Huang C, Neoh KG, Xu L, Kang ET, Chiong E. Polymeric nanoparticles with encapsulated superparamagnetic iron oxide and conjugated cisplatin for potential bladder cancer therapy. Biomacromolecules. 2012;13(8):2513–20.
104. Shen LF, Chen J, Zeng S, Zhou RR, Zhu H, Zhong MZ, et al. The superparamagnetic nanoparticles carrying the E1A gene enhance the radiosensitivity of human cervical carcinoma in nude mice. Mol Cancer Ther. 2010;9(7):2123–30.
105. Zhao D, Sun X, Tong J, Ma J, Bu X, Xu R, Fan R. A novel multifunctional nanocomposite C225-conjugated Fe3O4/Ag enhances the sensitivity of nasopharyngeal carcinoma cells to radiotherapy. Acta Biochim Biophys Sin. 2012;44(8):678–84.
106. Porcel E, Liehn S, Remita H, Usami N, Kobayashi K, Furusawa Y, et al. Platinum nanoparticles: a promising material for future cancer therapy? Nanotechnology. 2010;21(8):085103.
107. Hossain M, Su M. Nanoparticle location and material-dependent dose enhancement in X-ray radiation therapy. J Phys Chem C. 2012;116(43):23047–52.
108. Mallick S, Sun IC, Kim K, Yil DK. Silica coated gold nanorods for imaging and photothermal therapy of cancer cells. J Nanosci Nanotechnol. 2013;13(5):3223–9.
109. Elbialy N, Mohamed N, Monem AS. Preparation and characterization of SiO2-Au nanoshells: in vivo study of its photo-heat conversion. J Biomed Nanotechnol. 2013;9(2):158–66.
110. Li J, Jiang H, Yu Z, Xia H, Zou G, Zhang Q, Yu Y. Multifunctional uniform core-shell Fe3O4@mSiO2 mesoporous nanoparticles for bimodal imaging and photothermal therapy. Chem Asian J. 2013;8(2):385–91.
111. Klein S, Dell'Arciprete ML, Wegmann M, Distel LV, Neuhuber W, Gonzalez MC, Kryschi C. Oxidized silicon nanoparticles for radiosensitization of cancer and tissue cells. Biochem Biophys Res Commun. 2013;434(2):217–22.
112. Kargar S, Khoei S, Khoee S, Shirvalilou S, Mahdavi SR. Evaluation of the combined effect of NIR laser and ionizing radiation on cellular damages induced by IUdR-loaded PLGA-coated Nano-graphene oxide. Photodiagn Photodyn Ther. 2018;21:91–7.
113. Rezaie P, Khoei S, Khoee S, Shirvalilou S, Mahdavi SR. Evaluation of combined effect of hyperthermia and ionizing radiation on cytotoxic damages induced by IUdR-loaded PCL-PEG-coated magnetic nanoparticles in spheroid culture of U87MG glioblastoma cell line. Int J Radiat Biol. 2018;94(11):1027–37.
114. Jin C, Bai L, Wu H, Liu J, Guo G, Chen J. Paclitaxel-loaded poly(D,L-lactide-co-glycolide) nanoparticles for radiotherapy in hypoxic human tumor cells in vitro. Cancer Biol Ther. 2008;7(6):911–6.
115. Goswami N, Luo Z, Yuan X, Leong DT, Xie J. Engineering gold-based radiosensitizers for cancer radiotherapy. Mater Horizons. 2017;4(5):817–31.
116. Balasubramanian SK, Jittiwat J, Manikandan J, Ong C-N, Liya EY, Ong W-Y. Biodistribution of gold nanoparticles and gene expression changes in the liver and spleen after intravenous administration in rats. Biomaterials. 2010;31(8):2034–42.

117. Maeda H, Wu J, Sawa T, Matsumura Y, Hori K. Tumor vascular permeability and the EPR effect in macromolecular therapeutics: a review. J Control Release. 2000;65(1–2):271–84.
118. Wu J. The enhanced permeability and retention (EPR) effect: the significance of the concept and methods to enhance its application. J Personal Med. 2021;11(8)
119. Sykes EA, Chen J, Zheng G, Chan WC. Investigating the impact of nanoparticle size on active and passive tumor targeting efficiency. ACS Nano. 2014;8(6):5696–706.
120. Dimitriou NM, Tsekenis G, Balanikas EC, Pavlopoulou A, Mitsiogianni M, Mantso T, et al. Gold nanoparticles, radiations and the immune system: current insights into the physical mechanisms and the biological interactions of this new alliance towards cancer therapy. Pharmacol Ther. 2017;178:1–17.
121. Kumar R, Roy I, Ohulchanskky TY, Vathy LA, Bergey EJ, Sajjad M, Prasad PN. In vivo biodistribution and clearance studies using multimodal organically modified silica nanoparticles. ACS Nano. 2010;4(2):699–708.
122. De Jong WH, Hagens WI, Krystek P, Burger MC, Sips AJ, Geertsma RE. Particle size-dependent organ distribution of gold nanoparticles after intravenous administration. Biomaterials. 2008;29(12):1912–9.
123. Huang K, Ma H, Liu J, Huo S, Kumar A, Wei T, et al. Size-dependent localization and penetration of ultrasmall gold nanoparticles in cancer cells, multicellular spheroids, and tumors in vivo. ACS Nano. 2012;6(5):4483–93.
124. Ruan S, Cao X, Cun X, Hu G, Zhou Y, Zhang Y, et al. Matrix metalloproteinase-sensitive size-shrinkable nanoparticles for deep tumor penetration and pH triggered doxorubicin release. Biomaterials. 2015;60:100–10.
125. Zhang Y, Huang F, Ren C, Liu J, Yang L, Chen S, et al. Enhanced radiosensitization by gold nanoparticles with acid-triggered aggregation in cancer radiotherapy. Adv Sci. 2019;6(8):1801806.
126. Liu X, Chen Y, Li H, Huang N, Jin Q, Ren K, Ji J. Enhanced retention and cellular uptake of nanoparticles in tumors by controlling their aggregation behavior. ACS Nano. 2013;7(7):6244–57.
127. Miyamoto D, Oishi M, Kojima K, Yoshimoto K, Nagasaki Y. Completely dispersible PEGylated gold nanoparticles under physiological conditions: modification of gold nanoparticles with precisely controlled PEG-b-polyamine. Langmuir. 2008;24(9):5010–7.
128. Han HS, Martin JD, Lee J, Harris DK, Fukumura D, Jain RK, Bawendi M. Spatial charge configuration regulates nanoparticle transport and binding behavior in vivo. Angew Chem Int Ed. 2013;52(5):1414–9.
129. Zhang L, Cao Z, Li Y, Ella-Menye J-R, Bai T, Jiang S. Softer zwitterionic nanogels for longer circulation and lower splenic accumulation. ACS Nano. 2012;6(8):6681–6.
130. Liu X, Jin Q, Ji Y, Ji J. Minimizing nonspecific phagocytic uptake of biocompatible gold nanoparticles with mixed charged zwitterionic surface modification. J Mater Chem. 2012;22(5):1916–27.
131. Zhao J, Qin Z, Wu J, Li L, Jin Q, Ji J. Zwitterionic stealth peptide-protected gold nanoparticles enable long circulation without the accelerated blood clearance phenomenon. Biomater Sci. 2018;6(1):200–6.
132. Xuan M, Shao J, Dai L, He Q, Li J. Macrophage cell membrane camouflaged mesoporous silica nanocapsules for in vivo cancer therapy. Adv Healthc Mater. 2015;4(11):1645–52.
133. Xuan M, Shao J, Dai L, Li J, He Q. Macrophage cell membrane camouflaged Au nanoshells for in vivo prolonged circulation life and enhanced cancer photothermal therapy. ACS Appl Mater Interfaces. 2016;8(15):9610–8.
134. Baek S-K, Makkouk AR, Krasieva T, Sun C-H, Madsen SJ, Hirschberg H. Photothermal treatment of glioma; an in vitro study of macrophage-mediated delivery of gold nanoshells. J Neuro-Oncol. 2011;104(2):439–48.
135. Madsen SJ, Christie C, Hong SJ, Trinidad A, Peng Q, Uzal FA, Hirschberg H. Nanoparticle-loaded macrophage-mediated photothermal therapy: potential for glioma treatment. Lasers Med Sci. 2015;30(4):1357–65.

136. Oldenborg P-A, Zheleznyak A, Fang Y-F, Lagenaur CF, Gresham HD, Lindberg FP. Role of CD47 as a marker of self on red blood cells. Science. 2000;288(5473):2051–4.
137. Wang J, Dong Y, Li Y, Li W, Cheng K, Qian Y, et al. Designer exosomes for active targeted chemo-photothermal synergistic tumor therapy. Adv Funct Mater. 2018;28(18):1707360.
138. Zhu D-M, Xie W, Xiao Y-S, Suo M, Zan M-H, Liao Q-Q, et al. Erythrocyte membrane-coated gold nanocages for targeted photothermal and chemical cancer therapy. Nanotechnology. 2018;29(8):084002.
139. Dawidczyk CM, Russell LM, Searson PC. Nanomedicines for cancer therapy: state-of-the-art and limitations to pre-clinical studies that hinder future developments. Front Chem. 2014;2:69.
140. Paterson MC, Smith BP, Lohman PH, Anderson AK, Fishman L. Defective excision repair of gamma-ray-damaged DNA in human (ataxia telangiectasia) fibroblasts. Nature. 1976;260(5550):444–7.
141. Painter RB, Young BR. Radiosensitivity in ataxia-telangiectasia: a new explanation. Proc Natl Acad Sci USA. 1980;77(12):7315–7.
142. Grundy GJ, Parsons JL. Base excision repair and its implications to cancer therapy. Essays Biochem. 2020;64(5):831–43.
143. Caldecott KW. DNA single-strand break repair and human genetic disease. Trends Cell Biol. 2022;32(9):733–45.
144. Vogt A, He Y, Lees-Miller SP. How to fix DNA breaks: new insights into the mechanism of non-homologous end joining. Biochem Soc Trans. 2023;51(5):1789–800.
145. Elbakry A, Lobrich M. Homologous recombination subpathways: a tangle to resolve. Front Genet. 2021;12:723847.
146. Wood RD, Doublie S. DNA polymerase theta (POLQ), double-strand break repair, and cancer. DNA Repair (Amst). 2016;44:22–32.
147. Wagner SA, Oehler H, Voigt A, Dalic D, Freiwald A, Serve H, Beli P. ATR inhibition rewires cellular signaling networks induced by replication stress. Proteomics. 2016;16(3):402–16.
148. Nickoloff JA. Targeting replication stress response pathways to enhance genotoxic chemo- and radiotherapy. Molecules. 2022;27(15)
149. Reinhardt HC, Yaffe MB. Kinases that control the cell cycle in response to DNA damage: Chk1, Chk2, and MK2. Curr Opin Cell Biol. 2009;21(2):245–55.
150. McGowan CH, Russell P. Human Wee1 kinase inhibits cell division by phosphorylating p34cdc2 exclusively on Tyr15. EMBO J. 1993;12(1):75–85.
151. Mueller PR, Coleman TR, Kumagai A, Dunphy WG. Myt1: a membrane-associated inhibitory kinase that phosphorylates Cdc2 on both threonine-14 and tyrosine-15. Science. 1995;270(5233):86–90.
152. Chow JP, Poon RY, Ma HT. Inhibitory phosphorylation of cyclin-dependent kinase 1 as a compensatory mechanism for mitosis exit. Mol Cell Biol. 2011;31(7):1478–91.
153. Visconti R, Palazzo L, Della Monica R, Grieco D. Fcp1-dependent dephosphorylation is required for M-phase-promoting factor inactivation at mitosis exit. Nat Commun. 2012;3:894.
154. Zimmermann A, Zenke FT, Chiu LY, Dahmen H, Pehl U, Fuchss T, et al. A new class of selective ATM inhibitors as combination partners of DNA double-strand break inducing cancer therapies. Mol Cancer Ther. 2022;21(6):859–70.
155. Waqar SN, Robinson C, Olszanski AJ, Spira A, Hackmaster M, Lucas L, et al. Phase I trial of ATM inhibitor M3541 in combination with palliative radiotherapy in patients with solid tumors. Investig New Drugs. 2022;40(3):596–605.
156. Dok R, Glorieux M, Bamps M, Nuyts S. Effect of ATR inhibition in RT response of HPV-negative and HPV-positive head and neck cancers. Int J Mol Sci. 2021;22(4)
157. Dillon MT, Boylan Z, Smith D, Guevara J, Mohammed K, Peckitt C, et al. PATRIOT: a phase I study to assess the tolerability, safety and biological effects of a specific ataxia telangiectasia and Rad3-related (ATR) inhibitor (AZD6738) as a single agent and in combination with palliative radiation therapy in patients with solid tumours. Clin Transl Radiat Oncol. 2018;12:16–20.

158. Zenke FT, Zimmermann A, Sirrenberg C, Dahmen H, Kirkin V, Pehl U, et al. Pharmacologic inhibitor of DNA-PK, M3814, potentiates radiotherapy and regresses human tumors in mouse models. Mol Cancer Ther. 2020;19(5):1091–101.
159. Samuels M, Falkenius J, Bar-Ad V, Dunst J, van Triest B, Yachnin J, et al. A phase 1 study of the DNA-PK inhibitor Peposertib in combination with radiation therapy with or without cisplatin in patients with advanced head and neck tumors. Int J Radiat Oncol Biol Phys. 2024;118(3):743–56.
160. Loap P, Loirat D, Berger F, Rodrigues M, Bazire L, Pierga JY, et al. Concurrent Olaparib and radiotherapy in patients with triple-negative breast cancer: the phase 1 Olaparib and radiation therapy for triple-negative breast cancer trial. JAMA Oncol. 2022;8(12):1802–8.
161. Gani C, Coackley C, Kumareswaran R, Schutze C, Krause M, Zafarana G, Bristow RG. In vivo studies of the PARP inhibitor, AZD-2281, in combination with fractionated radiotherapy: an exploration of the therapeutic ratio. Radiother Oncol. 2015;116(3):486–94.
162. Zeng L, Nikolaev A, Xing C, Della Manna DL, Yang ES. CHK1/2 inhibitor Prexasertib suppresses NOTCH signaling and enhances cytotoxicity of cisplatin and radiation in head and neck squamous cell carcinoma. Mol Cancer Ther. 2020;19(6):1279–88.
163. Yang ES, Deutsch E, Mehmet A, Fayette J, Tao Y, Nabell L, et al. A phase 1b trial of prexasertib in combination with chemoradiation in patients with locally advanced head and neck squamous cell carcinoma. Radiother Oncol. 2021;157:203–9.
164. Yang L, Shen C, Pettit CJ, Li T, Hu AJ, Miller ED, et al. Wee1 kinase inhibitor AZD1775 effectively sensitizes esophageal cancer to radiotherapy. Clin Cancer Res. 2020;26(14):3740–50.
165. Chera BS, Sheth SH, Patel SA, Goldin D, Douglas KE, Green RL, et al. Phase 1 trial of adavosertib (AZD1775) in combination with concurrent radiation and cisplatin for intermediate-risk and high-risk head and neck squamous cell carcinoma. Cancer. 2021;127(23):4447–54.
166. Mateo J, Lord CJ, Serra V, Tutt A, Balmana J, Castroviejo-Bermejo M, et al. A decade of clinical development of PARP inhibitors in perspective. Ann Oncol. 2019;30(9):1437–47.
167. Zandarashvili L, Langelier MF, Velagapudi UK, Hancock MA, Steffen JD, Billur R, et al. Structural basis for allosteric PARP-1 retention on DNA breaks. Science. 2020;368(6486)
168. Kanev PB, Varhoshkova S, Georgieva I, Lukarska M, Kirova D, Danovski G, et al. A unified mechanism for PARP inhibitor-induced PARP1 chromatin retention at DNA damage sites in living cells. Cell Rep. 2024;43(5):114234.
169. Dungey FA, Loser DA, Chalmers AJ. Replication-dependent radiosensitization of human glioma cells by inhibition of poly(ADP-Ribose) polymerase: mechanisms and therapeutic potential. Int J Radiat Oncol Biol Phys. 2008;72(4):1188–97.
170. Verhagen CV, de Haan R, Hageman F, Oostendorp TP, Carli AL, O'Connor MJ, et al. Extent of radiosensitization by the PARP inhibitor olaparib depends on its dose, the radiation dose and the integrity of the homologous recombination pathway of tumor cells. Radiother Oncol. 2015;116(3):358–65.
171. Wurster S, Hennes F, Parplys AC, Seelbach JI, Mansour WY, Zielinski A, et al. PARP1 inhibition radiosensitizes HNSCC cells deficient in homologous recombination by disabling the DNA replication fork elongation response. Oncotarget. 2016;7(9):9732–41.
172. Lohse I, Kumareswaran R, Cao P, Pitcher B, Gallinger S, Bristow RG, Hedley DW. Effects of combined treatment with ionizing radiation and the PARP inhibitor Olaparib in BRCA mutant and wild type patient-derived pancreatic cancer xenografts. PLoS One. 2016;11(12):e0167272.
173. Parsels LA, Engelke CG, Parsels J, Flanagan SA, Zhang Q, Tanska D, et al. Combinatorial efficacy of Olaparib with radiation and ATR inhibitor requires PARP1 protein in homologous recombination-proficient pancreatic cancer. Mol Cancer Ther. 2021;20(2):263–73.
174. Wera AC, Lobbens A, Stoyanov M, Lucas S, Michiels C. Radiation-induced synthetic lethality: combination of poly(ADP-ribose) polymerase and RAD51 inhibitors to sensitize cells to proton irradiation. Cell Cycle. 2019;18(15):1770–83.

175. Karnak D, Engelke CG, Parsels LA, Kausar T, Wei D, Robertson JR, et al. Combined inhibition of Wee1 and PARP1/2 for radiosensitization in pancreatic cancer. Clin Cancer Res. 2014;20(19):5085–96.
176. Parsels LA, Karnak D, Parsels JD, Zhang Q, Velez-Padilla J, Reichert ZR, et al. PARP1 trapping and DNA replication stress enhance radiosensitization with combined WEE1 and PARP inhibitors. Mol Cancer Res. 2018;16(2):222–32.
177. Vance S, Liu E, Zhao L, Parsels JD, Parsels LA, Brown JL, et al. Selective radiosensitization of p53 mutant pancreatic cancer cells by combined inhibition of Chk1 and PARP1. Cell Cycle. 2011;10(24):4321–9.
178. Azad A, Bukczynska P, Jackson S, Haupt Y, Cullinane C, McArthur GA, Solomon B. Co-targeting deoxyribonucleic acid-dependent protein kinase and poly(adenosine diphosphate-ribose) polymerase-1 promotes accelerated senescence of irradiated cancer cells. Int J Radiat Oncol Biol Phys. 2014;88(2):385–94.
179. de Haan R, van den Heuvel MM, van Diessen J, Peulen HMU, van Werkhoven E, de Langen AJ, et al. Phase I and pharmacologic study of Olaparib in combination with high-dose radiotherapy with and without concurrent cisplatin for non-small cell lung cancer. Clin Cancer Res. 2021;27(5):1256–66.
180. Loap P, Loirat D, Berger F, Ricci F, Vincent-Salomon A, Ezzili C, et al. Combination of Olaparib and radiation therapy for triple negative breast cancer: preliminary results of the RADIOPARP phase 1 trial. Int J Radiat Oncol Biol Phys. 2021;109(2):436–40.
181. Prasad V, Zengerling F, Steinacker JP, Bolenz C, Beer M, Wiegel T, et al. First experiences with (177)Lu-PSMA therapy in combination with Pembrolizumab or after pretreatment with Olaparib in single patients. J Nucl Med. 2021;62(7):975–8.
182. Sheikh H, Ryder D, Bateman A, Chalmers A, Jackson A. Radiotherapy and olaparib in combination for carcinoma of the oesophagus: a phase I study. Clin Transl Radiat Oncol. 2023;40:100614.
183. Navran A, Al-Mamgani A, Elzinga H, Kessels R, Vens C, Tesselaar M, et al. Phase I feasibility study of Olaparib in combination with loco-regional radiotherapy in head and neck squamous cell carcinoma. Clin Transl Radiat Oncol. 2024;44:100698.
184. Chabot P, Hsia TC, Ryu JS, Gorbunova V, Belda-Iniesta C, Ball D, et al. Veliparib in combination with whole-brain radiation therapy for patients with brain metastases from non-small cell lung cancer: results of a randomized, global, placebo-controlled study. J Neuro-Oncol. 2017;131(1):105–15.
185. van de Ven AL, Tangutoori S, Baldwin P, Qiao J, Gharagouzloo C, Seitzer N, et al. Nanoformulation of Olaparib amplifies PARP inhibition and sensitizes PTEN/TP53-deficient prostate cancer to radiation. Mol Cancer Ther. 2017;16(7):1279–89.
186. Wu M, Liu J, Hu C, Li D, Yang J, Wu Z, et al. Olaparib nanoparticles potentiated radiosensitization effects on lung cancer. Int J Nanomedicine. 2018;13:8461–72.
187. Chan Wah Hak CML, Rullan A, Patin EC, Pedersen M, Melcher AA, Harrington KJ. Enhancing anti-tumour innate immunity by targeting the DNA damage response and pattern recognition receptors in combination with radiotherapy. Front Oncologia. 2022;12:971959.

Chapter 5
Role of Tumor-Initiating Stem Cells in Tumor Resistance and Relapse

Ali Amini and Babak Behnam

Abstract Tumor-initiating stem cells (TiSCs) are key drivers of tumor resistance, relapse, and immune evasion, with mitochondria playing a central role in regulating these processes. TiSCs and cancer stem cells (CSCs) rely heavily on oxidative phosphorylation (OXPHOS) for ATP production while maintaining metabolic plasticity, enabling them to switch between glycolysis and OXPHOS depending on nutrient availability and oxygen levels in the tumor microenvironment (TME). This adaptability enhances their ability to survive harsh conditions, resist chemotherapy and radiation, and fuel tumor progression. Mitochondria-generated reactive oxygen species (ROS) act as signaling molecules to support self-renewal, tumorigenesis, and DNA repair, further strengthening TiSC resilience to therapy.

Additionally, TiSCs evade apoptosis by tightly regulating cytochrome c release and caspase activation, making them more resistant to therapy-induced cell death. They also hijack mitochondria from immune cells, such as T cells and macrophages, impairing their function and promoting an immunosuppressive TME. Mitochondria with mtDNA mutations can be transferred to tumor-infiltrating lymphocytes (TILs), further dampening immune responses. Targeting mitochondrial pathways—including OXPHOS inhibition, ROS modulation, and disrupting mitochondrial transfer—could restore immune function, weaken TiSC defenses, and reduce tumor relapse. These therapeutic strategies represent promising avenues for improving cancer treatment outcomes.

Keywords Tumor-initiating stem cells (TiSCs) · Mitochondria · Oxidative phosphorylation · Immune evasion · Metabolic plasticity · Tumor relapse

---

A. Amini
American University, Washington, DC, USA

B. Behnam (✉)
American University, Washington, DC, USA

Avicenna Biotech Research, Clarksburg, MD, USA
e-mail: bbehnam@american.edu; babak.behnam@gmail.com

F. Taghizadeh-Hesary et al. (eds.), *From Radiobiology to Radiation Oncology*,
https://doi.org/10.1007/978-3-031-95724-6_5

5.1 Introduction

Imagine a fortress that repairs, renews, and evolves to withstand every attack. This is a perfect analogy to understand tumor-initiating stem cells (TiSCs) and cancer stem cells (CSCs), the amazing architects of cancer resilience, survival, and progression. These cells orchestrate the robust resistance to cancer treatment and tumor recurrence. TiSCs drive therapy resistance and tumor recurrence through cellular plasticity, metabolic adaptability, and dynamic interactions with the tumor microenvironment (TME), challenging therapeutic strategies.

Within the intricate landscape of cancer biology, TiSCs emerge as key players in tumor heterogeneity, capable of self-renewal and tumor initiation, and to our therapeutic endeavors. These cells, often referred to as CSCs, represent a subpopulation within tumors and a dynamic state that cells can enter and exit, challenging our traditional understanding of cancer cell hierarchies. Their ability to shift between different states, coupled with their extraordinary resistance mechanisms, makes them central figures in the story of cancer treatment failure and disease progression [1, 2].

This chapter takes you on a comprehensive journey through the world of TiSCs, structured to build a complete understanding of their role in cancer biology and therapy resistance. We begin by exploring the diverse origins of TiSCs, from oncogenic transformation of normal stem cells to the fascinating process of epithelial-mesenchymal transition (EMT). The chapter then surveys their biological properties, highlighting the remarkable metabolic plasticity that allows these cells to survive under extreme conditions and the sophisticated resistance mechanisms they employ to evade therapeutic intervention [3].

As we progress, we explore the intricate interplay between TiSCs and the TME, uncovering how these cells influence and are influenced by their microenvironment. This chapter explores therapeutic strategies, from conventional treatments like differentiation therapy to advanced innovations such as niche-modulating agents and targeted immunotherapies. In this regard, mitochondria play a essential role in the metabolic plasticity of TiSCs and CSCs, enabling them to adapt to extreme conditions and resist cancer therapies by supporting survival, energy production, and evasion of therapeutic interventions [4]. Understanding TiSC biology is reshaping cancer treatment, revealing new therapeutic possibilities. The chapter concludes by addressing key research priorities, including better characterization of TiSC heterogeneity, improved preclinical models, and optimized combination therapies—critical steps toward unlocking a new understanding of TiSCs and their role in cancer progression and treatment resistance. Overall, TiSCs are key drivers of tumor resistance and relapse, primarily due to their distinctive mitochondrial dynamics and metabolic flexibility [5].

The rest of this chapter explores into TiSCs'role in tumor resistance, relapse, and the key factors shaping their influence.

- Origins and Development of TiSCs.
- Biological Properties of TiSCs.

- Regulatory Mechanisms in Tumor Initiation, Metastasis, and Therapy Resistance via TiSC.
- The Role of Mitochondrial Biogenesis.
- The Role of Reactive Oxygen Species (ROS).
- The Role of Immune Modulation.
- Therapeutic Strategies of Targeting TiSCs.
- Contribution to Dynamics of Tumor Relapse.
- Conclusion and Future Directions.

5.2 Origins and Development of TiSC

Understanding how TiSCs arise and develop, including the important role of EMT in their formation, and the influence and role of TiSCs in the TME.

5.2.1 Multiple Origins of TiSC

TiSCs and CSCs are related but distinct concepts in cancer biology. TISCs are a subset of cells responsible for the initiation of tumors, particularly in experimental settings, and are crucial to understanding tumorigenesis and metastasis. However, CSCs have stem-like properties, including self-renewal and differentiation, which allow them to maintain tumor growth and contribute to relapse and resistance to therapy. Although TiSCs primarily drive tumor initiation, CSCs play a broader role in tumor maintenance and recurrence. Both contribute to cancer progression and serve as critical targets for the development of effective therapies.

Understanding the various cellular mechanisms through which TiSCs originate is essential, as these pathways can overlap with those that govern CSC behavior, further highlighting their interconnected roles in tumor evolution. These cells arise primarily through the oncogenic transformation of normal stem cells, although they can also develop through the reprogramming of differentiated cells under various stress conditions [6]. The hierarchical organization within tumors demonstrates different subpopulations with varying tumor-initiating potential, highlighting the complex nature of TiSC development [7]. The key characteristics of TiSCs include their ability to self-renew and differentiate into various tumor cell subtypes, properties that contribute significantly to tumor heterogeneity, and resistance to treatment. Recent studies using CellPhoneDB analysis have revealed sophisticated bidirectional communication between TiSCs and the TME, demonstrating how these cells adapt and influence their surroundings [8].

The multiple origins of TiSCs, which can arise from various normal stem cell populations, differentiated cells, or even non-tumorigenic cells, result in substantial variability in their mitochondrial content and function. This variability significantly affects their role in tumor resistance and relapse.

5.2.2 *Role of Epithelial-Mesenchymal Transition (EMT) in Cancer Stem Cell Development*

Epithelial-Mesenchymal Transition (EMT) plays a crucial role in the development of CSCs, contributing to tumor initiation, metastasis, and resistance to therapies. During EMT, cells lose epithelial traits and acquire mesenchymal properties, including increased motility and invasiveness, which are essential for metastasis. Mitochondria are central to this process, as they support the metabolic changes required for the survival and plasticity of CSC.

Mitochondrial reprogramming during EMT involves a shift from OXPHOS to glycolysis, particularly under hypoxic conditions, helping CSCs maintain their stem-like properties. Furthermore, mitochondrial dynamics, such as fission and fusion, enhance cell motility and invasiveness, critical features of metastasis [9–11]. On the other hand, while glycolysis is a common metabolic adaptation, some CSCs retain or even improve OXPHOS, depending on the type of cancer and microenvironment. Key EMT transcription factors, such as Snail, Twist, and ZEB family proteins, are activated during EMT and contribute to the stemness of CSCs. These transcription factors directly influence mitochondrial function, promoting metabolic reprogramming and mitochondrial remodeling. Snail, Twist, and ZEB proteins not only regulate the expression of genes involved in cell motility but also promote resistance to conventional therapies, such as chemotherapy and radiation. This resistance is due in part to altered mitochondrial bioenergetics, which enable CSCs to survive harsh conditions and regenerate tumors after treatment. In conclusion, EMT contributes to the development of CSCs through mitochondrial reprogramming and the activation of key transcription factors. However, these effects are highly context-dependent, varying according to tumor type, microenvironment, and external stimuli, highlighting the need for a personalized and individual approach to understanding CSC behavior.

EMT enhances the plasticity and adaptability of tumor cells, enabling them to resist therapeutic pressures and adapt to dynamic TME [12]. This plasticity, facilitated by "partial EMT" states, helps cancer cells maintain a balance between epithelial and mesenchymal characteristics, enhancing their metastatic potential. In partial EMT states, the cells retain both epithelial and mesenchymal traits, allowing cells to maintain some epithelial characteristics, such as adhesion molecules (e.g., E-cadherin), while acquiring mesenchymal traits like increased motility. These hybrid states contribute uniquely to metastasis by enabling cancer cells to circulate as clusters rather than as single cells, improving their survival in the bloodstream, and facilitating colonization at distant sites. In resistance to therapy, partial EMT allows cells to dynamically switch between epithelial and mesenchymal phenotypes, making them more adaptable to environmental stress, including chemotherapy and immune surveillance. This plasticity allows tumor persistence and relapse by allowing cells to evade targeted treatments that typically focus on epithelial or mesenchymal markers [13]. EMT also drives cancer cells to acquire stem cell-like properties, improving their tumorigenic capacity and the ability to evade therapies, although not all EMT-related changes necessarily lead to resistance. Some tumors may be more susceptible to targeted metabolic therapies post-EMT, addressing the context-dependent nature of therapy resistance [14].

Mitochondria play a key role in this process by supporting the metabolic shifts required for this plasticity, particularly through changes in mitochondrial dynamics and bioenergetics that sustain CSC characteristics. The interaction between EMT transcription factors and cellular plasticity is decisive for the development of CSCs and represents an important therapeutic target. Understanding these molecular mechanisms can help overcome resistance and improve clinical cancer treatment outcomes.

5.3 The Role of Biological Properties

5.3.1 Metabolic Adaptation: Plasticity, Reprogramming, and Pathway Alterations

Metabolic plasticity and reprogramming are indispensable characteristics of TiSCs and CSCs, allowing them to adapt to the ever-changing TME. These adaptive mechanisms are particularly significant in response to therapeutic pressures, such as radiation therapy, enabling tumor cells to survive, develop resistance, and contribute to relapse. By dynamically rewiring their metabolic pathways, TiSCs and CSCs optimize energy production, withstand oxidative stress, and adjust to fluctuations in nutrient availability, reinforcing their persistence and tumorigenic potential.

TiSCs and CSCs exhibit remarkable metabolic flexibility, enabling them to adapt their metabolic pathways to meet the increased energy and biosynthetic demands imposed by radiotherapy. Under stress conditions, these cells often shift their metabolic preferences, favoring glycolysis even in oxygen-rich environments, a phenomenon reminiscent of the Warburg effect. Concurrently, they may activate fatty acid oxidation and glutaminolysis, effectively generating ATP, producing biosynthetic precursors, and maintaining redox homeostasis, all of which are critical for their survival and proliferation. For instance, while bulk breast tumor cells typically display altered energetics characterized by aerobic glycolysis, breast CSCs demonstrate adaptive metabolic plasticity. This allows these tumor-initiating cells to dynamically switch between glycolysis and OXPHOS in response to various factors within the TME, such as hypoxia, ROS, and glucose availability [15].

5.3.2 *Resistance Mechanisms*

TiSCs employ sophisticated and multifaceted strategies to evade therapeutic interventions, demonstrating remarkable resilience in cancer treatment [6, 7]. These mechanisms collectively contribute to their extraordinary survival and resistance to treatment. For example, hepatocellular carcinoma (HCC), the most prevalent type of primary liver cancer, exhibits high molecular and cellular heterogeneity. The response to drugs and acquired resistance remains unpredictable under existing systematic guidelines based on mutation patterns and molecular biomarkers, leading to suboptimal treatment outcomes for many patients with atypical molecular profiles [16]. HCC remains one of the most challenging cancers to treat, underscoring the urgent need for more effective therapeutic strategies. Immunotherapy has emerged as a promising approach, benefiting approximately 15–20% of HCC patients. However, 3- and 5-year survival rates for HCC remain low, at 13–21% and 5%, respectively.

Despite therapeutic advances, TiSCs exhibit strong resistance to both conventional chemotherapy and immunotherapy, enabling them to persist as recurrent tumors or circulating tumor cells (CTCs). In particular, approximately 40% of the HCC cases originate from TiSCs, further emphasizing their pivotal role in cancer progression and resistance to treatment [17].

Immune Evasion via Mitochondrial Transfer and Tunneling Nanotubes

Recent studies have highlighted the role of mitochondrial transfer in cancer cells' resistance to therapy. Cancer cells can transfer mitochondria to T cells, leading to impaired anti-tumor activity [18]. Additionally, cancer cells induce T cell dysfunction through mitochondrial transfer, further contributing to immune evasion. It seems that the TiSCs and CSCs induce T cell dysfunction through this mitochondrial transfer [19]. The mitochondria-microtubule axis via the formed nanotubes facilitates mitochondrial transfer between TiSCs or CSCs and immune cells (T cells), contributing to cancer therapy resistance. Through tunneling nanotubes (TNTs), CSCs can transfer functional mitochondria to damaged or stressed neighboring cancer cells, enhancing their survival under therapeutic pressure. This transfer is guided by the microtubule network, which helps direct mitochondria along nanotubes to recipient cells. In addition, CSCs exploit nanotube connections to extract mitochondria from activated T cells, weakening their metabolic function and impairing immune responses (Fig. 5.1). By depleting T cells of mitochondria, CSCs induce T cell exhaustion, reducing their ability to mount an effective anti-tumor response. This exchange of mitochondria through the CSC-immune cell interface contributes to tumor resilience, immune evasion, and resistance to chemotherapy and radiation. Understanding this mechanism could provide therapeutic opportunities to disrupt mitochondrial transfer and restore immune function in cancer

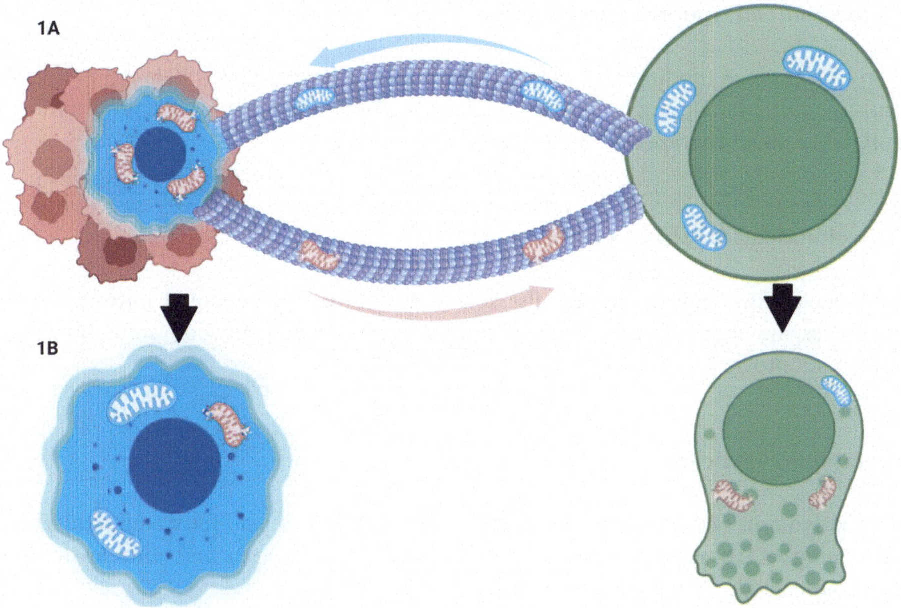

Fig. 5.1 Mitochondrial Hijacking by Tumor-Initiating Stem Cells (TiSCs) Leads to T Cell Dysfunction and Tumor Proliferation. This figure highlights the role of mitochondrial transfer in tumor immune evasion and underscores the potential of targeting mitochondrial dynamics as a therapeutic strategy. *Panel 1A:* Tumor-initiating stem cells (TiSCs) hijack intact mitochondria from T cells through tunneling nanotubes (TNTs). Simultaneously, TiSCs transfer dysfunctional mitochondria back to the T cell, impairing its metabolic and functional capacity. *Panel 1B:* Following mitochondrial transfer, the T cell becomes dysfunctional due to the loss of intact mitochondria and the acquisition of dysfunctional mitochondria. This metabolic and functional impairment suppresses the T cell's anti-tumor activity, while the tumor cell proliferates unchecked, promoting tumor progression and immune evasion. *Mitochondrial hijacking:* TiSCs exploit TNTs to acquire intact mitochondria from T cells. *Mitochondrial transfer:* Dysfunctional mitochondria are transferred to T cells, disrupting their metabolism. *T cell dysfunction:* Loss of intact mitochondria and accumulation of dysfunctional mitochondria impair T cell function. *Tumor proliferation:* TiSCs gain a survival advantage, leading to tumor growth and immune evasion

treatment. These findings accentuate the significance of mitochondrial dynamics in tumor progression and therapy resistance [20].

DNA Repair & Genomic Integrity

TiSCs possess highly efficient DNA repair mechanisms that rapidly resolve radiation- and chemotherapy-induced damage. By activating complex DNA repair pathways, these cells maintain genomic stability and evade therapeutic-induced cell death [11].

Mitochondrial Apoptotic Regulation

Through precise modulation of mitochondrial dynamics, TiSCs balance pro- and anti-apoptotic factors such as BCL-2 and BAX, allowing them to evade programmed cell death under therapeutic stress [10].

Mitochondrial Quality Control

Enhanced mitophagy facilitates the selective removal of damaged mitochondria, preserves metabolic fitness, reduces oxidative stress, and prevents mitochondria-driven apoptosis [9].

Metabolic Plasticity

TiSCs dynamically switch between glycolysis, OXPHOS, and fatty acid oxidation, sustaining ATP biogenesis and biosynthesis even under hypoxic and nutrient-limited conditions [15].

Drug Resistance Strategies

TiSCs develop pharmacological resistance by upregulating drug efflux transporters, activating detoxification pathways, adopting quiescent states, and leveraging immune-privileged niches. Their ability to adapt through these mechanisms presents a major challenge for effective cancer treatment.

5.3.3 Therapeutic Implications

As mentioned in the above sections, targeting tumor-initiating stem cell (TiSC)-specific resistance pathways, particularly those involving mitochondrial functions, offers promising therapeutic avenues in cancer treatment. TiSCs exhibit metabolic flexibility, switching between OXPHOS, glycolysis, and fatty acid oxidation to maintain energy production under stress inhibitors like metformin that have been explored to disrupt OXPHOS, aiming to impair TiSC energy metabolism and reduce their survival [21]. Additionally, TiSCs regulate mitochondrial dynamics, such as fission and fusion, to maintain their stemness and resist apoptosis. Compounds like Mdivi-1, a mitochondrial division inhibitor, have been investigated to disrupt these processes, potentially sensitizing TiSCs to conventional therapies [22]. TiSCs also exhibit high antioxidant capacities, allowing them to detoxify ROS and survive oxidative stress induced by treatments. Targeting their redox homeostasis by inhibiting antioxidant defenses could increase oxidative damage and sensitize TiSCs to therapy [23]. Furthermore, TiSCs can engage in mitochondrial transfer via tunneling nanotubes with immune cells, suppressing immune function and enhancing therapy resistance. Disrupting this process could restore immune surveillance and weaken TiSC survival [24]. By integrating mitochondrial-targeting strategies with existing treatments, it may be possible to overcome TiSC-mediated resistance, reduce cancer recurrence, and improve patient outcomes [8].

5.3.4 Organelles' Pivotal Role

Organelles such as mitochondria and microtubules are critical to the resistance mechanisms observed in TiSCs and CSCs, ultimately contributing to tumor relapse after conventional therapies. Mitochondria are central to metabolic reprogramming, enabling these cells to switch between oxidative OXPHOS and glycolysis. This metabolic flexibility allows for sustained ATP production even under hypoxic and nutrient-deprived conditions, ensuring cell survival during therapeutic stress [9]. In addition, recent studies have shown that mitochondrial transfer through tunneling nanotubes further bolsters the resistance capacity by enabling the sharing of functional mitochondria between cells, thereby promoting recovery from cytotoxic stress. Microtubules and cilia are equally important in this context, as they not only support cellular signaling and structural integrity but also play a leading role in the positioning and transport of mitochondria within the cell. This dynamic interplay facilitates rapid responses to environmental changes, enhancing cell motility, invasiveness, and immune evasion through EMT processes [10, 11]. Partial EMT states, characterized by a hybrid phenotype retaining both epithelial and mesenchymal features, further contribute to tumor cell plasticity and resistance by allowing CSCs to adapt to diverse microenvironments and evade targeted therapies [12].By understanding these interrelated mechanisms, it becomes apparent that targeting the mitochondria-microtubule axis may offer a promising therapeutic strategy. Inhibitors that disrupt mitochondrial bioenergetics, dynamics, or their intracellular transport could potentially reduce the resilience of TiSCs and CSCs, making them more susceptible to conventional treatments. This approach may not only prevent tumor relapse but also improve overall clinical outcomes in cancer patients. Integrating such targeted therapies with existing modalities could pave the way for more personalized and effective cancer treatment strategies.

5.4 Regulatory Mechanisms in Tumor Initiation, Metastasis, and Therapy Resistance via TiSCs

As discussed above, TiSCs and CSCs play influential roles in tumor initiation, metastasis, recurrence, and resistance by facilitating some key molecular pathways. The pathways include metabolic reprogramming, EMT, and immune evasion, which enable tumors to adapt to hostile environments and resist conventional therapies. TiSCs drive early tumor development, while CSCs maintain tumor growth and foster relapse after treatment. Their ability to dynamically switch between energy production modes and enhance survival underscores their importance in cancer progression. Targeting these regulatory mechanisms offers promising therapeutic strategies and is determining for development of effective personalized cancer treatments [9–11].

CSC-derived extracellular vesicles (ECVs) and non-coding RNAs are other mechanisms in regulating tumor initiation, metastasis, recurrence, and therapy resistance. ECVs serve as carriers of bioactive molecules—including proteins, microRNAs, and long non-coding RNAs—that modulate the TME, promote EMT, and enhance metastatic spread [25, 26]. Non-coding RNAs further regulate gene expression at the post-transcriptional level, maintaining CSC stemness and contributing to chemoresistance by modulating cell survival and drug efflux pathways [27, 28]. The interplay between CSC-derived ECVs and non-coding RNAs establishes a dynamic communication network that fosters tumor heterogeneity and adaptability under therapeutic stress. Disrupting these pathways is critical to dismantling cancer stem cell defenses, ultimately reducing relapse and improving survival outcomes.

5.5 The Role of Mitochondrial Biogenesis

TiSCs often exhibit higher mitochondrial mass and biogenesis, contributing to their enhanced bioenergetics and resistance to apoptosis. In addition, increased mitochondrial biogenesis plays a pivotal role in tumor resistance and relapse, particularly concerning TiSCs. TiSCs are a subpopulation of cancer cells endowed with self-renewal capabilities and are implicated in tumor initiation, metastasis, and recurrence. Elevated mitochondrial biogenesis within these cells enhances their metabolic flexibility, allowing them to switch efficiently between OXPHOS and glycolysis. This adaptability supports their survival under therapeutic stress and contributes to resistance against conventional treatments. Studies have demonstrated that TiSCs exhibit increased mitochondrial mass and activity, which are associated with heightened resistance to chemotherapy and radiotherapy. Targeting mitochondrial biogenesis pathways in TiSCs has been proposed as a potential therapeutic strategy to overcome resistance and prevent tumor relapse. By inhibiting key regulators of mitochondrial biogenesis, such as PGC-α 1, it may be possible to sensitize TiSCs to treatment and improve clinical outcomes [23, 29, 30].

Moreover, mtDNA stability is critical in tumor resistance and relapse, particularly concerning TiSCs. In addition, TiSCs exhibit mechanisms to preserve mtDNA integrity, which supports their survival and self-renewal capabilities. TiSCs are a subpopulation of cancer cells with self-renewal capabilities that drive tumor initiation and progression. Alterations in mtDNA, such as mutations or changes in copy number, can significantly impact TiSC function. For instance, a decrease in mtDNA copy number has been associated with the maintenance of stemness in cancer cells, leading to increased resistance to therapies. In squamous cell carcinoma of the esophagus, cells with reduced mtDNA exhibit enhanced stemness and prolonged cell cycle phases, contributing to therapeutic resistance. Similarly, mtDNA depletion in ovarian cancer cells upregulates genes related to proliferation and anti-apoptosis, further promoting resistance. These findings suggest that mtDNA stability is integral to TiSCs' resilience, and targeting mtDNA alterations may offer new avenues to overcome resistance to therapy and prevent tumor relapse [31].

5.6 The Role of Reactive Oxygen Species (ROS)

Regulation of ROS plays a major role in tumor resistance and relapse, particularly in TiSCs. These cells maintain low ROS levels through efficient mitochondrial respiration and robust antioxidant systems, preserving mitochondrial integrity, preventing DNA damage, and sustaining their stemness. As the primary sites of ROS generation, mitochondria produce reactive species during OXPHOS, with complexes I (NADH:ubiquinone oxidoreductase) and III (cytochrome bc1) being the major contributors. While excessive ROS can induce apoptosis or differentiation in TiSCs, thereby reducing their populations, moderate ROS levels drive adaptive responses that promote tumorigenesis, metastasis, and therapy resistance. TiSCs counteract oxidative stress by upregulating endogenous antioxidants and mitochondrial quality control mechanisms, allowing them to evade apoptosis and persist under therapeutic pressure. This precise ROS management enhances their survival, contributing to tumor recurrence. Recent studies by [32] and [33] underscore the critical role of mitochondrial ROS regulation in TiSC-driven tumor resilience and highlight novel therapeutic strategies targeting these pathways to overcome resistance and improve treatment efficacy [34, 35].

5.7 The Role of Immune Modulation

Recent studies highlight that TiSCs play a pivotal role in tumor resistance and relapse by employing sophisticated immune evasion mechanisms, many of which are intricately linked to mitochondrial dynamics. TiSCs often reside in immune-privileged niches, shielding them from immune surveillance while downregulating antigen presentation, secreting immunosuppressive cytokines (TGF-β, IL-10), and upregulating immune checkpoint molecules (PD-L1) to impair T-cell function. Recent sophisticated studies, including [32] and [33], underscore how disrupting these pathways could enhance the efficacy of immunotherapy. Targeting mitochondrial transfer and immune evasion mechanisms offers a promising strategy to overcome resistance to therapy and prevent tumor relapse. In addition, TiSCs engage in mitochondrial transfer through TNT with immune cells, particularly T cells, depleting them of functional mitochondria, thereby inducing exhaustion and weakening their anti-tumor response (Fig. 5.1) (see Sect. 3.3) [20].

TiSCs contribute to immune evasion and tumor resistance by creating an immune-privileged niche within the TME. One key mechanism involves the transfer of mitochondria that harbor mtDNA mutations from cancer cells to tumor-infiltrating lymphocytes (TILs), leading to metabolic and functional alterations that suppress anti-tumor immune responses [18, 19].

This mitochondrial transfer facilitates immune escape by reprogramming TILs toward a dysfunctional state, reducing their cytotoxic potential and contributing to tumor persistence and relapse. This strategic manipulation of mitochondrial dynamics allows TiSCs to evade immune-mediated destruction and persist after therapy.

5.8 Therapeutic Strategies Targeting TiSCs

Recent research underscores that targeting TiSCs through mitochondrial pathways holds significant promise in overcoming resistance to therapy and preventing tumor relapse. Secondarily, there are potentials for mitochondria-directed therapies to address resistance and minimize relapse in TiSCs. TiSC-specific mitochondrial pathways (e.g., OXPHOS inhibitors or ROS modulation) offer a potential strategy to overcome resistance and prevent relapse (mitochondria-directed therapies). TiSCs rely on robust mitochondrial OXPHOS and efficient antioxidant systems to maintain low ROS levels, thereby preserving mitochondrial integrity, preventing DNA damage, and maintaining stemness. Inhibiting OXPHOS with agents such as metformin or disrupting ROS modulation by targeting key antioxidant pathways impairs the metabolic flexibility of TiSCs, triggering apoptosis and enhancing treatment sensitivity. Emerging evidence from various studies—including [36, 37], and [38]—demonstrates that mitochondria-directed therapies significantly impair TiSC-driven tumor persistence. These innovative approaches are being explored in preclinical models and early-phase clinical trials, opening avenues for personalized cancer treatments that integrate metabolic inhibitors with conventional therapies. This innovative strategy offers a transformative approach to targeting TiSCs and ultimately improving long-term patient survival outcomes worldwide. Disrupting mitochondrial fission, fusion, or mitophagy can impair TiSC adaptability and survival.

5.9 Contribution to Dynamics of Tumor Relapse

5.9.1 Post-therapy Survival

Residual TiSCs surviving initial treatments serve as reservoirs for tumor regrowth, driven by their mitochondrial advantages. TiSCs and/or CSCs possess unique mitochondrial characteristics that enable their survival post-therapy, contributing to tumor relapse. These cells exhibit alterations in mitochondrial structure and function, enhancing their self-renewal and resistance to treatments. Such mitochondrial adaptations allow TiSCs to withstand therapeutic interventions, leading to tumor recurrence. TiSCs contribute to minimal residual disease (MRD) and the formation of metastatic niches. Their mitochondrial adaptations, such as enhanced OXPHOS and efficient energy production, enable survival post-therapy. These adaptations support TiSCs in evading treatment, leading to tumor relapse and metastasis [39].

Additionally, TiSCs can acquire mitochondria from surrounding cells through horizontal transfer mechanisms and tunneling nanotubes, further bolstering their energy production and survival capabilities. This mitochondrial acquisition supports TiSCs in evading therapy-induced cell death, thereby serving as a reservoir for tumor regrowth [18, 20]. Understanding these mitochondrial advantages is key for developing targeted therapies aimed at eradicating TiSCs and preventing tumor relapse.

5.9.2 Therapy-Induced Plasticity

Mitochondrial dynamics supports the phenotypic flexibility of TiSCs, enabling adaptation and relapse. TiSCs exhibit remarkable plasticity, enabling them to adapt to therapeutic pressures and contribute to tumor relapse. This adaptability is closely linked to mitochondrial dynamics. TiSCs can reprogram their mitochondrial metabolism, shifting between OXPHOS and glycolysis to meet energy demands and resist therapy-induced stress. This metabolic flexibility allows TiSCs to survive treatments and drive tumor recurrence [23]. Furthermore, TiSCs can modulate mitochondrial dynamics, such as fusion and fission processes, to maintain mitochondrial integrity and function under therapeutic stress. These adaptations enhance their survival and contribute to therapy resistance [40]. The interplay between TiSC plasticity and mitochondrial function is pivotal for developing targeted therapies aimed at eradicating these resilient cells and preventing tumor relapse.

5.10 Future Directions and Artificial Intelligence (AI) Integration

Future research should prioritize several key areas to advance our understanding and treatment of TiSCs. These priority areas include disrupting multiple TiSC survival pathways simultaneously, preventing mitochondrial adaptations that sustain resistance, developing combination treatments to counter metabolic plasticity, and inhibiting TiSC-driven tumor regeneration and metastasis. These interconnected approaches could lead to more effective therapeutic strategies. AI is revolutionizing cancer research, particularly in the field of CSCs investigation and treatment development. Through advanced visualization capabilities, AI enables researchers to rapidly model and analyze therapeutic interventions, significantly enhancing traditional methodologies [41]. Machine learning (ML)-based image analysis allows precise CSC identification, while transfer learning (TL) improves accuracy in medical imaging tasks like tumor detection [42].

Moreover, AI-driven radiogenomics integrates ML and deep learning (DL) approaches to enhance genomic classification in radiology, automating feature extraction for improved diagnostic precision. ML models, including random forests (RF), support vector machines (SVM), and XGBoost, have been widely utilized for radiomics, while deep learning architectures such as ResNet, VGG, and DenseNet provide automated feature learning for high-throughput genomic assessment [43]. These advancements position AI as a transformative tool in CSC research, optimizing therapeutic decision-making and accelerating the development of precision medicine strategies.

AI is the new electricity—just as electricity transformed nearly every industry a century ago [44], AI is revolutionizing healthcare today even in unexpected ways. When combined with cutting-edge medical advancements like organoid technology, AI has the potential to reshape human health. Recent research [45] shows liver cancer organoids from patient tissues effectively model cancer stem cell heterogeneity and enable personalized treatment testing. These 3D cultures preserve tumor characteristics, while AI enhances their utility through improved drug screening and response prediction. Combined with tools like CRISPR and microfluidics, AI-optimized organoid platforms offer promising solutions for treatment-resistant cancers, particularly hepatocellular carcinoma, by providing accurate patient-specific models.

References

1. Batlle E, Clevers H. Cancer stem cells revisited. Nat Med. 2017;23(10):1124–34. https://doi.org/10.1038/nm.4409.
2. Saygin C, Matei D, Majeti R, Reizes O. Targeting cancer stemness in the clinic: from hype to hope. Cell Stem Cell. 2019;24(1):25–40. https://doi.org/10.1016/j.stem.2018.11.017.

3. Viale A, Pettazzoni P, Lyssiotis CA, Ying H, Sánchez N, Marchesini M, et al. Oncogene ablation-resistant pancreatic cancer cells depend on mitochondrial function. Nature. 2014;514(7524):628–32. https://doi.org/10.1038/nature13611.
4. Peixoto J, Lima J, Metello J, Oliveira PJ. Mitochondrial metabolism and cancer stem cells: a complex interplay. Mol Asp Med. 2022;84:101058. https://doi.org/10.1016/j.mam.2021.101058.
5. Pastore N, Tasselli M, Hofer T. Mitochondria in cancer stem cells: a Janus-faced organelle. Front Cell Dev Biol. 2021;9:764217. https://doi.org/10.3389/fcell.2021.764217.
6. Li Y, Wang Z, Ajani JA, Song S. Drug resistance and cancer stem cells. Cell Commun Signal. 2021;15(19):19.
7. Abdullah LN, Chow EKH. Mechanisms of chemoresistance in cancer stem cells. Clin Transl Med. 2013 December;2:3.
8. He X, Smith SE, Chen S, Li H, Wu D, Meneses-Giles PI, et al. Tumor-initiating stem cell shapes its microenvironment into an immunosuppressive barrier and pro-tumorigenic niche. Cell Rep. 2021;36:109674.
9. Aguilar E, Marin de Mas I, Zodda E, Marin S, Morrish F, Selivanov V, et al. Metabolic reprogramming and dependencies associated with epithelial cancer stem cells independent of the epithelial-mesenchymal transition program. Stem Cells. 2016;34:1163–76.
10. Ma Y, Wang L, Jia R. The role of mitochondrial dynamics in human cancers. Am J Cancer Res. 2020;10:1278–93.
11. Sessions DT, Kashatus DF. Mitochondrial dynamics in cancer stem cells. Cell Mol Life Sci. 2021;78:3803–16.
12. Dave B, Mittal V, Tan NM, Kaestner L, Sheng S, Pestell R, et al. Epithelial-mesenchymal transition, cancer stem cells and treatment resistance. Breast Cancer Res. 2012;14:202.
13. Pradella D, Naro C, Sette C, Ghigna C, Valacca C. EMT and stemness: flexible processes tuned by alternative splicing in development and cancer progression. Mol Cancer. 2017;16:8.
14. Kong D, Li Y, Wang Z, Sarkar FH. Cancer stem cells and epithelial-to-mesenchymal transition (EMT)-phenotypic cells: are they cousins or twins? Cancers (Basel). 2011;3:716–29.
15. Walsh HR, Cruickshank BM, Brown JM, Marcato P. The flick of a switch: conferring survival advantage to breast cancer stem cells through metabolic plasticity. Front Oncol. 2019;9:753.
16. Chan YT, Zhang C, Wu J, et al. Biomarkers for diagnosis and therapeutic options in hepatocellular carcinoma. Mol Cancer. 2024;23:189. https://doi.org/10.1186/s12943-024-02101-z.
17. Machida K, Tahara SM. Immunotherapy and microbiota for targeting of liver tumor-initiating stem-like cells. Cancers. 2022;14:2381.
18. Ikeda H, Kawase K, Nishi T, et al. Immune evasion through mitochondrial transfer in the tumour microenvironment. Nature. 2025;638:225–36. https://doi.org/10.1038/s41586-024-08439-0.
19. AACR. Cancer cells induce T-cell dysfunction through mitochondrial transfer. Cancer Discovery—Research Watch; 2025 February.
20. Saha T, Dash C, Jayabalan R, Khiste S, Kulkarni A, Kurmi K, et al. Intercellular nanotubes mediate mitochondrial trafficking between cancer and immune cells. Nat Nanotechnol. 2022;17(1):98–106. https://doi.org/10.1038/s41565-021-01000-4.
21. Vasan K, Chandel NS. Molecular and cellular mechanisms underlying the failure of mitochondrial metabolism drugs in cancer clinical trials. J Clin Invest. 2024;134:e176736.
22. Armstrong JS. Mitochondrial medicine: pharmacological targeting of mitochondria in disease. Br J Pharmacol. 2007;151:1154–65.
23. Jin P, Jiang J, Zhou L, Huang Z, Nice EC, Huang C, et al. Mitochondrial adaptation in cancer drug resistance: prevalence, mechanisms, and management. J Hematol Oncol. 2022;15:97.
24. Li Y, Li Y, Li X, Wei L, Ye J. Advancements in mitochondrial-targeted nanotherapeutics: overcoming biological obstacles and optimizing drug delivery. Front Immunol. 2024;15:1451989. https://doi.org/10.3389/fimmu.2024.1451989.
25. Jiang Q, Tan XP, Zhang CH, Li ZY, Li D, Xu Y, et al. Non-coding RNAs of extracellular vesicles: key players in organ-specific metastasis and clinical implications. Cancers (Basel). 2022;14(22):5693. https://doi.org/10.3390/cancers14225693.

26. Kalluri R. The biology and function of exosomes in cancer. J Clin Invest. 2016;126:1208.
27. Qian Y, Shi L, Luo Z. Long non-coding RNAs in cancer: implications for diagnosis, prognosis, and therapy. Front Med (Lausanne). 2020;7:612393. https://doi.org/10.3389/fmed.2020.612393.
28. Wei L, Wang X, Lv L, et al. The emerging role of noncoding RNAs in colorectal cancer chemoresistance. Cell Oncol. 2019;42:757–68. https://doi.org/10.1007/s13402-019-00466-8.
29. De Luca A, Fiorillo M, Peiris-Pagès M, Ozsvari B, Smith DL, Sanchez-Alvarez R, et al. Mitochondrial biogenesis is required for the anchorage-independent survival and propagation of stem-like cancer cells. Oncotarget. 2015;6:14777–95.
30. Farnie G, Sotgia F, Lisanti MP. High mitochondrial mass identifies a sub-population of stem-like cancer cells that are chemo-resistant. Oncotarget. 2015;6:30472–86.
31. Zheng XX, Chen JJ, Sun YB, Chen TQ, Wang J, Yu SC. Mitochondria in cancer stem cells: Achilles heel or hard armor. Trends Cell Biol. 2023;33:708–27.
32. Tuy K, Rickenbacker L, Hjelmeland AB. Reactive oxygen species produced by altered tumor metabolism impacts cancer stem cell maintenance. Redox Biol. 2021;44:101953. https://doi.org/10.1016/j.redox.2021.101953.
33. Yuan S, Lu Y, Yang J, et al. Metabolic activation of mitochondria in glioma stem cells promotes cancer development through a reactive oxygen species-mediated mechanism. Stem Cell Res Ther. 2015;6:198. https://doi.org/10.1186/s13287-015-0174-2.
34. Zhou D, Shao L, Spitz DR. Reactive oxygen species in normal and tumor stem cells. Adv Cancer Res. 2014;122:1–67.
35. Hernansanz-Agustín P, Enríquez JA. Generation of reactive oxygen species by mitochondria. Antioxidants (Basel). 2021;10:415.
36. Behnam B, Fazilaty H, Ghadyani M, Fadavi P, Taghizadeh-Hesary F. Ciliated, mitochondria-rich Postmitotic cells are immune-privileged, and mimic immunosuppressive microenvironment of tumor-initiating stem cells: from molecular anatomy to molecular pathway. Front Biosci (Landmark Edition). 2023;28:261.
37. Behnam B, Taghizadeh-Hesary F. Mitochondrial metabolism: a new dimension of personalized oncology. Cancers (Basel). 2023;15:4058.
38. Taghizadeh-Hesary F, Ghadyani M, Kashanchi F, Behnam B. Exploring TSGA10 function: a crosstalk or controlling mechanism in the signaling pathway of carcinogenesis? Cancers (Basel). 2024 August;16:3044.
39. Garimella SV, Gampa SC, Chaturvedi P. Mitochondria in cancer stem cells: from an innocent bystander to a central player in therapy resistance. Stem Cell Cloning. 2023;16:19–41.
40. Genovese I, Carinci M, Modesti L, Aguiari G, Pinton P, Giorgi C. Mitochondria: insights into crucial features to overcome cancer chemoresistance. Int J Mol Sci. 2021;22:4770.
41. Abubakar M. Exploring the pivotal association of AI in cancer stem cells detection and treatment. Proc Anticancer Res. 2024:8.
42. Rana M, Bhushan M. Machine learning and deep learning approach for medical image analysis: diagnosis to detection. Multimed Tools Appl. 2023;82:26731–69.
43. Saxena S, Jena B, Gupta N, Das S, Sarmah D, Bhattacharya P, et al. Role of artificial intelligence in radiogenomics for cancers in the era of precision medicine. Cancers (Basel). 2022;14:2860.
44. Ng A, Why AI. Is the new electricity. Interview by Shana Lynch; 2017 March.
45. Giron-Michel J, Padelli M, Oberlin E, Guenou H, Duclos-Vallée JC. State-of-the-art liver cancer organoids: modeling cancer stem cell heterogeneity for personalized treatment. Bio. Drugs. 2025 Jan 18;39:237. https://doi.org/10.1007/s40259-024-00702-0. Epub ahead of print.

Chapter 6
Cancer Cell's Fates Upon Radiation

Bilal Siddiqui, Sakshi Tiwari, Syed Naved Quadri, and M. Z. Abdin

Abstract Cancer is still a major worldwide health concern, even with improvements in treatment approaches. The widespread use of radiotherapy, a well-established cytotoxic approach for many malignancies, has notably improved survival rates for patients. Traditionally, radiation is recognized for its ability to directly alter the DNA double helix structure, initiating DNA damage responses that can lead to necrosis, apoptosis, or cellular senescence. These processes can disrupt normal mitotic functions and ultimately modify the biological properties of cancer cells. According to recent research, radiation has a major impact on the immunogenicity, phenotypic, and surrounding tumor microenvironment of cancerous cells, leading to broader and more complex changes in cancer cell biology. It is noteworthy that some cancer cells can evade programmed cell death through the overactivation of pathways for DNA repair in the event of genomic damage. Furthermore, these surviving cells exposed to radiation often exhibit modification of the expression of significant oncogenes and tumor suppressor genes. This alteration can enhance various hallmarks of cancer, such as invasion, migration, and metastasis. Together with other epigenetic and structural cellular alterations, the induction of these genetic processes results in the phenomenon known as tumor radioresistance and is influenced by external factors from the tumor microenvironment. Understanding the intricate molecular and cellular mechanisms behind radioresistance is necessary, as it might unveil novel treatment objectives as well as diagnostic markers. This knowledge may ultimately contribute to improving radiation therapy (RT) outcomes through the establishment of novel therapeutic approaches, including the use of radiosensitizers and targeted approaches. This chapter discusses the mechanisms by which radiation affects cancer cells, the development of radioresistance, and potential therapeutic methods to improve RT's effectiveness.

Keywords Proliferation · Apoptosis · Autophagy · Metastasis · Senescence

---

B. Siddiqui · S. Tiwari · S. N. Quadri · M. Z. Abdin (✉)
Department of Biotechnology, Centre for Transgenic Plant Development, School of Chemical and Life Sciences, Jamia Hamdard, New Delhi, India

F. Taghizadeh-Hesary et al. (eds.), *From Radiobiology to Radiation Oncology*,
https://doi.org/10.1007/978-3-031-95724-6_6

6.1 Introduction

Cancer continues to be a serious global public health concern regardless progress in detection and management. Compared with 14.1 million cases reported and 8.2 million deaths from cancer in 2012, there were around 19 million cases of cancer, and nearly 10 million deaths globally were attributed to cancer in 2020 [1]. Over 34 million fresh cases of cancer are predicted to be diagnosed annually by 2070; lower-income nations are predicted to bear a disproportionate share of this burden, with a projected 400% spike in the incidence of cancer in those nations over the next fifty years [2]. As scientific research advances, numerous medications have been created to increase the effectiveness of treatment for particular cancer types. High-energy radiation is used in radiation therapy (RT) to destroy cancer cells and reduce the tumor size [3]. The process of tumor radiotherapy employs a variety of ionizing radiation techniques to halt and regulate the development, spread, and multiplication of cancerous tumor cells. Experimental methods and advancements in the field of molecular biology over the last few decades have helped to elucidate the impact of radiation on the biological attributes of tumor cells. Following the 1895 discovery of X-rays and Marie Curie's discoveries of radium and polonium toward the end of the nineteenth century, radiotherapy has been in use for over a century. According to estimates, RT is used by around fifty percent of cancer patients throughout medical treatment, which has significantly reduced death from cancer [4]. RT is thought to be a "two-edged sword" during cancer therapy because, apart from impacting neoplasms' development, spread, and other bodily processes, it can also genetically modify healthy tissues, causing injury to nontumor cells, that has an unforeseen adverse effect on the body. Because RT is capable of attacking both healthy and cancerous cells alike, damage to healthy tissue is minimized using segmented radiotherapy, which uses recurrently low doses for every part. As a result, for many tumor forms, a typical radiation treatment involving daily doses of about 2 Gy over 5–7 weeks is utilized [5]. The RT's primary goal is to destroy tumor cells by damaging their DNA. On the other hand, tumor cells identify the damage by means of DNA damage response (DDR) process, which induces the activation of several DNA repair processes by causing the cell cycle to stop and stimulating the engagement of cell cycle checkpoints. The three primary mechanisms for repairing radiation-induced DNA damage are homologous recombination (HR), non-homologous end joining (NHEJ), and base excision repair (BER) [6]. The immediate harm that ionizing radiation (IR) causes to DNA generates either single-strand breaks (SSBs) or double-strand breaks (DSBs). These incidents activate the DDR pathway, and while SSBs can be easily fixed, about 5% of DSBs cannot, which eventually results in cell death [5]. After DNA damage occurs, cell cycle checkpoints halt the cycle's progression and stop damaged DNA from replicating. The mechanisms that cause cell cycle arrest and radiosensitivity differ based on the cell cycle stage during infrared [7]. Cells experience a mitotic collapse if DNA damage happens before or during mitosis. As a result, cells get binucleated and fail to finish the cycle. These cells can inevitably age or undergo necrosis or apoptosis and die

[8]. The primary characteristic of tumor cells that sets them apart from normal cells in one's body is their capacity to survive death and continue to divide indefinitely. Numerous tumor cell fate decisions (CFD), including autophagy, apoptosis, pyroptosis, ferroptosis, cuproptosis, necroptosis, and programmed cell death (PCD), have been demonstrated to be induced by RT in recent investigations [8].

According to recent studies, radiation alters the immunogenicity of neoplasm cells as well as surrounding tissue by interfering with cell signal cascade and regulation in addition to disrupting the physical makeup of tumor cells, including the cell membrane and organelles. Furthermore, exposure to radiation causes changes in biology, including metabolic reconfiguration, epithelial-mesenchymal transition (EMT), broad variations in gene expression, chromatin remodeling, and irregularities in circadian cycles, among other factors [9]. Most radiation-induced changes promote the biological adaptation of malignant cells to their tumor microenvironment, which increases the tumors' ferocity and radioresistance and aggravates the invasion, movement, metastasis, and proliferation characteristics of cancer [10]. As the immune system and tumor stroma are better understood in influencing the evolution of tumors, novel radiosensitizer trials that target the stromal and immunological microenvironment in addition to important cell death pathways have attracted a lot of attention, especially when combined with stereotactic ablative body radiation (SABR). It is essential to comprehend how radiation affects cancer cells since it includes important consequences such as genomic instability, the development of radioresistance, and how the tumor microenvironment influences these reactions. This chapter's discussion of these interrelated processes advances the science of radiobiology and opens the door to the creation of novel approaches that maximize therapeutic effectiveness while minimizing side effects (Fig. 6.1).

6.2 Mechanism of Radiation on Cancer Cells

Gamma and electromagnetic radiation, such as X-rays, are the two forms of ionizing radiation used in RT treatments in clinical settings. This energy is absorbed by the impacted tissues, and its magnitude, expressed in gray (Gy) units, is based on the mass of the organs or tissues [4]. Several factors determine whether RT is used, such as the kind of cancer, and the dimension of the tumor, its location within the body, and the region of the body involved in the precision needed when applying the estimated dose of radiation. RT can be given intravenously (brachytherapy), superficially (teletherapy), or both. SSBs, DSBs, DNA cross-links, and DNA-protein cross-links are examples of direct DNA damage caused by radiation, while reactive oxygen species (ROS) and reactive nitrogen species (RNS) cause indirect DNA damage. DSBs are thought to be the most dangerous radiation-induced lesion among them since they are the triggering factor for rearrangements in chromosomes that expand in a linear-quadratic manner when exposed to radiation at high dose rates (HDR). The ability of radiation to regulate cancer, in other words, is mostly dependent on radiation-induced DNA damage (RIDD); consequently, the outcome

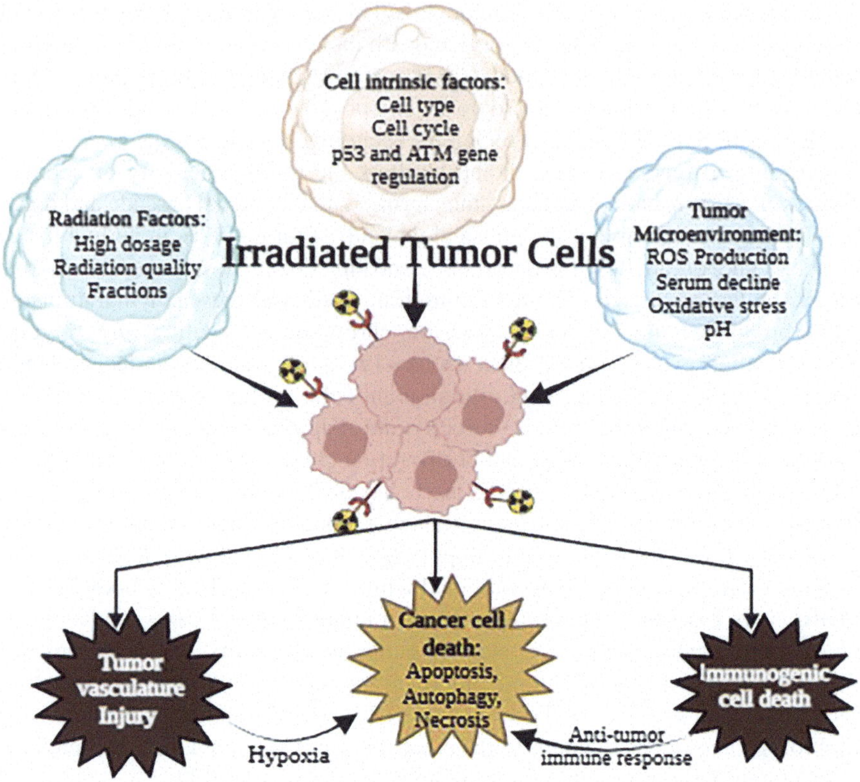

Fig. 6.1 Mechanisms of tumor cell response to radiation therapy: interplay of intrinsic, extrinsic, and microenvironmental factors

of cancer cells is largely determined by the DDR of cancerous cells and their capacity to repair DNA damage [11]. The kinases ataxia-telangiectasia mutant (ATM) and ataxia telangiectasia and Rad3-related protein (ATR) recognize RIDD and trigger downstream proteins to work together for initiating the DNA damage process [12]. DSBs are mostly repaired by two mechanisms during cell cycle arrest, which is brought on by NHEJ and HR. Despite being error-free, HR is only available within the late S and G2 phases because it needs an intact sister chromatid as the repair template, and NHEJ is active during the cell cycle but susceptible to errors.

In situations where DNA is damaged by ionizing radiation, one or two helix turns of the DNA will have multiple lesions. These locations, which also comprise DSB, are referred to be clustered damage sites since they are caused by a single radiation pathway. Ionizing radiation may be detected by clustered damage sites as opposed to single, endogenously caused lesions, which are often uniformly dispersed. Radiation-induced clustered damage sites were only identified in mammalian cells in the past decade, and when gamma radiation is used to produce DSB, non-DSB cluster yield is at least four to eight times higher than immediate DSB output [13]. Evidence suggests that most DSB caused by low LET radiation are repaired in 30–60 min, but a tiny percentage of DSB—typically less than 20%—are more difficult to repair in mammalian cells and some may last longer than 24 h. The percentage of slowly healed breaks rises with the LET of the radiation, and it is anticipated that the higher proportion of complicated breaks appear in the slower-repairing DSB [14]. DNA base damage (DBD), which is defined as alterations or damage to the bases inside the DNA molecule, is frequently caused by nucleotide oxidation, alkylation of deoxyribonucleotides, one base pair breakage, and base deletions or mismatches [15]. Numerous causes, including both inside and outside chemicals, IR, environmental influences, and metabolites generated by the cell itself, might result in these alterations or harm. One aspect of DNA damage caused by IR is DNA cross-linking (DCL), that mainly manifests as DNA–DNA cross-linkage and DNA–protein cross-linking [16]. DNA–DNA cross-linking is the process by which nucleotides on the identical strand of DNA or within the two strands can create a covalent bond. DNA–protein cross-linking is the process by which some proteins, including transcription and replication enzymes, regulatory proteins, nonhistone proteins in chromatin, and histones, form covalent bonds with DNA [17]. After being exposed to infrared radiation, this cross-linking is the chemical cause of the chromosomal defects observed under a microscope. Damage to cross-linking has the immediate effect of disrupting the current replication fork, which hinders cells' ability to replicate DNA normally. Moreover, DCL can increase the risk of cancer by causing mutations and encouraging malignant changes [18].

ROS levels are markedly increased both intracellularly and extracellularly by IR damage, which sets off cellular oxidative stress reactions [19]. Cells respond to oxidative stress by activating transcription factors and DNA repair mechanisms. Consequently, these variables alter the expression levels of enzymes that scavenge and process ROS [20]. For instance, elevated intracellular ROS levels alter one such transcription factor, Yap1, post-translationally, which in turn triggers the activation of many genes [21, 22]. Numerous biochemical mechanisms, including DNA damage, mitochondrial malfunction, and cell membrane oxidation, are involved in the generation of ROS brought on by IR. The prevention and treatment of RITI may be greatly aided by targeted management of ROS levels [23]. Furthermore, by intensifying inflammatory responses and promoting tumor cell proliferation, increased levels of ROS following tumor IR therapy can exacerbate tissue injury and tumor development [24].

6.2.1 Cellular Repair Pathways

Cells have developed a variety of mechanisms to respond to or repair DNA damage to preserve stability of the genome during ROS-induced stress. These processes include recombination (REC), HR, BER, NER, NHEJ, mismatch repair (MMR), translesion synthesis (TLS), and direct reversal [25]. BER, NER, MMR, and direct reversal are a few of the mechanisms that regulate the restoration of damaged DNA;

whereas, TLS and REC work to circumvent the damage so that replication can take place and the DNA lesion is tolerated. Although each of these pathways has a distinct function inside cell, each method can treat different types of DNA damage, and there are overlaps between them [26]. Every one of these mechanisms is essential to maintaining genomic integrity.

The main DNA repair mechanism for DSB is HR, which, by using identical sequences as repair models, is more precise and has a reduced mistake rate, such as sister chromatids, homologous chromosomes, and repetitive DNA sequences. Since sister chromatid homologous sequences constitute the selected template, HR repair mostly occurs during the S and G2 stages of the cell cycle [27]. One of the main mechanisms for DSB repair is NHEJ. This repair mechanism, which mostly takes place around the cell cycle's G0/G1 phases, is distinguished by its speed and error-proneness [28]. When a DSB is detected, the Ku70/Ku80 heterodimer immediately binds to the damaged DNA location, stabilizing and protecting the ends of the DNA [29]. The Ku70/Ku80 dimer then brings together two DNA-dependent protein kinase catalytic subunits (DNA–PKcs) to form the DNA–PK complex, which prevents end clipping and encourages the Ku heterodimer to move inward [30]. Thus, Artemis-DNA–PKcs complex acts as a major nuclease following the recruitment of other nucleases, including Artemis, CtIP, and EXO1. DNA polymerases μ and λ attach to the DNA after terminal clipping to promote synthesis; whereas, the NHEJ repair pathway is completed by DNA ligase IV [31].

A highly preserved biological mechanism, MMR, is necessary for mitotic recombination, cell death, and genomic stability. The template strand and the freshly synthesized strand are first separated by this method [32]. The mismatched nucleotide on the freshly produced DNA is then bound by the MutS protein dimer. Then, a specific cleavage is created by the cooperation of the MutL2 and MutH proteins [33]. Later, MutL2-MutS2-MutH complex moves to the mismatch from the incision site. The UvrD helicase unravels DNA's double helix structure in the process [34]. At the same time, the freshly manufactured DNA, which has the mismatched bases, is hydrolyzed by an exonuclease until it gets to where the mismatched bases are located. Lastly, the removed DNA sequence is synthesized again using DNA polymerase and DNA ligase.

BER mostly happens in the cell cycle's G1 phase. Xanthine and 7-methylguanosine are examples of alkylated or oxidized bases that are specifically targeted by this approach to fix minor flaws or breakage in the DNA strand [35]. An apurinic/apyrimidinic (AP) site is created when the damaged base is recognized and eliminated by DNA glycosylase, initiating the process [36]. AP endonuclease then breaks the phosphodiester bond at the AP site, resulting in the DNA splitting into two strands. The damaged area is then exposed to DNA polymerase, which is used to synthesize the missing base.

In mammals, NER is the main way of repairing severe DNA damage [37]. This procedure guarantees effectiveness, precision, manageability, ubiquity, low toxicity, and recovery by using nearby, intact DNA as a repair template. NER has two sub-pathways: transcription-coupled NER (TC-NER) and global genomic NER (GG-NER) [38]. GG-NER can fix damage that impacts transcribed and nontranscribed DNA strands in both active and dormant genes because it functions throughout the genome without relying on transcription [39]. Numerous "damage sensor" proteins, including the XPC–Rad23B complex and DNA binding, continually scan for helical aberrations in the genome.

6.2.2 IR Influences the Structure and Functioning of Cells

When cells are exposed to IR, their ROS levels increase as a result of cell membrane peroxidation, mitochondrial dysfunction, and damaged DNA. IR can directly disturb atomic structures when it is absorbed by live cells, resulting in chemical and biological alterations. By radiolyzing water, it can also have an indirect effect by producing reactive chemical species that can harm proteins, lipids, and nucleic acids [40]. When combined, radiation's direct and indirect impacts sequence of molecular and metabolic signals that may either repair the damage or cause irreversible changes in the body or cell death [41]. IR has a major effect on the structure and operation of mitochondria, resulting in increased ROS generation and mitochondrial malfunction. To begin, changes in the permeability of the inner membrane of the mitochondria disturb the balance of ions, which affects the mitochondrial membrane's potential and promotes the generation of ROS [42]. One major factor in the creation of ROS is excessive electron leakage during electron transport [43]. As a result, signal transmission and regular cellular metabolism are disturbed. ROS-mediated metabolic stress is linked to a number of illnesses, such as cancer, diabetes, and RITI. Therefore, strict regulation of the redox condition is necessary for homeostatic cellular activities [43]. Through controlling the expression of particular genes [44, 45], altering ion channel activity [46], and imitating or influencing signal transduction intermediates (such as second messengers) [47], ROS contribute to signaling pathways that sustain regular physiological processes at low levels. For instance, cells in the developmental cycle's G1 phase are strongly arrested when NAD(P)H-oxidases limit the generation of superoxide [48]. The effect must occur even in the absence of caused DNA damage since the ATM gene is necessary for the early detection and mending of ionizing radiation-induced damage to DNA [49].

Both short-term and long-term radiation-induced ROS/RNS can harm the mtDNA and/or nuclear DNA (nuDNA) coding for mitochondrial electron transfer chain (ETC) subunits, as well as the metabolic systems necessary for their proper expression and assembly [50]. It is interesting to note that radiation-induced ROS/RNS generation does not occur in rho(o) cells, which lack mitochondrial electron transport [51]. Compared to rho(+) cells, rho(o) cells show a decreased ability to recover past the G2 checkpoint and a delayed G2 arrest [52]. According to recent data, oxidative stress may cause mitochondria to undergo BER and MMR. Excessive oxidative stress, however, appears to focus on mitochondrial DNA polymerase γ (pol γ) activity, which is essential for mtDNA replication along with repair [53], lowering the total repair capacity [54]. It may be possible to uncover controlled pathways when metabolic alterations are present by revealing the relationship between these alterations and radiation-induced protective or harmful effects. The early and late responses of cells to ionizing radiation may be explained by different mechanisms that this technique may uncover. However, accurately describing the effects of radiation quality, dosage, and dose rate is crucial. For instance, low-dosage radiations may not always be mediated by the same molecular and metabolic processes that underpin high-dose radiations. Furthermore, it could be necessary to incorporate the metabolic rate of the system under investigation when extending findings across species.

6.2.3 Cellular Death Triggered by RT

IR has numerous useful applications, but excessive radiation exposure can be fatal. The two distinct groups into which the health effects of RT have been separated are Acute Radiation Syndrome (ARS) and Delayed Effects of Acute Radiation Exposure (DEARE). When a comparatively high dose (>2–5 Gy) strikes the organism as a whole or a significant portion of it is administered at an elevated rate, ARS is the resultant collection of health effects [55]. Depending on the organs' or tissues' level of radiosensitivity, ARS is typically classified into three types of syndromes: the Cardiovascular/Central Nervous System Syndrome (>50 Gy), the Gastrointestinal Syndrome (typically greater than 10 Gy), and the Hematopoietic Syndrome (which occurs following exposure to 0.7–10 Gy) [56]. DEARE syndrome, on the other hand, includes radiation-induced lung injury, delayed skin injury, delayed immunological reconstitution, protracted gastrointestinal consequences, and renal failure. These ailments might appear months or years after radiation exposure [57]. The practical application of RT for the eradication of cancer is frequently limited by damage to healthy tissue; protracted episodes of tissue damage, inflammation, atrophy, and fibrotic remodeling have been connected to all types of cell death along with accelerated senescence [58, 59].

Recent studies have shown that these proteins are crucial radiation targets that influence the pathways leading to cell death, challenging the traditional DNA-centric understanding of radiation harm. In certain instances, it is hypothesized that decreased DNA repair fidelity results in radiation-induced mortality through protein damage, which subsequently lowers cell viability [60]. Among prokaryotes, Qui et al. [61] hypothesized that *Shewanella oneidensis*'s radiosensitivity is caused by protein damage, while Daly et al. [62] suggested that *Deinococcus radiodurans*'s great radioresistance is caused by a number of defensive mechanisms that reduce protein oxidation. Because of their reduced protein oxidation, *Bdelloid rotifers* also exhibit resilience to radiation damage [62]. In traditional radiobiology, clonogenic tests are used to evaluate cell death, which can be characterized as the loss of replicative capacity (also known as replicative or reproductive death). Current knowledge of the mechanisms underpinning ionizing radiation-induced death of cells is predicated primarily on studies of immortal cancerous cell lines, which are not representative of normal cells' biological states that have not been immortalized or altered [63]. Even though their DNA is more susceptible to unrepaired damage because of their quicker rate of proliferation, cancer cells usually have numerous mutations that either inherently activate DNA repair pathways or enable them to endure damage that may render normal cells unviable [64].

Numerous processes of cell death, such as necrosis, apoptosis, or autophagy, were shown to occur when cells are exposed to radiation. Furthermore, radiation may cause rapid cellular senescence, a condition where a cell is still viable but no longer capable of proliferating due to changed functions [65]. In certain instances, it has been shown that greater IR dosages cause necrosis and change the cellular response from senescence to apoptosis and/or autophagy [66]. All cells do not, however, react in the same way to a certain radiation dose. For instance, primary human hematopoietic CD34+ cells undergo death, while pulmonary artery endothelial cells mostly experience expedited senescence. The same dose of radiation that induces senescence in one cell type can cause rapid apoptosis in others [66, 67]. More thorough classifications of radiation-induced cell death based on molecular mechanisms are given in this section. These include apoptosis, autophagy-dependent cellular death, necrosis (necroptosis, mitochondrial permeability transition, and MPT-driven necrosis), pyroptosis, ferroptosis, immunogenic cell death, and nonlethal processes (mitotic catastrophe and cellular senescence) (Fig. 6.2).

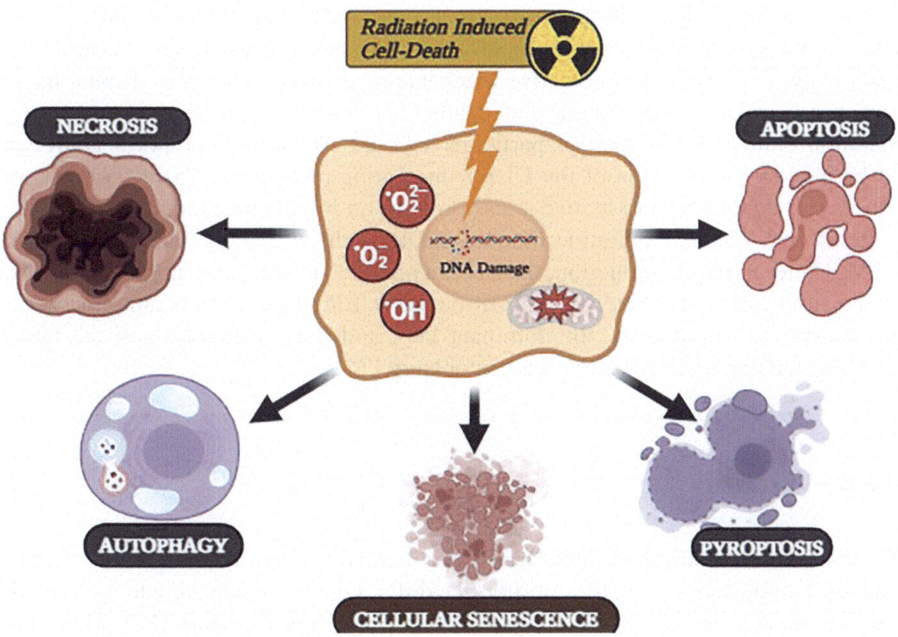

Fig. 6.2 Pathways of radiation-induced cell death: DNA damage and cellular fate determination

6.2.4 Apoptosis

Apoptosis is a gene-regulated, energy-dependent, active, and programmed process of cell death that is associated with contraction, pyknosis, and the growth of apoptotic bodies. IR, as a violent stimulation, causes irreversible microenvironmental disruptions of MCF-7 cells in a number of ways, mainly by acting on the mitochondrial and ER membranes to induce apoptotic signals [68], which can then cause massive accumulations of free radicals, such as ROS in IPEC-J2 cells, as well as destroying the plasma membrane or organelle membrane to alter intracellular ions, potentials, osmotic pressures, etc. [69]. All of these events can be detected by Bcl-2 family proteins located on the mitochondrial membrane, which activates the intrinsic apoptotic pathway [70]. Further evidence that radiation can induce cell apoptosis via activating P53-dependent pathways comes from the fact that radiation-sensitive tissues typically exhibit higher basic P53 mRNA expression and are more probable to produce apoptosis-inducing responses to radiation exposure. DNA damage caused by radiation can be repaired by activating the P53 pathway. Even when the damage is irreparable, P53 directly alters Bcl-2 and Bax by triggering apoptotic gene transcription, which controls the occurrence of apoptosis [71].

6.2.5 Autophagy

The process by which cellular stress causes damaged or aged cytoplasmic organelles to be sequestered within vesicles for lysosomal breakdown is known as autophagy. Usually a cytoprotective mechanism, an overabundance of autophagic response, results in autophagic cell death [72]. Autophagy can be activated by radiation-induced DNA damage, particularly DSBs. According to reports, DDR can be harmed by suppression of the ULK1 interacting protein FIP200, which in turn causes cell death under oxidative stress brought on by ionizing radiation [73]. The proteins that connect radiation-induced DDR with autophagy include P53 and ATM. Under radiation induction, the ER can operate via the following mechanisms: (1) by suppressing the Akt/TSC/MTOR pathway, ER stress triggers autophagy; (2) when exposed to radiation, the dominant Perk and IRE1 pathways activate Perk-eif2a, which results in ER stress and autophagy [74].

6.2.6 Necrosis

Necrosis is an uncontrolled, disorderly type of cell death that is brought on by unfavorable circumstances, such as abrupt pH shifts, energy depletion, and ion imbalance in the cell and its surroundings following radiation exposure [72]. There are three categories of cell necrosis: "necroptosis," "MPT (mitochondrial permeability transition)-driven necrosis," and "Parthanatos." Cell death, known as necroptosis, is brought on by disruptions in intracellular or extracellular homeostasis and is mostly reliant on MLKL, RIPK3, and RIPK1 kinase activity. According to experimental findings, under low-dose radiation, active caspase-8 causes cell apoptosis and prevents necrosis by cleaving RIP; under high-dose radiation, on the other hand, reduced caspase-8 activity causes the intracellular RIP1/RIP3 complex to form, which mediates necroptosis in tumor cells [75].

6.2.7 *Pyroptosis*

Pyroptosis is a type of cell death that is primarily triggered by inflammatory caspase-1 and caspase-11 and is highly dependent on members of the gasdermin protein family [76]. Through both conventional and nonclassical mechanisms, caspase1/11 can activate and cleave members of the gasdermin protein family, creating plasma membrane holes that result in cell enlargement and death. Under 10 and 20 Gy irradiation, studies have shown that bone marrow-derived macrophages (BMDMs) exhibit increased caspase-1 activity and pyroptosis proportion [77]; similarly, activation of NLRP3 inflammasomes not only mediates radiation-induced pyroptosis of BMDM but also causes pyroptosis of damaged lung tissue cells in mice [78].

6.2.8 *Ferroptosis*

Ferroptosis is an innovative form of death that is controlled by its own GPX4, brought on by oxidative stress, and prevented by lipophilic antioxidants and iron chelators [76]. It has just lately been identified as a key tumor suppression mechanism. Ferroptosis is essentially iron toxicity death brought on by a buildup of chronic lipid peroxidation, which is dependent on both iron and ROS generation. In a model of acute lung injury caused by radiation, ferroptosis has been documented. Transmission electron microscopy revealed ferroptotic distinctive alterations of mitochondria, and ferroptosis inhibitor treatment significantly reduced radiation-induced histopathological changes in mice's lungs as well as the levels of ROS and inflammatory cytokines (TNF-a, IL-6, IL-10) [79].

6.2.9 *Immunogenic Cell Death*

In the past ten years, the idea of ICD has surfaced to reinterpret cancer cell death from an immunological-functional perspective. ICD allows the dying cell to trigger anti-tumor adaptive immune responses through perimortem events [80]. An increasing number of preclinical and clinical studies, which combine RT with immunotherapy and demonstrate enhanced systemic tumor responses, also suggest radiation-induced ICD [81]. The release of DAMPs and antigenic variables to elicit cancer-specific immunity in these investigations is believed to be at least largely mediated through radiation-induced ICD, despite the indirect nature of the relationship. However, when some well-established immunosuppressive barriers are eliminated, for instance by combining immune checkpoint inhibitors (such as anti-CTLA4 or anti-PD-1) with local RT, the immunogenic effects of RT can result in immune-mediated tumor rejection [82]. Recent research has also shown that low-dose radiation can change macrophages to the iNOS+ or M1 phenotype, which allows them to attract T lymphocytes specific to tumors and encourage tumor rejection [83].

6.2.10 Cellular Senescence

Permanent cell cycle stoppage is referred to as senescence. Damage to DNA causes radiation-induced senescence, and the cell cycle block is caused by activation of the p53 and pRb pathways, but oxidative stress and other variables may also be significant triggers in the setting of radiation [84]. Because of several papers showing alterations in telomere control by TERT and other genes, radiation-induced senescence requires more consideration even if it is not dependent on the erosion and shortened erosion of chromosomal telomeres. P53 activation following radiation promotes cell survival by repairing DNA damage and inducing growth arrest. However, reports of expedited senescence without P53 or P21 and P16 have also surfaced, indicating that additional genes might be linked to controlling this mechanism [85].

6.3 Mitotic Catastrophe

A mechanism known as "mitotic catastrophe" controls cells that are incapable of completing mitosis by inducing mitotic arrest, which leads to regulated cell death (RCD) and senescence. Radiation-damaged cells can enter mitosis early with misrepaired DNA due to malfunctioning cell cycle checkpoints, a prevalent feature of cancer cells, which can result in mitotic catastrophe [76]. Uncertainty surrounds the precise processes that determine the ultimate destiny of cells and trigger the mitotic catastrophe. Solid tumors frequently exhibit delayed responses to RT because it may take multiple cell divisions before enough genetic damage is accumulated to cause mitotic death. Lethiferous mitotic catastrophe can be induced in two primary methods [86]: (1) P53 downregulation results in the suppression of G2 or M phase checkpoints, which causes irreparable DNA damage; and (2) centrosome hyperamplification [87], in which excessively enlarged centrosomes cause aberrant chromosome separation, multiploidy division, and cells with aberrant nuclei (megakaryon, binuclear, or multinuclear). It is possible to witness mitotic catastrophe-induced cell death in radiation or radioimmunotherapy treatment tests [88]. Cells undergo a brief G2 arrest upon exposure to radiation, proceed through mitosis, and then activate the mitotic checkpoint, also referred to as the spindle assembly checkpoint, to stop mitosis [89].

Another significant manner that radiation affects tumor cells is by controlling the secondary messengers and signaling molecules associated with the plasma membrane of cancer cells. According to Lammering [90] et al.'s in vitro tests, radiation can trigger downstream MAPK and PI3K pathways, being potent mediators of malignant development and proliferation, and stimulate EGFR via a nonligand-dependent mechanism. Park discovered that radiation activated the EGFR-mediated p38/Akt and PI3K/Akt signaling pathways, that in turn enhanced the glioma cell's capacity for invasion and metastasis by upregulating the production of matrix metalloproteinase 2 (MMP-2) [91]. On the one hand, radiation also encourages the growth, invasion, and spread of cancer cells by enhancing hypoxia inducible factor (HIF) and triggering the hepatocyte growth factor (HGF)/c-Met signal transduction pathway [93] or upregulating integrin, which is closely linked to the diverse biological behaviors of cancer cells [92]. On the other hand, damage to tumor cell membrane receptors brought on by ionizing radiation can lead to an imbalance in the downstream signal transduction pathway, which might confuse the cellular metabolic route and potentially trigger apoptosis or death.

6.4 Factors Influencing Cancer Cell Response to Radiation

Weinberg's team has identified 10 biological traits that are indicative of malignancies and have raised a lot of alarm [94]. Six damage scales connected to radiation are created by integrating the 10 properties according to the link between irradiation and cancerous cells: 1. Four of the ten characteristics of cell proliferation—perpetual expansion, escaping growth inhibition, and being resistant to cell death—are among the impacts of radiation on this scale. Two of the 10 characteristics—radiation effects on the infiltration and metasis scale are a result of the combination of angiogenesis induction, invasion stimulation, and metastasis. Radiation impacts on inflammation scales that promote cancer; the remaining three scales deal with immunogenicity, aberrant energy metabolism, and genomic instability [95]. Furthermore, recent research has demonstrated that tumor radiotherapy is intimately linked to immunological effects and that radiation indirectly influences tumor cell division or metastatic biological processes by influencing immunogenicity and the tumor microenvironment [96]. Magné et al. [97] discovered that one high radiation dosage administered to tumor cells triggered NF-κB signaling pathway, which in turn induced the development as well as preservation of the antigen-presenting cells, B cells, and immunological T cells, as well as controlling the expression of many apoptotic and antiapoptotic genes. It has been shown that local tumor radiation alters the immunogenicity of tumors and how they interact with the host, leading to the death of immunogenic antigenic tumor cells and the presentation of tumor-derived antigens by dendritic cells and T cells [98].

Although it is unclear if this is a direct result of irradiation, it may be enticing to hypothesize that ablative doses predispose cells to a distinct kind of cell death in comparison to normal doses per fraction. Establishing this connection involves separating the effect of impact of dosage per biologically effective dose (BED) fraction and taking into consideration temporary misconception brought on by dynamics of cell death and repopulation. It is important to note that high doses that are clinically irrelevant may be necessary to cause enough harm to nonnuclear cellular components to cause upfront necrosis [99]. The microenvironment of cells has a significant impact on the aforementioned characteristics. In a nutshell, oxygen tension reduces chromosomal abnormalities and the ability of cells exposed to hypoxic radiation to reproduce, hence modulating cell mortality following irradiation. This is thought to be caused by kinetic rivalry between chemical repair mechanisms and the oxygen "fixation" of DNA damage [100]. Cell mortality in hypoxic, nutrient-deficient in vivo environments will be impacted by the chemical characteristics of oxygen fixation and the indirect impact of hypoxia on biological processes, such as DNA repair, cell metabolism, and HIF-related survival mechanisms [101]. Regarding antigenicity, RT improves surface antigen processing and MHC-I-restricted presentation in a cell-autonomous approach, at least in part through the mTOR pathway [102]. More intriguingly, irradiation tumor cells in vitro exhibit a diversification of the epitope repertoire shown on MHC-I due to changed transcriptional activity and processing of antigen [103]. The first in-human demonstration of this idea was presented in a recent clinical trial that combined RT with immune checkpoint blockade and described a patient instance with proliferation of T cell clones toward a radiation-induced neoantigen [104].

6.5 Radiotherapy Optimization: Emerging Strategies for Improved Therapeutic Efficacy

Despite being one of the most successful cancer treatments, radiation nonetheless caused radioresistance in many people. When administered alone, ionizing radiation is more effective against a small number of cancers, such as lymphomas, head and neck cancer, prostate cancer, skin cancer, nonsmall cell lung cancer, cervical cancer, throat cancer, and skin cancer. Nevertheless, it is useless against a few kinds of cancer, including soft tissue carcinoma, progressive nonsmall-cell lung cancer, bladder-related cancer, glioblastoma, and breast cancer, possibly due to the inherent radioresistance of cancer cells [105]. Even with the development of new radiotherapy methods and the widespread use of new treatment regimens, radiation resistance remains a significant barrier to improve treatment outcomes, leading to RT failure, metastasis, recurrence of cancer, and a poor prognosis [106]. Because it is linked to many gene changes and the variation of the tumor and surrounding microenvironment, radiotherapy resistance can arise in a variety of ways [107, 108]. Therefore, evaluating the processes behind the emergence of solid tumor resistance to both novel and conventional RTs is crucial.

Personalized medicine has made it possible to accurately identify a patient's single-cell sequencing data for DNA and RNA. Additionally, modern technologies enabled it to precisely profile the immune system and protein expression. It is now possible to identify that tumor cells differ significantly among patients due to these technical advancements. Radiation oncologists can use these datasets to create new radiation sensitivity markers that will increase the effectiveness of treatment. Personalized cancer therapy has been successfully adopted through guided chemotherapy in chronic myelogenous leukemia utilizing imatinib, targeting the BCR-ABL oncoprotein [112]. Drugs that increase tumor cells' sensitivity to RT are known as radiosensitizers. These substances seem to aid in the molecular fixing of free radicals generated by radiation damage. The way it works is comparable to the oxygen effect, wherein the damaged molecules' metabolic responses prevent cells damaged by radiation from being repaired. The radiosensitizers absorb free radicals like OH+, which prevents the molecules from being repaired. Compounds known as radioprotectors are made to lessen the harm that radiation does to healthy tissues. Antioxidants can also serve as radioprotectors, but their efficiency depends on their presence prior to or during radiation exposure. It is possible to reduce toxicity after radiation had been administered by using additional substances known as mitigators [113].

RT is frequently used as a main treatment for cancer patients or as an adjuvant therapeutic method in conjunction with chemotherapy. Increased radiation dosages may destroy cancer cells, which would slow the spread of the disease and prevent recurrence [114]. Recent decades have seen the development of adjuvants or treatments based on nanoparticles, thanks to technological advancements in the field of nanomedicine. Due to their advantageous physical/chemical properties, high surface-to-volume ratio, and nanoscale size, nanocarriers are currently used in cancer therapy for efficient drug delivery [115]. Numerous radiosensitizing high-Z metal nanoparticles are useful for preventing the negative effects of RT and show a notable capacity to increase ionization-induced ROS production [114]. For many years, nanoparticles have been employed extensively in PET and a variety of therapeutic applications as contrast agents for improved imaging, radiation dosage enhancers, and radiation sensitizers. Recent developments in this field have placed more emphasis on modifying the tumor cell microenvironment to accomplish efficient therapy, in addition to the traditional function of improving dose enrichment in tumor lesion regions [115].

Radiation oncology's primary objectives are to use more secure and precise radiotherapy techniques to protect healthy tissue from dosage damage and increase the radiation dose delivered to the malignant tissue [117]. With time, new radiotherapy methods, such as proton radiotherapy, helical tomotherapy, and intensity-modulated radiation therapy (IMRT), have emerged as a result of the ongoing improvements in contemporary radiotherapy technology. There is still a limited therapeutic benefit even if these new radiation modalities can somewhat reduce radiation-induced damage [118]. Another cutting-edge radiotherapy method is FLASH-radiotherapy (FLASH-RT), which uses a linear electron accelerator (LINAC) to deliver ultra-high dose rate irradiation (UHDR) in less than 200 milliseconds at a mean dosage rate of over 40 Gy/s.

6.6 Synergistic Approaches in Radioimmunotherapy: A Novel Paradigm for Cancer Treatment

The integration of RT with immunotherapy has become a viable approach to improve promising strategy to enhance anti-tumor responses and improve clinical outcomes. RT not only induces direct DNA damage in cancer cells but also modulates the tumor microenvironment to enhance immune recognition [119]. A potentially effective approach involves combining RT with immune checkpoint inhibitors (ICIs), such as CTLA-4 and PD-1/PD-L1 inhibitors, which can amplify the abscopal effect by promoting T-cell priming and activation [120]. Furthermore, the use of radiosensitizing nanoparticles, including gold, hafnium, and iron oxide nanoparticles, has been explored to enhance radiation-induced ICD, thereby augmenting the immune response. In addition, recent studies highlight the role of Stimulator of Interferon Genes (STING) agonists in amplifying RT-induced innate immune activation, leading to enhanced Type I interferon responses and improved dendritic cell function [121]. Another promising approach is the combination of adoptive cell therapy, such as CAR-T cells or tumor-infiltrating lymphocytes (TILs), with RT, as radiation enhances tumor antigen expression and T-cell infiltration [122]. Additionally, epigenetic modulation using HDAC inhibitors or DNA methyltransferase (DNMT) inhibitors has been proposed to improve tumor antigen presentation, further sensitizing tumors to both radiation and immunotherapy [123]. Collectively, these multi-modal strategies hold immense potential to revolutionize cancer treatment by maximizing the immunogenic effects of radiotherapy while overcoming immune evasion mechanisms. Future studies should focus on optimal dosing, sequencing, and biomarker identification to enhance the efficacy of these synergistic radioimmunotherapy approaches.

6.7 Conclusion

RT is still a mainstay of cancer treatment; it works by causing DNA damage, oxidative damage, and disruption of the cell cycle, which ultimately causes cancer cells to undergo apoptosis, necrosis, or mitotic catastrophes. Radiation affects the TME in ways other than direct cytotoxicity, such as immunological feedback, vascular integrity, and stromal relations, all of which can improve or impair the effectiveness of treatment. To overcome resistance mechanisms and optimize radiation-based treatments, it is essential to comprehend these complex cellular and molecular responses.As molecular oncology has advanced, more sophisticated radiation techniques have been made possible by the combination of genomic, proteomic, and epigenetic knowledge. The significance of a precision medicine approach in radiation is highlighted by the discovery of biomarkers for radiosensitivity and resistance, as well as techniques, such as radioimmunotherapy, nanoparticle-assisted radiosensitization, and epigenetic reprogramming. The significance of a precision medicine approach in radiation is highlighted by the discovery of biomarkers for radiosensitivity and resistance, as well as tactics, such as radioimmunotherapy, nanoparticle-assisted radiosensitization, and epigenetic alterations. These molecular discoveries lessen off-target effects while simultaneously increasing therapy efficacy. Personalized treatment approaches, wherein individual tumor characteristics dictate radiation dosage, fractionation timelines, and combination treatments with immunotherapy and tailored medicines, are the way of the future for radiotherapy. The possibility of customizing RT to optimize tumor control while reducing normal tissue toxicity is further strengthened by the development of AI, radiomics, and predictive modeling. The use of our growing knowledge of how radiation exposure affects cancer cell destiny in clinical practice has enormous potential to enhance patient survival and oncological outcomes.

References

1. Ferlay J, Colombet M, Soerjomataram I, Parkin DM, Piñeros M, Znaor A, et al. Cancer statistics for the year 2020: an overview. Int J Cancer. 2021;149:778. https://doi.org/10.1002/ijc.33588.
2. Soerjomataram I, Bray F. Planning for tomorrow: global cancer incidence and the role of prevention 2020–2070. Nat Rev Clin Oncol. 2021;18(10):663–72.
3. Chen GZ, Zhu HC, Dai WS, Zeng XN, Luo JH, Sun XC. The mechanisms of radioresistance in esophageal squamous cell carcinoma and current strategies in radiosensitivity. J Thorac Dis. 2017;9(3):849.
4. Baskar R, Lee KA, Yeo R, Yeoh KW. Cancer and radiation therapy: current advances and future directions. Int J Med Sci. 2012;9(3):193.
5. Adjemian S, Oltean T, Martens S, Wiernicki B, Goossens V, Vanden Berghe T, et al. Ionizing radiation results in a mixture of cellular outcomes including mitotic catastrophe, senescence, methuosis, and iron-dependent cell death. Cell Death Dis. 2020;11(11):1003.
6. Minchom A, Aversa C, Lopez J. Dancing with the DNA damage response: next-generation anti-cancer therapeutic strategies. Therapeutic Adv Med Oncol. 2018;10:1758835918786658.
7. Huang X, Tran T, Zhang L, Hatcher R, Zhang P. DNA damage-induced mitotic catastrophe is mediated by the Chk1-dependent mitotic exit DNA damage checkpoint. Proc Natl Acad Sci. 2005;102(4):1065–70.
8. Hayes JD, Dinkova-Kostova AT, Tew KD. Oxidative stress in cancer. Cancer cell. 2020;38(2):167–97.
9. Bian L, Meng Y, Zhang M, Guo Z, Liu F, Zhang W, et al. ATM expression is elevated in established radiation-resistant breast cancer cells and improves DNA repair efficiency. Int J Biol Sci. 2020;16(7):1096.
10. Seibold P, Auvinen A, Averbeck D, Bourguignon M, Hartikainen JM, Hoeschen C, et al. Clinical and epidemiological observations on individual radiation sensitivity and susceptibility. Int J Radiat Biol. 2020;96(3):324–39.
11. Santivasi WL, Xia F. Ionizing radiation-induced DNA damage, response, and repair. Antioxid Redox Signal. 2014;21(2):251–9.
12. Maier P, Hartmann L, Wenz F, Herskind C. Cellular pathways in response to ionizing radiation and their targetability for tumor radiosensitization. Int J Mol Sci. 2016;17(1):102.
13. Sutherland BM, Bennett PV, Sutherland JC, Laval J. Clustered DNA damages induced by x rays in human cells. Radiat Res. 2002;157(6):611–6.

14. Schmid TE, Dollinger G, Beisker W, Hable V, Greubel C, Auer S, et al. Differences in the kinetics of γ-H2AX fluorescence decay after exposure to low and high LET radiation. Int J Radiat Biol. 2010;86(8):682–91.
15. Cadet J, Wagner JR. DNA base damage by reactive oxygen species, oxidizing agents, and UV radiation. Cold Spring Harb Perspect Biol. 2013;5(2):a012559.
16. Nakano T, Xu X, Salem AM, Shoulkamy MI, Ide H. Radiation-induced DNA–protein cross-links: mechanisms and biological significance. Free Radic Biol Med. 2017;107:136–45.
17. Caldecott K, Jeggo P. Cross-sensitivity of γ-ray-sensitive hamster mutants to cross-linking agents. Mutation Research/DNA Repair. 1991;255(2):111–21.
18. Richardson ME, Siemann DW. DNA damage in cyclophosphamide-resistant tumor cells: the role of glutathione. Cancer Res. 1995;55(8):1691–5.
19. Nie Q, Huan X, Kang J, Yin J, Zhao J, Li Y, Zhang Z. MG149 inhibits MOF-mediated p53 acetylation to attenuate X-ray radiation-induced apoptosis in H9c2 cells. Radiat Res. 2022;198(6):590–8.
20. Fry RC, Begley TJ, Samson LD. Genome-wide responses to DNA-damaging agents. Ann Rev Microbiol. 2005;59(1):357–77.
21. Delaunay A, Isnard AD, Toledano MB. H2O2 sensing through oxidation of the Yap1 transcription factor. EMBO J. 2000;19:5157.
22. Rodrigues-Pousada CA, Nevitt T, Menezes R, Azevedo D, Pereira J, Amaral C. Yeast activator proteins and stress response: an overview. FEBS Lett. 2004;567(1):80–5.
23. Muhammad W, Zhu J, Zhai Z, Xie J, Zhou J, Feng X, et al. ROS-responsive polymer nanoparticles with enhanced loading of dexamethasone effectively modulate the lung injury microenvironment. Acta Biomater. 2022;148:258–70.
24. Fu ZJ, Wang ZY, Xu L, Chen XH, Li XX, Liao WT, et al. HIF-1α-BNIP3-mediated mitophagy in tubular cells protects against renal ischemia/reperfusion injury. Redox Biol. 2020;36:101671.
25. Friedberg EC, Walker GC, Siede W, Wood RD, editors. DNA repair and mutagenesis. American Society for Microbiology Press; 2005.
26. Lindahl T, Wood RD. Quality control by DNA repair. Science. 1999;286(5446):1897–905.
27. Shamanna RA, Lu H, De Freitas JK, Tian J, Croteau DL, Bohr VA. WRN regulates pathway choice between classical and alternative non-homologous end joining. Nat Commun. 2016;7(1):13785.
28. Frock RL, Sadeghi C, Meng J, Wang JL. DNA end joining: G0-ing to the Core. Biomol Ther. 2021;11(10):1487.
29. Liu W, Miao C, Zhang S, Liu Y, Niu X, Xi Y, et al. VAV2 is required for DNA repair and implicated in cancer radiotherapy resistance. Signal Transduct Target Ther. 2021;6(1):322.
30. Dibitetto D, Marshall S, Sanchi A, Liptay M, Badar J, Lopes M, et al. DNA-PKcs promotes fork reversal and chemoresistance. Mol Cell. 2022;82(20):3932–42.
31. Guo M, Wang Y, Tang Y, Chen Z, Hou J, Dai J, et al. Mechanism of genome instability mediated by human DNA polymerase mu misincorporation. Nat Commun. 2021;12(1):3759.
32. Olave MC, Graham RP. Mismatch repair deficiency: the what, how and why it is important. Genes Chromosom Cancer. 2022;61(6):314–21.
33. Marinus MG. DNA mismatch repair. EcoSal Plus. 2012;5(1):10–1128.
34. Ahmad M, Tuteja R. Emerging importance of mismatch repair components including UvrD helicase and their cross-talk with the development of drug resistance in malaria parasite. Mutat Res—Fundam Mol Mech Mutagen. 2014;770:54–60.
35. Schermerhorn KM, Delaney S. A chemical and kinetic perspective on base excision repair of DNA. Acc Chem Res. 2014;47(4):1238–46.
36. Stratigopoulou M, van Dam TP, Guikema JE. Base excision repair in the immune system: small DNA lesions with big consequences. Front Immunol. 2020;11:1084.
37. Muñoz JC, Beckerman I, Choudhary R, Bouvier LA, Muñoz MJ. DNA damage-induced RNAPII degradation and its consequences in gene expression. Genes. 2022;13(11):1951.

38. Marteijn JA, Lans H, Vermeulen W, Hoeijmakers JH. Understanding nucleotide excision repair and its roles in cancer and ageing. Nat Rev Mol Cell Biol. 2014;15(7):465–81.
39. Yu S, Evans K, Van Eijk P, Bennett M, Webster RM, Leadbitter M, et al. Global genome nucleotide excision repair is organized into domains that promote efficient DNA repair in chromatin. Genome Res. 2016;26(10):1376–87.
40. Hall EJ, Giaccia AJ. Radiobiology for the. Radiologist. 2006;6:597.
41. Spitz DR, Azzam EI, Jian Li J, Gius D. Metabolic oxidation/reduction reactions and cellular responses to ionizing radiation: a unifying concept in stress response biology. Cancer Metastasis Rev. 2004;23:311–22.
42. He Z, Zhang C, Liang JX, Zheng FF, Qi XY, Gao F. Targeting mitochondrial oxidative stress: potential neuroprotective therapy for spinal cord injury. J Integr Neurosci. 2023;22(6):153.
43. Schafer FQ, Buettner GR. Redox environment of the cell as viewed through the redox state of the glutathione disulfide/glutathione couple. Free Radic Biol Med. 2001;30(11):1191–212.
44. Allen RG, Tresini M. Oxidative stress and gene regulation. Free Radic Biol Med. 2000;28(3):463–99.
45. Herrlich P, Böhmer FD. Redox regulation of signal transduction in mammalian cells. Biochem Pharmacol. 2000;59(1):35–41.
46. López-Barneo J, López-López JR, Ureña J, Gonzalez C. Chemotransduction in the carotid body: K+ current modulated by P O2 in type I chemoreceptor cells. Science. 1988;241(4865):580–2.
47. Schulze-Osthoff K, Bauer MK, Vogt M, Wesselborg S. Oxidative stress and signal transduction. Int J Vitam Nutr Res. Internationale Zeitschrift fur Vitamin-und Ernahrungsforschung. Journal international de vitaminologie et de nutrition. 1997;67(5):336–42.
48. Venkatachalam P, De Toledo SM, Pandey BN, Tephly LA, Carter AB, Little JB, et al. Regulation of normal cell cycle progression by flavin-containing oxidases. Oncogene. 2008;27(1):20–31.
49. Shiloh Y. ATM and related protein kinases: safeguarding genome integrity. Nat Rev Cancer. 2003;3(3):155–68.
50. Berg JM, Tymoczko JL, Stryer L. Biochemistry (loose-leaf). Macmillan; 2007.
51. Leach JK, Van Tuyle G, Lin PS, Schmidt-Ullrich R, Mikkelsen RB. Ionizing radiation-induced, mitochondria-dependent generation of reactive oxygen/nitrogen. Cancer Res. 2001;61(10):3894–901.
52. Kawamura S, Takai D, Watanabe K, Hayashi JI, Hayakawa K, Akashi M. Role of mitochondrial DNA in cells exposed to irradiation: generation of reactive oxygen species (ROS) is required for G2 checkpoint upon irradiation. J Health Sci. 2005;51(3):385–93.
53. Graziewicz MA, Day BJ, Copeland WC. The mitochondrial DNA polymerase as a target of oxidative damage. Nucleic Acids Res. 2002;30(13):2817–24.
54. Zastawny TH, Dabrowska M, Jaskolski T, Klimarczyk M, Kulinski L, Koszela A, et al. Comparison of oxidative base damage in mitochondrial and nuclear DNA. Free Radic Biol Med. 1998;24(5):722–5.
55. Waselenko JK, MacVittie TJ, Blakely WF, Pesik N, Wiley AL, Dickerson WE, et al. Medical management of the acute radiation syndrome: recommendations of the strategic National Stockpile Radiation Working Group. Ann Intern Med. 2004;140(12):1037–51.
56. Xiao M, Whitnall MH. Pharmacological countermeasures for the acute radiation syndrome. Curr Mol Pharmacol. 2009;2(1):122–33.
57. Coleman CN, Blakely WF, Fike JR, MacVittie TJ, Metting NF, Mitchell JB, et al. Molecular and cellular biology of moderate-dose (1–10 Gy) radiation and potential mechanisms of radiation protection: report of a workshop at Bethesda, Maryland, December 17–18, 2001. Radiat Res. 2003;159(6):812–34.
58. Friedman EJ. Immune modulation by ionizing radiation and its implications for cancer immunotherapy. Curr Pharm Des. 2002;8(19):1765–80.

59. Iglesias-Bartolome R, Patel V, Cotrim A, Leelahavanichkul K, Molinolo AA, Mitchell JB, Gutkind JS. mTOR inhibition prevents epithelial stem cell senescence and protects from radiation-induced mucositis. Cell Stem Cell. 2012;11(3):401–14.
60. Shuryak I, Brenner DJ. Mechanistic analysis of the contributions of DNA and protein damage to radiation-induced cell death. Radiat Res. 2012;178(1):17–24.
61. Qiu W, Leibowitz B, Zhang L, Yu J. Growth factors protect intestinal stem cells from radiation-induced apoptosis by suppressing PUMA through the PI3K/AKT/p53 axis. Oncogene. 2010;29(11):1622–32.
62. Krisko A, Leroy M, Radman M, Meselson M. Extreme anti-oxidant protection against ionizing radiation in bdelloid rotifers. Proc Natl Acad Sci. 2012;109(7):2354–7.
63. Güerci AM, Dulout FN, Grillo CA, Seoane AI. Differential response of two cell lines sequentially irradiated with low X-ray doses. Int J Radiat Biol. 2005;81(5):367–72.
64. Kaliberov SA, Buchsbaum DJ. Cancer treatment with gene therapy and radiation therapy. Adv Cancer Res. 2012;115:221–63.
65. Panganiban RAM, Snow AL, Day RM. Mechanisms of radiation toxicity in transformed and non-transformed cells. Int J Mol Sci. 2013;14(8):15931–58.
66. Panganiban RAM, Mungunsukh O, Day RM. X-irradiation induces ER stress, apoptosis, and senescence in pulmonary artery endothelial cells. Int J Radiat Biol. 2013;89(8):656–67.
67. Xiao M, Inal CE, Parekh VI, Chang CM, Whitnall MH. 5-androstenediol promotes survival of γ-irradiated human hematopoietic progenitors through induction of nuclear factor-κB activation and granulocyte colony-stimulating factor expression. Mol Pharmacol. 2007;72(2):370–9.
68. Yang S, Huang J, Liu P, Li J, Zhao S. Apoptosis-inducing factor (AIF) nuclear translocation mediated caspase-independent mechanism involves in X-ray-induced MCF-7 cell death. Int J Radiat Biol. 2017;93(3):270–8.
69. Kang R, Li R, Dai P, Li Z, Li Y, Li C. Deoxynivalenol induced apoptosis and inflammation of IPEC-J2 cells by promoting ROS production. Environ Pollut. 2019;251:689–98.
70. Luo X, Budihardjo I, Zou H, Slaughter C, Wang X. Bid, a Bcl2 interacting protein, mediates cytochrome c release from mitochondria in response to activation of cell surface death receptors. Cell. 1998;94(4):481–90.
71. Ruiter GA, Zerp SF, Bartelink H, van Blitterswijk WJ, Verheij M. Alkyl-lysophospholipids activate the SAPK/JNK pathway and enhance radiation-induced apoptosis. Cancer Res. 1999;59(10):2457–63.
72. Kroemer G, Levine B. Autophagic cell death: the story of a misnomer. Nat Rev Mol Cell Biol. 2008;9(12):1004–10.
73. Ying Y, Padanilam BJ. Regulation of necrotic cell death: p53, PARP1 and cyclophilin D-overlapping pathways of regulated necrosis? Cell Mol Life Sci. 2016;73:2309–24.
74. Gorbunov NV, Kiang JG. Up-regulation of autophagy in small intestine Paneth cells in response to total-body γ-irradiation. J Pathol. 2009;219(2):242–52.
75. Zhou Y, Liu L, Tao S, Yao Y, Wang Y, Wei Q, et al. Parthanatos and its associated components: promising therapeutic targets for cancer. Pharmacol Res. 2021;163:105299.
76. Galluzzi L, Vitale I, Aaronson SA, Abrams JM, Adam D, Agostinis P, et al. Molecular mechanisms of cell death: recommendations of the nomenclature committee on cell death 2018. Cell Death & Differentiation. 2018;25(3):486–541.
77. Liu YG, Chen JK, Zhang ZT, Ma XJ, Chen YC, Du XM, et al. NLRP3 inflammasome activation mediates radiation-induced pyroptosis in bone marrow-derived macrophages. Cell Death Dis. 2017;8(2):e2579–9.
78. Han R, Wu D, Deng S, Liu T, Zhang T, Xu Y. NLRP3 inflammasome induces pyroptosis in lung tissues of radiation-induced lung injury in mice. Xi bao yu fen zi Mian yi xue za zhi = Chinese J Cell Mol Immunol. 2017;33(9):1206–11.
79. Li X, Duan L, Yuan S, Zhuang X, Qiao T, He J. Ferroptosis inhibitor alleviates radiation-induced lung fibrosis (RILF) via down-regulation of TGF-β1. J Inflamm. 2019;16:1–10.
80. Kepp O, Senovilla L, Vitale I, Vacchelli E, Adjemian S, Agostinis P, et al. Consensus guidelines for the detection of immunogenic cell death. Onco Immunol. 2014;3(9):12472–124508.

81. Grassberger C, Ellsworth SG, Wilks MQ, Keane FK, Loeffler JS. Assessing the interactions between radiotherapy and antitumour immunity. Nat Rev Clin Oncol. 2019;16(12):729–45.
82. Golden EB, Demaria S, Schiff PB, Chachoua A, Formenti SC. An abscopal response to radiation and ipilimumab in a patient with metastatic non–small cell lung cancer. Cancer Immunol Res. 2013;1(6):365–72.
83. Klug F, Prakash H, Huber PE, Seibel T, Bender N, Halama N, et al. Low-dose irradiation programs macrophage differentiation to an iNOS+/M1 phenotype that orchestrates effective T cell immunotherapy. Cancer Cell. 2013;24(5):589–602.
84. Li M, You L, Xue J, Lu Y. Ionizing radiation-induced cellular senescence in normal, non-transformed cells and the involved DNA damage response: a mini review. Front Pharmacol. 2018;9:522.
85. Ou HL, Schumacher B. DNA damage responses and p53 in the aging process. Blood. 2018;131(5):488–95.
86. Vitale I, Galluzzi L, Castedo M, Kroemer G. Mitotic catastrophe: a mechanism for avoiding genomic instability. Nat Rev Mol Cell Biol. 2011;12(6):385–92.
87. Ye LF, Chaudhary KR, Zandkarimi F, Harken AD, Kinslow CJ, Upadhyayula PS, et al. Radiation-induced lipid peroxidation triggers ferroptosis and synergizes with ferroptosis inducers. ACS Chem Biol. 2020;15(2):469–84.
88. Eriksson D, Joniani HM, Sheikholvaezin A, Löfroth PO, Johansson L, Riklund Åhlström K, Stigbrand T. Combined low dose radio-and radioimmunotherapy of experimental HeLa Hep 2 tumours. Eur J Nucl Med Mol Imaging. 2003;30:895–906.
89. Eriksson D, Stigbrand T. Radiation-induced cell death mechanisms. Tumour Biol. 2010;31:363–72.
90. Lammering G, Valerie K, Lin PS, Hewit TH, Schmidt-Ullrich RK. Radiation-induced activation of a common variant of EGFR confers enhanced radioresistance. Radiother Oncol. 2004;72(3):267–73.
91. Park CM, Park MJ, Kwak HJ, Lee HC, Kim MS, Lee SH, et al. Ionizing radiation enhances matrix metalloproteinase-2 secretion and invasion of glioma cells through Src/epidermal growth factor receptor–mediated p 38/Akt and phosphatidylinositol 3-kinase/Akt signaling pathways. Cancer Res. 2006;66(17):8511–9.
92. Xu W, Luo T, Li P, Zhou C, Cui D, Pang B, et al. RGD-conjugated gold nanorods induce radiosensitization in melanoma cancer cells by downregulating αvβ3 expression. Int J Nanomedicine. 2012:915–24.
93. Schweigerer L, Rave-Fränk M, Schmidberger H, Hecht M. Sublethal irradiation promotes invasiveness of neuroblastoma cells. Biochem Biophys Res Commun. 2005;330(3):982–8.
94. Hanahan D, Weinberg RA. Hallmarks of cancer: the next generation. Cell. 2011;144(5):646–74.
95. Wang JS, Wang HJ, Qian HL. Biological effects of radiation on cancer cells. Mil Med Res. 2018;5:1–10.
96. Scheithauer H, Belka C, Lauber K, Gaipl US. Immunological aspects of radiotherapy. Radiat Oncol. 2014;9:1–4.
97. Magné N, Toillon RA, Bottero V, Didelot C, Van Houtte P, Gérard JP, Peyron JF. NF-κB modulation and ionizing radiation: mechanisms and future directions for cancer treatment. Cancer Lett. 2006;231(2):158–68.
98. Keisari Y, Hochman I, Confino H, Korenstein R, Kelson I. Activation of local and systemic anti-tumor immune responses by ablation of solid tumors with intratumoral electrochemical or alpha radiation treatments. Cancer Immunol Immunother. 2014;63:1–9.
99. Warters RL, Hofer KG, Harris CR, Smith JM. Radionuclide toxicity in cultured mammalian cells: elucidation of the primary site of radiation damage. Curr Top Radiat Res Q. 1978;12(1–4):389–407.
100. Stewart RD, Yu VK, Georgakilas AG, Koumenis C, Park JH, Carlson DJ. Effects of radiation quality and oxygen on clustered DNA lesions and cell death. Radiat Res. 2011;176(5):587–602.
101. Harris AL. Hypoxia—a key regulatory factor in tumour growth. Nat Rev Cancer. 2002;2(1):38–47.

102. Verbrugge I, Gasparini A, Haynes NM, Hagekyriakou J, Galli M, Stewart TJ, et al. The curative outcome of radioimmunotherapy in a mouse breast cancer model relies on mTOR signaling. Radiat Res. 2014;182(2):219–29.
103. Reits EA, Hodge JW, Herberts CA, Groothuis TA, Chakraborty M, Wansley E K, et al. Radiation modulates the peptide repertoire, enhances MHC class I expression, and induces successful antitumor immunotherapy. J Exp Med. 2006;203(5):1259–71.
104. Formenti SC, Rudqvist NP, Golden E, Cooper B, Wennerberg E, Lhuillier C, et al. Radiotherapy induces responses of lung cancer to CTLA-4 blockade. Nat Med. 2018;24(12):1845–51.
105. Begg AC, Stewart FA, Vens C. Strategies to improve radiotherapy with targeted drugs. Nat Rev Cancer. 2011;11(4):239–53.
106. Chandra RA, Keane FK, Voncken FE, Thomas CR. Contemporary radiotherapy: present and future. Lancet. 2021;398(10295):171–84.
107. Yard BD, Adams DJ, Chie EK, Tamayo P, Battaglia JS, Gopal P, et al. A genetic basis for the variation in the vulnerability of cancer to DNA damage. Nat Commun. 2016;7(1):11428.
108. Olivares-Urbano MA, Griñán-Lisón C, Marchal JA, Núñez MI. CSC radioresistance: a therapeutic challenge to improve radiotherapy effectiveness in cancer. Cells. 2020;9(7):1651.
109. Batlle E, Clevers H. Cancer stem cells revisited. Nat Med. 2017;23(10):1124–34.
110. Skvortsov S, Debbage P, Lukas P, Skvortsova I. Crosstalk between DNA repair and cancer stem cell (CSC) associated intracellular pathways. In: Seminars in cancer biology, vol. 31. Academic Press; 2015, April. p. 36–42.
111. Lei Y, Zhang D, Yu J, Dong H, Zhang J, Yang S. Targeting autophagy in cancer stem cells as an anticancer therapy. Cancer Lett. 2017;393:33–9.
112. Druker BJ, Lydon NB. Lessons learned from the development of an abl tyrosine kinase inhibitor for chronic myelogenous leukemia. J Clin Invest. 2000;105(1):3–7.
113. Citrin D, Cotrim AP, Hyodo F, Baum BJ, Krishna MC, Mitchell JB. Radioprotectors and mitigators of radiation-induced normal tissue injury. Oncologist. 2010;15(4):360–71.
114. Song G, Cheng L, Chao Y, Yang K, Liu Z. Emerging nanotechnology and advanced materials for cancer radiation therapy. Adv Mater. 2017;29(32):1700996.
115. Hong C, Wang D, Liang J, Guo Y, Zhu Y, Xia J, et al. Novel ginsenoside-based multifunctional liposomal delivery system for combination therapy of gastric cancer. Theranostics. 2019;9(15):4437.
116. Overchuk M, Zheng G. Overcoming obstacles in the tumor microenvironment: recent advancements in nanoparticle delivery for cancer theranostics. Biomaterials. 2018;156:217–37.
117. Diffenderfer ES, Verginadis II, Kim MM, Shoniyozov K, Velalopoulou A, Goia D, et al. Design, implementation, and in vivo validation of a novel proton FLASH radiation therapy system. Int J Radiat Oncol Biol Phys. 2020;106(2):440–8.
118. Pratx G, Kapp DS. Ultra-high-dose-rate FLASH irradiation may spare hypoxic stem cell niches in normal tissues. Int J Radiat Oncol Biol Phys. 2019;105(1):190–2.
119. Ngwa W, Irabor OC, Schoenfeld JD, Hesser J, Demaria S, Formenti SC. Using immunotherapy to boost the abscopal effect. Nat Rev Cancer. 2018;18(5):313–22.
120. Demaria S, Golden EB, Formenti SC. Role of local radiation therapy in cancer immunotherapy. JAMA Oncol. 2015;1(9):1325–32.
121. Corrales L, McWhirter SM, Dubensky TW, Gajewski TF. The host STING pathway at the interface of cancer and immunity. J Clin Invest. 2016;126(7):2404–11.
122. Rodriguez-Ruiz ME, Vitale I, Harrington KJ, Melero I, Galluzzi L. Immunological impact of cell death signaling driven by radiation on the tumor microenvironment. Nat Immunol. 2020;21(2):120–34.
123. Jones PA, Ohtani H, Chakravarthy A, De Carvalho DD. Epigenetic therapy in immune-oncology. Nat Rev Cancer. 2019;19(3):151–61.

Chapter 7
Overview of Radiotherapy Techniques and Modalities: Conventional and Modern Approaches

Mohadeseh Shahin, Zahra Siavashpour, Manijeh Beigi, and Seied Rabi Mahdavi

Abstract Radiotherapy has long been an essential element in the treatment of cancer patients, and more than half of the patients require radiotherapy for various purposes during their treatment. Radiotherapy has made tremendous progress in recent decades. In the past, two-dimensional treatments with large fields and fixed radiation angles were used, which led to high toxicity in normal tissues. After that, 3D radiotherapy was developed, which, by using imaging techniques like computed tomography (CT) scans, made it possible to identify tumors and normal tissues, resulting in reduced toxicity. Following these developments, newer methods were introduced, including intensity-modulated radiation therapy (IMRT), which created a revolution in radiotherapy by utilizing precise dynamic modulation of radiation intensity. Subsequently, volumetric modulated arc therapy (VMAT) increased the accuracy of treatment by enabling continuous arc radiation. Particle therapy has brought hope for the treatment of radio-resistant tumors, and the progress of radiotherapy continues.

M. Shahin (✉)
Department of Radiation Oncology, School of Medicine, Iran University of Medical Sciences, Tehran, Iran

Shohadaye Haftom-e-Tir Hospital, Iran University of Medical Sciences, Tehran, Iran
e-mail: Shahin.mo@iums.ac.ir

Z. Siavashpour
Department of Radiotherapy Oncology, Shohada-e Tajrish Educational Hospital, Medical School, Shahid Beheshti University of Medical Sciences, Tehran, Iran
e-mail: z_siavashpour@sbmu.ac.ir

M. Beigi
Department of Radiation Oncology, School of Medicine, Iran University of Medical Sciences, Tehran, Iran

S. R. Mahdavi
Departments of Medical Physics, Radio-oncology and Radiation Biology Research Center, Faculty of Medicine, Iran University of Medical Sciences, Tehran, Iran
e-mail: Mahdavi.r@iums.ac.ir

F. Taghizadeh-Hesary et al. (eds.), *From Radiobiology to Radiation Oncology*,
https://doi.org/10.1007/978-3-031-95724-6_7

Keywords 3D conformal radiotherapy · Intensity-modulated radiation therapy · Volumetric modulated arc therapy · Brachytherapy · Particle therapy

7.1 Advances in Radiotherapy: A Physics and Radiobiology Perspective

Radiotherapy has undergone significant advancements over several decades, evolving from traditional two-dimensional (2D) techniques to sophisticated modalities such as three-dimensional (3D) conformal radiotherapy, intensity-modulated radiation therapy (IMRT), volumetric modulated arc therapy (VMAT), and tomotherapy [16]. This chapter presents a summary of radiotherapy methods, highlighting their principles and applications.

7.2 From 2D to 3D Conformal Radiotherapy Techniques

Radiotherapy has been a cornerstone of cancer treatment, evolving from 2D to 3D techniques for improved precision and efficacy [8, 18]. Traditional 2D radiotherapy employs planar imaging, primarily X-rays, for treatment guidance. While quick to plan, this approach lacks detailed imaging data and uses fixed beam angles, resulting in larger treatment fields and higher radiation exposure to nearby healthy tissues, increasing side effects [12].

Conversely, the emergence of 3D conformal radiotherapy (3D-CRT) represented a significant advance in treatment techniques [17]. Utilizing sophisticated imaging techniques such as computed tomography (CT) and magnetic resonance imaging (MRI), 3D-CRT allows for accurate characterization of tumor shape, volume, and its relationship to adjacent organs [1]. The primary goal of 3D-CRT is to optimize radiation distribution, ensuring that maximal doses are delivered to the tumor while decreasing the dose received by adjacent healthy tissues. The methodology fundamentally relies on isodose curves, which graphically represent the distribution of radiation doses across the treated volume [1].

7.3 Intensity-Modulated Radiation Therapy (IMRT)

IMRT is a significant breakthrough in radiation oncology, building on the foundations of 3D-CRT while improving the capability to modulate dose delivery [21]. By allowing for the precise manipulation of the intensity of the radiation beams, IMRT enables clinicians to adapt the treatment plans based on the unique geometry of each patient's tumor, optimizing the therapeutic ratio by maximizing the dose to the targeted volume while minimizing the dose to adjacent healthy tissues [20]. IMRT provides versatility in treating patients through various distinct approaches, each employing unique dose delivery mechanisms [10].

Step-and-Shoot (SS) In this mode, the multi-leaf collimator (MLC) is programmed to create a series of discrete aperture shapes. Radiation is applied only when the MLC is stationary at each specific position. The outcome is a fluence pattern consisting of distinct intensity levels, with the number of levels matching the number of MLC steps.

Sliding Window (SW) Unlike SS, SW maintains the radiation beam while the MLC moves dynamically across the treatment aperture at varying speeds. This continuous motion "paints" a fluence pattern with smooth intensity variations.

Both methods employ a set of beam angles to accurately target the tumor with radiation while reducing the dose to healthy tissues.

The radiobiological implications of IMRT are profound. IMRT has the potential to improve tumor control rates significantly. Research indicates that IMRT can decrease toxicity levels, especially in head and neck cancers, significantly lowering the risk of long-term complications such as xerostomia [5] and dysphagia [2, 6]. The ability to escalate doses to tumors that are otherwise inoperable or located near sensitive structures provides opportunities for treatment that were previously unattainable. Moreover, IMRT applications have been extended to multiple cancer types, demonstrating improved patient outcomes and survival rates, such as those found in prostate cancer and breast cancer cases.

One of the critical considerations in the implementation of IMRT is its complexity, requiring well-trained personnel and considerable resources for treatment planning and delivery. As new techniques continue to emerge, the synergy between clinical practice and physics innovations remains essential to enhancing the efficacy of radiation therapy [3].

7.4 Volumetric Modulated Arc Therapy (VMAT)

VMAT emerges as a significant advancement in radiotherapy, capitalizing on the principles of intensity modulation while introducing the innovative concept of delivering radiation continuously in a partial or full arc. VMAT generates a continuous fluence pattern, similar to SW, but also leverages the additional dimensions of gantry motion and variable dose rates. This approach creates a fluence pattern across a broad arc, rather than being restricted to individual beam angles. By spreading the radiation dose over a larger volume, VMAT minimizes the dose per unit volume to normal tissues and improves dose distribution. In summary, while SS and SW offer precise control through static delivery methods, VMAT represents a leap forward by integrating dynamic motion and dose modulation, providing highly efficient and refined radiation therapy [10, 11, 15]. This continuous movement, in conjunction with precise control of radiation output, enables the delivery of variable doses to different regions of the tumor throughout a single treatment session. By enhancing spatial conformity and enabling the direct prescription of high doses to the tumor, VMAT allows for the potential escalation of doses in complex clinical cases, including locally advanced cancers. Several studies have confirmed that VMAT not only reduces treatment times but also results in superior dosimetric outcomes compared to traditional techniques, helping to mitigate the risks associated with prolonged radiation exposure [19].

7.4.1 Tomotherapy

Tomotherapy represents a pioneering approach in the field of radiation oncology, integrating cutting-edge imaging technologies with precision radiation delivery techniques to improve the accuracy and effectiveness of cancer treatment. This system integrates a linear accelerator with a helical CT scanner, combining advanced imaging and radiation delivery techniques to provide IMRT. Helical tomotherapy utilizes a fan-beam configuration combined with a rotating linear accelerator mounted on a gantry that continuously circles the patient and linear translational movement of the couch [7, 13]. This dynamic method of radiation delivery facilitates the creation of highly conformal dose distributions tailored to specific dose-volume constraints defined in the treatment planning phase. The megavoltage CT component enhances the system's ability to visualize the treatment area before and immediately after radiation delivery, allowing for precise monitoring. This capability facilitates the concept of adaptive radiotherapy, where the actual dose delivered on a daily basis is assessed and compared to the planned dose, enabling real-time adjustments to optimize treatment delivery. The helical configuration of tomotherapy offers a significant advantage in delivering isotropic radiation doses that closely conform to irregular tumor geometries [22, 24].

7.5 Image-Guided Radiation Therapy (IGRT)

IGRT represents a pivotal advancement aimed at enhancing the precision of radiation treatment through real-time imaging. By integrating advanced imaging modalities such as cone-beam CT (CBCT), X-ray, and ultrasound into the treatment process, IGRT enables clinicians to verify the tumor's position both before and during treatment delivery. The utilization of imaging technologies provides tremendous insights into patient anatomy, enabling corrections for anatomical differences that might arise between planning and treatment sessions [4]. The underlying physics of IGRT involves the acquisition and analysis of images that facilitate accurate patient positioning. Using fiducial markers and advanced imaging methods, the radiation oncology team can assess tumor location and make necessary adjustments.

These adjustments may include rotational corrections and translations, ensuring that the targeted area receives the prescribed radiation dose [23]. The implementation of real-time monitoring mitigates the risks associated with motion artifacts due to breathing or other movements, leading to improved precision in tumor targeting and a reduction in toxicity. Clinical studies have reported that IGRT correlates with reduced complication rates in various cancer types, emphasizing the importance of accuracy in radiation delivery [14].

IGRT introduces additional considerations such as increased treatment setup times and the need for robust quality assurance processes to validate imaging techniques. As technology continues to advance, integrating IGRT with other modalities like IMRT and VMAT enhances treatment adaptability and precision, ultimately leading to better patient outcomes [9].

7.5.1 Brachytherapy

Brachytherapy is a radiotherapy type through which a radiation source is implanted inside or near the target volume. "Brachy" refers to the meaning of a "short distance" Latin word. Due to the radiation inverse square law (i.e., the steep dose rate decreases as distance increases from the source), delivering very high radiation doses to a confined target volume over a short period is achievable by brachytherapy. Brachytherapy has been prescribed as a standalone treatment or in combination with EBRT and is commonly employed to deliver a boost dose in the management of various malignant and select benign tumors.

The very first roots of brachytherapy stem from the radioactivity discovery by Henri Becquerel in 1896 and the following Marie and Pierre Curie radium discovery by 1898. In 1934, the discovery of artificial radioactivity helped the use of man-made radioactive isotopes. In addition, remote afterloader development in 1960 allowed more staff radiation protection and flexibility to the treatment, and new horizons opened in brachytherapy technique [36].

Different classifications of brachytherapy techniques have been introduced, which include:

1. *Regarding the method of source implantation:*
   - *Interstitial:* The source is implanted directly within the target tissue.
   - *Intracavitary/Intraluminal:* The source is located inside the natural cavities of the patient's body close to the target tissue.
   - *Superficial:* Used primarily for treating skin cancers, the radiation source is positioned a short distance above the skin surface.
2. *Ocular:* Used for eye cancers, the source is applied as an eye plaque.
3. *Based on the residency duration of the radiation source in the body:*
   - *Temporary:* The source is removed after delivering the required dose, which occurs over a short period relative to the radioactive source's half-life.

- *Permanent:* The radioactive source persists in the patient's body for the rest of their life.

4. *Based on the method of loading the radioactive source:*

- *Preloading (hot loading):* Applicators preloaded with the radioactive source are directly used for treatment.
- *Afterloading:* Hollow applicators or catheters are first placed in the body, and the radioactive source is later inserted automatically using an afterloading device as per the treatment plan.

5. *Based on dose rate:*

- *Very Low-Dose Rate (VLDR):* < 0.4 Gy/h
- *Low-Dose Rate (LDR):* 0.4–2 Gy/h
- *Medium Dose Rate (MDR):* 2–12 Gy/h
- *High Dose Rate (HDR):* > 12 Gy/h
- *Pulsed Dose Rate (PDR):* Utilizes HDR hardware to deliver multiple low-dose fractions at short intervals.

6. *Regarding the source type:*

- *Radioactive materials:* Commonly used isotopes include iridium-192, cobalt-60, and ruthenium-106
- *Miniature X-ray generating devices (electronic brachytherapy):* An alternative to radioactive isotopes for certain applications [38, 48]

7.6 Stereotactic Radiosurgery and Radiotherapy

Stereotactic techniques have their roots in early neurosurgery, beginning with Horsley and Clarke's work on the monkey brain in 1908, followed by Spiegel and Weiss applying these methods to humans in 1947. A major milestone came in the 1950s when Lars Leksell developed the Gamma Knife, which paved the way for modern stereotactic radiosurgery (SRS) [32, 45].

The term "stereotactic" refers to treatments that rely on a three-dimensional reference system to localize the target lesion with exceptional accuracy. Initially, external reference frames were used, but advances in imaging have allowed the use of implanted markers visible in X-ray or CT scans [37].

Stereotactic radiotherapy, including stereotactic body radiotherapy (SBRT) and SRS, delivers highly focused radiation doses while minimizing exposure to surrounding healthy tissue. SBRT is particularly efficient for tumors that are less sensitive to conventional radiotherapy, enabling safe delivery of high doses in fewer sessions. This technique is widely selected to treat lesions of various body parts, such as the brain, lung, liver, and prostate. When treating intracranial or skull base lesions, the technique is specifically referred to as SRS [46].

SBRT and SRS are delivered using various advanced systems, including C-arm linear accelerator (LINAC) systems, robotic LINAC systems (such as CyberKnife), Gamma Knife systems, and helical tomotherapy systems. Using image-guided radiotherapy (IGRT) techniques is mandatory for SBRT and SRS treatments [28, 30, 31].

General Conditions for SRS/FSRT (fractionated stereotactic radiotherapy) [28, 44]:

1. *Maximum Lesion Size:*
   - For SRS: Lesions should be less than 3 cm.
   - For FSRT: Cumulative lesion size should be less than 6 cm.
2. *Lesion Shape:*
   - Lesions should be regular in shape for better targeting.
3. *Distance from Vital Structures:*
   - Brain lesions should be at least 3 mm away from critical areas.

Tables 7.1 and 7.2 outline the indications of SRS/FSRT for intracranial and extracranial lesions.

Table 7.1 Indications for SRS/FSRT (Intracranial and Skull Base Lesions)

Tumor type	Conditions
AVM	Suitable for SRS
Schwannomas	Includes acoustic and trigeminal schwannomas; typically, when surgical access is limited
Pituitary adenoma	For primary treatment or recurrence, particularly in inaccessible areas
Meningioma	For primary or recurrent meningiomas in regions difficult to operate on (e.g., skull base)
Craniopharyngioma	For non-operable cases or recurrent tumors
Chordoma/ chondrosarcoma	Suitable for treatment with SRS in appropriate anatomical conditions
Glioma	For gliomas under 5 cm or recurrent tumors in patients not suitable for surgery
Brain metastasis	For up to 10 lesions, the total volume is under 15 cm^3, with good performance status (KPS > 70)
Trigeminal neuralgia	For cases resistant to medical treatment
Intractable epilepsy	For refractory temporal lobe epilepsy when surgery is not an option

Table 7.2 Indications for SBRT (Extracranial)

Tumor type	Conditions
NSCLC	For tumors <5 cm, no distant spread, and when surgery is not feasible or the patient is unwilling
Spinal tumors	For primary lumbar spine tumors, even if previous radiotherapy has been given
Oligometastatic	For up to 3 lesions <5 cm and life expectancy over 6 months
Prostate cancer	For low-risk or intermediate-risk prostate cancer with good life expectancy, especially for tumors with Gleason score of 3 + 4
Hepatocellular carcinoma	For one to three tumoral lesions <4 cm (or a single lesion <6 cm) for non-candidates' patients for surgical procedures
Renal cell carcinoma (RCC)	For tumors ≤7 cm, non-operable, with good performance status (KPS ≥ 70%)
Pancreatic cancer	For localized advanced tumors with no distant spread and lesions <6 cm
Spinal metastasis	For metastases from melanoma, sarcoma, or RCC; ≤3 lesions <5 cm, life expectancy over 6 months

7.7 Particle Therapy.

Radiotherapy with particles is an external radiation therapy method employing charged particles, including protons, neutrons, or heavy ions (e.g., carbon), to hit the tumoral targets with high precision. Among these, proton therapy is the most commonly used technique owing to its unique dosimetric characteristics, particularly the Bragg Peak, where the maximum energy is deposited at a specific depth, allowing for delivering high doses to tumors while keeping surrounding healthy tissues safe, minimizing side effects compared to conventional photon-based radiotherapy [40, 43].

Key Features of Proton Therapy:

Energy Delivery:

- Protons deliver a minimal dose into the tissue before the Bragg Peak, and it can be planned so that they deposit their maximum energy at the target tissue.
- Clinical proton beam energies range from 60 to 300 MeV, enabling treatment of both superficial and deeply seated tumors.

Relative Biological Effectiveness (RBE):

- Protons have a higher RBE compared to photons due to their well-known high Linear Energy Transfer (LET), and because of this property, the absorbed dose in proton therapy is measured in terms of Gy (RBE). This effect enhances their effectiveness, especially for hypoxic tumors, which are resistant to photon therapy.

Beam Delivery Systems:

- Traditional systems use cyclotrons or synchrotrons to accelerate protons. Cyclotrons provide fixed-energy beams, while synchrotrons allow energy variation (superconducting synchrocyclotrons and compact synchrotrons are emerging).

Delivery Techniques:

- Passive Scattering: Uses mechanical devices to spread the proton beam.
- Pencil Beam Scanning (PBS): A highly precise technique where the proton beam is directed magnetically, allowing for Intensity-Modulated Proton Therapy (IMPT).

Clinical Applications:

- Adults: Ocular melanomas, chordomas, chondrosarcomas, hepatocellular carcinoma, and advanced head and neck tumors.
- Pediatrics: CNS tumors, Ewing's sarcoma, rhabdomyosarcoma, osteosarcoma, retinoblastoma, neuroblastoma, and tumors in genetically radiation-sensitive patients (e.g., NF1) [33, 41, 47].

Carbon ion therapy (CIRT) is almost more effective for radio-resistant tumors such as sarcomas and pancreatic cancer due to its ballistic and higher LET even than the protons. CIRT is two to three times higher relative to biological effectiveness (RBE) in facing targets with inherent radioresistance and improving tumor control [39, 40].

Neutron therapy is also a specialized radiotherapy that uses high-energy neutrons, effective for treating hypoxic or radio-resistant tumors due to their high LET. However, higher toxicity to normal tissues compared to proton therapy and its limited availability due to specialized equipment requirements are the challenges to its popularity [29, 47].

7.8 Motion Management

Tumors in the thoracic and abdominal regions often undergo displacement due to respiratory motion, creating significant uncertainties in accurate target localization. Conventional radiotherapy compensates for such motion by expanding treatment margins at the price of increasing the volume of healthy tissue exposure. This trade-off between treatment precision and healthy tissue protection is a critical consideration in radiotherapy planning [26, 35].

Target motion can be categorized into two types: intra-fractional motion, which occurs during a treatment session and is primarily caused by respiratory activity, and inter-fractional motion, which arises between sessions due to anatomical changes, such as weight loss, or daily setup variations [42].

Impact of motions and displacement on imaging, treatment planning, and dose delivery:

- Imaging Artifacts: Respiratory motion leads to blurring or artifacts in static 3D CT images, resulting in inaccurate tumor visualization. Solutions like 4D CT

account for respiratory motion during imaging, allowing phase-specific tumor data to be obtained and improving localization accuracy.
– Treatment Planning Limitations: Enlarged treatment margins to compensate for motion increase healthy tissue irradiation.
– Dose Delivery Challenges: Motion during treatment results in dose blurring, particularly at tumor edges, which deviates from the planned dose distribution. This effect is more pronounced in dynamic techniques such as IMRT or VMAT, where asynchronized multi-leaf collimator movements and tumor displacements can cause further treatment-delivering discrepancies [34, 35].

7.8.1 Motion Management Techniques During Radiotherapy

Motion management strategies include:

1. Breath-Hold Irradiation: Patients are guided to hold their breath, reducing respiratory motion during treatment.
2. Respiratory-Gated Irradiation: Motion is accounted for using external surrogates or 4D CT, with real-time tracking to synchronize treatment delivery with breathing cycles.
3. Immobilization Devices: Using specialized stereotactic body frames, often with abdominal compression, helps reduce respiratory movement.
4. Slow CT scans: These scans define the limits of tumor motion for accurate treatment planning.
5. Motion-Prediction Models: Using models to predict tumor trajectory based on respiratory patterns, helping adjust treatment delivery to predicted movements.
6. Real-Time Tumor Tracking (RTRT): Tracking tumor motion during irradiation for precise dose delivery by continuously adjusting the treatment beam to the tumor's position.
7. Surface Imaging: A method that uses non-invasive systems to monitor the patient's body surface and adjust the patient's position in real time [25, 27].

References

1. Beaton L, Bandula S, Gaze MN, Sharma RA. How rapid advances in imaging are defining the future of precision radiation oncology. Br J Cancer. 2019;120(8):779–90.
2. Bhide S, Newbold K, Harrington K, Nutting C. Clinical evaluation of intensity-modulated radiotherapy for head and neck cancers. Br J Radiol. 2012;85(1013):487–94.
3. Cheung K. Intensity-modulated radiotherapy: advantages, limitations and future developments. Biomed Imaging Interv J. 2006;2(1):e19.
4. De Los Santos J, Popple R, Agazaryan N, Bayouth JE, Bissonnette J-P, Bucci MK, Dieterich S, Dong L, Forster KM, Indelicato D. Image-guided radiation therapy (IGRT) technologies for radiation therapy localization and delivery. Int J Radiat Oncol Biol Phys. 2013;87(1):33–45.

5. Dirix P, Nuyts S, Van den Bogaert W. Radiation-induced xerostomia in patients with head and neck cancer: a literature review. Cancer. 2006;107(11):2525–34.
6. Dragan T, Beauvois S, Moreau M, Paesmans M, Vandekerkhove C, Cordier L, Van Gestel D. Clinical outcome and toxicity after simultaneous integrated boost IMRT in head and neck squamous cell cancer patients. Oral Oncol. 2019;98:132–40.
7. Escobar-Peralta E, Gil-García R, Quezada-Bautista AA, López-Riverol O. Helical tomotherapy: advanced radiotherapy technology. Revista médica del Hospital General de México. 2022;85(4):179–84.
8. Fang LC, Komaki R, Allen P, Guerrero T, Mohan R, Cox JD. Comparison of outcomes for patients with medically inoperable stage I non–small-cell lung cancer treated with two-dimensional vs. three-dimensional radiotherapy. Int J Radiat Oncol Biol Phys. 2006;66(1):108–16.
9. Grégoire V, Guckenberger M, Haustermans K, Lagendijk JJ, Ménard C, Pötter R, Slotman BJ, Tanderup K, Thorwarth D, Van Herk M. Image guidance in radiation therapy for better cure of cancer. Mol Oncol. 2020;14(7):1470–91.
10. Herman TDLF, Schnell E, Young J, Hildebrand K, Algan Ö, Syzek E, Herman T, Ahmad S. Dosimetric comparison between IMRT delivery modes: step-and-shoot, sliding window, and volumetric modulated arc therapy—for whole pelvis radiation therapy of intermediate-to-high risk prostate adenocarcinoma. J Med Phys. 2013;38(4):165–72.
11. Hunte SO, Clark CH, Zyuzikov N, Nisbet A. Volumetric modulated arc therapy (VMAT): a review of clinical outcomes—what is the clinical evidence for the most effective implementation? Br J Radiol. 2022;95(1136):20201289.
12. Khan FM, Gibbons JP. Khan's the physics of radiation therapy. Lippincott Williams & Wilkins; 2014.
13. Mackie TR, Holmes T, Swerdloff S, Reckwerdt P, Deasy JO, Yang J, Paliwal B, Kinsella T. Tomotherapy: a new concept for the delivery of dynamic conformal radiotherapy. Med Phys. 1993;20(6):1709–19.
14. Marinetti T, Dong L. Delivery of precision radiation therapy: IMRT, IGRT, proton beam therapy. Practical essentials of intensity modulated radiation. Therapy. 2013;1
15. Meyer J. IMRT, IGRT, SBRT: advances in the treatment planning and delivery of radiotherapy. Karger Medical and Scientific Publishers; 2011.
16. Mukherji A. Basics of planning and management of patients during radiation therapy. Springer; 2018.
17. Perez CA, Vijayakumar S, Purdy JA, Michalski JM, Bradley J, Vijayakumar S, Perez CA, Levitt SH. Three-dimensional treatment planning and conformal therapy. Technical Basis of Radiation Therapy: Practical Clinical Applications; 2006. p. 179–202.
18. Teo PM, Ma BB, Chan AT. Radiotherapy for nasopharyngeal carcinoma—transition from two-dimensional to three-dimensional methods. Radiother Oncol. 2004;73(2):163–72.
19. Teoh M, Clark C, Wood K, Whitaker S, Nisbet A. Volumetric modulated arc therapy: a review of current literature and clinical use in practice. Br J Radiol. 2011;84(1007):967–96.
20. Verhey LJ. Comparison of three-dimensional conformal radiation therapy and intensity-modulated radiation therapy systems. Semin Radiat Oncol. 1999;9:78.
21. Webb S. The physical basis of IMRT and inverse planning. Br J Radiol. 2003;76(910):678–89.
22. Welsh JS, Patel RR, Ritter MA, Harari PM, Mackie TR, Mehta MP. Helical tomotherapy: an innovative technology and approach to radiation therapy. Technol Cancer Res Treat. 2002;1(4):311–6.
23. Xing L, Thorndyke B, Schreibmann E, Yang Y, Li T-F, Kim G-Y, Luxton G, Koong A. Overview of image-guided radiation therapy. Med Dosim. 2006;31(2):91–112.
24. Zhu Z, Fu X. The radiation techniques of tomotherapy & intensity-modulated radiation therapy applied to lung cancer. Transl Lung Cancer Res. 2015;4(3):265.
25. Brandner ED, Chetty IJ, Giaddui TG, Xiao Y, Huq MS. Motion management strategies and technical issues associated with stereotactic body radiotherapy of thoracic and upper abdominal tumors: a review from NRG oncology. Med Phys. 2017;44(6):2595–612. https://doi.org/10.1002/mp.12227.

26. Chui CS, Yorke E, Hong L. The effects of intra-fraction organ motion on the delivery of intensity-modulated field with a multileaf collimator. Med Phys. 2003;30(7):1736–46. https://doi.org/10.1118/1.1578771.
27. Claridge Mackonis E, Stensmyr R, Poldy R, White P, Moutrie Z, Gorjiara T, et al. Improving motion management in radiation therapy: findings from a workshop and survey in Australia and New Zealand. Phys Eng Sci Med. 2024;47(3):813–20. https://doi.org/10.1007/s13246-024-01405-0.
28. Covington EL, Popple RA. Technical and quality considerations for stereotactic radiation treatment techniques. Cancer J. 2024;30(6):372–6. https://doi.org/10.1097/ppo.0000000000000756.
29. Gordon K, Gulidov I, Fatkhudinov T, Koryakin S, Kaprin A. Fast and furious: fast neutron therapy in cancer treatment. Int J Part Ther. 2022;9(2):59–69. https://doi.org/10.14338/ijpt-22-00017.
30. Halvorsen PH, Cirino E, Das IJ, Garrett JA, Yang J, Yin FF, Fairobent LA. AAPM-RSS medical physics practice guideline 9.A. For SRS-SBRT. J Appl Clin Med Phys. 2017;18(5):10–21. https://doi.org/10.1002/acm2.12146.
31. Hynes PR, Das JM. Stereotactic radiosurgery (SRS) and stereotactic body radiotherapy (SBRT). In: StatPearls. Treasure Island. FL: StatPearls Publishing; 2024. Copyright © 2024, StatPearls Publishing LLC.
32. Jensen RL, Stone JL, Hayne RA. Introduction of the human Horsley-Clarke stereotactic frame. Neurosurgery. 1996;38(3):563–7; discussion 567. https://doi.org/10.1097/00006123-199603000-00029.
33. José Santo R, Habraken SJM, Breedveld S, Hoogeman MS. Pencil-beam delivery pattern optimization increases dose rate for stereotactic FLASH proton therapy. Int J Radiat Oncol Biol Phys. 2023;115(3):759–67. https://doi.org/10.1016/j.ijrobp.2022.08.053.
34. Josipovic M, Persson GF, Dueck J, Bangsgaard JP, Westman G, Specht L, Aznar MC. Geometric uncertainties in voluntary deep inspiration breath hold radiotherapy for locally advanced lung cancer. Radiother Oncol. 2016;118(3):510–4. https://doi.org/10.1016/j.radonc.2015.11.004.
35. Keall PJ, Mageras GS, Balter JM, Emery RS, Forster KM, Jiang SB, et al. The management of respiratory motion in radiation oncology report of AAPM task group 7. 6. Med Phys. 2006;33(10):3874–900. https://doi.org/10.1118/1.2349696.
36. Kemikler G. History of brachytherapy. Turk Onkoloji Dergisi-Turkish J Oncol. 2019;34:1–10.
37. Lasak JM, Gorecki JP. The history of stereotactic radiosurgery and radiotherapy. Otolaryngol Clin N Am. 2009;42(4):593–9. https://doi.org/10.1016/j.otc.2009.04.003.
38. Lim YK, Kim D. Brachytherapy: a comprehensive review. PMP. 2021;32(2):25–39. https://doi.org/10.14316/pmp.2021.32.2.25.
39. Malouff TD, Mahajan A, Krishnan S, Beltran C, Seneviratne DS, Trifiletti DM. Carbon ion therapy: a modern review of an emerging technology. Front Oncol. 2020;10:82. https://doi.org/10.3389/fonc.2020.00082.
40. Orlandi E, Barcellini A, Vischioni B, Fiore MR, Vitolo V, Iannalfi A, et al. The role of carbon ion therapy in the changing oncology landscape-a narrative review of the literature and the decade of carbon ion experience at the Italian National Center for Oncological Hadrontherapy. Cancers (Basel). 2023;15(20) https://doi.org/10.3390/cancers15205068.
41. Overgaard CB, Reaz F, Ankjærgaard C, Andersen CE, Sitarz M, Poulsen P, et al. The proton RBE and the distal edge effect for acute and late normal tissue damage in vivo. Radiother Oncol. 2025;203:110668. https://doi.org/10.1016/j.radonc.2024.110668.
42. Park S, Lee SJ, Weiss E, Motai Y. Intra- and inter-fractional variation prediction of lung tumors using fuzzy deep learning. IEEE J Transl Eng Health Med. 2016;4:4300112. https://doi.org/10.1109/jtehm.2016.2516005.
43. Proton Beam Radiation Therapy (2024). Retrieved from https://www.uhcprovider.com/content/dam/provider/docs/public/policies/comm-medical-drug/proton-beam-radiation-therapy.pdf
44. Rohrer Bley C, Meier V, Turek M, Besserer J, Unterhirkhers S. Stereotactic radiation therapy planning, dose prescription and delivery in veterinary medicine: a systematic review on com-

pleteness of reporting and proposed reporting items. Vet Comp Oncol. 2024;22(4):457–69. https://doi.org/10.1111/vco.13011.
45. Schurr PH, Merrington WR. The Horsley–Clarke stereotaxic apparatus. Br J Surg. 1978;65(1):33–6. https://doi.org/10.1002/bjs.1800650110.
46. Smyth M, Williams M, Jones S, Woodley O, Bryant A, Matthams C, et al. Implementation of stereotactic ablative body radiotherapy (SABR) for spinal Oligometatases: initial experiences. Clin Oncol. 2018;30(6):e57. https://doi.org/10.1016/j.clon.2018.02.040.
47. Timoshchuk MA, Dekker P, Hippe DS, Parvathaneni U, Liao JJ, Laramore GE, Dillon JK. The efficacy of neutron radiation therapy in treating salivary gland malignancies. Oral Oncol. 2019;88:51–7. https://doi.org/10.1016/j.oraloncology.2018.11.006.
48. Wu N, Bu M, Jiang H, Mu X, Zhao H. Dose-effect relationship in external beam radiotherapy combined with brachytherapy for cervical cancer: a systematic review. J Contemp Brachytherapy. 2024;16(3):232–40. https://doi.org/10.5114/jcb.2024.140760.

Chapter 8
Recent Advances in Radiotherapy of Breast Cancer

Elham Khakshour, SaeidehSadat Shobeiri, James S. Welsh, and SeyedAlireza Javadinia

Abstract This chapter provides an overview of breast cancer management, emphasizing both conventional therapies and advanced techniques, including Intensity-Modulated Radiation Therapy (IMRT), Volumetric Modulated Arc Therapy (VMAT), and proton beam therapy. The fundamental principles of radiobiology including repair, redistribution, repopulation, and reoxygenation are explained in detail, alongside clinical data on the α/β ratio in the linear-quadratic model and the treatment fractionation schedules . In recent decades, attention has shifted to the interaction between radiation therapy (RT) and the immune system. Numerous studies have confirmed that radiation can elicit immune responses, recruiting immune cells to the tumor microenvironment, enhancing the production of tumor-specific antigens, and inducing immunological cell death. However, this combination has not yet fully eradicated primary and metastatic lesions . The abscopal effect, a phenomenon in which low-dose radiation modifies the tumor microenvironment, may help overcome treatment resistance. This chapter explores these evolving concepts in detail, highlighting technological advancements in radiotherapy, the biological mechanisms underlying treatment resistance, and the emerging potential of

E. Khakshour
Cellular and Molecular Research Center, Sabzevar University of Medical Sciences, Sabzevar, Iran
e-mail: elhamkhakshour1997@yahoo.com

S. Shobeiri
Clinical Research Development Unit, Vasei Hospital, Sabzevar University of Medical Sciences, Sabzevar, Iran
e-mail: Shobeiris1@medsab.ac.ir

J. S. Welsh
Department of Radiation Oncology Maywood, Loyola University Chicago Stritch School of Medicine, Edward Hines Jr., VA Hospital, Maywood, IL, USA
e-mail: James.Welsh@va.gov

S. Javadinia (✉)
Non-Communicable Diseases Research Center, Sabzevar University of Medical Sciences, Sabzevar, Iran
e-mail: javadinia.alireza@gmail.com

F. Taghizadeh-Hesary et al. (eds.), *From Radiobiology to Radiation Oncology*,
https://doi.org/10.1007/978-3-031-95724-6_8

combining RT with immunotherapy. By integrating physical and biological sciences with clinical innovation, the combination of RT and immunotherapy holds promise for delivering more effective and personalized care for breast cancer patients.

Keywords Breast cancer · Radiotherapy · Radiobiology · Treatment resistance · Immunotherapy

8.1 History of Radiotherapy in the Management of Breast Cancer

Radiotherapy has played a significant role in the treatment of breast cancer for over a century. Here's an overview of its historical development:

1. *Early Discoveries (Late Nineteenth Century)*
   - *1895*: Wilhelm Conrad Roentgen discovered X-rays, which paved the way for their use in medicine.
   - *1896*: The first therapeutic use of X-rays was reported by Emil Grubbe, who treated a patient with breast cancer.
2. *Initial Applications (1900s)*
   - *1900s*: Radiotherapy began to be used more widely for breast cancer, primarily as a palliative measure.
   - *1910s*: The introduction of radium therapy became popular, with radium being used for localized treatment of tumors.
3. *Advancements in Techniques (1940s–1960s)*
   - *1940s*: The use of cobalt-60 sources became common, providing more effective treatment options.
   - *1950s*: The development of fractionation techniques allowed for higher doses of radiation to be delivered safely over time, improving patient outcomes.
4. *Integration with Surgery (1970s–1980s)*
   - *1970s*: Radiotherapy began to be integrated with surgical approaches, such as lumpectomy, leading to the emergence of breast-conserving surgery.
   - *1980s*: Clinical trials demonstrated the efficacy of adjuvant radiotherapy following surgery, significantly improving local control rates.
5. *Technological Innovations (1990s–2000s)*
   - *1990s*: Advances in imaging technology (like CT and MRI) enhanced treatment planning and targeting of radiation.

- *2000s*: The introduction of intensity-modulated radiation therapy (IMRT) and image-guided radiation therapy (IGRT) allowed for more precise delivery of radiation, minimizing damage to surrounding healthy tissue.

6. *Current Trends and Future Directions (2010s-Present)*

- *2010s*: Research into hypofractionation (delivering higher doses in fewer sessions) gained traction, showing similar efficacy with reduced treatment times.
- *Present*: Ongoing studies focus on personalized radiotherapy approaches, integrating genomic profiling to tailor treatments based on individual tumor characteristics.

The history of radiotherapy in breast cancer management reflects significant advancements in technology and technique, improving patient outcomes and quality of life. As research continues to evolve, radiotherapy remains a cornerstone in the comprehensive treatment of breast cancer, with a promising future ahead.

8.2 Radiobiology of Radiotherapy in Breast Cancer

Conventional breast cancer treatment includes a standard therapy with a total dose of 40–50 Gy, which, depending on the patient's condition, can be delivered in 25–28 fractions, with about 1.8–2 Gy being delivered to the tumor site in each fraction [1]. The administration of this treatment regimen, which lasts 5–7 weeks, is based on the following four main radiobiological principles:

1. Repair: Because the tissues surrounding breast tumors, such as healthy breast tissue, blood vessels, and the heart (if the cancer is in the left breast), have low proliferative capacity and are more capable of repairing subcellular damage induced by radiation, reducing the prescribed dose per fraction and increasing the time between fractions allows healthy cells to repair this damage and reduce early complications such as erythema or skin inflammation and late complications such as fibrosis and necrosis. On the other hand, tumor cells have a lower ability than healthy cells to repair radiation damage. Fractionating the total radiation dose into repeated cycles causes an accumulation of radiation damage that can ultimately cause death. As a result, conventional treatment increases the ratio of tumor tissue damage to healthy tissue.
2. Redistribution: Since different phases of the cell cycle behave differently with radiation, during the radiation of tumor cells, those in the M and G2 phases (sensitive phases) are more likely to be destroyed by the damage caused, but cells in the G1 and S phases are more resistant to radiation and survive and repair the damage. Given that breast cancer cells have an average division rate, conventional methods adjust the time interval between irradiations according to the rate of division of these cells in such a way that they allow the remaining tumor cells in the resistant phase in the previous fraction to divide and enter the sensitive

phases and be destroyed by re-irradiation in the next fraction, increasing the efficiency of the treatment.

3. Repopulation: In conventional treatment, which has a longer duration and number of fractions, healthy tissue cells, especially those with a high division rate, such as skin, can divide quickly and compensate for lost tissue, and fewer complications are seen in healthy tissue. Although the conventional method prevents excessive tumor proliferation through the cumulative effect of radiation dose by dividing the total dose into daily fractions, the growth and proliferation rate of cancer cells is often faster, which can reduce the effectiveness of treatment, especially in invasive breast tumors.
4. Reoxygenation: Since the presence of oxygen during radiation can damage the DNA of cells by producing free radicals, in healthy cells, due to the strong repair system, these damages are repaired in the intervals between multiple fractions, but in tumor cells, due to the faster proliferation of tumor cells compared to blood vessels, the central areas of tumors are mostly hypoxic and, as a result, are more resistant to radiation-induced damage. After each fraction, the peripheral tumor cells, which are better oxygenated, are destroyed, and the inner cells are reoxygenated. This causes the tumor to become more sensitive to radiation in subsequent fractions gradually [2].

One of the mathematical radiobiological models that can predict the biological effects of radiation is the linear quadratic model, which can be used to estimate the percentage of cells that will survive at different doses without drawing a survival curve. Its equation is as follows:

$$S = e - \alpha D - \beta D2$$

In this equation, S is the fraction of cells that survive, D is the radiation dose, α is the linear component (representing direct acyls to DNA), and β is the quadratic component (representing cumulative damage). According to this equation, the ratio α/β is the dose at which the number of deaths caused by these two components is equal.

According to the graph below, which shows the ratio of survival fraction to dose for early- and late-responding tissues, early responding tissues such as skin and lymph nodes have a lower slope than late-responding tissues, and they are not very sensitive to the dose (fraction). Since these tissues have a high division rate and need time to repair, they are sensitive to the treatment time (interval between fractions). As a result, the α/β value occurs at a higher dose for these tissues. On the contrary, late-responding tissues such as nervous tissues and lungs have a steeper graph and are very sensitive to the dose. Because the division in these tissues is low, they do not respond much to the treatment time, and the α/β value occurs at a lower dose. In various studies, the α/β value in breast cancer has been measured at about 4–6 Gy, which is an average value. This value of α/β has led scientists toward hypofractionation treatments by keeping the biological effective dose (BED) constant. Conventional treatment with a gradual distribution of the total dose in breast cancer reduces the risk of acute and chronic side effects, and it provides more protection for healthy tissues around the breast tumor. However, the treatment comes with challenges, including the time-consuming treatment period and the high cost of treatment for the patient [3] (Fig. 8.1).

Due to the mentioned α/β ratio of breast cancer, its cancer cells can be more sensitive to changes in therapeutic dose [4]. Therefore, using a hypofractionation regimen that includes higher doses per fraction of about 2.6–3.2 Gy for 15–20 fractions can increase the irreparable DNA damage in tumor cells, while healthy cells can still repair this damage. Also, by reducing the time between fractions, tumor cells have less opportunity to repair the damage [5]. Several studies and clinical trials show that hypofractionation therapy has similar results in terms of cancer control and side effects as conventional therapy [6]. On the other hand, since the α/β ratio of breast cancer cells is similar to that of surrounding tissues, increasing the radiation dose can sharply increase the damage to surrounding healthy cells. Today, the use of modern radiotherapy technologies allows for more precise delivery of higher doses to tumor cells while reducing damage to healthy tissues surrounding the breast, such as the heart and lungs [7]. The most important radiotherapy technologies include the following:

1. Intensity-Modulated Radiation Therapy (IMRT)

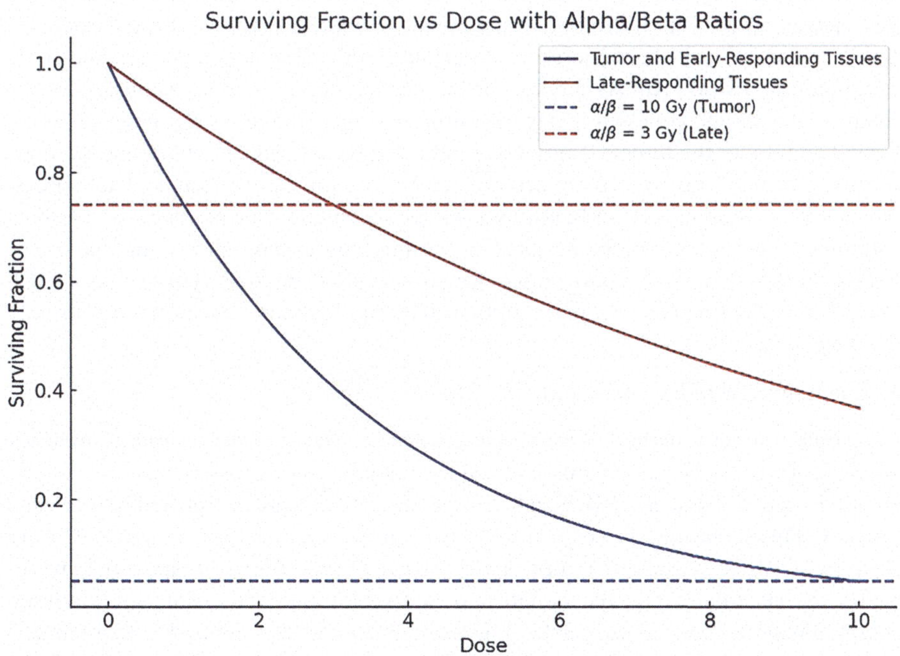

Fig. 8.1 Dose-response curve for early and late-responding tissues

This method is one of the advanced radiotherapy techniques that adjusts the radiation dose at different angles to a point where the tumor receives a more accurate dose, while the surrounding healthy tissues, such as the heart and lungs, receive less radiation in breast radiotherapy IMRT [8]. This method would permit tumors like those in the breast with an average α/β ratio to attain more accurate and effective doses at higher levels in a lesser number of fractions, with reduced acute and chronic complications in healthy tissues like dermatitis and fibrosis of soft tissues [9, 10].

2. Volumetric Modulated Arc Therapy (VMAT)

Another technique used in radiotherapy is the use of the VMAT technique, in which the radiation dose is continuously radiated by one or more arcs while the device rotates 360 degrees around the patient. During movement, the radiation intensity and its shape change dynamically, which can reduce the treatment time to about 2–4 min [11]. In addition, this method has high accuracy in distributing a homogeneous radiation dose to the tumor while at the same time exposing less radiation to the surrounding healthy tissues and can reduce the risk of patient movement errors [12]. Various studies have shown that this advanced 3D planning method can be very useful in planning a hypofractionation regimen for patients with breast cancer [13–15].

3. Proton Beam Therapy

This method is one of the most advanced techniques today that plays an important role in the treatment of breast cancer. In this method, positively charged particles such as protons are used; one of the physical characteristics of these particles is that they can release their energy at a certain depth (Bragg Peak) from the body tissue and then stop [16]. As a result of these interactions, a more accurate dose is given to the target area (tumor) while minimizing damage to the healthy tissue around and under the tumor. Using this method in breast cancer can reduce the dose delivered to the lungs, coronary arteries, and heart (in the treatment of left breast cancer) as well as late complications such as secondary cancers, serious cardiac, pulmonary, and other complications [17]. Various studies have shown that the use of proton therapy is a very good option for tumors that are near lymph nodes [18]. Overall, this method is a promising option that requires more research to accurately evaluate its benefits.

4. Image-guided radiation therapy (IGRT)

Another modern method of treating breast cancer that can deliver radiation doses more precisely to the tumor area and reduce damage to healthy tissue is the use of imaging such as CT scans, MRIs, and ultrasounds, which are performed before each fraction. This technique is very suitable for tumors such as breast cancer that are close to sensitive tissues or lymph nodes or even have a lot of respiratory movements so that it can adapt the radiation dose to the tumor area using the planning obtained with the help of images [19]. Despite the use of this method in many medical centers and the confirmation of the accuracy and reduction of complications of this method by clinical trials, its beneficial effect on overall survival or disease-free survival still requires more evidence [20, 21].

Key clinical studies on the switch from conventional to hypofractionated therapy for breast cancer have demonstrated that not only is the switch safe and effective, but it is also operationally and financially favorable for patients and healthcare systems. Some of the highlights of these studies include the following:

1. START Trials (A and B)

The START trial was one of the largest multicenter studies to have been undertaken in the UK. It consists of two parallel trials, START Trial A and START Trial B, comparing a hypofractionated schedule with a conventionally fractionated schedule for efficacy and safety in patients with early invasive breast cancer or in the post-mastectomy state. The details of the outline of these clinical trials are as follows: START-A compared the results of a hypofractionated regimen with a total dose of 41.6 Gy in 13 fractions with a dose per fraction of 3.2 Gy and a total dose of 39 Gy in 13 fractions with a dose per fraction of 3 Gy [22], and START-B compared a hypofractionated regimen with a total dose of 40 Gy in 15 fractions with a dose per fraction of 2.67 Gy with a conventional regimen with a total dose of 50 Gy in 25 fractions with a dose per fraction of 2 Gy [23]. Initial 5-year follow-up results confirm that a hypofractionated regimen, which delivers a lower total dose to the tumor area in fewer sessions, has rates similar to conventional therapy regarding local-regional tumor control and late side effects such as fibrosis, pain, and tissue stiffness; hypofractionated patients also had a better quality of life and shorter treatment. In addition, a 10-year follow-up showed that both studies had similar efficacy in local and regional tumor control compared with the conventional regimen. In START-A, the 39 Gy regimen showed a significant reduction in late complications such as fibrosis, telangiectasia, and breast edema, while the 41.6 Gy regimen had similar complications to 50 Gy. In START-B, the 40 Gy regimen was also associated with significantly reducing late complications, including reduced breast shrinkage, telangiectasia, and edema. In addition, the 40 Gy regimen was more comfortable for patients with a shorter treatment time (3 weeks vs. 5 weeks) [24]. These findings support the use of hypofractionation as a standard and effective option for the treatment of breast cancer.

2. FAST and Fast-Forward Trials

These two randomized controlled clinical trials were conducted in the UK for breast cancer patients who required radiotherapy after mastectomy, and their information and planning were carried out with the help of experience and information from clinical trials conducted in changing breast cancer treatment (START-A and START-B). These studies were conducted to investigate 5-fractionation regimens (ultrahypofractionation) in the adjuvant treatment of the whole breast. In the FAST trial, conventional radiotherapy (50 Gy in 25 fractions over 5 weeks) was used as the standard method, and two 5-fraction hypofractionation groups with a total dose of 30 Gy and 28.5 Gy while maintaining the 5-week treatment period (once a week) were also performed for comparison. The 10-year results of this trial showed that local tumor control was similar in all three study groups. Late complications such as breast firmness and changes were less than 30 Gy in the 28.5 Gy groups and comparable to the standard regimen, confirming that the hypofractionated regimen with a total dose of 28.5 Gy in 5 fractions over 5 weeks could be a potential alternative to the standard regimen [25]. The FAST-Forward study was conducted several years after a previous study that aimed to evaluate the efficacy and safety of ultra-hypofractionation, which delivers a total dose of 27 Gy or 26 Gy in 5 fractions over 1 week to the tumor area, compared with standard treatment, in which the hypofractionated regimen (40 Gy in 15 fractions over 3 weeks) was used as the standard treatment. The 5-year follow-up of this trial with Linke was similar in local tumor control in all groups, and the late complications studied in the 26 Gy ultrahypofraction regimen were similar to the standard hypofractionation regimen (40 Gy in 15 fractions), but the late complications of the 27 Gy regimen showed a higher rate than the standard hypofractionation group. Therefore, the 26 Gy regimen in 1 week, with a similar effect to the 40 Gy regimen (15 fractions) and reduced late complications, could be a new standard for treatment [26].

Advances in radiotherapy methods based on the modern science of radiobiology, including the shift toward hypofractionation and ultrahypofraction, have increased the effectiveness of treatment and provided a suitable platform for combining with more modern and advanced treatments such as immunotherapy. These new radiotherapy treatments have reduced damage to healthy tissues and attracted much attention by creating changes in the tumor microenvironment [27]. Studies have shown that short-term, concentrated doses of treatment stimulate the immunogenic death of tumor cells and activate the immune system by releasing tumor antigens [28–30]. In the following, we will further review the combination of radiotherapy and immunotherapy.

8.3 Advances in Combination of Immunotherapy and Radiotherapy

Immunotherapy has recently gained prominence as a therapeutic approach that activates the human immune system to target and eliminate tumor cells. The foundation of tumor immunotherapy lies in the concept of immune editing, which enhances the antigenic characteristics of tumor cells while simultaneously increasing the tumoricidal capabilities of immune cells. It also serves to inhibit immunosuppressive agents, restore immune system efficacy, enhance antitumor immune reactions, and restrict the proliferation of tumor cells [31].

Significant progress in the field of immunotherapy has been marked by the introduction of antibodies that specifically target tumor cells, immune checkpoint inhibitors, and chimeric antigen receptor T cell therapies [32]. Immune checkpoint inhibitors (ICIs) have notably altered the treatment paradigm for metastatic diseases in several cancers, including melanoma, non-small cell lung cancer, and renal cell carcinoma, with some patients achieving complete responses that endure for more than 5 years [33, 34]. Currently, therapies targeting the immune checkpoints CTLA-4 and PD-1/PD-L1 have been approved. The increasing precision of modern radiation therapy, in conjunction with immunotherapies, has minimized toxicity and broadened their application in clinical oncology [32].

1. *Anti-PD-1/PD-L1 therapy*

PD-1, which is a critical immunosuppressive and an immune checkpoint inhibitor receptor expressed on activated T cells, B cells, natural killer cells, dendritic cells, activated monocytes, myeloid cells, and a specific subset of thymocytes, serves to limit autoimmunity by regulating the activity of effector T cells in response to inflammatory stimuli [31, 35]. PD-L1, the ligand for PD-1, acts as an immunosuppressive signal and is upregulated in response to pro-inflammatory signals such as interferon-γ [35]. PD-L1 is found on tumor cells, macrophages, activated T cells, and cancer-associated fibroblasts [36]. The expression of PD-1 occurs on the surface of T cells following the activation of the T cell receptor [37]. The binding of PD-1 to PD-L1 activates SH2 protein tyrosine phosphatase 2 (SHP2), leading to the downregulation of the PI3K/AKT pathway. This process inhibits the proliferation of T and B cells in peripheral tissues, promotes programmed cell death in T cells, and helps prevent autoimmune disorders, thus contributing to the maintenance of immune homeostasis. However, this mechanism also facilitates the potential for immune evasion by tumor cells [31]. Anti-PD-1/PD-L1 antibodies inhibit the immune checkpoint pathway, thereby reactivating the function of activated T cells in their response to tumors [38]. PD-1 is highly expressed in breast cancer cells, particularly in triple-negative breast cancer (TNBC) cells, with the level of expression being indicative of the malignancy's severity [39]. The activation of PD-1 is intrinsically linked to PD-L1. Conventional cancer treatment modalities frequently enhance PD-L1 activation. An expanding body of literature supports the notion that blockade of the PD-1/PD-L1 axis may offer a promising therapeutic avenue, particularly for TNBC [40]. The United States Food and Drug Administration (FDA) has granted approval for several PD-1/PD-L1 inhibitors [41]. Notable examples of anti-PD-1 antibodies include cemiplimab, pembrolizumab, and nivolumab; anti-PD-L1 antibodies comprise avelumab, atezolizumab, and durvalumab [40]. A recent study showed that the combination of PD-1/PD-L1 and extracellular signal-regulated kinase (ERK1/2) inhibitors presents a promising therapeutic strategy, yielding more favorable outcomes in TNBC cell lines than in their non-TNBC counterparts [42]. The KEYNOTE-012 trial (NCT01848834) was a multicenter, nonrandomized phase Ib investigation that evaluated the effects of pembrolizumab, administered intravenously at a dose of 10 mg/kg biweekly, in patients suffering from advanced TNBC with PD-L1 positivity. Results from this phase I clinical trial revealed that pembrolizumab exhibited a commendable safety profile and an objective response rate (ORR) in the TNBC patient population [43]. In cohort A of the phase II KEYNOTE-086 trial, the efficacy of pembrolizumab was assessed as a second or subsequent line of therapy for individuals with metastatic triple-negative breast cancer (mTNBC). The results indicated that pembrolizumab monotherapy exhibited sustained antitumor effects in a specific group of patients who had undergone prior treatment for mTNBC while also maintaining an acceptable safety profile [44]. In a phase 1 clinical trial (JAVELIN Solid Tumor; NCT01772004), patients diagnosed with metastatic breast cancer who were either refractory to or had progressed following standard treatment received intravenous avelumab at a dosage of 10 mg/kg biweekly. The administration of avelumab demonstrated a tolerable safety profile along with clinical efficacy in a specific group of metastatic breast cancer patients. The overall confirmed ORR was recorded at 3.0%, while in patients diagnosed with TNBC, the ORR was noted to be 5.2%. Furthermore, the expression of PD-L1 in immune cells associated with tumors may correlate with an increased likelihood of clinical response to avelumab in individuals with metastatic breast cancer [45]. In the phase III randomized trial IMpassion-130 (NCT02425891), patients diagnosed with metastatic TNBC and exhibiting a favorable performance status (0–1) were assigned to receive either nab-paclitaxel (100 mg/m^2 on days 1, 8, and 15) in combination with atezolizumab (800 mg on days 1 and 15) or a placebo, administered in 28-day cycles [35, 46–48]. Prior treatments, including radiotherapy and chemotherapy (taxanes included), were permissible provided they occurred at least 12 months prior to randomization. Patients with asymptomatic central nervous system (CNS) metastases that had been treated were also included in the study. Stratification factors included whether patients received neoadjuvant or adjuvant taxane therapy, the presence or absence of liver metastases at baseline, and the expression of PD-L1 at baseline, categorized as positive or negative. Atezolizumab in combination with nab-paclitaxel has been shown to extend progression-free survival in patients diagnosed with metastatic triple-negative breast cancer, applicable to both the intention-to-treat cohort and the subgroup expressing PD-L1. The adverse events observed were in line with the established safety profiles of the individual agents involved [47]. The administration of anti-PD-1 or anti-PD-L1 agents, or a combination of both, appears to significantly influence survival rates, even among patients who did not receive immunotherapy as their initial treatment. Further investigations involving patients lacking a positive immune infiltrate are warranted [35].

2. *Anti-CTLA-4 therapy*

CTLA-4 is present in various cell types, including T cells, non-lymphoid cells, B cells, dendritic cells, stromal cells, and tumors. This molecule plays a critical role in the negative regulation of T effector cell activity by competing with CD28/CD80 co-stimulatory receptors for their common ligand, B7–1 [49, 50]. The binding of CTLA-4 to B7–1 occurs with greater avidity and affinity, leading to a reduction in T cell activation. Consequently, the interaction of CTLA-4 with B7–1 diminishes the co-stimulatory signal, resulting in an overall immunosuppressive effect [49, 51]. CTLA-4 plays a crucial role in inhibiting T cell proliferation, cell cycle progression, interleukin-2 (IL-2) production, and T cell differentiation. As such, it serves as a specific modulator of the CD4+ T cell response, contributing to the maintenance of T cell homeostasis. Research has examined CTLA-4 expression across various breast cancer subtypes using data from The Cancer Genome Atlas (TCGA), revealing that CTLA-4 expression is notably elevated in TNBC [31, 52]. CTLA-4 blockade may consequently enhance or reinstate T cell functionality. Humanized monoclonal antibodies targeting CTLA-4, such as ipilimumab and tremelimumab, have shown clinical efficacy across various malignancies, including melanoma, lung cancer, and renal cell carcinoma; however, their effectiveness in breast cancer remains unproven [31, 53, 54]. The antitumor efficacy of CTLA-4 monoclonal antibodies is considered suboptimal, and the high toxicity associated with anti-CTLA-4 monotherapy in breast cancer has prompted investigations into combination therapies as a potential alternative treatment strategy [31, 55].

3. *Combination of immunotherapy and radiotherapy*

Significant progress has been made in the treatment of breast cancer through chemotherapy, endocrine therapy, and targeted therapy; however, radiotherapy remains a crucial component in the management of TNBC. Radiotherapy facilitates the recruitment of immune cells into the tumor microenvironment (TME) through various mechanisms [31, 56]. Radiotherapy exhibits significant immunomodulatory capabilities by facilitating the production of tumor-specific antigens and enhancing the initiation and activation of cytotoxic T cells, which are crucial for tumor eradication during immune surveillance [57]. Furthermore, it can trigger immunogenic cell death through the release of cytokines, inflammatory mediators, and various immune-related molecules. Although activated CD8+ T cells and other immune-stimulatory cells can migrate to and infiltrate metastatic sites to exert antitumor effects, the upregulation of immune-suppressive cells due to inflammatory factors may counteract these effects and contribute to tumor progression. This indicates that radiotherapy by itself is inadequate for the complete eradication of both primary and metastatic tumor lesions [57, 58]. Moreover, radiotherapy is primarily administered for local control of tumors; however, it can also induce systemic effects on distant and non-irradiated tumor sites, a phenomenon known as the abscopal effect. The perception of radiotherapy as merely a localized treatment has significantly evolved in recent years, leading to a broader recognition that it can elicit a systemic immune response. This understanding provides a compelling rationale for integrating radiotherapy with immunotherapy. Additionally, the concept of the "radscopal effect," which involves the application of low-dose radiation to modify the tumor microenvironment, may enhance the likelihood of the abscopal effect and overcome the resistance of radiotherapy and immunotherapy combination. Collectively, these insights position radiotherapy as a catalyst for systemic antitumor immune responses, suggesting that, when combined with immunotherapy, it can serve as a comprehensive and radical treatment option [59]. Preclinical studies have demonstrated that the combination of radiotherapy with ICIs not only enhances antitumor efficacy but also elicits immune responses beyond the irradiated area. In a mouse model of TNBC characterized by low immunogenicity, radiotherapy was found to upregulate the expression of genes associated with immunogenic mutations. Vaccination using new epitopes derived from these genes successfully stimulated CD8+ and CD4+ T cell responses, thereby bolstering the overall immune response [31, 60]. A comparative study involving radiotherapy, anti-CTLA-4 antibody treatment, and their combination in a low-immunogenicity metastatic mouse model of breast cancer (4 T1 cells) indicated that the combined approach with ICIs and chemotherapy resulted in increased T cell infiltration, delayed growth of the primary irradiated tumor, improved survival rates, and reduced lung metastasis [61]. A phase 2 clinical trial examined the effectiveness and safety of combining radiotherapy with pembrolizumab in individuals diagnosed with mTNBC. The findings indicated that within the intention-to-treat group, the ORR for the unselected PD-L1 population was 17.6%, which surpassed the ORR observed in mTNBC patients who had previously undergone ICI monotherapy. The most frequently observed grade 1–2 toxicity was dermatitis, occurring in 29% of patients. Grade 3 adverse events were linked to pembrolizumab, including fatigue, lymphopenia, and infection. No grade 4 adverse events or treatment-related fatalities were reported. The combination of pembrolizumab and radiotherapy was deemed safe and exhibited promising efficacy in patients with poor-prognosis, metastatic, triple-negative breast cancer, regardless of PDL1 expression status [62]. In the TONIC trial, which is an adaptive, non-comparative phase 2 study, patients diagnosed with metastatic triple-negative breast cancer (TNBC) were randomized into several treatment groups. These included nivolumab administered either without prior induction or following a two-week low-dose induction, as well as combinations with (i) irradiation (3 × 8 Gy), (ii) cyclophosphamide, (iii) cisplatin, or (iv) doxorubicin, all of which were subsequently followed by nivolumab. The overall objective ORR was found to be 20%. Notably, the highest response rates were recorded in the cisplatin (ORR 23%) and doxorubicin (ORR 35%) groups. Furthermore, following the administration of doxorubicin and cisplatin, an upregulation of immune-related genes associated with the PD-1-PD-L1 axis and T cell cytotoxicity pathways was observed. Collectively, the clinical and translational findings from this study suggest that short-term treatment with doxorubicin and cisplatin may enhance the tumor microenvironment, thereby increasing the potential for a favorable response to PD-1 blockade in patients with TNBC [63]. The synergistic use of immunotherapy and radiotherapy may yield both benefits and drawbacks. The modifications brought about by ICIs can shift the balance between immune resistance and tolerance, resulting in heightened tumor recognition and increased immune responses in non-cancerous tissues. The application of radiotherapy can lead to the release of various antigens, both tumor-specific and non-tumor-specific, into the TME, which may activate autoreactive T cells that could attack normal tissues. As a result, this combination therapy may intensify the adverse effects experienced by patients. Therefore, it is imperative that future large-scale clinical trials are conducted to determine the optimal strategies for integrating radiotherapy with immunotherapy [31].

References

1. Yarnold J, Ashton A, Bliss J, Homewood J, Harper C, Hanson J, et al. Fractionation sensitivity and dose response of late adverse effects in the breast after radiotherapy for early breast cancer: long-term results of a randomised trial. Radiother Oncol. 2005;75(1):9–17.
2. Joiner MC, van der Koge JA. Basic clinical radiobiology. CRC Press; 2025.
3. McMahon SJ. The linear quadratic model: usage, interpretation and challenges. Phys Med Biol. 2018;64(1):01TR.
4. Alzibdeh A, Abuhijlih R, Abuhijla F. Breast cancer radiobiology: The renaissance of whole breast radiation fractionation (review). Mol Clin Oncol. 2024;21(6):97.
5. Tutt A, Yarnold J. Radiobiology of breast cancer. Clin Oncol (Royal College of Radiologists (Great Britain)). 2006;18(3):166–78.
6. Whelan TJ, Pignol J-P, Levine MN, Julian JA, MacKenzie R, Parpia S, et al. Long-term results of hypofractionated radiation therapy for breast cancer. N Engl J Med. 2010;362(6):513–20.
7. Alzibdeh A, Abuhijlih R, Abuhijla F. Breast cancer radiobiology: the renaissance of whole breast radiation fractionation. Mol Clin Oncol. 2024;21(6):97.
8. Chang JS, Chang JH, Kim N, Kim YB, Shin KH, Kim K. Intensity modulated radiotherapy and volumetric modulated arc therapy in the treatment of breast cancer: an updated review. J Breast Cancer. 2022;25(5):349.
9. Jagsi R, Griffith KA, Moran JM, Matuszak MM, Marsh R, Grubb M, et al. Comparative effectiveness analysis of 3D-conformal radiation therapy versus intensity modulated radiation therapy (IMRT) in a prospective multicenter cohort of patients with breast cancer. Int J Radiat Oncol Biol Phys. 2022;112(3):643–53.
10. Allali S, Kirova Y. Radiodermatitis and fibrosis in the context of breast radiation therapy: a critical review. Cancers. 2021;13(23):5928.
11. Racka I, Majewska K, Winiecki J. Three-dimensional conformal radiotherapy (3D-CRT) vs. volumetric modulated arc therapy (VMAT) in deep inspiration breath-hold (DIBH) technique in left-sided breast cancer patients—comparative analysis of dose distribution and estimation of projected secondary cancer risk. Strahlenther Onkol. 2023;199(1):90–101.
12. Jingu K, Ito K, Sato K, Umezawa R, Yamamoto T, Takahashi N, et al. VMAT with DIBH in hypofractionated radiotherapy for left-sided breast cancer after breast-conserving surgery: results of a non-inferiority clinical study. J Radiat Res. 2024;65(1):87–91.
13. Kim N, Chang JS, Shah C, Shin H, Keum KC, Suh CO, et al. Hypofractionated volumetric-modulated arc therapy for breast cancer: a propensity-score-weighted comparison of radiation-related toxicity. Int J Cancer. 2021;149(1):149–57.
14. De Rose F, Fogliata A, Franceschini D, Navarria P, Villa E, Iftode C, et al. Phase II trial of hypofractionated VMAT-based treatment for early stage breast cancer: 2-year toxicity and clinical results. Radiat Oncol. 2016;11:1–9.
15. Tang L, Ishikawa Y, Ito K, Yamamoto T, Umezawa R, Jingu K. Evaluation of DIBH and VMAT in hypofractionated radiotherapy for left-sided breast cancers after breast-conserving surgery: a planning study. Technol Cancer Res Treat. 2021;20:15330338211048706.
16. Newhauser WD, Zhang R. The physics of proton therapy. Phys Med Biol. 2015;60(8):R155.
17. Wu X-Y, Chen M, Cao L, Li M, Chen J-Y. Proton therapy in breast cancer: a review of potential approaches for patient selection. Technol Cancer Res Treat. 2024;23:15330338241234788.
18. Holt F, Probert J, Darby SC, Haviland JS, Coles CE, Kirby AM, et al. Proton beam therapy for early breast cancer: a systematic review and meta-analysis of clinical outcomes. Int J Radiat Oncol Biol Phys. 2023;117(4):869–82.
19. Lemanski C, Thariat J, Ampil FL, Bose S, Vock J, Davis R, et al. Image-guided radiotherapy for cardiac sparing in patients with left-sided breast cancer. Front Oncol. 2014;4:257.
20. Liang JA, Lee PC, Ku CP, Chen WT, Chung CY, Kuo YC, et al. Effectiveness of image-guided radiotherapy in adjuvant radiotherapy on survival for localized breast cancer: a population-based analysis. Cancer Manag Res. 2021;13:3465–72.

21. Van Parijs H, Cecilia-Joseph E, Gorobets O, Storme G, Adriaenssens N, Heyndrickx B, et al. Lung-heart toxicity in a randomized clinical trial of hypofractionated image guided radiation therapy for breast cancer. Front Oncol. 2023;13:1211544.
22. Group ST, et al. Lancet Oncol. 2008;9(4):331–41.
23. Group ST, The UK. Standardisation of Breast Radiotherapy (START) Trial B of radiotherapy hypofractionation for treatment of early breast cancer: a randomised trial. Lancet. 2008;371(9618):1098–107.
24. Haviland JS, Owen JR, Dewar JA, Agrawal RK, Barrett J, Barrett-Lee PJ, et al. The UK Standardisation of Breast Radiotherapy (START) trials of radiotherapy hypofractionation for treatment of early breast cancer: 10-year follow-up results of two randomised controlled trials. Lancet Oncol. 2013;14(11):1086–94.
25. Brunt AM, Haviland JS, Sydenham M, Agrawal RK, Algurafi H, Alhasso A, et al. Ten-year results of FAST: a randomized controlled trial of 5-fraction whole-breast radiotherapy for early breast cancer. J Clin Oncol. 2020;38(28):3261–72.
26. Brunt AM, Haviland JS, Wheatley DA, Sydenham MA, Alhasso A, Bloomfield DJ, et al. Hypofractionated breast radiotherapy for 1 week versus 3 weeks (FAST-Forward): 5-year efficacy and late normal tissue effects results from a multicentre, non-inferiority, randomised, phase 3 trial. Lancet. 2020;395(10237):1613–26.
27. Wang Y. Advances in hypofractionated irradiation-induced immunosuppression of tumor microenvironment. Front Immunol. 2021;11:612072.
28. Zhu S, Wang Y, Tang J, Cao M. Radiotherapy induced immunogenic cell death by remodeling tumor immune microenvironment. Front Immunol. 2022;13(1074477):2022.
29. Zhu M, Yang M, Zhang J, Yin Y, Fan X, Zhang Y, et al. Immunogenic cell death induction by ionizing radiation. Front Immunol. 2021;12:705361.
30. Ko EC, Benjamin KT, Formenti SC. Generating antitumor immunity by targeted radiation therapy: role of dose and fractionation. Adv Radiat Oncol. 2018;3(4):486–93.
31. Liu Y, Hu Y, Xue J, Li J, Yi J, Bu J, et al. Advances in immunotherapy for triple-negative breast cancer. Mol Cancer. 2023;22(1):145.
32. Jagodinsky JC, Harari PM, Morris ZS. The promise of combining radiation therapy with immunotherapy. Int J Radiat Oncol Biol Phys. 2020;108(1):6–16.
33. Topalian SL, Hodi FS, Brahmer JR, Gettinger SN, Smith DC, McDermott DF, et al. Five-year survival and correlates among patients with advanced melanoma, renal cell carcinoma, or non–small cell lung cancer treated with nivolumab. JAMA Oncol. 2019;5(10):1411–20.
34. Hamid O, Robert C, Daud A, Hodi F, Hwu W, Kefford R, et al. Five-year survival outcomes for patients with advanced melanoma treated with pembrolizumab in KEYNOTE-001. Ann Oncol. 2019;30(4):582–8.
35. Dey A, Ghosh S, Jha S, Hazra S, Srivastava N, Chakraborty U, et al. Recent advancement in breast cancer treatment using CAR T cell therapy: a review. Adv Cancer Biol Metastasis. 2023;7:100090.
36. Ahmed FS, Gaule P, McGuire J, Patel K, Blenman K, Pusztai L, et al. PD-L1 protein expression on both tumor cells and macrophages are associated with response to neoadjuvant durvalumab with chemotherapy in triple-negative breast cancer. Clin Cancer Res. 2020;26(20):5456–61.
37. Bullock KK, Richmond A. Beyond anti-PD-1/PD-L1: improving immune checkpoint inhibitor responses in triple-negative breast cancer. Cancers. 2024;16(12):2189.
38. Sato H, Okonogi N, Nakano T. Rationale of combination of anti-PD-1/PD-L1 antibody therapy and radiotherapy for cancer treatment. Int J Clin Oncol. 2020;25:801–9.
39. Mittendorf EA, Philips AV, Meric-Bernstam F, Qiao N, Wu Y, Harrington S, et al. PD-L1 expression in triple-negative breast cancer. Cancer Immunol Res. 2014;2(4):361–70.
40. Ye F, Dewanjee S, Li Y, Jha NK, Chen Z-S, Kumar A, et al. Advancements in clinical aspects of targeted therapy and immunotherapy in breast cancer. Mol Cancer. 2023;22(1):105.
41. Hargadon KM, Johnson CE, Williams CJ. Immune checkpoint blockade therapy for cancer: an overview of FDA-approved immune checkpoint inhibitors. Int Immunopharmacol. 2018;62:29–39.

42. Bräutigam K, Kabore-Wolff E, Hussain AF, Polack S, Rody A, Hanker L, et al. Inhibitors of PD-1/PD-L1 and ERK1/2 impede the proliferation of receptor positive and triple-negative breast cancer cell lines. J Cancer Res Clin Oncol. 2021;147(10):2923–33.
43. Nanda R, Chow LQ, Dees EC, Berger R, Gupta S, Geva R, et al. Pembrolizumab in patients with advanced triple-negative breast cancer: phase Ib KEYNOTE-012 study. J Clin Oncol. 2016;34(21):2460–7.
44. Adams S, Schmid P, Rugo H, Winer E, Loirat D, Awada A, et al. Pembrolizumab monotherapy for previously treated metastatic triple-negative breast cancer: cohort A of the phase II KEYNOTE-086 study. Ann Oncol. 2019;30(3):397–404.
45. Dirix LY, Takacs I, Jerusalem G, Nikolinakos P, Arkenau H-T, Forero-Torres A, et al. Avelumab, an anti-PD-L1 antibody, in patients with locally advanced or metastatic breast cancer: a phase 1b JAVELIN Solid Tumor study. Breast Cancer Res Treat. 2018;167:671–86.
46. Flies DB, Han X, Higuchi T, Zheng L, Sun J, Ye JJ, et al. Coinhibitory receptor PD-1H preferentially suppresses CD4+ T cell–mediated immunity. J Clin Invest. 2014;124(5):1966–75.
47. Schmid P, Adams S, Rugo HS, Schneeweiss A, Barrios CH, Iwata H, et al. Atezolizumab and nab-paclitaxel in advanced triple-negative breast cancer. N Engl J Med. 2018;379(22):2108–21.
48. Emens LA, Molinero L, Loi S, Rugo HS, Schneeweiss A, Diéras V, et al. Atezolizumab and nab-paclitaxel in advanced triple-negative breast cancer: biomarker evaluation of the IMpassion130 study. JNCI J Natl Cancer Inst. 2021;113(8):1005–16.
49. Jama M, Tabana Y, Barakat KH. Targeting cytotoxic lymphocyte antigen 4 (CTLA-4) in breast cancer. Eur J Med Res. 2024;29(1):353.
50. Lüönd F, Tiede S, Christofori G. Breast cancer as an example of tumour heterogeneity and tumour cell plasticity during malignant progression. Br J Cancer. 2021;125(2):164–75.
51. Rudd CE, Taylor A, Schneider H. CD28 and CTLA-4 coreceptor expression and signal transduction. Immunol Rev. 2009;229(1):12–26.
52. Peng Z, Su P, Yang Y, Yao X, Zhang Y, Jin F, et al. Identification of CTLA-4 associated with tumor microenvironment and competing interactions in triple negative breast cancer by co-expression network analysis. J Cancer. 2020;11(21):6365.
53. Darvin P, Toor SM, Sasidharan Nair V, Elkord E. Immune checkpoint inhibitors: recent progress and potential biomarkers. Exp Mol Med. 2018;50(12):1–11.
54. Weber JS, O'Day S, Urba W, Powderly J, Nichol G, Yellin M, et al. Phase I/II study of ipilimumab for patients with metastatic melanoma. J Clin Oncol. 2008;26(36):5950–6.
55. Liu Y, Zheng P. Preserving the CTLA-4 checkpoint for safer and more effective cancer immunotherapy. Trends Pharmacol Sci. 2020;41(1):4–12.
56. Charpentier M, Spada S, Van Nest SJ, Demaria S. Radiation therapy-induced remodeling of the tumor immune microenvironment. Semin Cancer Biol. 2022;86:737–47. Elsevier.
57. Gao Z, Zhao Q, Xu Y, Wang L. Improving the efficacy of combined radiotherapy and immunotherapy: focusing on the effects of radiosensitivity. Radiat Oncol. 2023;18(1):89.
58. Chicas-Sett R, Zafra-Martin J, Morales-Orue I, Castilla-Martinez J, Berenguer-Frances MA, Gonzalez-Rodriguez E, et al. Immunoradiotherapy as an effective therapeutic strategy in lung cancer: from palliative care to curative intent. Cancers. 2020;12(8):2178.
59. Zhang Z, Liu X, Chen D, Yu J. Radiotherapy combined with immunotherapy: the dawn of cancer treatment. Signal Transduct Target Ther. 2022;7(1):258.
60. Lhuillier C, Rudqvist N-P, Yamazaki T, Zhang T, Charpentier M, Galluzzi L, et al. Radiotherapy-exposed CD8+ and CD4+ neoantigens enhance tumor control. J Clin Invest. 2021;131(5)
61. Song H-N, Jin H, Kim J-H, Ha I-B, Kang K-M, Choi H-S, et al. Abscopal effect of radiotherapy enhanced with immune checkpoint inhibitors of triple negative breast cancer in 4T1 mammary carcinoma model. Int J Mol Sci. 2021;22(19):10476.

62. Ho AY, Barker CA, Arnold BB, Powell SN, Hu ZI, Gucalp A, et al. A phase 2 clinical trial assessing the efficacy and safety of pembrolizumab and radiotherapy in patients with metastatic triple-negative breast cancer. Cancer. 2020;126(4):850–60.
63. Voorwerk L, Slagter M, Horlings HM, Sikorska K, van de Vijver KK, de Maaker M, et al. Immune induction strategies in metastatic triple-negative breast cancer to enhance the sensitivity to PD-1 blockade: the TONIC trial. Nat Med. 2019;25(6):920–8.

Chapter 9
Advances in Radiotherapy of Gastrointestinal Cancers

Tahreen Taj, Fathima Beegum, M. P. Mohammed Ameen, Biplab Debnath, Sumel Ashique, Debajit Dewan, Raneet Das, Chowdhury Mobaswar Hossain, Shaghayegh Rezvani Nezhad, and Neeraj Mishra

Abstract One of the most widely used treatments for cancer is radiation therapy. It is crucial in managing gastrointestinal (GI) cancers and enhancing treatment outcomes when integrated with multimodal therapies. Advancements in technology like brachytherapy, external beam radiotherapy, and adaptive radiotherapy offer superior accuracy by improving tumor control and decreasing radiation exposure to neighboring healthy tissues. The emerging role of radiotherapy is discovered by recent clinical trials and evidence-based advances, aiming to overcome challenges related to radiation resistance in GI tumors to progress therapeutic conclusions by demanding innovative approaches. Overcoming resistance involves targeting tumor-specific factors such as hypoxia, DNA repair mechanisms, and the immunosuppressive tumor microenvironment. Managing side effects and toxicity is essential due to the critical location of GI tumors near radiosensitive organs such as the bowel, liver, and kidneys. Incorporating molecular biomarkers for therapeutic customization is the main focus of future developments in treating GI cancers. To overcome resistance, integrating radiotherapy with systemic therapies, such as checkpoint inhibitors and targeted drugs, shows potential outcomes. Precision and effectiveness are

T. Taj · F. Beegum · M. P. Mohammed Ameen
Department of Pharmacology, Yenepoya Pharmacy College and Research Centre, Yenepoya (Deemed to be) University, Mangalore, Karnataka, India

B. Debnath · D. Dewan · R. Das · S. Ashique
Department of Pharmaceutical Technology, Bharat Technology, Uluberia, West Bengal, India

C. M. Hossain
Department of Pharmaceutical Technology, Maulana Abul Kalam Azad University of Technology, Haringhata, Nadia, West Bengal, India

S. R. Nezhad
Department of Radiation Oncology, School of Medicine, Shahid Beheshti University of Medical Sciences, Tehran, Iran

N. Mishra (✉)
Amity Institute of Pharmacy, Amity University Madhya Pradesh, Gwalior, Madhya Pradesh, India

F. Taghizadeh-Hesary et al. (eds.), *From Radiobiology to Radiation Oncology*,
https://doi.org/10.1007/978-3-031-95724-6_9

also being enhanced by developments in imaging and treatment planning. Research into epigenetic and genomic factors driving resistance is critical to developing novel therapeutic targets. These combined efforts aim to optimize radiotherapy for GI tumors, improving both survival rates and long-term patient outcomes.

Keywords Radiotherapy · Gastrointestinal cancers · Multimodal therapies · Radiation resistance · Toxicity management

9.1 Introduction

Radiotherapy has undergone major progressions over the past few spans by hardening its role as a keystone in the management of gastrointestinal (GI) cancers. It is employed in numerous clinical scenarios, comprising definitive local therapy, adjuvant treatment following surgery, and more lately, as a neoadjuvant method to boost surgical conclusions [1]. The radiotherapy integration into multidisciplinary cancer care is well recognized, along with the recent imaging techniques and also innovative planning and delivery approaches permitting for more accurate tumor targeting while decreasing destruction to surrounding healthy tissues. Depending on tumor type and location, radiotherapy can be directed preoperatively, postoperatively, or even intraoperatively [2]. With patients having metastatic gastric cancer, this radiotherapy plays a vital role in symptom palliation, particularly for managing bone pain and controlling bleeding from the primary tumor. Studies have constantly validated that neoadjuvant radiotherapy contributes to tumor downstaging and enhanced local disease control. Furthermore, radiotherapy remains a crucial element in the beneficial treatment of elderly patients with GI malignancies [3]. GI cancers are among the most prevalent malignancies which affect both men and women. To improve clinical outcomes as well as deliver optimal patient care, a multidisciplinary treatment approach is essential. However, for rare digestive cancers, the deficiency of established treatment protocols and the low incidence rates present substantial trials, delaying research progress and leading to poorer prognoses [4]. Recent advancements in imaging, radiotherapy planning, and delivery combined with the integration of novel therapeutic agents have led to improved oncological outcomes, better quality of life, reduced adverse effects, and greater potential for organ preservation [5]. Despite these advancements, a significant proportion of patients experience disease recurrence following curative resection, and nearly 50% present with advanced-stage disease at diagnosis. GI cancers display substantial heterogeneity, both within individual tumors and across different patients, further complicating treatment strategies. Addressing these challenges requires continued innovation in radiotherapy and a collaborative, multidisciplinary approach to care [6].

9.2 Techniques and Modalities of Radiotherapy

For cancer, one of the most tremendously used treatments is radiation therapy. Over time, substantial improvements in radiotherapy techniques have allowed for further accurate targeting of tumors while diminishing harm to surrounding healthy tissues. This approach employs high-energy X-rays to terminate cancer cells by disrupting their ability to multiply [7]. With continuous technological progress, radiotherapy methods have evolved rapidly, particularly in Asia. The development of advanced radiotherapy technologies has greatly enhanced the accuracy, efficiency, and safety of cancer treatments. Radiotherapy is often integrated with surgery and systemic therapies as part of a comprehensive multimodal treatment approach. This integrated approach aims to optimize tumor control, improve patient outcomes, and preserve organ function while minimizing side effects [8]. The treatment can be administered through two primary approaches: external beam radiation therapy (EBRT) and internal radiation therapy (brachytherapy). EBRT includes directing high-energy rays at the tumor from an external machine, whereas brachytherapy involves placing radioactive implants (seeds) inside or near the affected area to deliver targeted radiation [9]. Furthermore, various innovative radiotherapy modalities have been introduced to enhance cancer treatment efficacy. These advancements continue to shape the landscape of oncology, proposing patients further effective and personalized treatment options [10]. Numerous systems in the treatment of cancer, specifically radiotherapy modalities, are as follows:

9.2.1 Brachytherapy (Internal Radiotherapy)

Brachytherapy is also recognized as internal radiotherapy, a specialized form of radiotherapy that delivers radiation directly to cancerous tissue using sealed radioactive sources. Unlike EBRT, which requires radiation to pass through healthy tissue to reach the tumor, brachytherapy provides a highly localized dose, minimizing exposure to surrounding structures [9]. Standalone treatment or in combination, it can be used with EBRT, surgery, or chemotherapy. The advancement of brachytherapy depends on emerging technologies that address existing challenges and enhance its clinical application.

9.2.2 External Beam Radiotherapy (EBRT)

EBRT is a widely utilized and effective approach for managing advanced localized malignancies. It encompasses various specialized techniques, each tailored to enhance precision, improve tumor targeting, and minimize damage to surrounding healthy tissues [11]. These modalities include:

Three-Dimensional Conformal Radiotherapy (3DCRT)

3DCRT utilizes advanced imaging techniques to shape radiation beams precisely according to the tumor's dimensions. It employs multiple cross-firing beams with carefully contoured fields, offering a significant improvement over traditional 2D radiotherapy. This technique is extensively used in treating diverse malignancies, including GI cancers [12].

Intensity-Modulated Radiotherapy (IMRT)

IMRT optimizes radiation delivery by adjusting the intensity of each beam, ensuring precise targeting of the tumor while sparing surrounding healthy tissues. A computer-assisted optimization algorithm aids in this process. IMRT has been largely applied in patients with GI cancers [13].

Volumetric Modulated Arc Therapy (VMAT)

VMAT transports radiation in an endless arc, allowing for dynamic dose modulation and optimal tumor coverage. By adjusting the dose rate and multileaf collimator (MLC) positions during a 360° rotation, VMAT achieves superior dose distribution while reducing treatment time. It is frequently used for different malignancies [14]. Emerging evidence has demonstrated the clinical benefits of VMAT over IMRT in patients with gastric [15], rectal [16], and pancreatic cancer [17].

Image-Guided Radiotherapy (IGRT)

IGRT integrates real-time imaging during treatment to enhance precision by accounting for tumor motion, patient positioning, and anatomical variations. This technique has significantly improved the accuracy of radiation delivery and is widely applied in gastric cancers [18].

Stereotactic Body Radiotherapy (SBRT)

SBRT delivers high-precision radiation doses by directing multiple beams from various angles to converge on the tumor site. This method is highly effective for treating prostate, head and neck, spinal, and renal oligometastases. It has also shown promising outcomes in patients with pancreatic and liver cancer [19, 20].

9.2.3 Adaptive Radiotherapy (ART)

Adaptive radiotherapy (ART) is an advanced progression in radiation oncology that influences progressive imaging to accommodate the dynamic variations in patient anatomy throughout the treatment of cancer [21]. This current progress in radiotherapy technology has also improved treatment personalization in the setting of GI cancers [22].

9.2.4 Radiopharmaceutical Therapy (Molecular Radiotherapy)

Radiopharmaceutical therapy utilizes radionuclides conjugated to tumor-targeting agents to selectively transport cytotoxic radiation to cancer cells. Its effectiveness depends on achieving therapeutic radiation doses to tumors while minimizing toxicity to normal tissues. This approach has been emergently applied in patients [23], colon [24], and pancreatic cancers [25].

9.3 Integration with Radiotherapy Multimodal Therapies

Radiotherapy (RT) plays a fundamental role in multidisciplinary cancer care, contributing precise tumor targeting while lessening destruction to surrounding healthy tissues. Treatment accuracy has been significantly improved by innovative planning techniques, advances in imaging technologies, and enhanced computational resources in radiation oncology. The seamless integration of multimodal imaging into radiation therapy workflows has further refined tumor localization and treatment planning, leading to improved clinical conclusions. Integration of radiotherapy with immunotherapy has appeared as an auspicious method for enhancing therapeutic efficacy and reducing tumor recurrence, which can modulate the tumor microenvironment, increase antigen presentation, and improve immune cell infiltration, thereby overcoming tumor immune tolerance mechanisms.

Numerous preclinical and clinical studies on GI cancers have demonstrated that combining radiotherapy with immune checkpoint inhibitors [26], monoclonal antibodies [27], and immunomodulating therapies [28] can enhance anti-tumor responses and improve patient consequences. Adoptive cell therapy (ACT) is a personalized management strategy that includes the infusion of immune cells engineered to target tumors. ACT includes tumor-infiltrating lymphocytes (TILs), lymphokine-activated killer (LAK) cells, cytokine-induced killer (CIK) cells, dendritic cells (DCs), natural killer (NK) cells, chimeric antigen receptor T (CAR-T) cells, and T cell receptor-engineered T (TCR-T) cells [29]. Among these, CAR-T cell therapy has gained significant attention due to its potential for durable tumor control. A preclinical study on glioblastoma has demonstrated that combining sub-therapeutic doses of localized RT with NKG2D-based CAR-T cells can improve CAR-T cell operation at the site of the tumor, thereby achieving a synergistic antitumor effect [30]. Further studies are required to demonstrate whether this benefit is also true in GI cancers.

Cancer vaccines alone often fail to generate a strong and sustained immune response capable of eliminating tumors. However, when combined with radiotherapy, they can induce a more robust anti-tumor effect by priming the immune system and enhancing antigen release. Clinical trials investigating the synergistic potential of RT and cancer vaccines are ongoing, with promising preliminary results in multiple cancer types [31]. As research continues to advance, scientists are discovering new ways to make cancer treatment more effective by combining radiotherapy with innovative approaches.

9.4 Radiotherapy in Specific Gastrointestinal Cancers

Radiotherapy is a crucial component in the treatment of GI cancers, utilized either as a standalone therapy or as part of a comprehensive approach that includes surgery and chemotherapy. It employs high-energy radiation to eliminate cancer cells, thereby reducing tumor size and enhancing treatment outcomes. While surgery remains the primary curative option for most GI cancers, radiotherapy is frequently used in neoadjuvant, adjuvant, or palliative settings to maximize therapeutic benefits [32]. The role of radiotherapy in GI cancers varies depending on the type, stage, and location of the tumor. For example, neoadjuvant radiotherapy in rectal cancer improves the likelihood of successful surgical resection and reduces the risk of local recurrence, while in esophageal and gastric cancers, it is often combined with chemotherapy to improve treatment efficacy. Advances in radiotherapy techniques, such as VMAT and SBRT, have enhanced precision, allowing targeted delivery of radiation to tumors while minimizing harm to nearby healthy tissues and reducing side effects [15, 19].

Radiotherapy reduces tumor size, aiding in surgical resection or alleviating symptoms in cases where surgery is not an option. In cancers such as esophageal and rectal malignancies, preoperative radiotherapy has demonstrated the ability to downstage tumors, making curative surgery more feasible [33, 34]. In advanced stages, radiotherapy is effective in alleviating symptoms such as pain, bleeding, and obstruction [35]. Despite its benefits, radiotherapy in GI cancers poses challenges due to the anatomical complexity of the GI tract, the sensitivity of surrounding organs (e.g., liver, kidneys, and spinal cord), and the diverse nature of tumor biology, necessitating meticulous planning and execution [36]. Innovative approaches, including proton beam therapy and combination therapies integrating radiotherapy with immunotherapy, are under active research and show promise in improving survival rates and minimizing treatment-related toxicity [37, 38].

9.5 Personalized and Adaptive Radiotherapy in Gastrointestinal Cancers

9.5.1 Definition and Principles of Personalized Radiotherapy for GI Cancers

Personalized radiotherapy involves customizing radiation therapy to the specific characteristics of an individual patient and their tumor. This approach considers genetic, molecular, and anatomical features of the tumor, alongside patient-specific factors such as age, gender, comorbidities, and overall health status. Unlike traditional, standardized treatment protocols, personalized radiotherapy leverages advanced diagnostic tools, genomic profiling, and cutting-edge imaging technologies to achieve an optimal balance between effective tumor control and minimal harm to healthy tissues [10]. The fundamental principles of personalized radiotherapy include:

Tailored Treatment Planning

Designing radiation doses and techniques to suit the tumor's type, size, location, and radiosensitivity [39].

Biological Targeting

Employing biomarkers and genomic insights to identify tumors that are most likely to respond to radiotherapy or combination therapies [40].

Dynamic Adaptation

Utilizing adaptive radiotherapy techniques to modify treatment plans in real-time as tumors respond to therapy or anatomical changes occur [41].

Integration with Multimodal Therapies

Combining radiotherapy with systemic treatments such as chemotherapy, targeted drugs, or immunotherapy based on the cancer's specific molecular profile [42].

9.5.2 Importance of Tailoring Treatment to Individual Patient Characteristics

Customizing radiotherapy is crucial for improving outcomes in GI cancers, given their significant heterogeneity. Personalization addresses the unique characteristics of both the tumor and the patient, resulting in.

Enhanced Effectiveness

By focusing on specific molecular pathways and accounting for individual radiosensitivity, personalized radiotherapy increases treatment success rates [43].

Minimized Toxicity

Accurate radiation delivery spares nearby healthy tissues, reducing side effects and enhancing the patient's quality of life [44].

Improved Prognostic Accuracy

Biomarkers and genomic profiling help predict treatment responses, facilitating better decision-making and preventing over- or undertreatment [45].

Flexibility to Tumor Changes

Adaptive radiotherapy adjusts treatments as tumors shrink or change during therapy, ensuring ongoing precision and effectiveness [22].

For example, in rectal cancer, patients with specific genetic or molecular characteristics (e.g., microsatellite instability) can benefit from personalized approaches that combine radiotherapy with immunotherapy [46]. Likewise, in esophageal cancer, tailoring treatments based on tumor location and histology can substantially improve clinical outcomes [47]. Ultimately, personalized radiotherapy enhances survival rates while aligning with the goals of precision oncology, delivering treatments uniquely designed for each patient [48].

9.5.3 Principles of Adaptive Radiotherapy in GI Cancers

ART is an advanced, flexible approach to radiation therapy that continuously updates and optimizes the treatment plan in response to real-time changes in the tumor's size, shape, or position, as well as alterations in the anatomy of surrounding healthy tissues [21]. In GI cancers, ART is particularly vital due to the anatomical complexity of the GI tract and the high likelihood of organ motion caused by breathing, digestion, or tumor shrinkage during treatment [22].

Real-Time Monitoring

ART utilizes state-of-the-art imaging techniques, such as cone-beam computed tomography (CBCT), magnetic resonance imaging (MRI)-guided radiotherapy, and 4D-CT imaging, to continuously monitor tumor and organ movement during treatment sessions. This ensures that radiation is delivered with high accuracy despite anatomical changes [49].

Customized Treatment Adjustments

The treatment plan is dynamically revised based on variations in tumor volume, patient weight, or the position of nearby critical organs like the liver, kidneys, and intestines. These modifications are crucial for GI cancers, where even small deviations can significantly affect treatment outcomes [50].

Management of Tumor and Organ Motion

Tumors in the GI tract are prone to movement caused by normal physiological activities such as peristalsis, stomach filling, and respiratory motion. ART incorporates motion management strategies, including respiratory gating, breath-hold techniques, and real-time tracking, to ensure precise radiation delivery [51].

Minimizing Margins and Toxicity

ART reduces the need for large safety margins by accounting for daily anatomical changes, thereby ensuring that healthy tissues receive minimal radiation exposure. This approach decreases the risk of treatment-related side effects and enhances patient safety [52].

Incorporation of Advanced Technologies

ART often integrates innovative methods such as IMRT, SBRT, and proton therapy to deliver highly targeted doses while minimizing radiation to adjacent organs [53].

Dynamic Adaptation to Tumor Response

Tumor shrinkage or anatomical changes during treatment can alter the initial radiation plan. ART dynamically adjusts the treatment approach to maintain optimal dose delivery, which is especially important in cancers like rectal, pancreatic, and esophageal cancers, where significant tumor regression is common [54].

9.5.4 Significance of Adaptive Radiotherapy in GI Cancers

Increased Accuracy

ART precisely targets tumors while minimizing radiation to critical structures such as the spinal cord, liver, and intestines [55].

Enhanced Clinical Outcomes

By adapting to tumor response, ART improves local tumor control and overall therapeutic success [56].

Lower Toxicity Levels

Limiting radiation exposure to healthy tissues reduces side effects such as GI discomfort, diarrhea, and radiation-induced organ injury [57].

9.6 Clinical Trials and Evidence-Based Advances of Radiotherapy in Gastrointestinal Cancers

Clinical trials and evidence-based advances in radiotherapy have significantly improved treatment outcomes for GI cancers. Innovations in radiotherapy techniques, coupled with clinical trial data, have led to better survival rates and enhanced quality of life for patients [58, 59].

9.6.1 Intensity-Modulated Radiotherapy

In comparison to conventional 2D and 3D-conformal techniques, conventional radiotherapy techniques like IMRT and VMAT allow for more precise radiation delivery, targeting tumors while minimizing damage to surrounding healthy tissue.

This is especially crucial in GI cancers, where critical organs like the bowel, liver, and kidneys are located nearby. IMRT, in particular, has proven beneficial in reducing side effects such as GI toxicity by providing highly targeted doses of radiation [13].

9.6.2 Stereotactic Body Radiation Therapy

SBRT, also known as stereotactic ablative radiotherapy (SABR), delivers high doses of radiation in fewer treatment sessions. It has become increasingly popular for treating localized tumors in the liver, pancreas, and other GI regions where surgery may not be an option. Clinical trials have shown that SBRT is particularly effective for small, isolated tumors, including those that are inoperable or have recurred after surgery [2, 60].

9.6.3 Combination with Chemotherapy

Evidence has shown that combining radiotherapy with chemotherapy enhances treatment outcomes, particularly for esophageal, rectal, and pancreatic cancers. For instance, chemoradiation has significantly improved survival rates in locally advanced esophageal cancer, especially when used before surgery (neoadjuvant chemoradiation). Landmark trials such as the CROSS trial [61] and the RTOG 8501 trial [62] have established chemoradiation as a cornerstone of treatment.

9.6.4 Neoadjuvant and Adjuvant Radiotherapy

Neoadjuvant radiotherapy, given before surgery, has been shown to shrink tumors, making them more resectable and improving survival outcomes, particularly in rectal cancer. Meanwhile, adjuvant radiotherapy, administered after surgery, helps reduce recurrence risks, especially in cancers such as pancreatic and colorectal cancer [63].

9.6.5 Immunotherapy and Radiotherapy Combinations

Emerging research suggests that combining radiotherapy with immunotherapy may offer a synergistic approach, as radiation can stimulate the immune system to better target cancer cells, known as the 6th R of Radiobiology: Immune Reactivation [64]. Early-phase clinical trials investigating this combination, particularly in hepatocellular carcinoma (liver cancer) [65] and gastric cancer [66], have shown promise, although further evidence is needed to confirm its efficacy.

9.6.6 Radiotherapy for Metastatic Disease

Radiotherapy is commonly used to treat metastatic GI cancers, particularly when metastases are confined to a few locations, i.e., the oligometastatic disease. SBRT has shown great promise in treating oligometastatic disease. This application has demonstrated clinical promise in patients with metastatic colorectal [67], gastric [68], esophageal [69], and pancreatic cancers [70].

9.6.7 Personalized Radiotherapy

Advances in genomics and molecular imaging are enabling more personalized radiotherapy approaches. By identifying specific genetic mutations and tumor characteristics, clinicians can better predict which patients will benefit most from particular radiation therapies. Ongoing trials are exploring the potential for biomarker-driven strategies to refine radiotherapy regimens for individual patients [59].

9.6.8 Advanced Imaging and Treatment Planning

The development of advanced imaging technologies, such as PET/CT scans and MRI, has significantly improved the precision of radiotherapy planning. These imaging techniques allow for more accurate tumor localization and monitoring of tumor changes during treatment, leading to more effective dose delivery while minimizing damage to surrounding healthy tissue [58].

9.6.9 Index Clinical Trials and Research

The LAPC-1 Trial (Pancreatic Cancer)

This study evaluated the use of SBRT in patients with locally advanced pancreatic cancer and found that it could be an effective treatment option for patients who were not candidates for surgery [62].

POET Trial [Gastroesophageal Junction (GEJ) Cancers]

A randomized control trial was conducted in Germany in which a comparison of neoadjuvant chemotherapy alone to neoadjuvant chemoradiotherapy in patients with gastroesophageal junction (GEJ) cancers was performed, which showed a statistically significant increase in survival rate in the neoadjuvant chemoradiotherapy group [71].

Stockholm III Trial (Rectal Cancer)

This trial focused on determining the most effective radiotherapy regimen for rectal cancer, comparing short-course radiation therapy with long-course radiation therapy, and helping refine preoperative treatment strategies [67].

The PREOPANC Trial (Pancreatic Cancer)

The randomized control trial examines the role of neoadjuvant chemoradiation (gemcitabine-based CRT regimen) in resectable and borderline resectable pancreatic cancer [71].

CROSS Trail (Oesophageal and Gastroesophageal Cancer)

In the Netherlands, it was conducted and evaluated neoadjuvant chemoradiotherapy followed by surgery versus surgery alone in esophageal and gastroesophageal cancer and observed an increase in survival rate in the neoadjuvant chemoradiotherapy arm [72].

TOPGEAR Trial (Gastric and Gastroesophageal Cancer)

This is a phase III trial conducted to evaluate survival rates in patients with gastric and gastroesophageal cancer, but it does not get significant change by adding radiotherapy to chemotherapy and compared with chemotherapy alone. This study did not demonstrate a survival benefit for the radiotherapy-included arm [73].

INT-0116 Trial (Gastric Adenocarcinoma)

A randomized clinical trial (US-based) with stage $\geq$ T3 or node-positive gastric adenocarcinoma showed significant improvement in survival with adjuvant chemoradiation compared with surveillance [74].

The IMRT Trials in Esophageal Cancer

Studies on IMRT in esophageal cancer have demonstrated its ability to provide effective treatment with reduced toxicity compared to traditional 3D-CRT, leading to improved local control and quality of life for patients [75].

9.7 Overcoming Radiation Resistance in Gastrointestinal Tumors: Detailed Approaches

Colorectal, gastric, esophageal, pancreatic, and liver cancers are among the GI cancers that pose a significant global health burden. Radiation therapy is still an essential component of the treatment plan for many GI malignancies, even with improvements in detection and treatment [3]. However, radiation resistance continues to be a significant challenge in treating these tumors, leading to treatment failure, recurrence, and poor prognosis in some patients. Understanding and overcoming radiation resistance in GI tumors is essential for improving clinical outcomes [66]. This overview explores multiple strategies being investigated to overcome mechanisms of radiation resistance in GI tumors [4]. It's critical to comprehend the basic mechanisms underlying radiation resistance in GI tumors before delving into methods for overcoming it. The way radiation therapy works is by causing DNA damage, mostly double-strand breaks (DSBs), which, if left untreated, can kill cells. However, tumor cells, including those in GI cancers, have developed mechanisms to repair this damage and survive, which contributes to radiation resistance [67].

9.7.1 Mechanisms Behind Radiation Resistance

Damage to DNA Repair Mechanisms

The capacity of cancer cells to repair radiation-induced DNA damage, particularly DSBs, is a defining characteristic of radiation resistance. DNA damage repair mechanisms are necessary for cellular integrity under normal circumstances, but cancer cells frequently take advantage of these pathways to evade radiation treatment. Homologous recombination (HR) is a repair pathway which uses an intact sister chromatid to repair DSBs accurately. In certain GI tumors, HR may be overactive, enabling them to effectively repair radiation-induced damage and survive. Repression of DNA Damage Checkpoints: Cells typically arrest in specific phases of the cell cycle in response to DNA damage, allowing time for repair before mitosis. Tumor cells can bypass these checkpoints, allowing them to continue dividing even in the presence of DNA damage. This checkpoint abrogation contributes to radiation resistance [68].

Tumor Hypoxia

Due to their rapid growth and insufficient blood supply, many GI tumors have hypoxic (low oxygen) areas. Because oxygen intensifies the DNA-damaging effects of ionizing radiation, hypoxic conditions decrease the effectiveness of radiation therapy. Hypoxia-inducible factors (HIFs), transcription factors that support cellular adaptation to low oxygen levels, are activated in response to hypoxia. HIFs lead to the expression of genes that protect cells from radiation-induced death by promoting cell survival, angiogenesis, and metabolic adaptation. Tumors with extensive hypoxic regions are often more resistant to radiation and less responsive to therapy [69].

Tumor Microenvironment

Radiation resistance in GI malignancies is significantly influenced by the tumor microenvironment (TME). To support tumor cell survival, repair, and proliferation, TME constituents such as immune cells, endothelial cells, and cancer-associated fibroblasts (CAFs) can interact with tumor cells. For example, macrophages and myeloid-derived suppressor cells (MDSCs) are often present in the TME of GI tumors and can contribute to immune suppression and radiation resistance by secreting anti-inflammatory cytokines and promoting cell survival pathways [70].

9.7.2 Strategies to Overcome Radiation Resistance in GI Tumors

Targeting DNA Repair Pathways

One of the most promising approaches to overcoming radiation resistance in GI tumors involves targeting the DNA repair mechanisms that enable cancer cells to survive radiation-induced damage. Inhibiting key components of the NHEJ repair pathway, such as DNA-PKcs (DNA-dependent protein kinase catalytic subunit), can prevent tumor cells from efficiently repairing DSBs. Inhibition of NHEJ has shown promise in preclinical models, enhancing the efficacy of radiation therapy [71].

Hypoxia Targeting

Given that tumor hypoxia is a significant contributor to radiation resistance, strategies aimed at mitigating its effects are crucial in overcoming resistance. Hypoxia-activated prodrugs (HAPs) are inactive under normal oxygen levels but become cytotoxic in hypoxic conditions. Tirapazamine is one example of a hypoxia-activated drug that has been shown to increase tumor cell death when combined with radiation. HIF-1 plays a central role in the adaptation of tumor cells to hypoxic conditions. Inhibiting HIF-1 can reduce the tumor's ability to survive under hypoxia, making it more sensitive to radiation. Agents that target HIF-1 or its downstream pathways, such as PX-478 and EZN-2968, are being explored in combination with radiation therapy [72].

Modification of the Tumor Microenvironment

The TME is crucial to the survival and development of cancer cells, as well as their ability to withstand radiation. CAFs secrete growth factors and extracellular matrix proteins that promote tumor growth and radiation resistance. This might be achieved by altering the TME to make it less conducive to tumor survival. TGF-β drugs, such as fresolimumab, can change the TME and lessen radiation therapy resistance by blocking CAF activity. Tumor-associated macrophages and MDSCs can contribute to immune suppression and resistance to radiation. Using agents like anti-PD-1/PD-L1 inhibitors (e.g., pembrolizumab) or immune-modulating drugs that repolarize macrophages from a pro-tumor phenotype to an anti-tumor phenotype could help overcome radiation resistance and enhance the immune response against the tumor [70].

Advanced Radiation Technique

To overcome resistance, new radiation delivery techniques increase precision and tumor. By concentrating the radiation dose on tumor cells and causing the least amount of harm to nearby healthy tissue, SBRT can overcome resistance by precisely delivering high doses of radiation. Liver, pancreas, and colorectal malignancies are among the GI tumors that SBRT has demonstrated potential in treating [29, 63]. Particle therapy (also known as proton or heavy ion therapy): These treatments deliver radiation using charged particles, like protons or carbon ions, rather than conventional X-rays. Particle therapy has a different biological effect, including a higher relative biological effectiveness (RBE) at the Bragg peak, making it potentially more effective in overcoming radiation resistance [73].

9.8 Side Effects and Toxicity Management in Gastrointestinal Tumor Treatment

Among the most prevalent cancers in the world are GI cancers, which include colorectal, esophageal, pancreatic, gastric, and liver cancers [2]. Usually, surgery, chemotherapy, radiation therapy, and, more recently, immunotherapy are used to treat these malignancies [39]. Even though these treatments have greatly increased survival rates, they also have several toxicities and adverse effects, some of which can be crippling. Controlling these adverse effects is essential for enhancing treatment adherence and patient quality of life (QoL). This article describes frequent toxicities and side effects related to the treatment of GI tumors, with an emphasis on chemotherapy, radiation therapy, and immunotherapy. We will also talk about several approaches to dealing with these adverse effects [74].

The management of toxicities from chemotherapy, radiation therapy, and immunotherapy requires a multidisciplinary approach that includes pharmacological treatment, supportive care, and regular monitoring. Some important supportive care strategies include. (i) GI cancer patients often experience malnutrition, weight loss, and difficulty eating due to the disease or side effects of treatment. Nutritional counseling and enteral or parenteral feeding may be necessary for those unable to eat normally [76]. A balanced diet tailored to the patient's needs can help maintain strength, improve immune function, and enhance overall recovery. (ii) Pain management: Side effects of cancer treatments or the disease itself can cause excruciating pain. Opioids, nonsteroidal anti-inflammatory medicines (NSAIDs), and adjuvant therapies like physical therapy or nerve blocks are among the medications used to treat pain. It's critical to evaluate the patient's discomfort frequently and modify treatment regimens as necessary [77]. Palliative care professionals can enhance comfort and assist in better managing chronic pain. (iii) Psychosocial Support: It's critical to address the psychological and emotional effects of cancer treatment. Patient outcomes can be greatly impacted by how well a patient handles the stress, worry, and sadness that frequently accompany a cancer diagnosis. Patients can manage these difficulties with the use of mental health professionals, support groups, and counseling. A network of friends and family is also essential for offering emotional support throughout therapy [78]. (iv) Managing Fatigue: One of the most frequent adverse effects of cancer treatment is fatigue. Patients may feel emotionally and physically exhausted, which can greatly affect their day-to-day activities. Cognitive-behavioral therapy (CBT), energy-saving techniques, and regular exercise can all help manage cancer-related fatigue. (v) Prevention and Screening: Regular follow-up care is essential for detecting and managing long-term toxicities, secondary cancers, and recurrence of GI cancers. Preventative measures, such as vaccinations, regular imaging, and blood tests, help ensure early intervention and improve outcomes. Screening for potential complications like secondary malignancies is especially important in patients who have undergone radiation or certain chemotherapy regimens [79].

9.9 Conclusions and Future Directions

The treatment of GI tumors has advanced significantly with the use of IMRT that reduces damage to healthy tissue while enabling more accurate tumor targeting by varying the radiation beams' intensity. Its promise in GI tumors can still be increased, though, and future studies will try to improve treatment strategies and reduce toxicity associated with them. Efforts are focused on improving IMRT planning and delivery. Adaptive radiotherapy, which adjusts doses based on tumor shrinkage or changes in patient anatomy, is a promising area of study. Researchers are also working to refine dose distribution, especially for tumors near critical organs like the liver, intestines, and kidneys. SBRT delivers highly focused, high-dose radiation in a small number of sessions. It is increasingly used for treating primary and metastatic GI tumors, particularly in the liver and pancreas, offering shorter treatment times and reduced exposure of healthy tissues to radiation [19]. Advancing SBRT accuracy is crucial for tumors in challenging locations, such as the pancreas. Combining SBRT with therapies like immunotherapy or chemotherapy may improve outcomes for metastatic GI cancers. Further studies on optimal fractionation schemes and long-term effects on healthy tissues will refine its role in treatment. Ongoing studies are evaluating the cost-effectiveness of proton therapy and its role in treating locally advanced or recurrent GI tumors. Additionally, researchers are improving treatment planning methods and investigating how they may be combined with other therapies like immunotherapy and chemotherapy [56]. Real-time imaging and tumor tracking during radiotherapy enhance precision, especially for moving organs like the stomach, liver, and pancreas. Technologies such as cone-beam computed tomography (CBCT) and magnetic resonance imaging (MRI) are being integrated into treatment planning and delivery. Innovations in real-time imaging aim to improve tumor tracking and account for organ motion, such as respiratory motion in abdominal tumors. MRI-guided radiotherapy, which provides superior soft tissue visualization and real-time tumor response assessment, is an area of significant interest [18, 49].

Traditional radiotherapy is limited by variability in patient responses. Some tumors are highly sensitive to radiation, while others are resistant due to molecular alterations. Personalized approaches that integrate genetic and molecular profiling are critical to improving outcomes in GI cancer treatment. Radiogenomics examines the genetic factors that influence a patient's response to radiotherapy [78]. By identifying specific genetic mutations and variations, clinicians can customize treatment plans. Researchers are working to identify biomarkers that predict tumor response to radiation, along with patient-specific factors like DNA repair capacity and tumor microenvironment characteristics. These findings will enable the development of predictive models for more tailored treatment plans. Certain molecular pathways contribute to radiation resistance in GI tumors. For example, mutations in DNA repair genes like BRCA1/BRCA2 can reduce radiation effectiveness. Targeting these pathways to increase radiosensitivity is a promising strategy. There is ongoing research into creating radiosensitizers that target particular molecular pathways, like the PI3K-AKT-mTOR pathway [80]. To improve treatment effectiveness, clinical trials are looking into combining targeted medicines with radiation. Hypoxia, or low oxygen levels in the tumor microenvironment, is a major factor in radiation resistance. Hypoxic cells are less sensitive to radiation because of reduced oxygen-induced DNA damage. Strategies to address hypoxia could improve radiotherapy outcomes. Researchers are exploring hypoxia-targeting agents, such as tirapazamine, to sensitize hypoxic tumors. Other approaches, like normalizing blood vessels or increasing oxygen delivery to tumors, may also help overcome resistance [81].

Conflict of Interest None.

Funding NA

References

1. Kumar R, Rosati LM, Herman JM. Advances in radiation therapy for gastrointestinal cancers. In: Textbook of gastrointestinal oncology; 2019. p. 421–43.
2. Hajj C, Goodman KA. Role of radiotherapy and newer techniques in the treatment of GI cancers. J Clin Oncol. 2015;33(16):1737–44.
3. Smith GL, Smith BD. Radiation treatment in older patients: a framework for clinical decision making. J Clin Oncol. 2014;20:2669.
4. Bordry N, Astaras C, Ongaro M, Goossens N, Frossard JL, Koessler T. Recent advances in gastrointestinal cancers. World J Gastroenterol. 2021;27(28):4493.
5. Chong IY, Chau I. Is there still a place for radiotherapy in gastric cancer? Curr Opin Pharmacol. 2023;68:102325.
6. Berardi R, Morgese F, Rinaldi S, Torniai M, Mentrasti G, Scortichini L, et al. Benefits and limitations of a multidisciplinary approach in cancer patient management. Cancer Manag Res. 2020;12:9363.
7. Akino Y, Tohyama N, Akita K, Ishikawa M, Kawamura R, Kurooka M. Modalities and techniques used for stereotactic radiotherapy, intensity-modulated radiotherapy, and image-guided radiotherapy: a 2018 survey by the Japan Society of Medical Physics. Phys Med. 2019;64:182.
8. Hosseinzadeh E, Banaee N, Ali NH. Cancer and treatment modalities. Curr Cancer Ther Rev. 2017;13(1):17–27.
9. Cheng T, Peng R, Qu A, Wang H. High-dose rate endorectal brachytherapy for rectal cancer: a state-of-the-art review. Cancer Sci. 2023;114(11):4145–56.
10. Manem VSK, Taghizadeh-Hesary F. Advances in personalized radiotherapy. BMC Cancer. 2024;24(1):556.
11. Koka K, Verma A, Dwarakanath BS, Papineni RV. Technological advancements in external beam radiation therapy (EBRT): an indispensable tool for cancer treatment. Cancer Manag Res. 2022;11:1421–9.
12. Ren F, Li S, Zhang Y, Zhao Z, Wang H, Cui Y, et al. Efficacy and safety of intensity-modulated radiation therapy versus three-dimensional conformal radiation treatment for patients with gastric cancer: a systematic review and meta-analysis. Radiat Oncol. 2019;14(1):84.
13. Yeh CY. Integral dose comparison of VMAT and IMRT for rectal cancer treatment: a dosimetric analysis. Radiat Phys Chem. 2024;221:111736.
14. Hunte SO, Clark CH, Zyuzikov N, Nisbet A. Volumetric modulated arc therapy (VMAT): a review of clinical outcomes-what is the clinical evidence for the most effective implementation? Br J Radiol. 2022;95(1136):20201289.
15. Li Z, Zeng J, Wang Z, Zhu H, Wei Y. Dosimetric comparison of intensity modulated and volumetric arc radiation therapy for gastric cancer. Oncol Lett. 2014;8(4):1427–34.
16. Shi J, Yang F, Ju X, Huang Y, Jiang F. Comparative study on dosimetry of VMAT and IMRT in assisted radiotherapy after radical resection of rectal cancer. Oncol Lett. 2017;13(5):2971–4.
17. Nabavizadeh N, Simeonova AO, Waller JG, Romer JL, Monaco DL, Elliott DA, et al. Volumetric-modulated arc radiotherapy for pancreatic malignancies: Dosimetric comparison

with sliding-window intensity-modulated radiotherapy and 3-dimensional conformal radiotherapy. Med Dosim. 2014;39(3):256–60.
18. Shen J, Lian X, Guan Q, He L, Zhang F, Shen J. Neoadjuvant chemo-radiation using IGRT in patients with locally advanced gastric cancer. Curr Oncol. 2022;29(10):7450–60.
19. Shouman MA, Fuchs F, Walter F, Corradini S, Westphalen CB, Vornhülz M, et al. Stereotactic body radiotherapy for pancreatic cancer – a systematic review of prospective data. Clin Transl Radiat Oncol. 2024;45:100738.
20. Matsuo Y. Stereotactic body radiotherapy for hepatocellular carcinoma: a brief overview. Curr Oncol. 2023;30(2):2493–500.
21. Lavrova E, Garrett MD, Wang YF, Chin C, Elliston C, Savacool M, et al. Adaptive radiation therapy: a review of CT-based techniques. Radiol Imaging Cancer. 2023;5(4):e230011.
22. Yariv O, Elamir A, Al Mutar S, Kazmi S, Beg SM, Porembka M, et al. Personalized Ultrafractionated stereotactic adaptive radiotherapy (PULSAR) for the treatment of gastrointestinal cancers. Int J Radiat Oncol Biol Phys. 2024;120(2):e502–3.
23. Du H, Hao X, Lin B, Tang M, Wang D, Yang X, et al. ^{177} Lu-labeled Anticlaudin 6 monoclonal antibody for targeted therapy in esophageal cancer. J Nucl Med. 2025;66(3):377–84.
24. Alkatheeri A, Salih S, Kamil N, Alnuaimi S, Abuzar M, Abdelrahman SS. Nano-radiopharmaceuticals in colon cancer: current applications, challenges, and future directions. Pharmaceuticals. 2025;18(2):257.
25. Calistri S, Ottaviano G, Ubaldini A. Radiopharmaceuticals for pancreatic cancer: a review of current approaches and future directions. Pharmaceuticals. 2024;17(10):1314.
26. Jin Y, Jiang J, Mao W, Bai M, Chen Q, Zhu J. Treatment strategies and molecular mechanism of radiotherapy combined with immunotherapy in colorectal cancer. Cancer Lett. 2024;591:216858.
27. Velenik V, Ocvirk J, Oblak I, Anderluh F. Neoadjuvant cetuximab, capecitabine, and radiotherapy (RT) in locally advanced resectable rectal cancer: results of a phase II trial. JCO. 2009;27(15_suppl):e15029.
28. Corrò C, Dutoit V, Koessler T. Emerging trends for radio-immunotherapy in rectal cancer. Cancers (Basel). 2021;13(6):1374.
29. Yu S, Wang Y, He P, Shao B, Liu F, Xiang Z. Effective combinations of immunotherapy and radiotherapy for cancer treatment. Front Oncol. 2022; https://doi.org/10.3389/fonc.2022.809304.
30. Weiss T, Weller M, Guckenberger M, Sentman CL, Roth P. NKG2D-based CAR T cells and radiotherapy exert synergistic efficacy in glioblastoma. Cancer Res. 2018;78(4):1031–43.
31. Tharmalingam H, Hoskin PJ. The optimism surrounding stereotactic body radiation therapy and immunomodulation. Chin Clin Oncol. 2017;6(Suppl 2):S9.
32. Gunderson LL, Tepper JE. Clinical radiation oncology. 5th ed. Elsevier; 2020.
33. Huang R, Qiu Z, Zheng C, Zeng R, Chen W, Wang S, et al. Neoadjuvant therapy for locally advanced esophageal cancers. Front Oncol. 2022;12:734581.
34. Mehraj A, Baba AA, Khan B. Predictors of pathological complete response following neoadjuvant chemoradiotherapy for rectal cancer. J Cancer Res Ther. 2022;18(Supplement):S391–6.
35. Cameron MG, Kersten C, Vistad I, Fosså S, Guren MG. Palliative pelvic radiotherapy of symptomatic incurable rectal cancer – a systematic review. Acta Oncol. 2014;53(2):164–73.
36. Shinde A, Hou WH, Han C, Glaser S, Chao J, Woo Y, et al. Reduced acute and late toxicities with intensity-modulated radiation therapy compared to three-dimensional conformal radiation therapy in post-operative gastric cancer. J Radiat Oncol. 2019;8(1):73–80.
37. Fok M, Toh S, Easow J, Fowler H, Clifford R, Parsons J, et al. Proton beam therapy in rectal cancer: a systematic review and meta-analysis. Surg Oncol. 2021;38:101638.
38. Sartorius D, Blume ML, Fleischer JR, Ghadimi M, Conradi LC, De Oliveira T. Implications of rectal cancer radiotherapy on the immune microenvironment: allies and foes to therapy resistance and patients' outcome. Cancers (Basel). 2023;15(21):5124.
39. Baumann M, Krause M, Overgaard J. Radiotherapy in individualized cancer treatment. Nat Rev Cancer. 2016;16(4):234–49.

40. Taghizadeh-Hesary F. "Reinforcement" by tumor microenvironment: the seventh "R" of radiobiology. Int J Radiat Oncol Biol Phys. 2024;119(3):727–33.
41. Limkin EJ, Sun R, Dercle L. Promising radiomics features for early prediction of response to adaptive radiotherapy. Eur J Cancer. 2020;130:108–17.
42. Begg AC, Stewart FA, Vens C. Strategies to improve radiotherapy with targeted drugs. Nat Rev Cancer. 2011;11(4):239–53.
43. Hall EJ, Giaccia AJ. Radiobiology for the radiologist. 8th ed. Lippincott Williams & Wilkins; 2018.
44. Stojanovic-Rundic S, Marinkovic M, Stanojevic A, Gavrilovic D, Jankovic R, Maksimovic N, et al. Dose escalation in neoadjuvant chemoradiotherapy for rectal cancer: short-term efficacy and toxicity of VMAT–SIB vs. 3D-CRT. Medicina. 2025;61(3):483.
45. Badic B, Hatt M, Durand S, Jossic-Corcos CL, Simon B, Visvikis D, et al. Radiogenomics-based cancer prognosis in colorectal cancer. Sci Rep. 2019;9(1):9743.
46. Nguyen NP, Mohammadianpanah M, SunMyint A, Page BR, Vinh-Hung V, Gorobets O, et al. Immunotherapy and radiotherapy for older patients with locally advanced rectal cancer unfit for surgery or decline surgery: a practical proposal by the International Geriatric Radiotherapy Group. Front Oncol. 2024;14:1325610.
47. Shah MA, Altorki N, Patel P, Harrison S, Bass A, Abrams JA. Improving outcomes in patients with oesophageal cancer. Nat Rev Clin Oncol. 2023;20(6):390–407.
48. Chong LM, Wang P, Lee VV, Vijayakumar S, Tan HQ, Wang FQ, Yeoh TD, Truong AT, Tan LW, Tan SB, Senthil Kumar K. Radiation therapy with phenotypic medicine: towards N-of-1 personalization. Br J Cancer. 2024;131(1):1–10.
49. Dasnoy-Sumell D. Towards real-time image-guided radiotherapy: 3D patient-specific breathing motion models driven by 2D cine-MRI. Doctoral dissertation, UCL-Université Catholique de Louvain.
50. Joshi SS, Badgwell BD. Current treatment and recent progress in gastric cancer. CA Cancer J Clin. 2021;71(3):264–79.
51. Lucchini R, Panizza D, Colciago RR, Vernier V, Daniotti MC, Faccenda V, Arcangeli S. Treatment outcome and compliance to dose-intensified linac-based SBRT for unfavorable prostate tumors using a novel real-time organ-motion tracking. Radiat Oncol. 2021;16:1–9.
52. Lee J, Lin JB, Weng CS, Chen SJ, Chen TC, Chen YJ. Impact of reduced margin pelvic radiotherapy on gastrointestinal toxicity and outcome in gynecological cancer. Clin Transl Radiat Oncol. 2023;43:100671.
53. Diwanji TP, Mohindra P, Vyfhuis M, Snider JW III, Kalavagunta C, Mossahebi S, Yu J, Feigenberg S, Badiyan SN. Advances in radiotherapy techniques and delivery for non-small cell lung cancer: benefits of intensity-modulated radiation therapy, proton therapy, and stereotactic body radiation therapy. Transl Lung Cancer Res. 2017;6(2):131.
54. Clairambault J, Pouchol C. A survey of adaptive cell population dynamics models of emergence of drug resistance in cancer, and open questions about evolution and cancer. Biomath. 2019;8(1):23.
55. Bulens P, Thomas M, Deroose CM, Haustermans K. PET imaging in adaptive radiotherapy of gastrointestinal tumors. Q J Nucl Med Mol Imaging. 2018;62(4):385–403.
56. Bakker A, van der Zee J, van Tienhoven G, Kok HP, Rasch CR, Crezee H. Temperature and thermal dose during radiotherapy and hyperthermia for recurrent breast cancer are related to clinical outcome and thermal toxicity: a systematic review. Int J Hyperth. 2019;36(1):1023–38.
57. Araujo IK, Munoz-Guglielmetti D, Molla M. Radiation-induced damage in the lower gastrointestinal tract: clinical presentation, diagnostic tests and treatment options. Best Pract Res Clin Gastroenterol. 2020;48:101707.
58. Boustani J, Huguet F, Vendrely V. Practice-changing clinical trials in radiation oncology for gastrointestinal malignancies in 2021–2023. Cancer/Radiothérapie. 2023;27(8):768–77.
59. Bowen JM, Gibson RJ, Coller JK, Blijlevens N, Bossi P, Al-Dasooqi N, Bateman EH, Chiang K, de Mooij C, Mayo B, Stringer AM. Systematic review of agents for the management of

cancer treatment-related gastrointestinal mucositis and clinical practice guidelines. Support Care Cancer. 2019;27:4011–22.
60. Kimura T, Fujiwara T, Kameoka T, Adachi Y, Kariya S. The current role of stereotactic body radiation therapy (SBRT) in hepatocellular carcinoma (HCC). Cancers. 2022;14(18):4383.
61. Ikeda T, Toyama S, Harada T, Noma K, Hamada M, Kitagawa T. Effectiveness of prehabilitation during neoadjuvant therapy for patients with esophageal or gastroesophageal junction cancer: a systematic review. Esophagus. 2024;21(3):283–97.
62. Teriaca MA, Loi M, Suker M, Eskens FA, van Eijck CH, Nuyttens JJ. A phase II study of stereotactic radiotherapy after FOLFIRINOX for locally advanced pancreatic cancer (LAPC-1 trial): long-term outcome. Radiother Oncol. 2021;155:232–6.
63. Popek S, Tsikitis VL. Neoadjuvant vs adjuvant pelvic radiotherapy for locally advanced rectal cancer: which is superior? World J Gastroenterol. 2011;17(7):848–54.
64. Boustani J, Grapin M, Laurent PA, Apetoh L, Mirjolet C. The 6th R of radiobiology: reactivation of anti-tumor immune response. Cancers (Basel). 2019;11(6):860.
65. Chen L, Zhang R, Lin Z, Tan Q, Huang Z, Liang B. Radiation therapy in the era of immune treatment for hepatocellular carcinoma. Front Immunol. 2023;14:1100079.
66. Donlon NE, Power R, Hayes C, Davern M, Reynolds JV, Lysaght J. Radiation and immunotherapy in upper gastrointestinal cancers: the current state of play. Int J Mol Sci. 2021;22(3):1071.
67. Erlandsson J, Holm T, Pettersson D, Berglund Å, Cedermark B, Radu C, Johansson H, Machado M, Hjern F, Hallböök O, Syk I. Optimal fractionation of preoperative radiotherapy and timing to surgery for rectal cancer (Stockholm III): a multicentre, randomised, non-blinded, phase 3, non-inferiority trial. Lancet Oncol. 2017;18(3):336–46.
68. Krishnatry R, Gudi S, Ostwal V, Shrikhande S, Engineer R. P-201 using SBRT in treating oligorecurrences of carcinoma of stomach. Ann Oncol. 2022;33:S321.
69. Liu Q, Zhu Z, Chen Y, Deng J, Ai D, Liu Q, et al. Phase 2 study of stereotactic body radiation therapy for patients with Oligometastatic esophageal squamous cell carcinoma. Int J Radiat Oncol Biol Phys. 2020;108(3):707–15.
70. Jiang L, Ye Y, Feng Z, Liu W, Cao Y, Zhao X, et al. Stereotactic body radiation therapy for the primary tumor and oligometastases versus the primary tumor alone in patients with metastatic pancreatic cancer. Radiat Oncol. 2024;19(1):111.
71. Versteijne E, van Dam JL, Suker M, Janssen QP, Groothuis K, Akkermans-Vogelaar JM, Besselink MG, Bonsing BA, Buijsen J, Busch OR, Creemers GJ. Neoadjuvant chemoradiotherapy versus upfront surgery for resectable and borderline resectable pancreatic cancer: long-term results of the Dutch randomized PREOPANC trial. J Clin Oncol. 2022;40(11):1220–30.
72. van Hagen P, Hulshof MC, Van Lanschot JJ, Steyerberg EW, Henegouwen MV, Wijnhoven BP, Richel DJ, Nieuwenhuijzen GA, Hospers GA, Bonenkamp JJ, Cuesta MA. Preoperative chemoradiotherapy for esophageal or junctional cancer. N Engl J Med. 2012;366(22):2074–84.
73. Leong T, Smithers BM, Michael M, Haustermans K, Wong R, Gebski V, O'Connell RL, Zalcberg J, Boussioutas A, Findlay M, Willis D. Preoperative chemoradiotherapy for resectable gastric cancer. N Engl J Med. 2024;391(19):1810–21.
74. Smalley SR, Benedetti JK, Haller DG, Hundahl SA, Estes NC, Ajani JA, Gunderson LL, Goldman B, Martenson JA, Jessup JM, Stemmermann GN. Updated analysis of SWOG-directed intergroup study 0116: a phase III trial of adjuvant radiochemotherapy versus observation after curative gastric cancer resection. J Clin Oncol. 2012;30(19):2327–33.
75. Xu D, Li G, Li H, Jia F. Comparison of IMRT versus 3D-CRT in the treatment of esophagus cancer: a systematic review and meta-analysis. Medicine. 2017;96(31):e7685.
76. Cencioni C, Trestini I, Piro G, Bria E, Tortora G, Carbone C, Spallotta F. Gastrointestinal cancer patient nutritional management: from specific needs to novel epigenetic dietary approaches. Nutrients. 2022;14(8):1542.
77. Magee DJ, Jhanji S, Poulogiannis G, Farquhar-Smith P, Brown MR. Nonsteroidal anti-inflammatory drugs and pain in cancer patients: a systematic review and reappraisal of the evidence. Br J Anaesth. 2019;123(2):e412–23.

78. Lin L, Lin H, Zhou R, Liu B, Liu K, Jiang R. Surviving and thriving: assessing quality of life and psychosocial interventions in mental health of head and neck cancer patients. Asian J Surg. 2024;48:1634.
79. Fabi A, Bhargava R, Fatigoni S, Guglielmo M, Horneber M, Roila F, Weis J, Jordan K, Ripamonti CI. ESMO guidelines committee. Cancer-related fatigue: ESMO clinical practice guidelines for diagnosis and treatment. Ann Oncol. 2020;31(6):713–23.
80. Rubini D, Gagliardi F, Menditti VS, D'Ambrosio L, Gallo P, D'Onofrio I, Pisani AR, Sardaro A, Rubini G, Cappabianca S, Nardone V. Genetic profiling in radiotherapy: a comprehensive review. Front Oncol. 2024;14:1337815.
81. Kao TW, Bai GH, Wang TL, Shih IM, Chuang CM, Lo CL, Tsai MC, Chiu LY, Lin CC, Shen YA. Novel cancer treatment paradigm targeting hypoxia-induced factor in conjunction with current therapies to overcome resistance. J Exp Clin Cancer Res. 2023;42(1):171.

Chapter 10
Recent Advances in Radiotherapy of Lung Cancer

Maria Protopapa, Myrsini Balafouta, Georgia Lymperopoulou, Andreas Koulouris, and Maria Tolia

Abstract Modern radiotherapy techniques achieve better tumor coverage and greater sparing of organs at risk (OARs), reinforcing the role of radiotherapy in the treatment of lung cancer. Intensity-modulated radiation therapy (IMRT) and volumetric modulated arc therapy (VMAT) deliver radiation to the targeted volumes with better conformity compared to 3D conformal radiotherapy (3DCRT). Coupled with image-guided radiation therapy (IGRT), IMRT and VMAT ensure high precision in radiotherapy delivery. Motion and breathing management techniques have found a particularly useful application in early lung cancer treatment with the use of stereotactic body radiotherapy (SBRT). These techniques enable the irradiation of a target that continuously moves during breathing, while limiting excess irradiation of the healthy lung parenchyma, permitting the delivery of high, stereotactic doses to the lung primary tumor. SBRT improves rates of local control (LC) and overall survival (OS) in inoperable stage I non-small cell lung cancer (NSCLC) compared to conventional radiotherapy. SBRT is an established alternative to surgery, in early lung cancer. Radiotherapy can also be used with curative intent postoperatively for selected patients, as part of a radical chemoradiation approach for locally advanced disease, and as a metastasis-directed treatment for selected oligometastatic patients.

M. Protopapa
Department of Radiotherapy and Stereotactic Radiosurgery, Mediterraneo Hospital, Athens, Greece

M. Balafouta
Department of Biomedical Science, University of West Attica, Athens, Greece

G. Lymperopoulou
1st Department of Radiology, Medical School, National and Kapodistrian University of Athens, Athens, Greece

A. Koulouris
Department of Oncology-Pathology, Karolinska Institutet/Thoracic Oncology Center, Karolinska University Hospital, Stockholm, Sweden

M. Tolia (✉)
Department of Radiotherapy, School of Medicine, University of Crete, Heraklion, Greece
e-mail: mariatolia@uoc.gr

F. Taghizadeh-Hesary et al. (eds.), *From Radiobiology to Radiation Oncology*,
https://doi.org/10.1007/978-3-031-95724-6_10

The synergy of modern radiotherapy with immunotherapy has increased the OS of locally advanced lung cancer patients, and its combination with systemic treatments is currently being further investigated. In the palliative setting, advanced radiotherapy techniques offer improved quality of life. Radiation oncologists are an integral part of the multidisciplinary team for the management of lung cancer at all stages of the disease.

Keywords Stereotactic body radiotherapy (SBRT) · Lung cancer · Radiotherapy · Intensity-modulated radiation therapy (IMRT) · Volumetric modulated arc therapy (VMAT)

10.1 Introduction

Lung cancer is the most prevalent and morbid malignant disease worldwide. It is responsible for almost one in eight new cancer diagnoses each year, and for a fifth of all cancer-related deaths. The incidence of lung cancer steadily drops in developed countries, mainly following the decline in smoking, the main driver of carcinogenesis of lung cancer [11]. Advances in radiotherapy, surgery, and chemotherapy, and the introduction of immunotherapy have improved the survival of patients with lung cancer over the last years [82]. Still, the prognosis remains dismal, with the majority of patients dying of their disease [77].

Radiotherapy is part of the treatment for most lung cancer patients. In the very early stages, stereotactic treatments of the lung can primarily achieve durable local control and OS, being an alternative to surgery, especially for medically inoperable patients. Locally advanced disease is optimally addressed with concurrent chemoradiotherapy and immunotherapy maintenance. Radiotherapy has a role in the metastatic setting both as a metastasis-directed treatment for oligometastatic disease (to prolong survival and gain time off systemic treatments) and as a palliative treatment both at the lung and at distal sites of the disease. Recently, systemic treatments, namely tyrosine kinase inhibitors (TKIs) and immune checkpoint inhibitors (ICIs), have revolutionized lung cancer treatment for the subset of patients responsive to these agents. In this changing landscape, the role of local treatments is enhanced, as the scope of potentially curative disease broadens [86].

10.2 Modern Radiotherapy Techniques

10.2.1 Planning Techniques

The aim of radiation therapy is to cover the planning target volume (PTV), comprising the clinical target volume (CTV) plus a margin to compensate for geometrical uncertainties, with the prescribed dose, while sparing organs at risk (OARs), i.e., healthy organs, from clinically significant radiation dose, protecting them from acute or long-term adverse effects [35]. Three-dimensional conformal radiation therapy (3DCRT) development relies on technical advances in linear accelerators. Multileaf collimators (MLCs), with each leaf moving independently, were a key step toward better conformity of the radiation beam to the target volume [20]. Modern radiotherapy relies on computational optimization techniques that enable the formation of complex, intensity-modulated fields. With the use of intensity-modulated radiotherapy (IMRT), a very high coverage of the PTV and a steep dose fall-off toward the directions of important OARs can be achieved [35]. The main dose-limiting OARs in lung cancer radiotherapy include the lungs, heart, esophagus, and spinal cord. The volume of OARs irradiated with clinically significant doses has been markedly reduced with IMRT [5].

Secondary analyses comparing IMRT to 3DCRT are based on the dose-escalation RTOG0617 RCT where the patient population was stratified by radiotherapy technique. This showed significantly improved quality of life (QOL) outcomes [52], lesserious pneumonitis (7.9% vs 3.5%, $p = 0.039$), and lower heart doses ($p < 0.05$) with IMRT [16]. With a median [IQR] follow-up of 5.2 [4.8–6.0] years, IMRT resulted in significantly less grade 3 or higher pneumonitis (8 [3.5%] vs 21 [8.2%]; $p = 0.03$) and in significantly decreased 20% or more volume of the heart receiving 40Gy (V40 ≥ 20%) (16.5% vs 20.5%; $p < 0\ 0.001$). The latter was associated with better OS in multivariate analysis [17]. A considerable 10% reduction in acute esophagitis has also been observed in the years following the adoption of modern radiotherapy techniques [5]. The most important long-term complication of RT, radiation-induced lung fibrosis (occasionally leading to hospitalization), is also diminished with modern techniques [44].

10.2.2 Imaging Techniques

The achievement of precise target volume irradiation with modern planning techniques is mainly ensured through imaging. In newer linear accelerators, verification of treatment delivery is done by means of a cone-beam CT (CBCT), a three-dimensional image resembling that of a CT scan, depicting both bones and soft tissue [19]. A CBCT can also detect major anatomic changes, such as atelectasis formation and body weight loss, that result in clinically unacceptable dose distribution differences, prompting the need to repeat a CT simulation to replan the patient with the new anatomy (Adaptive Radiotherapy) [28]. Even better soft tissue visualization has been made possible with linear accelerators that use an MRI scanner as imaging instead of a CT scanner. Thoracic MRI is seldom used in lung cancer delineation, but it could help define the extent of chest wall invasion, the presence of mediastinal lymph nodes metastasis, and the closeness with structures such as proximal bronchial tree and brachial plexus, which are commonly of concern in stereotactic ablative treatments. More frequent use of SBRT and MRI LINACS may increase the use of MRI in lung cancer radiotherapy in the future [15].

10.2.3 Motion and Breathing Management Techniques

The technology implemented to address issues of intrafraction and interfraction movement falls under the umbrella of image-guided radiotherapy (IGRT). IGRT encompasses all forms of ionizing and nonionizing radiation that monitor positioning and movement. Commonly, CBCT is performed prior to treatment to verify correct patient positioning and relative position of lung tumor and OARs. During treatment, there are different systems to monitor patient movement, usually surface-guided radiotherapy (SGRT), tracking body surface movements with nonionizing radiation, to prompt a new CBCT if the motion detected exceeds a prespecified accepted range of motion [18, 45].

Gating is a method of limiting treatment only at a particular phase of the breathing cycle. In deep inspiration breath hold (DIBH) technique, irradiation is delivered only at a narrow window during inspiration [12]. Another gating technique is based on a 4D-CT, a CT scan that incorporates a time factor based on the patient's breathing cycle, generating different sets of images, each corresponding to the different phases of a breathing cycle. The kinetics of the tumor during a typical breathing cycle are visualized to guide the decision on treatment to the whole respiratory cycle within certain upper and lower limits (amplitude gating) or to a specific respiratory phase (phase gating) for limited and significant tumor mobility, respectively. Tracking refers to the continuous irradiation of the moving target, predicting its anticipated position real time by means of mathematical models. These models are based on data from the direct tracking of the tumor itself or of fiducials placed in or near the tumor, or from tracking external surrogates through the use of a series of radiographic images acquired by two orthogonal cameras [8, 12].

10.3 Curative Intent Radiotherapy

10.3.1 Stereotactic Radiotherapy for Early Lung Cancer

SBRT, also referred to as stereotactic ablative radiotherapy (SABR), is a high-precision radiotherapy technique that delivers ablative doses to the target with a very steep dose gradient to spare neighboring OARs from clinically relevant dose. Its precision is based on all the planning, imaging, and motion and breathing management advances of modern radiotherapy, combined with robust immobilization systems that result in highly reproducible patient positioning. Ablative radiotherapy prescribes doses with a biological effective dose (BED) > $100Gy_{10}$. Tumors treated with SBRT are usually less than 5 cm in their largest diameter to limit excessive healthy lung tissue radiation exposure and, therefore, avoid high rates of radiation pneumonitis [32]. Multiple doses and fractionations are used in lung SBRT. Typical prescribed doses for peripheral lung tumors range from 50 to 60 Gy in 3 to 5 fractions. Central lung tumors, i.e., tumors within 2 cm of the proximal bronchial tree, or abutting mediastinal structures, which are considered more challenging, and prescriptions include 50Gy in 5 fractions and 60Gy in 8 fractions [92]. Single-fraction SBRT is also an option for patients with small, peripheral lesions, as was shown during the COVID-19 pandemic period [96].

In lung cancer, SBRT is an established alternative to surgery for all early-stage patients who are poor surgical candidates or refuse surgery. The first prospective robust evidence that SBRT can improve local control over conventional radiotherapy was based on the results of the phase 2 RTOG 0236 trial on early-stage inoperable NSCLC. At 3 years, primary tumor control was 97.6% (95% confidence interval [CI], 84.3–99.7%) and locoregional control 87.2% (95% CI, 71.0–94.7%). At 5 years, disease-free survival (DFS) and OS were 25.5% (95% CI, 14.9–37.4%) and 40.0% (95% CI, 27.1–52.5%), respectively. Adverse events (AEs) ≥grade 3, mainly radiation pneumonitis and rib fractures, were reported in almost a third of the patients, but there were no lethal events and only rare cases of grade 4 AE [79, 80]. These results were also confirmed by the JCOG0403 phase II trial, which also included operable patients. The published 3, 5, and 10-year LC rates were 85.4% (95% CI: 73.8–92.1%), 85.4% (95% CI: 73.8–92.1%), and 85.4% (95% CI: 73.8–92.1%), respectively. Grade 3 AEs were 12.3%, with no grade 4 or 5 AEs [53].

SBRT is under investigation as an alternative to surgery for operable patients. Lobectomy remains the mainstay of treatment for most stage I NSCLC. Segmentectomy is reserved for lung tumors up to 2 cm with a consolidation-to-tumor ratio >0.5. Nonanatomic resections have an inferior oncologic outcome and are reserved for patients who are not fit for an anatomic resection [34, 65]. Two randomized controlled trials (RCTs), ROSEL and STARS, comparing surgery with SBRT for operable stage I NSCLC patients were terminated early because of poor accrual. A pooled analysis of these trials combined showed a significant 3-year OS benefit in the SBRT group compared to the surgical group (79% vs 95%, $p = 0.037$), with no significant difference in recurrence-free survival, but higher clinically significant toxicity in the surgery group. Even if the pooled analysis was underpowered, it demonstrated that SBRT is a promising technique that merits further investigation [13]. STABLE-MATES (NCT02468024) and VALOR (NCT02984761) are ongoing trials comparing SBRT to sublobar and anatomical resection, respectively.

10.3.2 Definitive Chemoradiotherapy for NSCLC

Radical concurrent chemoradiotherapy (CRT) with durvalumab maintenance is the preferred treatment for unresectable lung tumors. Standard dose is 60–66Gy delivered with conventional fractionation in a continuous regimen, as was established by the Radiation Therapy Oncology Group (RTOG) 7301 almost half a century ago. OS was very disappointing at the time, with a two-year survival rate of 25% for those treated with the established regimen [58]. In the modern era, patients should be treated by making use of all available techniques to respect dose constraints to OARs, and IGRT, including motion and breathing management techniques, to ensure accurate radiation delivery, aiming to minimize toxicity. With current radiotherapy techniques and systemic treatments, the five-year OS of unresectable stage III NSCLC patients treated with concurrent chemoradiotherapy either alone or with durvalumab maintenance has risen to 32.1% and 42.9%, respectively [9, 73].

Radical RT should always be delivered based on an FDG-PET, a functional imaging technique, used both for staging and for delineation purposes. The use of FDG-PET is indispensable in staging lung cancer, resulting in upstaging and downstaging in almost half of the patients, which often leads to changes in their management [40, 69]. Because of a sensitivity of 58–94% and a specificity of 76–96% for mediastinal lymph node staging, selectively targeting FDG-avid lymph nodes, a method called involved-field radiotherapy (IF-RT), has been made possible [67]. Prophylactic mediastinal irradiation, referred to as elective nodal irradiation (ENI), the standard RT technique before the generalized use of FDG-PET, has been abandoned as it is associated with a statistically significant increase in ≥grade 4 toxicities, while IF-RT does not compromise the oncological outcomes [24, 66]. FDG-PET can also be used to differentiate tumor from atelectasis and help in primary tumor delineation. The safety and efficacy of FDG-PET-based IF-RT have been shown in an open-label, multicenter RCT, where a tendency toward improved LC was shown in the IF-RT arm in spite of the more restricted PTV [54]. Efforts to improve results of CRT with induction chemotherapy, dose escalation, altered fractionation, or preoperative radiotherapy have all failed. There is no level 1 evidence in support of hypofractionation for the radical treatment of locally advanced lung cancer as all RCTs were not powered for equivalence and were negative for superiority [97]. Evidence is even more scarce on accelerated hyperfractionation, with only one RCT, which failed to prove a benefit from the experimental arm [98]. A phase III RCT comparing radical CRT with preoperative CRT followed by resection for N2 tumors showed no difference in OS (HR 0.87, CI 0.70–1.10, $p = 0.24$), despite a progression-free survival (PFS) benefit in the surgical arm. As pneumonectomy patients had a high treatment-related mortality rate, an exploratory analysis comparing lobectomy with radical CRT was undertaken. In this matched cohort analysis, patients treated with the combination of preoperative CRT and lobectomy had an OS benefit over patients treated with definitive CRT [1].

In the modern era of 3DCRT, hopes for better results with dose-escalation were revived, but the RTOG 0617 trial, a phase III study was also negative, with worse survival rates in the dose-escalation group (HR 1·38, 95% CI 1·09–1·76; $p = 0{\cdot}004$) [10]. With a median follow-up of 5 years, there was a statistically significant increase in ≥grade 3 dysphagia and in esophagitis, without difference in radiation pneumonitis. In multivariate analysis, OS was improved with standard prescription dose, tumor location in the upper lobes, institutions with higher accrual volume, decreased esophagitis and dysphagia, lower planning target volume, and lower heart V5 [9]. A further negative dose escalation trial, RTOG 1106, incorporated a mid-therapy FDG-PET to monitor response and dose-escalated in the interventional arm only in those areas that had FDG-avid remnant disease. This PET-guided dose-escalation approach was safe but did not improve results [43]. A phase II trial of a mid-treatment PET-guided approach has shown better LC results in the experimental arm compared to the standard treatment, and this approach merits further investigation [85].

10.3.3 Adjuvant Radiotherapy for NSCLC

Complete surgical resection is the gold standard for all resectable locally advanced NSCLC patients. Resectability of lung cancer and medical operability of the patient should always be determined by a thoracic surgeon experienced in lung cancer. In general, tumors smaller than T4 with up to single-station N2 disease and T4 tumors with N0–1 lymph nodes are considered operable [36]. Even after complete resection, patients present a high risk of locoregional recurrence. Perioperative systemic treatment is recommended for all stage IIB-III patients and can be considered for stage IB-IIA patients with high-risk features [91].

Adjuvant radiotherapy is not supported by robust evidence, especially for N0–1 disease. As for N2 disease, the significance of adjuvant radiotherapy is still a matter of debate, but it cannot be recommended for all N2 patients. A meta-analysis of 11 clinical trials published in 1998 correlated postoperative radiotherapy with a worse survival outcome [61]. Possible explanations included the presence of low-risk stage I-II patients who would not benefit much from adjuvant treatments, the outdated radiotherapy techniques of the time that had significant cardiac and pulmonary toxicities, and the absence of systemic treatment which led to a high rate of distant metastases.

Three RCTs were initiated in the modern era to address these limitations using modern RT techniques and limiting the trial population to stage III (N2) patients. Two phase III RCTs found no clinically significant improvement in disease-free survival (DFS) or OS with postoperative radiotherapy in a population of stage IIIA (N2) patients after complete resection. Patients were prescribed 50–54 Gy with conventional fractionation and were treated with IMRT and 3DCRT techniques, but IGRT was performed only if available. In the Lung-ART trial, where patient selection was based on FDG-PET to better exclude more advanced disease, a nonsignificant increase in DFS of 15% was observed. This tendency of an increase in DFS shows that PORT with the use of modern RT techniques can potentially improve the outcome for carefully selected patients [37, 50, 57]. A third trial was closed early due to slow accrual [59]. A meta-analysis of all three RCTs confirms the lack of benefit from adjuvant RT in N2 disease, even though there is a decrease in local recurrences (odds ratio = 0.56, 95% confidence interval [CI]: 0.40–0.76) [89]. The debate is still not closed as a current meta-analysis of over 21,000 patients from 33 studies shows improvement not only in LR but also in OS with the use of PORT with IMRT and 3DCRT [94].

10.3.4 SCLC

Small cell lung cancer (SCLC) comprises about 11% of lung cancer diagnoses and is more aggressive than NSCLC, with two-thirds of new cases being diagnosed with distant metastases [68]. Although TNM staging is also used, the most common classification system for SCLC is the Veterans Administration Lung Study Group

(VALSG), which separates patients into limited-stage (LS-SCLC), with a median OS of 25–30 months, and extensive-stage (ES-SCLC), with a median OS of less than 13 months [25]. Initially, when radiation was the primary treatment of SCLC, this distinction was based upon the possibility of encompassing all disease sites within one radiation portal. In the modern era of concurrent systemic treatments, this simplified staging is still used for its practicality in guiding treatment management as it distinguishes between patients treated with curative intent or not [23, 39]. Chemoradiotherapy is the standard of care for LS-SCLC. The addition of radiotherapy to platinum-based chemotherapy, usually combined with etoposide, has been shown to improve OS and LC [60, 90]. Once-daily conventional fractionation is usually prescribed, although RCTs support the use of hyperfractionated accelerated radiotherapy [2, 41]. A twice-daily TRT regimen of 45Gy in 30 fractions confers a statistically significant survival benefit over conventional fractionation of 45Gy in daily fractions of 1.8Gy, both given concurrently with chemotherapy [81], albeit at the expense of a moderate grade 3 esophagitis increase. In the era of modern radiotherapy techniques, two RCTs failed to prove the superiority of a higher dose delivered daily over the same hyperfractionated regimen. Although not powered to assess noninferiority, oncological outcomes and toxicity were comparable between the conventional and the hyperfractionated groups in both studies [7, 22]. A phase II RCT suggests that a higher dose of 60Gy, delivered twice-daily, can offer a significant survival advantage, without an increase in adverse events [30]. PET-CT-based staging and contouring limit toxicity rates, with the same oncological outcomes [29].

Surgery, for long considered to have no place in SCLC, is gaining grounds for the few patients diagnosed with very early SCLC, especially for stage I disease, along with adjuvant chemotherapy, as it confers a significant survival advantage over chemoradiotherapy [31, 74, 87]. CCRT and SBRT can be used as alternative treatments for stage I patients when medically inoperable and when complete resection is at stake. Considerable variability in practice patterns is observed, with surgery being preferred for peripheral and T1 tumors [48]. A multidisciplinary tumor board should decide the treatment management of each patient. Postoperative radiotherapy is recommended in case of pN2 disease and in incomplete resections [95]. CCRT is the treatment of choice for LS-SCLC from stage II and above. Immunotherapy plus EP is currently the first-line treatment for ES-SCLC [14, 27, 47, 64]. The addition of TRT of 30Gy in ten fractions decreases progression (HR 0.73, $p = 0.001$) and improves the two-year OS from 3% without TRT to 13% ($p = 0.004$) [72].

10.3.5 PCI

Both NSCLC and SCLC have a high propensity to metastasize the brain. In locally advanced NSCLC, prophylactic cranial irradiation (PCI) has been largely abandoned following the results of several RCTs that showed decrease in brain metastasis (BM)-free survival, but with no OS improvement [62]. In the current era of ICI and TKIs with substantial intracranial activity, the incidence of BM is significantly lower, further supporting PCI omission. On the contrary, LS-SCLC patients in complete remission after chemoradiotherapy receive prophylactic cranial irradiation, as it lowers the risk of developing brain metastases (BM) and improves OS. Yet, the impact of PCI in patients' quality of life and neurocognitive function has not been adequately studied [3, 49]. Similarly, a study by Slotman et al. showed that PCI prevents BM and prolongs OS also in ES-SCLC patients responding to chemotherapy [71]. More recent studies introducing MRI surveillance in the observation group failed to confirm a BM risk reduction benefit from PCI in ES-SCLC patients, as well as in patients with a complete response to chemotherapy [55, 75, 78]. Modern radiotherapy techniques can selectively avoid high doses to the hippocampi with Hippocampal Avoidance PCI (HA-PCI), which can possibly preserve memory function, as the irradiation of the hippocampi is known to have a detrimental effect on cognition. Results from clinical trials on HA-PCI are promising regarding its safety, as there was no increase in BM incidence, but contradictory regarding its effect on neurocognition [6, 63].

10.4 Radiotherapy for Stage IV Patients

10.4.1 Radiotherapy for Oligometastatic Patients

According to a multidisciplinary consensus report, synchronous oligometastatic NSCLC is defined as a metastatic disease (excluding mediastinal lymph nodes) of up to five metastases to three sites maximum. It is considered that this state represents an entity in between locoregional disease and polymetastatic state. All disease sites should be treated with radical intent in order to achieve long-term disease control [21]. A multicenter phase III RCT from China is the first large-scale trial to show an LC and OS benefit from the addition of thoracic radiotherapy to standard TKI treatment compared to TKIs alone for oligometastatic EGFR mutant NSCLC patients. All patients in the experimental arm were treated with 60Gy to the lung primary and positive regional lymph nodes, while metastasis-directed radiotherapy was at the discretion of the treating physician. Both PFS (hazard ratio [HR], 0.57; $P = 0.004$) and OS (HR, 0.62; $P = 0.029$) were statistically improved from thoracic radiotherapy (TRT) addition [76]. On the other hand, a phase II/III trial, NRG-LU002, randomized NSCLC patients with up to three metastases and stable disease after four cycles of ICI, to either continue with maintenance therapy alone or together with local consolidative therapy, surgery and/or radiotherapy, and it was closed after the planned interim analysis, which found an increase in toxicity without a PFS or OS benefit [38]. These two trials had two major differences, the site, primary or metastatic, treated by radiotherapy, and the use of different systemic agents, which could explain their different outcomes.

10.4.2 Palliative Radiotherapy

There is a changing landscape concerning palliation in cancer patients as metastatic patients survive longer. Goals of treatment include durable and rapid palliation and limiting systemic treatment breaks. The possibilities for retreatment should also be considered. Advanced radiation techniques, such as SBRT, are appropriate if these goals are achieved without compromising safety [83]. All patients should be considered for palliative treatments at symptomatic sites. Based on expert consensus, TRT should not be offered to stage IV lung cancer patients unless for palliation. Hypofractionated radiotherapy should be administered concurrently with chemotherapy for stage III incurable patients who are fit for chemotherapy, have a PS 0–2, and a prognosis of more than 3 months [51].

As lung cancer treatments improve patients' survival, the incidence of BM increases, as the brain-blood barrier hinders the penetration of most systemic agents to the brain parenchyma. Up to 50% of patients will develop BM, weighing heavily upon patients' prognosis and quality of life. TKIs and ICI show promising intracranial activity for lung cancer patients. Still, local treatments, surgery and, mainly, radiotherapy and stereotactic radiosurgery (SRS), remain the standard treatment modalities for intracranial metastases. SRS gains ground over WBRT for limited intracranial metastases on the basis of neurocognitive improvement without compromising OS [62].

10.5 Synergy of Radiotherapy with Systemic Treatments

There is level 1 evidence that concurrent CRT can significantly improve outcomes over sequential CRT for NSCLC unresectable stage III patients, albeit at the expense of an acceptable toxicity increase, and, therefore, should be offered to all patients fit to receive the concurrent regimen [4]. Sequential CRT is preferred over RT alone, improving the OS [20]. All patients who respond to concurrent or sequential CRT should then be offered consolidation therapy with durvalumab, mainly in PD-L1-positive patients, or osimertinib, a TKI, in EGFR-mutant patients, following the results of the phase III RCTs PACIFIC and LAURA, respectively [91]. The main concern with the combination of ICI with radiotherapy is an increase in symptomatic pneumonitis, which does not compromise oncological outcomes [84]. Similarly, maintenance with durvalumab after radical concurrent chemoradiotherapy for LS-SCLC is the standard of care as it improves OS and PFS, according to the ADRIATIC phase III RCT [15]. Both once- and twice-daily radiotherapy fractionations were allowed in the ADRIATIC trial, with no significantly different impact on outcomes (HR, 0.68 vs 0.72 for twice-daily vs once-daily) [56].

10.6 Future Directions

The future of lung cancer radiotherapy lies in novel ways to minimize toxicity, enabling dose escalation. Combination of SBRT to the primary and standard dose IMRT to regional lymph nodes is studied in ongoing trials, such as NRG-LU008. SBRT is used as a means of dose escalation to the primary, the most common site of treatment failure, while potentially limiting toxicity as it is a more conformal technique [70]. Proton treatments in lung cancer have a theoretical advantage because of the steep dose fall-off that characterizes protons. Intensity-modulation techniques have also been applied to proton treatments (IMPT), and their comparative benefit to photon IMRT is the subject of an ongoing trial, RTOG 1308 [26]. Ultra-central early lung cancer patients are at increased toxicity risks when treated with SBRT. Future improvements are awaited with the use of IMPT, MR-LINACS, and online adaptive radiotherapy [93].

One suggested approach to minimize radiation pneumonitis is by avoiding irradiation of functional areas of the lung based on functional imaging techniques, and by incorporating predictive models of radiation pneumonitis for the different lung areas [99, 100]. Techniques to avoid excessive esophagitis and decrease radiation-induced heart disease are actively being investigated [46, 88]. FLASH-RT, delivering radiotherapy at an ultra-high dose rate, promises to revolutionize radiotherapy by significantly reducing normal tissue toxicity, and a first in-human clinical trial is underway [42]. Novel combinations in the multidisciplinary management of lung cancer are also investigated to improve outcomes. The incorporation of radiation therapy in the neoadjuvant setting concurrently with immunotherapy to facilitate clear margin resection of stage III NSCLC is being investigated in an ongoing phase II clinical trial [33]. Lung cancer still poses a great challenge and further advances in all related disciplines of medicine are required. The future lies in a multidisciplinary approach focused on better patient selection to personalize treatments based not only on staging, but also on molecular and imaging biomarkers and comorbidities.

References

1. Albain KS, Swann RS, Rusch VR, Turrisi AT, Shepherd FA, Smith C, Chen Y, Livingston RB, Feins R, Gandara DR, Fry WA, Darling G, Johnson DH, Green MR, Miller RC, Ley J, Sause WT, Cox JD. Radiotherapy plus chemotherapy with or without surgical resection for stage III non-small cell lung cancer. Lancet. 2009;374(9687):379–86. https://doi.org/10.1016/S0140-6736(09)60737-6.
2. Allen SG, Dragovic AF, Yin HM, Bryant AK, Paximadis PA, Matuszak MM, Schipper MJ, Dess RT, Hayman JA, Dominello MM, Kestin LL, Movsas B, Jolly S, Bergsma DP, Michigan Radiation Oncology Quality Consortium. Prospective evaluation of limited-stage small cell lung cancer radiotherapy fractionation regimen usage and acute toxicity in a large statewide quality collaborative. Pract Radiat Oncol. 2023;13(5):444–53. https://doi.org/10.1016/j.prro.2023.04.007.

3. Arriagada R, Le Chevalier T, Borie F, Rivière A, Chomy P, Monnet I, Tardivon A, Viader F, Tarayre M, Benhamou S. Prophylactic cranial irradiation for patients with small-cell lung cancer in complete remission. J Natl Cancer Inst. 1995;87(3):183–90. https://doi.org/10.1093/jnci/87.3.183.
4. Aupérin A, Le Péchoux C, Rolland E, Curran WJ, Furuse K, Fournel P, Belderbos J, Clamon G, Ulutin HC, Paulus R, Yamanaka T, Bozonnat M-C, Uitterhoeve A, Wang X, Stewart L, Arriagada R, Burdett S, Pignon J-P. Meta-analysis of concomitant versus sequential radiochemotherapy in locally advanced non-small-cell lung cancer. J Clin Oncol Off J Am Soc Clin Oncol. 2010;28(13):2181–90. https://doi.org/10.1200/JCO.2009.26.2543.
5. Baker S, Dahele M, Lagerwaard FJ, Senan S. A critical review of recent developments in radiotherapy for non-small cell lung cancer. Radiat Oncol (London). 2016;11(1):115. https://doi.org/10.1186/s13014-016-0693-8.
6. Belderbos JSA, De Ruysscher DKM, De Jaeger K, Koppe F, Lambrecht MLF, Lievens YN, Dieleman EMT, Jaspers JPM, Van Meerbeeck JP, Ubbels F, Kwint MH, Kuenen MA, Deprez S, De Ruiter MB, Boogerd W, Sikorska K, Van Tinteren H, Schagen SB. Phase 3 randomized trial of prophylactic cranial irradiation with or without hippocampus avoidance in SCLC (NCT01780675). J Thorac Oncol. 2021;16(5):840–9. https://doi.org/10.1016/j.jtho.2020.12.024.
7. Bogart J, Wang X, Masters G, Gao J, Komaki R, Gaspar LE, Heymach J, Bonner J, Kuzma C, Waqar S, Petty W, Stinchcombe TE, Bradley JD, Vokes E. High-dose once-daily thoracic radiotherapy in limited-stage small-cell lung cancer: CALGB 30610 (alliance)/RTOG 0538. J Clin Oncol Off J Am Soc Clin Oncol. 2023;41(13):2394–402. https://doi.org/10.1200/JCO.22.01359.
8. Botticella A, Levy A, Auzac G, Chabert I, Berthold C, Pechoux CL. Tumour motion management in lung cancer: a narrative review. Transl Lung Cancer Res. 2021;10(4):10.21037/tlcr-20-856.
9. Bradley JD, Hu C, Komaki RR, Masters GA, Blumenschein GR, Schild SE, Bogart JA, Forster KM, Magliocco AM, Kavadi VS, Narayan S, Iyengar P, Robinson CG, Wynn RB, Koprowski CD, Olson MR, Meng J, Paulus R, Curran WJ, Choy H. Long-term results of NRG oncology RTOG 0617: standard- versus high-dose chemoradiotherapy with or without cetuximab for unresectable stage III non-small-cell lung cancer. J Clin Oncol Off J Am Soc Clin Oncol. 2020;38(7):706–14. https://doi.org/10.1200/JCO.19.01162.
10. Bradley JD, Paulus R, Komaki R, Masters G, Blumenschein G, Schild S, Bogart J, Hu C, Forster K, Magliocco A, Kavadi V, Garces YI, Narayan S, Iyengar P, Robinson C, Wynn RB, Koprowski C, Meng J, Beitler J, et al. Standard-dose versus high-dose conformal radiotherapy with concurrent and consolidation carboplatin plus paclitaxel with or without cetuximab for patients with stage IIIA or IIIB non-small-cell lung cancer (RTOG 0617): a randomised, two-by-two factorial phase 3 study. Lancet Oncol. 2015;16(2):187–99. https://doi.org/10.1016/S1470-2045(14)71207-0.
11. Bray F, Laversanne M, Sung H, Ferlay J, Siegel RL, Soerjomataram I, Jemal A. Global cancer statistics 2022: GLOBOCAN estimates of incidence and mortality worldwide for 36 cancers in 185 countries. CA Cancer J Clin. 2024;74(3):229–63. https://doi.org/10.3322/caac.21834.
12. Caillet V, Booth JT, Keall P. IGRT and motion management during lung SBRT delivery. Phys Med. 2017;44:113–22. https://doi.org/10.1016/j.ejmp.2017.06.006.
13. Chang JY, Mehran RJ, Feng L, Balter P, McRae S, Berry DA, Roth JA, STARS Clinical Trials Group. Stereotactic ablative radiotherapy in operable stage I NSCLC patients: long-term results of the expanded STARS clinical trial. J Clin Oncol. 2021;39(15_suppl):8506. https://doi.org/10.1200/JCO.2021.39.15_suppl.8506.
14. Cheng Y, Han L, Wu L, Chen J, Sun H, Wen G, Ji Y, Dvorkin M, Shi J, Pan Z, Shi J, Wang X, Bai Y, Melkadze T, Pan Y, Min X, Viguro M, LI X, Zhao Y, et al. Effect of first-line Serplulimab vs placebo added to chemotherapy on survival in patients with extensive-stage small cell lung cancer: the ASTRUM-005 randomized Clinical trial. JAMA. 2022;328(12):1223–32. https://doi.org/10.1001/jama.2022.16464.

15. Cheng Y, Spigel DR, Cho BC, Laktionov KK, Fang J, Chen Y, Zenke Y, Lee KH, Wang Q, Navarro A, Bernabe R, Buchmeier EL, Chang JW-C, Shiraishi Y, Goksu SS, Badzio A, Shi A, Daniel DB, Hoa NTT, et al. Durvalumab after chemoradiotherapy in limited-stage small-cell lung cancer. N Engl J Med. 2024;391(14):1313–27. https://doi.org/10.1056/NEJMoa2404873.
16. Chun SG, Hu C, Choy H, Komaki RU, Timmerman RD, Schild SE, Bogart JA, Dobelbower MC, Bosch W, Galvin JM, Kavadi VS, Narayan S, Iyengar P, Robinson CG, Wynn RB, Raben A, Augspurger ME, MacRae RM, Paulus R, Bradley JD. Impact of intensity-modulated radiation therapy technique for locally advanced non-small-cell lung cancer: a secondary analysis of the NRG oncology RTOG 0617 randomized clinical trial. J Clin Oncol Off J Am Soc Clin Oncol. 2017;35(1):56–62. https://doi.org/10.1200/JCO.2016.69.1378.
17. Chun SG, Hu C, Komaki RU, Timmerman RD, Schild SE, Bogart JA, Dobelbower MC, Bosch W, Kavadi VS, Narayan S, Iyengar P, Robinson C, Rothman J, Raben A, Augspurger ME, MacRae RM, Paulus R, Bradley JD. Long-term prospective outcomes of intensity modulated radiotherapy for locally advanced lung cancer: a secondary analysis of a randomized clinical trial. JAMA Oncol. 2024;10(8):1111–5. https://doi.org/10.1001/jamaoncol.2024.1841.
18. Darréon J, Massabeau C, Geffroy C, Maroun P, Simon L. Surface-guided radiotherapy overview: technical aspects and clinical applications. Cancer Radiother. 2023;27(6–7):504–10. https://doi.org/10.1016/j.canrad.2023.07.003.
19. de Crevoisier R, Lafond C, Mervoyer A, Hulot C, Jaksic N, Bessières I, Delpon G. Image-guided radiotherapy. Cancer Radiother. 2022;26(1–2):34–49. https://doi.org/10.1016/j.canrad.2021.08.002.
20. Dillman RO, Herndon J, Seagren SL, Eaton WL, Green MR. Improved survival in stage III non-small-cell lung cancer: seven-year follow-up of cancer and leukemia group B (CALGB) 8433 trial. J Natl Cancer Inst. 1996;88(17):1210–5. https://doi.org/10.1093/jnci/88.17.1210.
21. Dingemans A-MC, Hendriks LEL, Berghmans T, Levy A, Hasan B, Faivre-Finn C, Giaj-Levra M, Giaj-Levra N, Girard N, Greillier L, Lantuéjoul S, Edwards J, O'Brien M, Reck M, Smit EF, Van Schil P, Postmus PE, Ramella S, Lievens Y, et al. Definition of synchronous oligometastatic non-small cell lung cancer-a consensus report. J Thorac Oncol. 2019;14(12):2109–19. https://doi.org/10.1016/j.jtho.2019.07.025.
22. Faivre-Finn C, Snee M, Ashcroft L, Appel W, Barlesi F, Bhatnagar A, Bezjak A, Cardenal F, Fournel P, Harden S, Pechoux CL, McMenemin R, Mohammed N, O'Brien M, Pantarotto J, Surmont V, Meerbeeck JPV, Woll PJ, Lorigan P, Blackhall F. Concurrent once-daily versus twice-daily chemoradiotherapy in patients with limited-stage small-cell lung cancer (CONVERT): an open-label, phase 3, randomised, superiority trial. Lancet Oncol. 2017;18(8):1116–25. https://doi.org/10.1016/S1470-2045(17)30318-2.
23. Farago AF, Keane FK. Current standards for clinical management of small cell lung cancer. Transl Lung Cancer Res. 2018;7(1):69–79. https://doi.org/10.21037/tlcr.2018.01.16.
24. Fernandes AT, Shen J, Finlay J, Mitra N, Evans T, Stevenson J, Langer C, Lin L, Hahn S, Glatstein E, Rengan R. Elective nodal irradiation (ENI) vs. involved field radiotherapy (IFRT) for locally advanced non-small cell lung cancer (NSCLC): a comparative analysis of toxicities and clinical outcomes. Radiother Oncol. 2010;95(2):178–84. https://doi.org/10.1016/j.radonc.2010.02.007.
25. García-Campelo R, Sullivan I, Arriola E, Insa A, Juan Vidal O, Cruz-Castellanos P, Morán T, Reguart N, Zugazagoitia J, Dómine M. SEOM-GECP Clinical guidelines for diagnosis, treatment and follow-up of small-cell lung cancer (SCLC) (2022). Clin Transl Oncol. 2023;25(9):2679–91. https://doi.org/10.1007/s12094-023-03216-3.
26. Giaddui T, Chen W, Yu J, Lin L, Simone CB, Yuan L, Gong YUT, Wu QJ, Mohan R, Zhang X, Bluett JB, Gillin M, Moore K, O'Meara E, Presley J, Bradley JD, Liao Z, Galvin J, Xiao Y. Establishing the feasibility of the dosimetric compliance criteria of RTOG 1308: phase III randomized trial comparing overall survival after photon versus proton radiochemotherapy for inoperable stage II-IIIB NSCLC. Radiat Oncol (London). 2016;11:66. https://doi.org/10.1186/s13014-016-0640-8.

27. Goldman JW, Dvorkin M, Chen Y, Reinmuth N, Hotta K, Trukhin D, Statsenko G, Hochmair MJ, Özgüroğlu M, Ji JH, Garassino MC, Voitko O, Poltoratskiy A, Ponce S, Verderame F, Havel L, Bondarenko I, Każarnowicz A, Losonczy G, et al. Durvalumab, with or without tremelimumab, plus platinum-etoposide versus platinum-etoposide alone in first-line treatment of extensive-stage small-cell lung cancer (CASPIAN): updated results from a randomised, controlled, open-label, phase 3 trial. Lancet Oncol. 2021;22(1):51–65. https://doi.org/10.1016/S1470-2045(20)30539-8.
28. Gonzalez Y, Visak J, Overman L, Liao C-Y, Yen A, Zhuang T, Cai B, Godley A, Zhang Y, Timmerman R, Iyengar P, Westover K, Parsons D, Lin M-H. Beyond conventional bounds: surpassing system limits for stereotactic ablative (SAbR) lung radiotherapy using CBCT-based adaptive planning system. J Appl Clin Med Phys. 2024;25(8):e14375. https://doi.org/10.1002/acm2.14375.
29. Graabak G, Grønberg BH, Killingberg KT, Halvorsen TO. Effect of FDG PET-CT for staging and radiotherapy planning—a comparison of cohorts from two randomized trials of thoracic radiotherapy in limited-stage SCLC. JTO Clin Res Rep. 2024;5(9):100688. https://doi.org/10.1016/j.jtocrr.2024.100688.
30. Grønberg BH, Killingberg KT, Fløtten Ø, Brustugun OT, Hornslien K, Madebo T, Langer SW, Schytte T, Nyman J, Risum S, Tsakonas G, Engleson J, Halvorsen TO. High-dose versus standard-dose twice-daily thoracic radiotherapy for patients with limited stage small-cell lung cancer: an open-label, randomised, phase 2 trial. Lancet Oncol. 2021;22(3):321–31. https://doi.org/10.1016/S1470-2045(20)30742-7.
31. Guo X, Wang B, Sun J, Li J, Jia W, Zhu H, Guo H. Role of surgery in potentially resectable small-cell lung cancer based on the eighth edition of the TNM classification: a population study of the US SEER database. Clin Respir J. 2024;18(10):e70024. https://doi.org/10.1111/crj.70024.
32. Hadziahmetovic M, Loo BW, Timmerman RD, Mayr NA, Wang JZ, Huang Z, Grecula JC, Lo SS. Stereotactic body radiation therapy (stereotactic ablative radiotherapy) for stage I non-small cell lung cancer—updates of radiobiology, techniques, and clinical outcomes. Discov Med. 2010;9(48):411–7.
33. Hamada A, Soh J, Hata A, Nakamatsu K, Shimokawa M, Yatabe Y, Oizumi H, Tsuboi M, Horinouchi H, Yoshino I, Tanahashi M, Toyooka S, Okada M, Yokomise H, Yamashita M, Nishimura Y, Yamamoto N, Nakagawa K, Mitsudomi T. Phase II study of neoadjuvant concurrent chemo-immuno-radiation therapy followed by surgery and adjuvant immunotherapy for resectable stage IIIA-B (discrete N2) non-small-cell lung cancer: SQUAT trial (WJOG 12119L). Clin Lung Cancer. 2021;22(6):596–600. https://doi.org/10.1016/j.cllc.2021.04.006.
34. Hattori A, Suzuki K, Takamochi K, Wakabayashi M, Sekino Y, Tsutani Y, Nakajima R, Aokage K, Saji H, Tsuboi M, Okada M, Asamura H, Nakamura K, Fukuda H, Watanabe S-I, Japan Clinical Oncology Group, West Japan Oncology Group. Segmentectomy versus lobectomy in small-sized peripheral non-small-cell lung cancer with radiologically pure-solid appearance in Japan (JCOG0802/WJOG4607L): a post-hoc supplemental analysis of a multicentre, open-label, phase 3 trial. Lancet Respir Med. 2024;12(2):105–16. https://doi.org/10.1016/S2213-2600(23)00382-X.
35. Hodapp N. Der ICRU-Report 83: Verordnung, Dokumentation und Kommunikation der fluenzmoduliertenPhotonenstrahlentherapie (IMRT). Strahlenther Onkol. 2012;188(1):97–100. https://doi.org/10.1007/s00066-011-0015-x.
36. Houda I, Bahce I, Dickhoff C, Kroese TE, Kroeze SGC, Mariolo AV, Tagliamento M, Moliner L, Brandão M, Pretzenbacher Y, Edwards J, Opitz I, Brunelli A, Guckenberger M, van Schil PE, Popat S, Blum T, Faivre-Finn C, de Ruysscher D, et al. An international and multidisciplinary EORTC survey on resectability of stage III non-small cell lung cancer. Lung Cancer. 2025;199:108061. https://doi.org/10.1016/j.lungcan.2024.108061.
37. Hui Z, Men Y, Hu C, Kang J, Sun X, Bi N, Zhou Z, Liang J, Lv J, Feng Q, Xiao Z, Chen D, Wang Y, Li J, Wang J, Gao S, Wang L, He J. Effect of postoperative radiotherapy for patients with pIIIA-N2 non-small cell lung cancer after complete resection and adjuvant che-

motherapy: the phase 3 PORT-C randomized clinical trial. JAMA Oncol. 2021;7(8):1178–85. https://doi.org/10.1001/jamaoncol.2021.1910.
38. Iyengar P, Hu C, Gomez DR, Timmerman RD, Simone CB, Robinson CG, Gerber DE, Waqar SN, Donington J, Swisher S, Weldon M, Wu J, Faller BA, Rashdan S, Stephans KL, Sampson P, Higgins KA, Nowak RK, Lyness J, Bradley JD. NRG-LU002: randomized phase II/III trial of maintenance systemic therapy versus local consolidative therapy (LCT) plus maintenance systemic therapy for limited metastatic non-small cell lung cancer (NSCLC). J Clin Oncol. 2024;42(16_suppl):8506. https://doi.org/10.1200/JCO.2024.42.16_suppl.8506.
39. Kalemkerian GP. Staging and imaging of small cell lung cancer. Cancer Imaging. 2012;11(1):253–8. https://doi.org/10.1102/1470-7330.2011.0036.
40. Kaseda K. Recent and current advances in FDG-PET imaging within the field of clinical oncology in NSCLC: a review of the literature. Diagnostics. 2020;10(8):561. https://doi.org/10.3390/diagnostics10080561.
41. Kazemi M, Ladbury C, Liu J, Glaser S, Williams T, Amini A. Thoracic radiation in limited stage small cell lung cancer: trends in radiation fractionation. Clin Lung Cancer. 2023;24(4):322–8. https://doi.org/10.1016/j.cllc.2023.02.001.
42. Kinj R, Gaide O, Jeanneret-Sozzi W, Dafni U, Viguet-Carrin S, Sagittario E, Kypriotou M, Chenal J, Duclos F, Hebeisen M, Falco T, Geyer R, Gonçalves Jorge P, Moeckli R, Bourhis J. Randomized phase II selection trial of FLASH and conventional radiotherapy for patients with localized cutaneous squamous cell carcinoma or basal cell carcinoma: a study protocol. Clin Transl Radiat Oncol. 2024;45:100743. https://doi.org/10.1016/j.ctro.2024.100743.
43. Kong F-MS, Hu C, Pryma DA, Duan F, Matuszak M, Xiao Y, Ten Haken R, Siegel MJ, Hanna L, Curran WJ, Dunphy M, Gelblum D, Piert M, Jolly S, Robinson CG, Quon A, Loo BW, Srinivas S, Videtic GM, et al. Primary results of NRG-RTOG1106/ECOG-ACRIN 6697: a randomized phase II trial of individualized adaptive (chemo)radiotherapy using midtreatment 18F-fluorodeoxyglucose positron emission tomography/computed tomography in stage III non-small cell lung cancer. J Clin Oncol Off J Am Soc Clin Oncol. 2024;42(33):3935–46. https://doi.org/10.1200/JCO.24.00022.
44. Konkol M, Śniatała P, Milecki P. Radiation-induced lung injury—what do we know in the era of modern radiotherapy? Rep Pract Oncol Radiother. 2022;27(3):552–65. https://doi.org/10.5603/RPOR.a2022.0046.
45. Lefkopoulos D, Ferreira I, Isambert A, Le Péchoux C, Mornex F. Présent et avenir de la radiothérapieguidée par l'image (IGRT) et ses applications possibles dans le traitement des cancers bronchiques. Cancer/Radiothér. 2007;11(1):23–31. https://doi.org/10.1016/j.canrad.2006.10.001.
46. Lim HS, Yang K, Noh JM, Pyo H, Kim JM, Kwon D, Ju SG, Ahn YC. Reduced frequency and severity of radiation esophagitis without marginal failure risk by contralateral esophagus sparing IMRT in stage III non-small cell lung cancer patients undergoing definitive concurrent chemoradiotherapy. Radiother Oncol. 2024;199:110436. https://doi.org/10.1016/j.radonc.2024.110436.
47. Liu SV, Reck M, Mansfield AS, Mok T, Scherpereel A, Reinmuth N, Garassino MC, De Castro Carpeno J, Califano R, Nishio M, Orlandi F, Alatorre-Alexander J, Leal T, Cheng Y, Lee J-S, Lam S, McCleland M, Deng Y, Phan S, Horn L. Updated overall survival and PD-L1 subgroup analysis of patients with extensive-stage small-cell lung cancer treated with Atezolizumab, carboplatin, and etoposide (IMpower133). J Clin Oncol Off J Am Soc Clin Oncol. 2021;39(6):619–30. https://doi.org/10.1200/JCO.20.01055.
48. Malakouti-Nejad B, Moore S, Wheatley-Price P, Tiberi D. Management of very early small cell lung cancer: a Canadian survey study. Curr Oncol. 2023;30(7):6006–18. https://doi.org/10.3390/curroncol30070449.
49. Meert AP, Paesmans M, Berghmans T, Martin B, Mascaux C, Vallot F, Verdebout JM, Lafitte JJ, Sculier JP. Prophylactic cranial irradiation in small cell lung cancer: a systematic review of the literature with meta-analysis. BMC Cancer. 2001;1:5. https://doi.org/10.1186/1471-2407-1-5.

50. Men Y, Bao Y, Bi N, Zhou Z, Liang J, Lv J, Feng Q, Xiao Z, Wang Y, Li J, Wang J, Gao S, Wang LH, He J, Hui Z. Long-term outcomes of postoperative radiotherapy for patients with pIIIA-N2 non–small cell lung cancer after complete resection and adjuvant chemotherapy: the phase 3 PORT-C randomized clinical trial. Int J Radiat Oncol Biol Phys. 2024;120(2):S84–5. https://doi.org/10.1016/j.ijrobp.2024.07.149.
51. Moeller B, Balagamwala EH, Chen A, Creach KM, Giaccone G, Koshy M, Zaky S, Rodrigues G. Palliative thoracic radiation therapy for non-small cell lung cancer: 2018 update of an American Society for Radiation Oncology (ASTRO) evidence-based guideline. Pract Radiat Oncol. 2018;8(4):245–50. https://doi.org/10.1016/j.prro.2018.02.009.
52. Movsas B, Hu C, Sloan J, Bradley J, Komaki R, Masters G, Kavadi V, Narayan S, Michalski J, Johnson DW, Koprowski C, Curran WJ, Garces YI, Gaur R, Wynn RB, Schallenkamp J, Gelblum DY, MacRae RM, Paulus R, Choy H. Quality of life analysis of a radiation dose-escalation study of patients with non-small-cell lung cancer: a secondary analysis of the radiation therapy oncology group 0617 randomized clinical trial. JAMA Oncol. 2016;2(3):359–67. https://doi.org/10.1001/jamaoncol.2015.3969.
53. Nagata Y, Hiraoka M, Shibata T, Onishi H, Kokubo M, Karasawa K, Shioyama Y, Onimaru R, Kunieda E, Ishikura S. A phase II trial of stereotactic body radiation therapy for operable T1N0M0 non-small cell lung cancer; Japan clinical oncology group (JCOG0403); long term follow-up results. Int J Radiat Oncol Biol Phys. 2018;102(3):S9. https://doi.org/10.1016/j.ijrobp.2018.06.115.
54. Nestle U, Schimek-Jasch T, Kremp S, Schaefer-Schuler A, Mix M, Küsters A, Tosch M, Hehr T, Eschmann SM, Bultel Y-P, Hass P, Fleckenstein J, Thieme A, Stockinger M, Dieckmann K, Miederer M, Holl G, Rischke HC, Gkika E, et al. Imaging-based target volume reduction in chemoradiotherapy for locally advanced non-small-cell lung cancer (PET-Plan): a multicentre, open-label, randomised, controlled trial. Lancet Oncol. 2020;21(4):581–92. https://doi.org/10.1016/S1470-2045(20)30013-9.
55. Nosaki K, Seto T, Shimokawa M, Takahashi T, Yamamoto N. Is prophylactic cranial irradiation (PCI) needed in patients with extensive-stage small cell lung cancer showing complete response to first-line chemotherapy? Radiother Oncol. 2018;127(3):344–8. https://doi.org/10.1016/j.radonc.2018.04.010.
56. Pan B, Shen B. The ADRIATIC study: revolutionizing the standard treatment paradigm for concurrent chemoradiotherapy in limited-stage small cell lung cancer. J Natl Cancer Center. 2024;5:1. https://doi.org/10.1016/j.jncc.2024.12.003.
57. Pechoux CL, Pourel N, Barlesi F, Lerouge D, Antoni D, Lamezec B, Nestle U, Boisselier P, Dansin E, Paumier A, Peignaux K, Thillays F, Zalcman G, Madelaine J, Pichon E, Larrouy A, Lavole A, Argo-Leignel D, Derollez M, et al. Postoperative radiotherapy versus no postoperative radiotherapy in patients with completely resected non-small-cell lung cancer and proven mediastinal N2 involvement (Lung ART, IFCT 0503): an open-label, randomised, phase 3 trial. Lancet Oncol. 2022;23(1):104–14. https://doi.org/10.1016/S1470-2045(21)00606-9.
58. Perez CA, Stanley K, Rubin P, Kramer S, Brady L, Perez-Tamayo R, Brown GS, Concannon J, Rotman M, Seydel HG. A prospective randomized study of various irradiation doses and fractionation schedules in the treatment of inoperable non-oat-cell carcinoma of the lung. Preliminary report by the Radiation Therapy Oncology Group. Cancer. 1980;45(11):2744–53. https://doi.org/10.1002/1097-0142(19800601)45:11<2744::aid-cncr2820451108>3.0.co;2-u.
59. Perry MC, Kohman LJ, Bonner JA, Gu L, Wang X, Vokes EE, Green MR. A phase III study of surgical resection and paclitaxel/carboplatin chemotherapy with or without adjuvant radiation therapy for resected stage III non-small-cell lung cancer: Cancer and Leukemia Group B 9734. Clin Lung Cancer. 2007;8(4):268–72. https://doi.org/10.3816/CLC.2007.n.005.
60. Pignon J-P, Arriagada R, Ihde DC, Johnson DH, Perry MC, Souhami RL, Brodin O, Joss RA, Kies MS, Lebeau B, Onoshi T, Østerlind K, Tattersall MHN, Wagner H. A meta-analysis of thoracic radiotherapy for small-cell lung cancer. N Engl J Med. 1992;327(23):1618–24. https://doi.org/10.1056/NEJM199212033272302.

61. Postoperative radiotherapy in non-small-cell lung cancer: systematic review and meta-analysis of individual patient data from nine randomised controlled trials. PORT Meta-analysis Trialists Group. Lancet (London). 1998;352(9124):257–63.
62. Protopapa M, Kouloulias V, Nikoloudi S, Papadimitriou C, Gogalis G, Zygogianni A. From whole-brain radiotherapy to immunotherapy: a multidisciplinary approach for patients with brain metastases from NSCLC. J Oncol. 2019;2019:3267409. https://doi.org/10.1155/2019/3267409.
63. Rodríguez de Dios N, Couñago F, Murcia-Mejía M, Rico-Oses M, Calvo-Crespo P, Samper P, Vallejo C, Luna J, Trueba I, Sotoca A, Cigarral C, Farré N, Manero RM, Durán X, Gispert JD, Sánchez-Benavides G, Rognoni T, Torrente M, Capellades J, et al. Randomized phase III trial of prophylactic cranial irradiation with or without hippocampal avoidance for small-cell lung cancer (PREMER): a GICOR-GOECP-SEOR study. J Clin Oncol Off J Am Soc Clin Oncol. 2021;39(28):3118–27. https://doi.org/10.1200/JCO.21.00639.
64. Rudin CM, Awad MM, Navarro A, Gottfried M, Peters S, Csőszi T, Cheema PK, Rodriguez-Abreu D, Wollner M, Yang JC-H, Mazieres J, Orlandi FJ, Luft A, Gümüş M, Kato T, Kalemkerian GP, Luo Y, Ebiana V, Pietanza MC, et al. Pembrolizumab or placebo plus etoposide and platinum as first-line therapy for extensive-stage small-cell lung cancer: randomized, double-blind, phase III KEYNOTE-604 study. J Clin Oncol Off J Am Soc Clin Oncol. 2020;38(21):2369–79. https://doi.org/10.1200/JCO.20.00793.
65. Saji H, Okada M, Tsuboi M, Nakajima R, Suzuki K, Aokage K, Aoki T, Okami J, Yoshino I, Ito H, Okumura N, Yamaguchi M, Ikeda N, Wakabayashi M, Nakamura K, Fukuda H, Nakamura S, Mitsudomi T, Watanabe S-I, et al. Segmentectomy versus lobectomy in small-sized peripheral non-small-cell lung cancer (JCOG0802/WJOG4607L): a multicentre, open-label, phase 3, randomised, controlled, non-inferiority trial. Lancet (London). 2022;399(10335):1607–17. https://doi.org/10.1016/S0140-6736(21)02333-3.
66. Schild SE, Fan W, Stinchcombe TE, Vokes EE, Ramalingam SS, Bradley JD, Kelly K, Pang HH, Wang X. Toxicity related to radiotherapy dose and targeting strategy: a pooled analysis of cooperative group trials of combined modality therapy for locally advanced non–small cell lung cancer. J Thorac Oncol. 2019;14(2):298–303. https://doi.org/10.1016/j.jtho.2018.09.021.
67. Schimmer C, Neukam K, Elert O. Staging of non-small cell lung cancer: clinical value of positron emission tomography and mediastinoscopy. Interact Cardiovasc Thorac Surg. 2006;5(4):418–23. https://doi.org/10.1510/icvts.2006.129478.
68. SEER*Explorer. An Interactive Website for SEER Cancer Statistics [Internet]. Surveillance Research Program, National Cancer Institute; 2024 [Updated: 2024 Nov 5; Cited 2025 Jan 5]. Available from: https://Seer.Cancer.Gov/Statistics-Network/Explorer/. Data Source(s): SEER Incidence Data, November 2023 Submission (1975–2021), SEER 22 Registries; (n.d.).
69. Seltzer MA, Yap CS, Silverman DH, Meta J, Schiepers C, Phelps ME, Gambhir SS, Rao J, Valk PE, Czernin J. The impact of PET on the management of lung cancer: the referring physician's perspective. J Nucl Med. 2002;43(6):752–6.
70. Simone CB, Hu C, Heinzerling JH, Mileham K, Higgins KA, Lin L, Abazeed M, Ohri N, Bradley JD. NRG LU008: phase III prospective randomized trial of primary lung tumor stereotactic body radiation therapy (SBRT) followed by concurrent mediastinal chemoradiation for locally-advanced non-small cell lung cancer (LA-NSCLC). Int J Radiat Oncol Biol Phys. 2024;120(2):e63. https://doi.org/10.1016/j.ijrobp.2024.07.1917.
71. Slotman B, Faivre-Finn C, Kramer G, Rankin E, Snee M, Hatton M, Postmus P, Collette L, Musat E, Senan S, EORTC Radiation Oncology Group and Lung Cancer Group. Prophylactic cranial irradiation in extensive small-cell lung cancer. N Engl J Med. 2007;357(7):664–72. https://doi.org/10.1056/NEJMoa071780.
72. Slotman BJ, van Tinteren H, Praag JO, Knegjens JL, El Sharouni SY, Hatton M, Keijser A, Faivre-Finn C, Senan S. Use of thoracic radiotherapy for extensive stage small-cell lung cancer: a phase 3 randomised controlled trial. Lancet (London). 2015;385(9962):36–42. https://doi.org/10.1016/S0140-6736(14)61085-0.

73. Spigel DR, Faivre-Finn C, Gray JE, Vicente D, Planchard D, Paz-Ares L, Vansteenkiste JF, Garassino MC, Hui R, Quantin X, Rimner A, Wu Y-L, Özgüroğlu M, Lee KH, Kato T, de Wit M, Kurata T, Reck M, Cho BC, et al. Five-year survival outcomes from the PACIFIC trial: durvalumab after chemoradiotherapy in stage III non–small-cell lung cancer. J Clin Oncol. 2022;40(12):1301–11. https://doi.org/10.1200/JCO.21.01308.
74. Stokes M, Berfeld N, Gayle A, Descoteaux A, Rohrmoser O, Franks A. A systematic literature review of real-world treatment outcomes of small cell lung cancer. Medicine. 2022;101(26):e29783. https://doi.org/10.1097/MD.0000000000029783.
75. Sun A, Hu C, Wong SJ, Gore E, Videtic G, Dutta S, Suntharalingam M, Chen Y, Gaspar LE, Choy H. Prophylactic cranial irradiation vs observation in patients with locally advanced non-small cell lung cancer: a long-term update of the NRG oncology/RTOG 0214 phase 3 randomized clinical trial. JAMA Oncol. 2019;5(6):847–55. https://doi.org/10.1001/jamaoncol.2018.7220.
76. Sun H, Li M, Huang W, Zhang J, Wei S, Yang Y, Wang Z, Ye S, Gong H, Zhang Y, Li J, Song H, Wang L, Chen X, Lin H, Ding G, Li H, Zheng A, Ma X, et al. Thoracic radiotherapy improves the survival in patients with EGFR-mutated oligo-organ metastatic non-small cell lung cancer treated with epidermal growth factor receptor-tyrosine kinase inhibitors: a multicenter, randomized, controlled, phase III trial. J Clin Oncol Off J Am Soc Clin Oncol. 2025;43(4):412–21. https://doi.org/10.1200/JCO.23.02075.
77. Sung H, Ferlay J, Siegel RL, Laversanne M, Soerjomataram I, Jemal A, Bray F. Global cancer statistics 2020: GLOBOCAN estimates of incidence and mortality worldwide for 36 cancers in 185 countries. CA Cancer J Clin. 2021;71(3):209–49. https://doi.org/10.3322/caac.21660.
78. Takahashi T, Yamanaka T, Seto T, Harada H, Nokihara H, Saka H, Nishio M, Kaneda H, Takayama K, Ishimoto O, Takeda K, Yoshioka H, Tachihara M, Sakai H, Goto K, Yamamoto N. Prophylactic cranial irradiation versus observation in patients with extensive-disease small-cell lung cancer: a multicentre, randomised, open-label, phase 3 trial. Lancet Oncol. 2017;18(5):663–71. https://doi.org/10.1016/S1470-2045(17)30230-9.
79. Timmerman RD, Hu C, Michalski JM, Bradley JC, Galvin J, Johnstone DW, Choy H. Long-term results of stereotactic body radiation therapy in medically inoperable stage I non–small cell lung cancer. JAMA Oncol. 2018;4(9):1287–8. https://doi.org/10.1001/jamaoncol.2018.1258.
80. Timmerman R, Paulus R, Galvin J, Michalski J, Straube W, Bradley J, Fakiris A, Bezjak A, Videtic G, Johnstone D, Fowler J, Gore E, Choy H. Stereotactic body radiation therapy for inoperable early stage lung cancer. JAMA. 2010;303(11):1070–6. https://doi.org/10.1001/jama.2010.261.
81. Turrisi AT, Kim K, Blum R, Sause WT, Livingston RB, Komaki R, Wagner H, Aisner S, Johnson DH. Twice-daily compared with once-daily thoracic radiotherapy in limited small-cell lung cancer treated concurrently with cisplatin and etoposide. N Engl J Med. 1999;340(4):265–71. https://doi.org/10.1056/NEJM199901283400403.
82. Tzschoppe T, Ohlinger J, Vordermark D, Bedir A, Medenwald D. Population based study on the progress in survival of primarily metastatic lung cancer patients in Germany. Sci Rep. 2024;14(1):16005. https://doi.org/10.1038/s41598-024-66307-3.
83. Udovicich C, Lo SS, Guckenberger M, Sahgal A. Shifting the landscape of spine and non-spine bone metastases: a review of stereotactic body radiotherapy. Cancer J (Sudbury). 2024;30(6):385–92. https://doi.org/10.1097/PPO.0000000000000755.
84. Vansteenkiste JF, Naidoo J, Faivre-Finn C, Özgüroğlu M, Villegas A, Daniel D, Murakami S, Hui R, Lee KH, Cho BC, Kubota K, Broadhurst H, Wadsworth C, Newton M, Thiyagarajah P, Antonia SJ. Symptomatic pneumonitis with durvalumab after concurrent chemoradiotherapy in unresectable stage III NSCLC. JTO Clin Res Rep. 2024;5(3):100638. https://doi.org/10.1016/j.jtocrr.2024.100638.
85. Vera P, Thureau S, Le Tinier F, Chaumet-Riffaud P, Hapdey S, Kolesnikov-Gauthier H, Martin E, Berriolo-Riedinger A, Pourel N, Broglia JM, Boissellier P, Guillemard S, Salem N, Brenot-Rossi I, Le Péchoux C, Berthold C, Giroux-Leprieur E, Moreau D, Guillerm S,

et al. Adaptive radiotherapy (up to 74 Gy) or standard radiotherapy (66 Gy) for patients with stage III non-small-cell lung cancer, according to [18F]FDG-PET tumour residual uptake at 42 Gy (RTEP7-IFCT-1402): a multicentre, randomised, controlled phase 2 trial. Lancet Oncol. 2024;25(9):1176–87. https://doi.org/10.1016/S1470-2045(24)00320-6.

86. Visa MA, Abazeed ME, Avella Patino D. Integrative approaches in non-small cell lung cancer management: the role of radiotherapy. J Clin Med. 2024;13(15):4296. https://doi.org/10.3390/jcm13154296.
87. Wakeam E, Acuna SA, Leighl NB, Giuliani ME, Finlayson SRG, Varghese TK, Darling GE. Surgery versus chemotherapy and radiotherapy for early and locally advanced small cell lung cancer: a propensity-matched analysis of survival. Lung Cancer. 2017;109:78–88. https://doi.org/10.1016/j.lungcan.2017.04.021.
88. Walls GM, Bergom C, Mitchell JD, Rentschler SL, Hugo GD, Samson PP, Robinson CG. Cardiotoxicity following thoracic radiotherapy for lung cancer. Br J Cancer. 2024;132:311. https://doi.org/10.1038/s41416-024-02888-0.
89. Wang S-F, Mao N-Q, Zhao W-H, Pan X-B. Postoperative radiotherapy in pIIIA-N2 non-small cell lung cancer after complete resection and adjuvant chemotherapy: a meta-analysis. Medicine. 2022;101(28):e29550. https://doi.org/10.1097/MD.0000000000029550.
90. Warde P, Payne D. Does thoracic irradiation improve survival and local control in limited-stage small-cell carcinoma of the lung? A meta-analysis. J Clin Oncol Off J Am Soc Clin Oncol. 1992;10(6):890–5. https://doi.org/10.1200/JCO.1992.10.6.890.
91. Watanabe S-I, Yotsukura M, Miyoshi T, Hattori A, Isaka T, Maniwa T, Isaka M, Yoshioka H, Endo M, Mimae T, Tsutani Y, Nakagawa K, Aokage K, Lung Cancer Surgical Study Group (LCSSG) of the Japan Clinical Oncology Group (JCOG). Updated review of perioperative treatment for non-small-cell lung cancer in the new era of immune checkpoint inhibitors: past, present, and future. Jpn J Clin Oncol. 2024;54(12):1244–53. https://doi.org/10.1093/jjco/hyae106.
92. Wolf A, Loo BW, Mak RH, Liptay M, Pettiford B, Rocco G, Lanuti M, Merritt RE, Keshavarz H, Suh RD, Brunelli A, Criner GJ, Mazzone PJ, Walsh G, Eileen Wafford Q, Murthy S, Blair Marshall M, Tong B, Luketich J, et al. Systematic review of stereotactic ablative radiotherapy (SABR)/stereotactic body radiation therapy (SBRT) for treatment of high-risk patients with stage I non-small cell lung cancer. Semin Thorac Cardiovasc Surg. 2024;S1043-0679(24):00105–9. https://doi.org/10.1053/j.semtcvs.2024.10.005.
93. Yan M, Louie AV, Kotecha R, Ashfaq Ahmed M, Zhang Z, Guckenberger M, Kim M-S, Lo SS, Scorsetti M, Tree AC, Sahgal A, Slotman BJ. Stereotactic body radiotherapy for ultra-central lung tumors: a systematic review and meta-analysis and international stereotactic radiosurgery society practice guidelines. Lung Cancer. 2023;182:107281. https://doi.org/10.1016/j.lungcan.2023.107281.
94. Yao B, Zhang H, Liang X, Zhao T, Hu J, Ge J, Han Y, Wang W, Xu W, He J. Efficacies of different postoperative radiotherapy techniques in patients with N2 non-small cell lung cancer: a meta-analysis. Am J Transl Res. 2024;16(11):7016–35. https://doi.org/10.62347/JGIB9696.
95. Zhang C, Zhao G, Wu H, Jiang J, Duan W, Fan Z, Wang Z, Wang R. Application of post-operative adjuvant radiotherapy in limited-stage small cell lung cancer: a systematic review and meta-analysis. Radiother Oncol. 2024;193:110123. https://doi.org/10.1016/j.radonc.2024.110123.
96. Salem, Alfaifi V., Pareek J., Kim S., Rathod W., Hunter A., Leylek N., Ahmed S., Venkataraman N., Venugopal A., Chowdhury A., Dubey S., Kakumanu B., Bashir Moving towards single fraction peripheral lung stereotactic body radiation therapy: patient care during and after the global COVID-19 pandemic Lung Cancer Management 10.1080/17581966.2024.2367369
97. May N., Tsao Y., Ung P., Cheung I., Poon Alexander V., Louie (2024). A Systematic Review of Phase II/III Trials of Hypofractionated versus Conventionally Fractionated Radiation Therapy in Stage III Non-Small Cell Lung Cancer Patients Cancers 16(19) 3384-10.3390/cancers16193384

98. Badr Id, Said Y., Geng Shahed N., Badiyan A., Bang A., Bezjak Kevin L.M., Chua C., Faivre-Finn Feng-Ming, Kong D., Przybysz Paul M., Putora P., Munoz-Schuffenegger S., Siva M., Xu-Welliver F., McDonald A., Louie Stephen G., Chun (2025). Accelerated Hypofractionated Radiotherapy for Locally Advanced NSCLC: A Systematic Review From the International Association for the Study of Lung Cancer Advanced Radiation Technology Subcommittee Journal of Thoracic Oncology 20(1) 39-51 10.1016/j.jtho.2024.09.1437
99. Vincent, Bourbonne F., Lucia V., Jaouen J., Bert O., Pradier D., Visvikis U., Schick (2022). VMAT-Based Planning Allows Sparing of a Spatial Dose Pattern Associated with Radiation Pneumonitis in Patients Treated with Radiotherapy for a Locally Advanced Lung Cancer Cancers 14(15) 3702-10.3390/cancers14153702
100. Donna M., Edwards Caitlin A., Schonewolf John D., Rice M., Schipper Randall K. Ten, Haken Martha, Matuszak J., Balter D., Jarema Douglas A., Arenberg M., Piert A., Qin Gregory P., Kalemkerian Bryan J., Schneider N., Ramnath Christina H., Chapman David A., Elliott Theodore S., Lawrence J., Hearn James A., Hayman S., Jolly (2024). Phase 2 Trial Assessing Toxicity of Personalized Response-Based Radiation Treatment in Patients With Locally Advanced Non-Small Cell Lung Cancer International Journal of Radiation Oncology*Biology*Physics 120(5) 1332-1343 10.1016/j.ijrobp.2024.06.018

Chapter 11
Advances in Radiotherapy of Genitourinary Cancers

Debasmita Paul, Utpal Bhui, Joy Das, Biplab Debnath, Mohini Mondal, Sumel Ashique, and Shayan Sheikhmiri

Abstract Genitourinary cancers including prostate, bladder, renal, and testicular cancers are among the most prevalent malignancies worldwide, accounting for over 13% of total cancer incidence in 2022. Radiation therapy is considered a key component in the treatment of these cancers. Innovations such as intensity-modulated radiation therapy, volumetric-modulated arc therapy, and image-guided radiation therapy have redefined prostate cancer radiotherapy. Stereotactic body radiotherapy has become an established frontline alternative to long-course treatment schedules for localized prostate cancer. Radioligand therapy has emerged as a standard treatment option for managing metastatic prostate cancer. For muscle-invasive bladder cancer, chemoradiation has proven to be an effective alternative to radical surgery. Stereotactic ablative radiotherapy is now regarded as a viable treatment option for both primary and metastatic renal cell carcinoma. This chapter summarizes recent advancements in the radiotherapeutic management of genitourinary malignancies.

Keywords Genitourinary cancer · Prostate cancer · Bladder cancer · Kidney cancer · Urothelial carcinoma · Testicular cancer · Radiotherapy

D. Paul · B. Debnath · M. Mondal · S. Ashique
Department of Pharmaceutical Technology, Bharat Technology, Uluberi, West Bengal, India

U. Bhui
NSHM Institute of Pharmaceutical Technology, NSHM Knowledge Campus, Durgapur, West Bengal, India

J. Das
School of Pharmaceutical Sciences, Lovely Professional University, Phagwara, Punjab, India

S. Sheikhmiri (✉)
Cancer Research Institute, Shahid Beheshti University of Medical Sciences, Tehran, Iran

F. Taghizadeh-Hesary et al. (eds.), *From Radiobiology to Radiation Oncology*,
https://doi.org/10.1007/978-3-031-95724-6_11

11.1 Introduction

Genitourinary (GU) cancers include a broad group of malignancies of the urinary and reproductive organs, including the prostate, bladder, kidneys, ureters, urethra, penis, and testicles. The incidence of these cancers continues to rise [1]. Collectively, they accounted for over 13% of the total cancer incidence and over 8% of the total cancer mortality worldwide in 2022, making them the most frequently occurring cancers and surpassing female breast cancer in mortality [2]. Smoking and aging are significant risk factors, along with other established factors such as germline genetic mutations, family history, and exposure to chemicals [3–6].

Radiation therapy (RT) is a cornerstone in the treatment of GU cancers, especially prostate cancer. Although the principles have largely remained unchanged, there have been significant advancements in the application and techniques of RT over the last two decades, particularly in the field of prostate cancer. Advanced conformal techniques including intensity-modulated RT (IMRT) and volumetric-modulated RT (VMAT), image-guided RT (IGRT), moderately hypofractionated RT (MHFRT), and ultrahypofractionated RT (UHFRT) have significantly transformed RT for prostate cancer. Neoadjuvant, concurrent, and adjuvant androgen deprivation therapy (ADT) have markedly enhanced oncological outcomes when combined with prostate cancer RT, achieving results unparalleled by any other intervention. Radioligand therapy (RLT) has become a significant breakthrough in the management of metastatic prostate cancer. Chemoradiation is now an alternative standard of care (SoC) to radical surgery in the management of muscle-invasive bladder cancer (MIBC). Stereotactic Body Radiation Therapy (SBRT), Stereotactic Ablative Body Radiation Therapy (SABR), and Stereotactic Radiosurgery (SRS) have emerged as effective treatment modalities for managing metastatic disease originating from various genitourinary (GU) malignancies, including renal cell carcinoma (RCC). This chapter provides a brief summary of these advancements in the radiotherapeutic management of genitourinary cancers.

11.2 Prostate Cancer Radiation Therapy

Over the past two decades, few advancements in prostate RT have been as transformative as the adoption of advanced conformal techniques, including IMRT and VMAT. These advanced techniques, combined with IGRT, have become the SoC in prostate cancer RT, leaving three-dimensional conformal RT (3DCRT) as a technique primarily used in resource-limited settings. These methods enable dose escalation while sparing organs at risk, improving treatment outcomes, and reducing acute and late toxicities [7–9]. VMAT may offer shorter treatment times and better dosimetric parameters compared to IMRT [10]; however, no significant improvement in toxicity or cancer outcomes has been demonstrated.

MR-guided adapted RT is an innovative advancement in radiation oncology, offering real-time imaging for precise tumor targeting and improved organ sparing [11]. Its application in prostate cancer is constrained by limited long-term evidence, prolonged workflow per fraction, and restricted field size [12]. A single-center phase 3 randomized trial demonstrated reduced gastrointestinal (GI) and GU toxicity with the use of MR-guided RT in prostate SBRT [13]. Until more evidence on long-term oncological superiority and cost-effectiveness becomes available, MR-guided RT remains an investigational approach in the treatment of prostate cancer.

Biodegradable spacers are a recent innovation in mitigating radiation-induced side effects during prostate cancer RT. These tools help protect the rectum by creating a temporary space between it and the prostate. Systematic reviews and meta-analyses have demonstrated that the use of spacers is associated with lower rectal radiation doses, resulting in reduced early and late rectal and GI toxicities, as well as decreased late GU toxicities [14–17]. These benefits have been observed in conventional RT, SBRT, and brachytherapy. The use of spacers is associated with well-documented complications [18], which should be carefully considered when weighing their relatively modest benefit in reducing radiation toxicities.

High-dose, conventionally fractionated RT (HD-CFRT), delivering escalated doses in the range of 74–80 Gy, has been a major advancement in prostate cancer treatment over the past two decades. Supported by multiple randomized trials demonstrating improved biochemical relapse-free survival (bRFS) [19, 20], it has firmly established itself as an SoC in prostate RT. However, with the introduction of MHFRT, which offers comparable outcomes and toxicity [21, 22], its use has been increasingly limited to select clinical scenarios. MHFRT, defined as fraction sizes of 2.5–3.4 Gy, has become a widely adopted approach in modern prostate RT, consistently demonstrating equal outcomes compared to HD-CFRT [23, 24], which it has largely replaced in routine practice. MHFRT is also gaining supporting evidence in the postoperative setting, where CFRT has long been the standard of care. The NRG-GU003 trial has now established postoperative mHFRT as a new SoC, proving its non-inferiority to CFRT [25].

The effort to escalate doses using external beam RT (EBRT) has recently centered on selective dose intensification, often incorporating MRI guidance and simultaneous integrated boost (SIB). The FLAME trial was the first phase 3 randomized study to show an improvement in biochemical disease-free survival (bDFS) and comparable toxicities with the use of focal boost [26]. Recent systematic reviews and meta-analyses [27, 28] have provided level 1 evidence supporting this approach over whole-prostate HD-CFRT in terms of bDFS, while highlighting the need for further research into focal boost within the contexts of MHFRT and UHFRT.

The advancements in conformal techniques, such as IMRT and VMAT, have paved the way for the emergence of ultrahypofractionation, also known as SBRT, defined as fraction sizes exceeding 3.4 Gy. Over the past decade, this approach has revolutionized prostate cancer RT and significantly influenced clinical practice patterns. Accumulating evidence [29, 30] has demonstrated that UHFRT is a safe and effective treatment option, offering comparable oncological outcomes across all clinical risk categories of prostate cancer, although it is supported by a lower category of evidence in the high-risk group [31]. SBRT has also emerged as a valuable approach in the management of locally recurrent and oligometastatic disease, establishing itself as a highly effective option in this context [32–37]. SBRT is also a feasible and effective option for patients presenting with nodal-only recurrence [38, 39]. Recently, reirradiation using SBRT has gained attention, with a growing body of evidence supporting its use [40].

Although the theoretical rationale for using protons or heavy particles is sound, current evidence does not clearly demonstrate their superiority in terms of toxicity or oncological outcomes compared to standard-of-care advanced conformal photon-based techniques [41–43]. Given the high costs and logistical challenges [44], the incorporation of these techniques into the routine management of prostate cancer remains premature and is unlikely to be implemented in the foreseeable future.

Likely, no intervention aside from the classic standards of care—RT and radical prostatectomy—has had a greater impact on oncological outcomes of prostate cancer than the integration of ADT with RT. The SoC now includes a short duration of four to 6 months for intermediate-risk disease and a longer duration of 2–3 years for high-risk diseases. While data on overall survival (OS) benefits for intermediate-risk disease remains somewhat inconsistent [45], the addition of ADT has consistently demonstrated improvements in biochemical control (BCR) and metastasis-free survival (MFS) [46–48]. Sequencing ADT with RT, whether as neoadjuvant or adjuvant therapy, likely has minimal impact on oncological outcomes, with both approaches being considered acceptable [49]. If a neoadjuvant approach is chosen, only a short course of ADT should be administered before RT, with the remaining duration given concurrently during RT or as part of adjuvant therapy [47].

Prostate RT in the context of metastatic castration-sensitive prostate cancer (mCSPC) represents a significant breakthrough from the last decade, extending the use of EBRT to the metastatic setting. Prostate RT, largely supported by findings from the landmark STAMPEDE trial [50], has now been established as a recommended SoC for patients with low-volume mCSPC. Although the STAMPEDE trial demonstrated improved OS in this setting, subsequent meta-analyses [51] and findings from the recent PEACE-1 trial [52] have not validated these results when novel hormonal therapies are employed in the first-line setting. Nevertheless, prostate RT improves radiographic progression-free survival (rPFS) and reduces the risk of high-grade GU events in this setting [52].

Lastly, there is the case for RLT, a form of brachytherapy, which has improved outcomes in metastatic castration-resistant prostate cancer (mCRPC) and, more recently, in the setting of mCSPC [53]. Lutetium-177, a β-emitting radionuclide with a favorable half-life of 6.6 days and an energy range of 100–200 KeV, has become the preferred radionuclide. Actinium-225 has also been explored, yielding promising results [54, 55]. Lutetium-177 is now an FDA-approved option for the treatment of mCRPC, both in taxane-naïve [56, 57] and taxane-resistant [58] cases, following progression on a novel hormone therapy agent. While favorable outcomes have been observed in terms of response, PFS, and toxicity, no OS benefit has been demonstrated to date [59].

11.3 Urothelial Cancers Radiation Therapy

Arguably the most notable advancement in the field of RT for urothelial cancers over the past two decades is the emergence of chemoradiation as a viable alternative SoC to radical cystectomy (RC), which has long been the traditional standard treatment for MIBC. In the absence of robust randomized data, meta-analyses have consistently demonstrated the safety and efficacy of chemoradiation as part of a trimodality approach, with some analyses reporting outcomes comparable to those of RC [60–62]. Whether employing a continuous or split-course technique for EBRT, trimodality therapy (TMT) has been consistently associated with excellent 5-year OS [63]. Achieving optimal outcomes with TMT depends significantly on appropriate patient selection. NCCN guidelines state that "Optimal candidates for bladder preservation with chemoradiotherapy include patients with tumors that present without moderate/severe hydronephrosis, are without concurrent extensive or multifocal CIS, and are <6 cm". A hypofractionated regimen of 55 Gy in 20 fractions appears to be at least as effective as, and potentially superior to, the conventional regimen of 64 Gy in 32 fractions [64].

Integrating immunotherapy with standard chemoradiation protocols may improve bladder preservation outcomes. Multiple ongoing trials are investigating this approach and their results are eagerly awaited [65–67]. A recently completed phase II trial combining durvalumab and tremelimumab with chemoradiation as part of TMT demonstrated an impressive complete response rate of 81% [68]. Currently, major guidelines do not recommend concurrent immunotherapy as an option; however, this approach is anticipated to become a standard in the near future, given the encouraging outcomes observed in neoadjuvant and adjuvant settings.

For non-muscle invasive bladder cancer (NMIB), while major guidelines do not currently include RT as a treatment option, emerging evidence suggests that high-risk tumors may be effectively controlled with RT. A recent meta-analysis [69], encompassing studies conducted between 1986 and 2019 and including only one RCT from 2007 [70], reported acceptable outcomes with RT. Five-year relapse-free survival (RFS), cancer-specific survival (CSS), and OS were 54%, 86%, and 72%, respectively. Notably, only 13% of patients required salvage RC, and only 9% developed metastatic disease. Despite the low quality of current evidence, with just one randomized trial available, the role of RT in managing NMIBC warrants renewed consideration, particularly for BCG-unresponsive cases where non-radical treatment options remain limited.

For urothelial carcinomas of the upper tract, radiation therapy (RT) has traditionally played a limited role. Radical nephroureterectomy remains the standard treatment, with adjuvant RT reserved for select cases. Recent systematic reviews and meta-analyses have consistently concluded that adjuvant RT offers no survival benefit for upper tract urothelial carcinoma [71–73]. Adjuvant RT may improve locoregional control [72], particularly in margin-positive cases; however, this potential benefit must be weighed against the risk of increased toxicity and associated complications.

11.4 Renal Cell Carcinoma Radiation Therapy

RCC has long been recognized for its resistance to conventional radiation schedules. However, the introduction of extreme hypofractionation (SBRT, SABR, and SRS) has established radiation as a viable option in the management of both primary and metastatic RCC.

SABR is now a recognized treatment option for medically inoperable, localized cT1a and cT1b (≤7 cm) tumors. Common recommended dose fractionation regimens include 25–26 Gy in a single fraction for tumors less than 4–5 cm in size, and 42–48 Gy in three fractions for larger tumors [74]. A recent meta-analysis demonstrated excellent short-term outcomes, with 2-year local control rates of 97% and overall survival (OS) rates of 85%, accompanied by a very low incidence of high-grade toxicities [75]. Preliminary studies have indicated a potential role for SABR in the neoadjuvant management of locally advanced cases with inferior vena cava tumor thrombus, a challenging clinical scenario [76].

Oligometastatic and oligoprogressive disease represent additional contexts in which SBRT/SABR have gained acceptance as recommendable treatment options with favorable treatment outcomes [77, 78]. Radiosurgery is now standard care for brain metastases, including RCC. Studies have demonstrated the safety and efficacy of integrating these treatments with systemic therapy standards of care [79–81].

Although the kidneys are traditionally considered fixed organs, significant respiration-associated motion has been documented [82], posing challenges to the delivery of high-dose ablative therapies. Consequently, implementing motion control measures is essential when utilizing SABR for primary renal tumors [83]. MR-guided adaptive RT (SMART) is an advanced technique with early studies demonstrating its feasibility and safety in the application of SABR for renal tumors [84]. With adaptive daily planning and real-time tracking, this high-precision technique has the potential to reduce adverse events and improve oncologic outcomes.

11.5 Testicular Cancer Radiation Therapy

Historically, RT has played a limited role in the treatment of testicular germ cell tumors (GCTs), with the exception of stage I and II seminoma [85, 86]. Whole-brain RT (WBRT) continues to be a standard approach for managing brain metastases associated with GCTs. There have been no significant advancements in the radiotherapeutic management of testicular germ cell tumors (GCTs). Nonetheless, two topics deserve mentioning.

The first is the increasing interest in utilizing proton radiation for the radiotherapeutic treatment of GCTs. GCTs primarily affect young individuals. While RT is highly effective, its use presents concerns regarding secondary malignancies, due to the high curability of GCTs and the extended life expectancy of most patients. Proton therapy (PT) offers potential benefits, including improved organ sparing and a reduced integral radiation dose, which contribute to lower acute and late toxicities as well as a decreased risk of secondary cancers [87–92]. PT has emerged as a viable and increasingly preferred alternative to 3DCRT and IMRT/VMAT in the management of GCTs, particularly in the treatment of stage I and II seminoma, which generally involves a significantly larger treatment area compared to, for instance, primary brain GCTs.

The second is the emergence of stereotactic radiotherapy (SRT) as a potentially viable option for managing metastases from germ cell tumors (GCTs), particularly brain metastases originating from non-seminomatous germ cell tumors (NSGCTs), as well as cases where surgical intervention is either technically challenging or medically not recommended [93–97].

11.6 Conclusions

Over the past two decades, radiotherapy advancements for genitourinary cancers have primarily focused on enhancing precision, with the integration of high-dose stereotactic techniques becoming a significant milestone in the last decade. Hypofractionated and stereotactic treatments have firmly established themselves as standards of care across diverse clinical scenarios, including in the metastatic setting. The integration of systemic therapies, including androgen deprivation therapy, chemotherapy, and immunotherapy, represents another significant advancement in the radiotherapeutic management of prostate and bladder cancer. Radioligand therapy has emerged as a viable option in the management of metastatic prostate cancer. In conclusion, while novel systemic therapies have transformed the management of advanced genitourinary cancers, particularly prostate and bladder cancer, radiotherapy continues to be a cornerstone in the treatment of both early and advanced disease. Advancing precision, expanding therapeutic indications, and developing personalized treatment strategies are expected to remain central areas of focus in the years ahead.

References

1. Tian YQ, Yang JC, Hu JJ, Ding R, Ye DW, Shang JW. Trends and risk factors of global incidence, mortality, and disability of genitourinary cancers from 1990 to 2019: systematic analysis for the Global Burden of Disease Study 2019. Front Public Health. 2023;11:1119374.
2. Bray F, Laversanne M, Sung H, Ferlay J, Siegel RL, Soerjomataram I, Jemal A. Global cancer statistics 2022: GLOBOCAN estimates of incidence and mortality worldwide for 36 cancers in 185 countries. CA Cancer J Clin. 2024;74(3):229–63.
3. Cumberbatch MGK, Jubber I, Black PC, Esperto F, Figueroa JD, Kamat AM, et al. Epidemiology of bladder cancer: a systematic review and contemporary update of risk factors in 2018. Eur Urol. 2018;74(6):784–95.

4. Al-Bayati O, Hasan A, Pruthi D, Kaushik D, Liss MA. Systematic review of modifiable risk factors for kidney cancer. Urol Oncol. 2019;37(6):359–71.
5. Bergengren O, Pekala KR, Matsoukas K, Fainberg J, Mungovan SF, Bratt O, et al. 2022 update on prostate cancer epidemiology and risk factors-a systematic review. Eur Urol. 2023;84(2):191–206.
6. Hemminki K, Kiemeney LA, Morgans AK, Ranniko A, Pichler R, Hemminki O, et al. Hereditary and familial traits in urological cancers and their underlying genes. Eur Urol Open Sci. 2024;69:13–20.
7. Yu T, Zhang Q, Zheng T, Shi H, Liu Y, Feng S, et al. The effectiveness of intensity modulated radiation therapy versus three-dimensional radiation therapy in prostate cancer: a meta-analysis of the literatures. PLoS One. 2016;11(5):e0154499.
8. Guo W, Sun YC, Zhang LY, Yin XM. Gastrointestinal/genitourinary adverse event after intensity modulated versus three-dimensional primary radiation therapy in the treatment of prostate cancer: systematic review and meta-analysis. J Cancer. 2023;14(15):2878–88.
9. de Crevoisier R, Bayar MA, Pommier P, Muracciole X, Pêne F, Dudouet P, et al. Daily versus weekly prostate cancer image guided radiation therapy: phase 3 multicenter randomized trial. Int J Radiat Oncol Biol Phys. 2018;102(5):1420–9.
10. Ren W, Sun C, Lu N, Xu Y, Han F, Liu YP, Dai J. Dosimetric comparison of intensity-modulated radiotherapy and volumetric-modulated arc radiotherapy in patients with prostate cancer: a meta-analysis. J Appl Clin Med Phys. 2016;17(6):254–62.
11. Leeman JE, Shin KY, Chen YH, Mak RH, Nguyen PL, D'Amico AV, Martin NE. Acute toxicity comparison of magnetic resonance-guided adaptive versus fiducial or computed tomography-guided non-adaptive prostate stereotactic body radiotherapy: a systematic review and meta-analysis. Cancer. 2023;129(19):3044–52.
12. Tocco BR, Kishan AU, Ma TM, Kerkmeijer LGW, Tree AC. MR-guided radiotherapy for prostate cancer. Front Oncol. 2020;10:616291.
13. Kishan AU, Ma TM, Lamb JM, Casado M, Wilhalme H, Low DA, et al. Magnetic resonance imaging-guided vs computed tomography-guided stereotactic body radiotherapy for prostate cancer: the MIRAGE randomized clinical trial. JAMA Oncol. 2023;9(3):365–73.
14. Miller LE, Efstathiou JA, Bhattacharyya SK, Payne HA, Woodward E, Pinkawa M. Association of the Placement of a perirectal hydrogel spacer with the clinical outcomes of men receiving radiotherapy for prostate cancer: a systematic review and meta-analysis. JAMA Netw Open. 2020;3(6):e208221.
15. Armstrong N, Bahl A, Pinkawa M, Ryder S, Ahmadu C, Ross J, et al. SpaceOAR hydrogel spacer for reducing radiation toxicity during radiotherapy for prostate cancer. A systematic review. Urology. 2021;156:e74–85.
16. Payne HA, Pinkawa M, Peedell C, Bhattacharyya SK, Woodward E, Miller LE. SpaceOAR hydrogel spacer injection prior to stereotactic body radiation therapy for men with localized prostate cancer: a systematic review. Medicine (Baltimore). 2021;100(49):e28111.
17. Wong CH, Ko IC, Leung DK, Yuen SK, Siu B, Yuan C, Teoh JY. Does biodegradable peri-rectal spacer mitigate treatment toxicities in radiation therapy for localised prostate cancer-a systematic review and meta-analysis. Prostate Cancer Prostatic Dis. 2025; https://doi.org/10.1038/s41391-025-00961-0.
18. Aminsharifi A, Kotamarti S, Silver D, Schulman A. Major complications and adverse events related to the injection of the SpaceOAR hydrogel system before radiotherapy for prostate cancer: review of the manufacturer and user facility device experience database. J Endourol. 2019;33(10):868–71.
19. Viani GA, Stefano EJ, Afonso SL. Higher-than-conventional radiation doses in localized prostate cancer treatment: a meta-analysis of randomized, controlled trials. Int J Radiat Oncol Biol Phys. 2009;74(5):1405–18.
20. Kissel M, Krhili SL, Minsat M, El Ayachy R, Bringer S, Lahmi L, et al. Dose-escalation in prostate cancer: results of randomized trials. Cancer/Radiothérapie. 2022;26(6):899–904.

21. Ferella L, Limoncin E, Vittorini F, Chalaszczyk A, Sorce C, Grimaldi G, et al. Are we ready for a paradigm shift from high-dose conventional to moderate hypofractionated radiotherapy in intermediate-high risk prostate cancer? A systematic review of randomized controlled trials with trial sequential analysis. Crit Rev Oncol Hematol. 2019;139:75–82.
22. Sinzabakira F, Brand V, Heemsbergen WD, Incrocci L. Acute and late toxicity patterns of moderate hypo-fractionated radiotherapy for prostate cancer: a systematic review and meta-analysis. Clin Transl Radiat Oncol. 2023;40:100612.
23. Lilleby O, Meidahl Petersen P, Daugaard G, Anne Perell K. Current evidence for moderate and ultra-hypofractionated radiation therapy in prostate cancer: a summary of the results from phase 3 randomised trials. Scand J Urol. 2023;58:21–7.
24. Niazi T, Nabid A, Malagon T, Tisseverasinghe S, Bettahar R, Dahmane R, et al. Hypofractionated dose escalation radiotherapy for high-risk prostate cancer: the survival analysis of the prostate cancer Study-5 (PCS-5), a GROUQ-led phase III trial. Eur Urol. 2025;87(3):314–23.
25. Buyyounouski MK, Pugh SL, Chen RC, Mann MJ, Kudchadker RJ, Konski AA, et al. Noninferiority of Hypofractionated vs conventional Postprostatectomy radiotherapy for genitourinary and gastrointestinal symptoms: the NRG-GU003 Phase 3 Randomized Clinical Trial. JAMA Oncol. 2024;10(5):584–91.
26. Kerkmeijer LGW, Groen VH, Pos FJ, Haustermans K, Monninkhof EM, Smeenk RJ, et al. Focal boost to the Intraprostatic tumor in external beam radiotherapy for patients with localized prostate cancer: results from the FLAME randomized phase III trial. J Clin Oncol. 2021;39(7):787–96.
27. Poon DMC, Yuan J, Yang B, Kerkmeijer LGW, Kishan AU, Murthy V, et al. Magnetic resonance imaging–guided focal boost to Intraprostatic lesions using external beam radiotherapy for localized prostate cancer: a systematic review and meta-analysis. Eur Urol Oncol. 2023;6(2):116–27.
28. Dornisch AM, Zhong AY, Poon DMC, Tree AC, Seibert TM. Focal radiotherapy boost to MR-visible tumor for prostate cancer: a systematic review. World J Urol. 2024;42(1):56.
29. Cushman TR, Verma V, Khairnar R, Levy J, Simone CB 2nd, Mishra MV. Stereotactic body radiation therapy for prostate cancer: systematic review and meta-analysis of prospective trials. Oncotarget. 2019;10(54):5660–8.
30. Jackson WC, Silva J, Hartman HE, Dess RT, Kishan AU, Beeler WH, et al. Stereotactic body radiation therapy for localized prostate cancer: a systematic review and meta-analysis of over 6,000 patients treated on prospective studies. Int J Radiat Oncol Biol Phys. 2019;104(4):778–89.
31. Foerster R, Zwahlen DR, Buchali A, Tang H, Schroeder C, Windisch P, et al. Stereotactic body radiotherapy for high-risk prostate cancer: a systematic review. Cancers (Basel). 2021;13(4):759.
32. Connor MJ, Smith A, Miah S, Shah TT, Winkler M, Khoo V, Ahmed HU. Targeting Oligometastasis with stereotactic ablative radiation therapy or surgery in metastatic hormone-sensitive prostate cancer: a systematic review of prospective clinical trials. Eur Urol Oncol. 2020;3(5):582–93.
33. Viani GA, Arruda CV, Hamamura AC, Faustino AC, Freitas Bendo Danelichen A, Guimarães FS. Stereotactic body radiotherapy for Oligometastatic prostate cancer recurrence: a meta-analysis. Am J Clin Oncol. 2020;43(2):73–81.
34. Yan M, Moideen N, Bratti VF, Moraes FY. Stereotactic body radiotherapy (SBRT) in metachronous oligometastatic prostate cancer: a systematic review and meta-analysis on the current prospective evidence. Br J Radiol. 2020;93(1116):20200496.
35. Marvaso G, Volpe S, Pepa M, Augugliaro M, Corrao G, Biffi A, et al. Oligorecurrent prostate cancer and stereotactic body radiotherapy: where are we now? A systematic review and meta-analysis of prospective studies. Eur Urol Open Sci. 2021;27:19–28.
36. Persson AE, Hallqvist A, Bjørn Larsen L, Rasmussen M, Scherman J, Nilsson P, et al. Stereotactic body radiotherapy as metastasis-directed therapy in oligometastatic prostate

cancer: a systematic review and meta-analysis of randomized controlled trials. Radiat Oncol. 2024;19(1):173.
37. Meng Y, Liu J, Shen B, Xu H, Wu D, Ying Y. Evaluation of the safety and efficacy of stereotactic body radiotherapy for radio-recurrent prostate cancer: a systematic review and meta-analysis. Prostate Cancer Prostatic Dis. 2024; https://doi.org/10.1038/s41391-024-00927-8.
38. Zamagni A, Bonetti M, Buwenge M, Macchia G, Deodato F, Cilla S, et al. Stereotactic radiotherapy of nodal oligometastases from prostate cancer: a prisma-compliant systematic review. Clin Exp Metastasis. 2022;39(6):845–63.
39. Aman A, Akram A, Akram B, Husnain A, Akram A, Akram S, et al. Salvage stereotactic radiotherapy for nodal oligo-recurrent prostate cancer: a systematic review and meta-analysis of randomized controlled trials. Clin Genitourin Cancer. 2024;22(6):102239.
40. Schröder C, Tang H, Lenffer B, Buchali A, Zwahlen DR, Förster R, Windisch P. Re-irradiation to the prostate using stereotactic body radiotherapy (SBRT) after initial definitive radiotherapy – a systematic review and meta-analysis of recent trials. Clin Transl Radiat Oncol. 2024;48:100806.
41. Li M, Li X, Yao L, Han X, Yan W, Liu Y, et al. Clinical efficacy and safety of proton and carbon ion radiotherapy for prostate cancer: a systematic review and meta-analysis. Front Oncol. 2021;11:709530.
42. Corrao G, Marvaso G, Mastroleo F, Biffi A, Pellegrini G, Minari S, et al. Photon vs proton hypofractionation in prostate cancer: a systematic review and meta-analysis. Radiother Oncol. 2024;195:110264.
43. Wisdom AJ, Yeap BY, Michalski JM, Horick NK, Zietman AL, Christodouleas JP, et al. Setting the stage: feasibility and baseline characteristics in the PARTIQoL trial comparing proton therapy versus intensity modulated radiation therapy for localized prostate cancer. Int J Radiat Oncol Biol Phys. 2025;121(3):741–51.
44. Schroeck FR, Jacobs BL, Bhayani SB, Nguyen PL, Penson D, Hu J. Cost of new Technologies in Prostate Cancer Treatment: systematic review of costs and cost effectiveness of robotic-assisted laparoscopic prostatectomy, intensity-modulated radiotherapy, and proton beam therapy. Eur Urol. 2017;72(5):712–35.
45. Krauss DJ, Karrison T, Martinez AA, Morton G, Yan D, Bruner DW, et al. Dose-escalated radiotherapy alone or in combination with short-term androgen deprivation for intermediate-risk prostate cancer: results of a phase III multi-institutional trial. J Clin Oncol. 2023;41(17):3203–16.
46. Nabid A, Carrier N, Vigneault E, Van Nguyen T, Vavassis P, Brassard MA, et al. Androgen deprivation therapy and radiotherapy in intermediate-risk prostate cancer: a randomised phase III trial. Eur J Cancer. 2021;143:64–74.
47. Kishan AU, Sun Y, Hartman H, Pisansky TM, Bolla M, Neven A, et al. Androgen deprivation therapy use and duration with definitive radiotherapy for localised prostate cancer: an individual patient data meta-analysis. Lancet Oncol. 2022;23(2):304–16.
48. Kishan AU, Wang X, Sun Y, Romero T, Michalski JM, Ma TM, et al. High-dose radiotherapy or androgen deprivation therapy (HEAT) as treatment intensification for localized prostate cancer: an individual patient-data network meta-analysis from the MARCAP consortium. Eur Urol. 2022;82(1):106–14.
49. Malone S, Roy S, Eapen L, Choan E, MacRae R, Perry G, et al. Sequencing of androgen-deprivation therapy with external-beam radiotherapy in localized prostate cancer: a phase III randomized controlled trial. J Clin Oncol. 2020;38(6):593–601.
50. Parker CC, James ND, Brawley CD, Clarke NW, Ali A, Amos CL, et al. Radiotherapy to the prostate for men with metastatic prostate cancer in the UK and Switzerland: long-term results from the STAMPEDE randomised controlled trial. PLoS Med. 2022;19(6):e1003998.
51. Roy S, Fervaha G, Spratt DE, Sun Y, Kishan AU, Loblaw A, et al. Prostate radiotherapy in Low-volume metastatic hormone-sensitive prostate cancer: a network meta-analysis. Eur Urol. 2024;86(1):10–7.

52. Bossi A, Foulon S, Maldonado X, Sargos P, MacDermott R, Kelly P, et al. Efficacy and safety of prostate radiotherapy in de novo metastatic castration-sensitive prostate cancer (PEACE-1): a multicentre, open-label, randomised, phase 3 study with a 2 × 2 factorial design. Lancet. 2024;404(10467):2065–76.
53. Laudicella R, Bauckneht M, Burger IA, Cacciola A, Fanti S, Farolfi A, et al. The role of PSMA-based radioligand therapy in hormone-sensitive prostate cancer. Eur J Nucl Med Mol Imaging. 2025;52:2723.
54. Sathekge MM, Lawal IO, Bal C, Bruchertseifer F, Ballal S, Cardaci G, et al. Actinium-225-PSMA radioligand therapy of metastatic castration-resistant prostate cancer (WARMTH Act): a multicentre, retrospective study. Lancet Oncol. 2024;25(2):175–83.
55. Parida GK, Panda RA, Bishnoi K, Agrawal K. Efficacy and safety of Actinium-225 prostate-specific membrane antigen Radioligand therapy in metastatic prostate cancer: a systematic review and Metanalysis. Med Princ Pract. 2023;32(3):178–91.
56. Morris MJ, Castellano D, Herrmann K, de Bono JS, Shore ND, Chi KN, et al. ¹⁷⁷Lu-PSMA-617 versus a change of androgen receptor pathway inhibitor therapy for taxane-naive patients with progressive metastatic castration-resistant prostate cancer (PSMAfore): a phase 3, randomised, controlled trial. Lancet. 2024;404(10459):1227–39.
57. Almeida LS, García Megías I, Etchebehere E, Calapaquí Terán AK, Herrmann K, Giammarile F, et al. Assessment of the therapeutic efficacy of [(177)Lu]Lu-PSMA-X compared to taxane chemotherapy in taxane-chemo-naïve patients with metastatic castration-resistant prostate cancer: a systematic review and meta-analysis. Eur J Nucl Med Mol Imaging. 2025;52(3):936–54.
58. Sartor O, Bono J, Chi KN, Fizazi K, Herrmann K, Rahbar K, et al. Lutetium-177–PSMA-617 for metastatic castration-resistant prostate cancer. N Engl J Med. 2021;385(12):1091–103.
59. Ciccarese C, Bauckneht M, Zagaria L, Fornarini G, Beccia V, Lanfranchi F, et al. Defining the position of [(177)Lu]Lu-PSMA Radioligand therapy in the treatment landscape of metastatic castration-resistant prostate cancer: a meta-analysis of clinical trials. Target Oncol. 2025;20(1):103–12.
60. Fahmy O, Khairul-Asri MG, Schubert T, Renninger M, Malek R, Kübler H, et al. A systematic review and meta-analysis on the oncological long-term outcomes after trimodality therapy and radical cystectomy with or without neoadjuvant chemotherapy for muscle-invasive bladder cancer. Urol Oncol. 2018;36(2):43–53.
61. Matsukawa A, Yanagisawa T, Miszczyk M, Kardoust Parizi M, Fazekas T, Tsuboi I, et al. Trimodality therapy versus radical cystectomy for muscle-invasive bladder cancer: a systematic review and meta-analysis of matched cohort studies. Eur Urol Focus. 2024; https://doi.org/10.1016/j.euf.2024.11.003.
62. Fong KY, Lim EJ, Wong HC, Tay KJ, Gan VHL, Ho HSS, et al. Trimodality therapy versus radical cystectomy for muscle-invasive bladder cancer: a systematic review and meta-analysis. Urol Oncol. 2025;43:412.
63. Arcangeli G, Arcangeli S, Strigari L. A systematic review and meta-analysis of clinical trials of bladder-sparing trimodality treatment for muscle-invasive bladder cancer (MIBC). Crit Rev Oncol Hematol. 2015;94(1):105–15.
64. Choudhury A, Porta N, Hall E, Song YP, Owen R, MacKay R, et al. Hypofractionated radiotherapy in locally advanced bladder cancer: an individual patient data meta-analysis of the BC2001 and BCON trials. Lancet Oncol. 2021;22(2):246–55.
65. Singh P, Efstathiou JA, Tangen C, Jhavar SG, Hahn NM, Costello BA, et al. INTACT (S/N1806) phase III randomized trial of concurrent chemoradiotherapy with or without atezolizumab in localized muscle-invasive bladder cancer: safety update on first 73 patients. J Clin Oncol. 2021;39(6_suppl):428.
66. Joshi M, Kim SE, Solanki AA, Miyamoto DT, Degraff D, Zou JW, et al. EA8185: phase 2 study of bladder-sparing chemoradiation (chemoRT) with durvalumab in clinical stage III, node-positive urothelial carcinoma (INSPIRE), an ECOG-ACRIN/NRG collaboration. J Clin Oncol. 2022;40(16_suppl):TPS4617-TPS.

67. Tissot G, Xylinas E. Efficacy and safety of Pembrolizumab (MK-3475) in combination with Chemoradiotherapy versus Chemoradiotherapy alone in muscle-invasive bladder cancer: the MK-3475-992/KEYNOTE-992 trial. Eur Urol Focus. 2023;9(2):227–8.
68. Garcia-Del-Muro X, Valderrama BP, Medina-Colmenero A, Etxaniz O, Gironés Sarrió R, Juan-Fita MJ, et al. Bladder preservation with Durvalumab plus Tremelimumab and concurrent radiotherapy in patients with localized muscle-invasive bladder cancer (IMMUNOPRESERVE): a phase II Spanish oncology GenitoUrinary group trial. Clin Cancer Res. 2025;31(4):659–66.
69. Rodrigues Pessoa R, Mueller AC, Boxley P, Flaig TW, Piper C, Konety B, et al. Systematic review and meta-analysis of radiation therapy for high-risk non-muscle invasive bladder cancer. Urol Oncol. 2021;39(11):786.e1–e8.
70. Harland SJ, Kynaston H, Grigor K, Wallace DM, Beacock C, Kockelbergh R, et al. A randomized trial of radical radiotherapy for the management of pT1G3 NXM0 transitional cell carcinoma of the bladder. J Urol. 2007;178(3 Pt 1):807–13. discussion 13
71. Iwata T, Kimura S, Abufaraj M, Janisch F, Karakiewicz PI, Seebacher V, et al. The role of adjuvant radiotherapy after surgery for upper and lower urinary tract urothelial carcinoma: a systematic review. Urol Oncol. 2019;37(10):659–71.
72. Zalay O, Yan M, Sigurdson S, Malone S, Vera-Badillo FE, Mahmud A. Adjuvant radiotherapy for upper tract urothelial carcinoma: systematic review and meta-analysis. Curr Oncol. 2022;30(1):19–36.
73. Zhou X, Luo G. Is radiotherapy helpful for the prognosis of upper urinary tract urothelial carcinoma? A meta-analysis. Urol Oncol. 2022;40(10):424–33.
74. Siva S, Louie AV, Kotecha R, Barber MN, Ali M, Zhang Z, et al. Stereotactic body radiotherapy for primary renal cell carcinoma: a systematic review and practice guideline from the International Society of Stereotactic Radiosurgery (ISRS). Lancet Oncol. 2024;25(1):e18–28.
75. Suleja A, Bilski M, Laukhtina E, Fazekas T, Matsukawa A, Tsuboi I, et al. Stereotactic body radiotherapy (SBRT) for the treatment of primary localized renal cell carcinoma: a systematic review and meta-analysis. Cancers (Basel). 2024;16(19):3276.
76. Margulis V, Freifeld Y, Pop LM, Manna S, Kapur P, Pedrosa I, et al. Neoadjuvant SABR for renal cell carcinoma inferior vena cava tumor thrombus-safety Lead-in results of a phase 2 trial. Int J Radiat Oncol Biol Phys. 2021;110(4):1135–42.
77. Villafuerte CJQ, Swaminath A. Stereotactic body radiotherapy for renal cell carcinoma-a review of use in the primary, Cytoreductive and Oligometastatic settings. Cancers (Basel). 2024;16(19):3334.
78. Le Guevelou J, Sargos P, Siva S, Ploussard G, Ost P, Gillessen S, Zilli T. The emerging role of Extracranial stereotactic ablative radiotherapy for metastatic renal cell carcinoma: a systematic review. Eur Urol Focus. 2023;9(1):114–24.
79. Zhang S, Xiong X, Xie N, Zheng W, Li Y, Lin T, et al. The efficacy and safety of stereotactic body radiotherapy combined with systematic therapy for metastatic renal cell carcinoma: a systematic review and meta-analysis. MedComm (2020). 2024;5(5):e544.
80. Khan M, Zhao Z, Arooj S, Liao G. Impact of Tyrosine Kinase Inhibitors (TKIs) combined with radiation therapy for the Management of Brain Metastases from Renal Cell Carcinoma. Front Oncol. 2020;10:1246.
81. Podder V, Ranjan T, Margolin K, Maharaj A, Ahluwalia MS. Evaluating the safety of immune checkpoint inhibitors and combination therapies in the Management of Brain Metastases: a comprehensive review. Cancers (Basel). 2024;16(23)
82. Pham D, Kron T, Foroudi F, Schneider M, Siva S. A review of kidney motion under free, deep and forced-shallow breathing conditions: implications for stereotactic ablative body radiotherapy treatment. Technol Cancer Res Treat. 2014;13(4):315–23.
83. Barbour AB, Kirste S, Grosu AL, Siva S, Louie AV, Onishi H, et al. The judicious use of stereotactic ablative radiotherapy in the primary Management of Localized Renal Cell Carcinoma. Cancers (Basel). 2023;15(14)

84. Yim K, Hsu SH, Nolazco JI, Cagney D, Mak RH, D'Andrea V, et al. Stereotactic magnetic resonance-guided adaptive radiation therapy for localized kidney cancer: early outcomes from a prospective phase 1 trial and supplemental cohort. Eur Urol Oncol. 2024;7(1):147–50.
85. Petrelli F, Coinu A, Cabiddu M, Ghilardi M, Borgonovo K, Lonati V, Barni S. Surveillance or adjuvant treatment with chemotherapy or radiotherapy in stage I seminoma: a systematic review and meta-analysis of 13 studies. Clin Genitourin Cancer. 2015;13(5):428–34.
86. Giannatempo P, Greco T, Mariani L, Nicolai N, Tana S, Farè E, et al. Radiotherapy or chemotherapy for clinical stage IIA and IIB seminoma: a systematic review and meta-analysis of patient outcomes. Ann Oncol. 2015;26(4):657–68.
87. Pasalic D, Prajapati S, Ludmir EB, Tang C, Choi S, Kudchadker R, Frank SJ. Outcomes and toxicities of proton and photon radiation therapy for testicular seminoma. Int J Particle Ther. 2020;7(2):11–20.
88. Xiang M, Chang DT, Pollom EL. Second cancer risk after primary cancer treatment with three-dimensional conformal, intensity-modulated, or proton beam radiation therapy. Cancer. 2020;126(15):3560–8.
89. Lui A, Zeng J, Chen J, Weg ES, Ellis W, Psutka SP, et al. Proton radiation therapy for stage IIA/IIB testicular seminoma. Int J Radiat Oncol Biol Phys. 2023;117(2):e411–e2.
90. Maxwell R, Chang Y, Paul C, Vaughn DJ, Christodouleas JP. Cancer control, toxicity, and secondary malignancy risks of proton radiation therapy for stage I-IIB testicular seminoma. Adv Radiat Oncol. 2023;8(5):101259.
91. Rønde HS, Kronborg C, Høyer M, Hansen J, Bak ME, Agergaard SN, et al. Dose comparison of robustly optimized intensity modulated proton therapy (IMPT) vs IMRT and VMAT photon plans for testicular seminoma. Acta Oncol. 2023;62(10):1222–9.
92. Pursley J, Remillard K, Depauw N, Lee G, Grassberger C, Paganetti H, et al. Radiation therapy for stage IIA/B seminoma: modeling secondary cancer risk for protons and VMAT versus 3D photons. Cancers (Basel). 2024;16(4)
93. Laflamme P, Doucet C, Sirois C, Kopek N, Vanhuyse M. Stereotactic radiation therapy for residual chemorefractory primary mediastinal non-seminomatous germ cell tumor after surgical thoracotomy. Pract Radiat Oncol. 2017;7(4):260–3.
94. Satyanarayan A, Mooney R, Bhanvadia RR, Iyengar P, Margulis V, Desai NB, Bagrodia A. Stereotactic ablative radiotherapy (SAbR) in the setting of metastatic Nonseminomatous germ cell tumor of testis. Clin Genitourin Cancer. 2019;17(4):e768–e71.
95. Khalifa J, Fléchon A, Chevreau C. Brain metastases from germ cell tumor: time to reconsider radiotherapy? Crit Rev Oncol Hematol. 2020;150:102946.
96. Gui C, Funt S, Boe LA, Yu Y, Zinovoy M, Chen L, et al. Stereotactic radiosurgery and whole brain radiotherapy for germ cell tumor brain metastases: practice patterns and outcomes. Int J Radiat Oncol Biol Phys. 2024;120(2):e233–e4.
97. Akhavan-Sigari A, Hori YS, Harary PM, Persad AR, Kassu R, Tayag A, et al. Stereotactic radiosurgery for Choriocarcinoma brain metastases: illustrative case presentation and systematic review. World Neurosurg. 2025;194:123387.

Chapter 12
Advances in Radiotherapy of Brain Tumors

Sanjesh Kumar, Mansi Singh, M. V. N. L. Chaitanya, Devesh Kapoor, Mohini Mondal, Biplab Debnath, Sumel Ashique, and Tahreen Taj

Abstract The complexity, heterogeneity, and poor prognosis of brain tumors present a substantial challenge in oncology, necessitating the continuous development of therapeutic modalities. With developments in intensity-modulated radiotherapy (IMRT), volumetric modulated arc therapy (VMAT), and stereotactic radiosurgery (SRS), radiotherapy (RT) remains the foundation of brain tumor treatment. Personalized techniques, which are driven by radiogenomics, integrate genetic profiling with radiotherapy to predict individual tumor activity and optimize treatment planning. This is demonstrated by the utilization of biomarkers such as IDH mutations and MGMT methylation levels. Advancements in imaging technologies, such as functional MRI (fMRI), diffusion tensor imaging (DTI), PET, and hybrid modalities like PET-MRI, have significantly altered the visualization of malignancies. This has enabled real-time adaptive interventions with exceptional spatial accuracy. Nevertheless, these developments, the underlying radioresistance, and tumor microenvironmental problems of recurrent and metastatic brain cancers still provide a treatment challenge. Emerging approaches such as tumor-treating fields (TTFs), AI-powered adaptive radiotherapy, and RT combined with immunotherapy provide promising options. These advancements underscore the transformative potential of radiotherapy in mitigating therapeutic limitations, although significant challenges

S. Kumar · M. Singh
Institute of Pharmaceutical Research, GLA University, Mathura, Uttar Pradesh, India

M. V. N. L. Chaitanya
School of Pharmaceutical Sciences, Lovely Professional University, Phagwara, Punjab, India

D. Kapoor
Dr. Dayaram Patel Pharmacy College, Surat, Gujarat, India

M. Mondal · B. Debnath · S. Ashique (✉)
Department of Pharmaceutical Technology, Bharat Technology, Uluberia, West Bengal, India

T. Taj
Department of Pharmacology, Yenepoya Pharmacy College and Research Centre, Yenepoya (Deemed to be) University, Mangalore, Karnataka, India

F. Taghizadeh-Hesary et al. (eds.), *From Radiobiology to Radiation Oncology*,
https://doi.org/10.1007/978-3-031-95724-6_12

persist in accessibility, cost, and treatment resistance. Overcoming these obstacles and releasing the full possibilities of tailored radiation depend on the combination of translational research and multidisciplinary approaches. This chapter presents a thorough examination of technology advances, clinical insights, and future perspectives in radiation for brain tumors, stressing its critical role in increasing patient survival and quality of life.

Keywords Brain tumors · Radiotherapy · Radiogenomics · Advanced imaging · Personalized medicine

12.1 Introduction

Brain tumors are a critical global health concern, with substantial morbidity and mortality stemming from their aggressive behavior, complexity, and limited treatment options. The treatment of brain tumors is exceptionally difficult due to their molecular and clinical heterogeneity, invasive nature, and location within vital neurological structures [1]. The responses of these tumors to treatment exhibit considerable variability, and their repercussions extend beyond traditional survival; they can induce substantial functional and cognitive impairments that diminish quality of life (QOL). The intricacies involved highlight the necessity for advanced therapeutic strategies that are tailored to the distinct tumor characteristics and individual patient needs [2]. Radiotherapy (RT) is currently a critical component of brain tumor treatment, and it is employed either independently or in conjunction with surgery and systemic treatments, such as chemotherapy and immunotherapy. The results have been significantly enhanced by its ability to precisely target tumor cells while avoiding healthy cerebral tissue. The efficacy of RT is frequently impeded by biological variables, such as genetic diversity, complicated microenvironments, and inherent tumor resistance [3]. Moreover, clinical challenges such as radiation-induced toxicity, including edema, necrosis, and delayed neurological sequelae, continue to pose substantial hurdles. These challenges compel a review of conventional RT techniques, including new knowledge and modern technologies to maximize therapeutic index. The emergence of customized medicine has transformed brain tumor care, transitioning from standardized protocols to tailored treatment approaches [4]. By use of radiogenomics, predictive and prognostic biomarkers including IDH mutations, MGMT promoter methylation, and TERT promoter mutations have been identified, therefore enabling exact prediction of therapy responses and disease trajectories. Concurrent developments in computational techniques have improved our ability to dynamically simulate tumor behavior, therefore enabling adaptive treatment plans that change with therapy. Advances in imaging technology, where modalities include functional MRI (fMRI), tractography, and PET imaging, provide high-resolution observation of both tumor margins and functioning brain pathways, complement these advances [5]. Real-time imaging capabilities, supported by innovations like MR-guided radiotherapy systems, offer opportunities for intra-treatment adjustments that significantly improve therapeutic precision [6]. The application of radiotherapy for brain tumours has been further redefined by technological advancements in delivery systems. Different modalities of radiation therapy, including intensity-modulated radiotherapy (IMRT), volumetric modulated arc therapy (VMAT), and stereotactic approaches including stereotactic radiosurgery (SRS) and stereotactic body radiotherapy (SBRT), can precisely target and administer radiation, thus safeguarding vital structures to the greatest extent possible. Brain cancers that come back or spread may be able to be cured with new treatments like proton therapy (PT) and tumor-treating fields (TTFs) [7]. Despite these advancements, significant challenges remain, particularly in the treatment of metastatic or recurrent malignancies. This is because there aren't many systemic alternatives and the cancer is becoming more resistant to traditional treatments, which makes it harder to keep the cancer under control [8]. Recent research seeks to address these issues by integrating radiotherapy with systemic targeted therapies, such as immunotherapies and radiosensitizers. The exploration of artificial intelligence (AI) and machine learning is underway to improve the selection process, treatment planning, and outcome prediction. An entirely new subfield of precision oncology has developed [9]. This chapter thoroughly explores developments in radiation treatment for brain cancers, including their therapeutic impacts, challenges, and future possibilities. By using cutting-edge technology and gaining a better understanding of tumor biology, the field is well positioned to make significant strides in improving patient outcomes and quality of life.

12.2 Technical Advancements in Radiation Therapy in Brain Tumors

When it comes to treating brain tumors, radiation therapy is a must. All parts of radiotherapy (RT) have benefited from technological improvements in the previous several decades, including imaging, patient immobilization, treatment planning, and delivery. Modern radiation oncology and imaging techniques have allowed for more precise tumor location and dosage delivery, reducing the amount of healthy brain tissue exposed to harmful radiation levels [10]. Radiation may cause long-term harm, but normal tissue sparing aims to reduce that risk while keeping the treatment effective [11]. In the treatment of brain tumors, radiation therapy is essential. The use of intensity-modulated radiotherapy (IMRT), volumetric-modulated arc treatment (VMAT), and stereotactic radiosurgery (SRS) is examples of recent developments in radiation technology [12]. When contrasted with traditional three-dimensional conformal radiotherapy (3D-CRT), all of these methods increase radiation doses delivered to the target volume while decreasing the danger of damage to healthy tissues [13]. Because proton therapy delivers a more uniform dosage to the target volume, it could be a viable alternative to photon irradiation as a therapeutic option. Recent advances in radiation treatment and their practical implications for the care of patients diagnosed with brain tumors are outlined in this study [14].

12.2.1 Intensity-Modulated Radiotherapy (IMRT)

Offering enhanced radiation treatment capabilities, intensity-modulated radiation therapy (IMRT) has been rapidly embraced globally since its inception in 1997. This method employs advanced imaging and computational technology to customize radiation therapy according to the distinct morphology and characteristics of each cancer [15]. IMRT allows for the exact modulation of the form and intensity of individual radiation beams, facilitating the delivery of greater radiation doses more effectively while reducing exposure to adjacent healthy tissues. In order to create a three-dimensional (3D) digital envision of the tumor, a radiation oncologist first combines data from several imaging modalities, such as computed tomography (CT), magnetic resonance imaging (MRI), and positron emission tomography (PET). This 3D mapping allows for accurate tumor targeting while preserving healthy tissue. By varying the beam directions, forms, and intensities to maximize therapeutic efficacy, a specialized computer program subsequently optimizes the patient-specific therapy plan [16]. In patients diagnosed with high-grade gliomas, a comparative study assessed the effectiveness of IMRT and three-dimensional conformal radiation therapy (3D-CRT). Fused CT and MRI scans were implemented to develop treatment plans for 20 patients. The results suggested that IMRT considerably enhanced target coverage, resulting in a 31% reduction in the brain volume exposed to radiation doses exceeding 45 Gy. Furthermore, the percentage of brain tissue that was exposed to dosages exceeding 18 Gy, 24 Gy, and 45 Gy was significantly reduced by 10%, 14%, and 40%, respectively. Furthermore, IMRT minimized the exposure of the optic chiasm to doses exceeding 45 Gy. The technique enhanced tumor control probability while decreasing the likelihood of radiation-induced complications in healthy brain and brainstem tissue. Advances in cancer imaging and the development of more effective systemic therapies are expanding the treatment possibilities for high-grade gliomas [17]. Applying localized radiation promoter to the tumor bed, which had previously shown strong local tumor control, was the goal of a study by the National Cancer Institute to replicate the effective treatment results achieved for medulloblastoma. The study included 33 patients newly diagnosed with medulloblastoma who underwent an IMRT-based tumor-bed boost following craniospinal irradiation (CSI). Most patients concurrently received vincristine along with standard adjuvant chemotherapy. Among patients classified as standard risk, the five-year progression-free survival (PFS) rate was recorded at 81.4%, while the overall survival (OS) rate was 88.4% following CSI doses of either 23.4 Gy or 36 Gy. There was an 87.5% PFS and OS rate for patients at high risk. It is worth mentioning that no cases of posterior fossa failures were documented. Six percent of the thirty-one individuals had Grade 3 hearing loss throughout the median 19-month follow-up. These findings show that IMRT-based tumor-bed boosts are very good at controlling local tumors while keeping the mean radiation dose to the ear low. This lowers the risk of ototoxicity [18].

12.2.2 Volumetric Modulated Arc Treatment (VMAT)

The treatment of glioblastoma multiforme (GBM) involves a trimodality regimen, including maximum resection, concomitant chemoradiation, and adjuvant chemotherapy. Radiation conformality has improved, leading to more sparing of normal brain tissue due to developments in patient immobilization, linear accelerator technology, and the introduction of magnetic resonance imaging (MRI) [19]. Volumetric modulated arc therapy (VMAT) is employed due to its decreased treatment delivery time and improved sparing of normal tissues compared to IMRT or 3-dimensional conformal radiation [20]. VMAT has shown potential advantages over IMRT, such as increasing dose conformality and reducing dose to the brain stem, optic chiasm, hippocampi, or cochleae. A study by Navarria et al. evaluated 341 patients (74% with GBM) who were treated with either VMAT or 3-dimensional conformal radiation therapy. The study found no significant difference in PTV coverage between the two techniques, but at a median follow-up time of 1.3 years, VMAT was associated with a statistically significant PFS (15 months) and OS (19 months) benefit compared with 3-dimensional conformal radiation therapy. Patients treated with VMAT also had a 10% absolute reduction in all recurrences. However, VMAT has potential for lower mean and minimum doses to the PTV or the volume of dose receiving 95% of the prescribed dose, even among plans that are considered acceptable for target coverage.

There are a number of logistical benefits to VMAT that could increase patient comfort while decreasing expenses and resource use. A thorough evaluation of the possible effects of utilizing a new technology within the framework of current standards should precede the practical shift to VMAT-based treatment delivery, despite numerous reports of a comparable or even better dosimetric profile with VMAT in addition to reduced treatment durations. Be cautious while thinking about volume definitions, dose distributions with various beam configurations, and treatment planning objectives [21].

12.2.3 Stereotactic Radiosurgery (SRS)

Traditionally, whole-brain radiation therapy (WBRT) was the standard approach for treating brain metastases, regardless of their size or number. In comparison with WBRT alone, early clinical trials showed that stereotactic radiosurgery (SRS) in conjunction with WBRT enhanced local tumor control [22]. However, studies indicated that patients with 1–3 or 1–4 metastases measuring less than 3–4 cm in diameter experienced inferior local and intracranial control when treated with SRS alone.

Additionally, combining WBRT with SRS was associated with neurocognitive decline, reduced quality of life, and increased mortality.

SRS alone is advised by current recommendations for patients with 1–4 brain metastases, making it the recommended course of treatment in both definitive and postoperative contexts. SRS is increasingly used postoperatively for larger tumors to preserve cognitive function [23]. However, this approach is linked to a higher incidence of nodular meningeal recurrences compared to WBRT. Preoperative SRS is one approach to perhaps reduce this risk. Providing a basis for therapeutic decision-making, the Radiosurgery Society has created case-based recommendations defining ideal treatment regimens for brain metastases controlled with SRS [24]. With an increasing number of patients qualifying for SRS, a comprehensive evaluation of patient selection criteria and treatment parameters is essential. Future advancements in SRS include innovations such as multiple lesion therapy, masked SRS, alternative fractionation regimens, and the integration of systemic therapies that may complement or substitute SRS. The purpose of these guidelines is to provide a method for the systematic and easy management of SRS-related metastatic brain tumors [25]. In patients with numerous brain metastases, comparative studies comparing SRS and WBRT indicate that the median survival difference after SRS is negligible, at about 0.8 months. It is important to note that 90% of patients suffer from systemic disease rather than neurological deterioration [26]. Outcomes such as neurological mortality, local recurrence rates, repeat SRS for newly developed lesions, and treatment-related complications were similar between groups receiving SRS or WBRT. WBRT affords temporary control over the formation of new lesions, usually for 6–8 months, whereas SRS alone offers the possibility of later salvage SRS in situations of new metastases [27]. The long-term advantages of WBRT are limited because many patients with numerous metastases survive for more than a year.

Although clinical practice often changes ahead of randomized studies, current evidence-based guidelines recommend SRS for patients with 1–4 metastatic brain lesions. The North American Gamma Knife Consortium conducted a prospective randomized study to compare the neurocognitive outcomes of WBRT with SRS at baseline and six months. Findings indicated that patients with five or more metastatic lesions did not experience increased SRS-related complications compared to those with fewer lesions. To ensure the efficacy of SRS alone in patients with 10 or more metastases, randomized controlled trials (RCTs) are required. Ultimately, patient-specific factors play a major role in determining the optimal treatment plan [28].

12.2.4 IGRT (Image-Guided Radiotherapy)

Due to the low survival rates and potential risk of brain injury associated with radiation therapy, the treatment of glioblastoma multiforme (GBM) and brain metastases remains a significant challenge. Despite the use of concurrent chemotherapy and radiation dose escalation, local recurrence remains the predominant failure pattern in GBM, likely attributed to the repopulation of cancer stem cells [29]. Radiotherapy plays a crucial role in targeting radioresistant tumors, such as melanoma and renal cell carcinoma. However, for patients with brain metastases, disease progression at the systemic level often leads to mortality. Maintaining the quality of life (QOL) for these patients is a major concern. Image-guided radiotherapy (IGRT) has shown potential in preserving neurocognitive function while ensuring local tumor control. Moreover, IGRT can reduce treatment duration in GBM patients without causing excessive toxicity [30]. Future prospective studies should evaluate the impact of IGRT on patient QOL in cases of brain tumors and metastases. The primary objective of therapy for patients with GBM and brain metastases is to enhance their QOL, given the poor long-term survival rates. Advances in radiation oncology, particularly IGRT, have facilitated shorter treatment durations while maintaining safety. Magnetic resonance imaging (MRI) is now an integral part of radiation planning, improving precision. Additionally, advanced imaging modalities such as diffusion tensor imaging (DTI) and positron emission tomography (PET) enhance tumor delineation and minimize radiation exposure to normal tissues [31]. IGRT delivery systems are categorized into radiation-based (kV and MV) and non-radiation-based (ultrasound, electromagnetic) technologies. These systems utilize fiducial markers implanted within tumors or enable direct tumor visualization. Daily imaging reduces the risk of marginal misses, and the comparison of pre- and post-treatment images aids in refining radiation planning [32]. The choice of IGRT technology depends on factors such as tumor type, clinician preference, and financial considerations. Radiation-based systems produce high-quality 2D and 3D images through kV imaging, while non-radiation-based approaches use ultrasound, electromagnetic transponders, and combined MRI-cobalt linear accelerator techniques for tumor localization. Further clinical research is required to validate these technologies' effectiveness [33].

IGRT is a valuable tool in radiotherapy, offering improved tumor visualization, daily imaging, and precise dose administration. Compared to three-dimensional conformal radiation therapy (3D-CRT), IGRT enables dose escalation while minimizing toxicity to normal brain tissues. Accelerated radiation therapy has the potential to enhance QOL and extend survival for GBM patients [34]. Furthermore, IGRT allows for a reduction in the planning target volume due to the accuracy achieved through daily imaging. This is particularly advantageous when tumors are situated near critical radiosensitive structures, including the optic chiasm, optic nerves, and brainstem. IGRT is also applicable for re-irradiation of recurrent tumors following standard chemoradiotherapy, with median survival durations ranging from 7 to 11 months, with or without chemotherapy [35]. Integrating fluoroethyl tyrosine (FET)-PET into radiotherapy planning may further enhance local tumor control and survival outcomes. Additional research is necessary to explore IGRT's full potential in the management of GBM [36]. IGRT has shown promise in reducing treatment duration for patients with brain metastases, and its combination with radiosurgery may help avoid delays in chemotherapy initiation. However, the risk of tumor recurrence in untreated regions of the brain remains a limitation. Whole-brain IGRT with simultaneous integrated boost (SIB) has demonstrated favorable outcomes, ensuring local control in patients with one to six brain metastases within 1–2 weeks of radiotherapy [37]. To mitigate neurocognitive decline in long-term survivors following whole-brain radiation therapy (WBRT), the protection of the limbic system from excessive radiation exposure should be considered. Technically, IGRT enables hippocampal and neural stem cell (NSC) compartment sparing without compromising target volume coverage. Preclinical studies have confirmed the feasibility of NSC-sparing intensity-modulated radiation therapy (IMRT). In patients with brain metastases, the local tumor control may be enhanced, while neurocognitive function is preserved by combining whole-brain IGRT with SIB and hippocampal sparing. Nevertheless, radiation dose escalation for brain tumors and metastases should only be pursued with proper image guidance to prevent increased neurotoxicity. IGRT should be evaluated in future clinical trials to see whether or not it is feasible to improve quality of life for individuals who have brain metastases and GBM [38].

12.2.5 Proton RT

Proton therapy (PT) is a treatment for brain tumors due to its precise radiation delivery and no exit dose. It reduces neurocognitive toxicity and improves quality of life, especially in children. PT has around 150,000 patients, but no level I evidence exists. Future trials should include protons or compare photons to protons [39]. PT has additional costs and toxicity, but it may reduce radiation-induced toxicities and tumor recurrences, ultimately optimizing the photon/proton financial ratio. Because proton therapy (PT) delivers radiation precisely and does not produce an exit dosage, it is an effective treatment for brain tumours. To a lesser extent, this therapy may enhance quality of life and lessen neurocognitive toxicity, particularly in children. However, high-quality data is required since PT lacks level I evidence. There are extra expenses associated with PT that might increase to the rising cost of cancer care. The possibility for dose escalation, conformal radiation delivery, nontarget brain irradiation, and other factors should be taken into account when analyzing these expenses in comparison to photons. PT has the potential to optimize the photon/proton financial ratio by reducing radiation-induced toxicity and tumor recurrences [40]. In the era of evidence-based medicine (EBM), high-quality data is crucial to justify the additional cost factor associated with proton therapy for brain and non-CNS tumors. Randomised controlled trials (RCTs) are of paramount importance in assessing the value of any treatment modality in cancer care. Clinical validation of proton therapy can be achieved with a non-RCT paradigm, specifically a model-based/driven validation approach [38]. A combination of these trial strategies will best generate data that creates scientifically sound evidence on how to select patients for protons and increase the therapeutic balance of various malignancies, including CNS tumors, for value-based cancer management. A number of RCTs and prospective Phase II trials have been proposed and are currently accruing patients worldwide. The most common brain tumors for these trials were chordoma or chondrosarcoma, meningioma, and low-grade glioma [41]. A study comparing photon and proton radiation therapy (PT) for treating brain tumors found that PT can reduce doses to normal brain structures, potentially benefiting cognition. The study analyzed 23 adult patients with proton and photon plans for primary brain tumor treatment. Cognitive outcomes were predicted using converted equivalent dose (EQD2) and risks of cognitive decline on two specific tests. Results showed that PT reduced the probability of impairment in specific neurocognitive test scores after PT [42]. The mean dose reduction to the left temporal lobe reduced the probability of impairment in the COWAT from 6.8% to 5.4%. The same reduction was observed with the normal brain, reducing the probability of impairment in the WAIS-IV Coding test from 5% to 4.1%. The study concluded that PT can reduce doses to brain substructures, potentially decreasing the probability of impairment in neurocognitive test scoring. Research comparing photon and proton methods for treating primary brain tumors found that proton therapy resulted in reduced radiation exposure for most brain substructures. So, this may lessen the likelihood of neurocognitive deterioration due to RT [41]. Time required to develop changes in cognition may vary greatly, and neurocognitive toxicity encompasses a spectrum of toxicities. Research on the potential associations between neurocognitive deterioration following RT and certain brain substructures is limited. Radiation oncologists and neuro-oncologists are showing a growing interest in treating patients with protons since PT (proton therapy) has the ability to lessen long-term toxicity following radiation therapy [41]. Photon RT alone is still used by many nations, including Canada, since there is a dearth of data and PT therapy devices are expensive. It is possible to measure the relative dosimetric advantages of proton and photon therapies using comparative dosimetry. Proton therapy was associated with reduced dosages to certain brain substructures, including the left temporal lobe and normal brain, according to the research [43].

12.2.6 Adoptive Radiotherapy

The current gold standard for treating glioblastoma (GBM) is chemotherapy; however, significant advances in radiation dosage have not been made for decades. Conventional wisdom is that treatment should focus on a volume (CTV) 1.5–3.0 cm beyond the GTV [43]. Incorporating functional and conventional MRI sequences into treatment workflow has been normal during the last 15 years. With an emphasis on the advantages and disadvantages of adaptive workflows, both online and offline, this paper delves into margin methods and the possibility of adaptive RT for GBM. There is also discussion of potential opportunities to use noninvasive imaging biomarkers to physiologically guide adaptive RT [43]. Advanced imaging methods can help define margins and create personalized treatment plans for GBM. Hyperperfused tumors are linked to poorer progression and chemoradiation resistance, so functional methods can generate target areas for dose escalation. Multiparametric DCE- and DWI-guided radiation treatment is a viable option for detecting resistant subregions in GBM patients. Proton magnetic resonance spectroscopic imaging (MRSI) can predict progression factor and overall survival in GBM patients [44]. Further research is needed to identify the most predictive functional MRI maps and consider tissue displacement from tumor growth or shrinking for accurate recurrence forecasts. By incorporating MRI with standard RT procedures, it is now possible to see the dynamics of a GBM tumor while it undergoes chemoradiation. It is possible to enhance the therapeutic ratio by combining adaptive RT techniques (online or offline) with CTV margin reduction using MR image guidance. Novel dosage techniques and functional imaging integration are now being studied within the adaptive RT framework [43]. The MR-linac offers the potential to visualize anatomical changes during radiotherapy, allowing for optimal treatment planning. However, there is a trade-off between time spent adapting the treatment plan and dosimetric gain. This study explored various daily plan adaptation methods for various tumor sites, aiming to provide insight into the behavior of state-of-the-art 1.5 T MRI-guided online adaptive radiotherapy methods [45]. The study simulated online plan adaptation for five cases with varying levels of interfraction motion, regions of interest, and target sizes. Results showed that the time needed for plan adaptation ranged between 17 and 485 s, with more advanced methods resulting in more plans meeting clinical dose criteria. Full online replanning was found to be the most robust adaptive planning method for Unity, feasible for specific sites in clinically acceptable times. The clinical trial focuses on the effectiveness of ATS-based approaches and ATP-based plan adjustments. ATS validates daily MRI against treatment planning requirements, but it increases time due to online contour inspection or modification. Error tolerance is a deciding factor between ATP methods. Offline replanning is a potential solution for reducing treatment dosage violations. Both ATS and ATP can be integrated into the Unity process, allowing rapid adaptation in case of alterations during preparation. More research is needed on intra-fraction plan adaptation [46].

12.3 Radiogenomics and Personalized Radiotherapy for Brain Tumors

Radiogenomics is a breakthrough convergence of radiation oncology, genomics, and bioinformatics, providing a precision-focused methodology for irradiation (RT) in brain malignancies. Gliomas and glioblastomas have diverse cell and molecular profiles. This variation underpins the different responses to standard therapies like radiation therapy [47]. Radiogenomics seeks to dissect these variations by correlating genetic and epigenetic features with radiosensitivity, resistance, and normal tissue response, enabling truly personalized treatment strategies that optimize therapeutic efficacy while minimizing toxicity [48]. Radiation sensitivity varies among brain cancers because of intrinsic genetic markers, tumor microenvironment, and DNA repair capacity. For example, glioblastoma (GBM), recognized as the most aggressive primary brain tumor, is influenced by intricate genetic changes, such as mutations in IDH1/IDH2, TERT promoter, and TP53, among others [49]. Radiogenomics provides insights into these genetic bases, thus enabling personalized RT regimens by linking molecular characteristics to therapeutic responses. The use of targeted treatments along with radiation, immunotherapy, and alkylating drugs is made easier by genomic analysis, which also improves the future prospects [56]. The resistance to temozolomide and RT is associated with key biomarkers such as O6-methylguanine-DNA methyltransferase (MGMT), a DNA repair enzyme. The methylation of the MGMT promoter leads to the silencing of gene expression, which in turn leads to an increase in tumor radiosensitivity and impaired DNA damage repair [50].

This biomarker, which serves as a foundation for the identification of combined modality therapies for glioblastoma, enables tailored adjustments to fractionation schedules and temozolomide dosing. Likewise, alterations in IDH1/IDH2 lead to modifications in cellular metabolism and the accumulation of 2-hydroxyglutarate, a metabolic oncometabolite that enhances genomic stability and interferes with epigenetic regulation. Because of their diminished DNA repair capacity, tumors with IDH mutations have a more radiosensitive phenotype, allowing lower dosage regimens that protect nearby brain tissues without affecting oncologic outcomes [51]. TERT mutations, common in glioblastomas and oligodendrogliomas, extend replicative potential by maintaining telomere length but are associated with treatment resistance and poor prognosis. The mechanistic link between TERT activity and radiation response offers potential for dose escalation or combination therapies targeting telomere stabilization mechanisms [52]. The classification of glioblastomas into proneural, mesenchymal, neural, and classical subtypes has been made possible by advancements in transcriptional profiling. Each subtype demonstrates unique immune evasion and radiosensitivity mechanisms. Mesenchymal subtypes, marked by NF-κB and STAT3 activation, are more resistant to RT due to enhanced DNA repair and inflammatory signaling, necessitating intensification of therapeutic regimens or synergistic approaches with radiosensitizers [53]. The response of both tumor and normal tissue to radiation therapy is influenced by single-nucleotide polymorphisms (SNPs) in genes that are involved in DNA damage recognition, cell-cycle control (e.g., ATM, TP53), and apoptosis (e.g., FAS, BCL2). These discoveries facilitate the estimation of acute and late toxicities in critical structures such as the hippocampus and medulla [54]. As the therapy progresses, ART provides a dynamic approach by allowing RT regimens to adapt to account for intratumoral and structural changes. Technologies like magnetic resonance-guided linear accelerators (MR-LINACs) allow for real-time imaging and dose adjustment. For example, tumor volume reductions during treatment, often seen in glioblastoma, are counteracted by re-optimized dose distributions to maintain efficacy in resistant regions [55]. Hypofractionation capitalizes on the radiobiological advantages of rapidly progressing malignancies, thereby minimizing the treatment duration while administering greater doses per fraction, resulting in an enhanced biological effect. To enable accurate dose customization, advanced BED models integrate patient-specific radiosensitivity factors such as MGMT status or DNA damage response proficiency [56]. To measure tumor response, modern dose-response models combine radiogenomic data with details about the body's structure. Computational platforms leverage biomarkers, including homologous recombination repair defects and microenvironmental hypoxia levels, for fine-tuning radiation fields and dose escalation zones [57]. Though it has great potential, radiogenomics is hampered by technical problems obtaining tumor specimens, lack of longitudinal genomic data, and high cost of sequencing equipment. The diversity of intracranial tumors, which encompasses ependymomas and metastatic lesions, complicates standardized approaches. The application of AI within bioinformatics breakthroughs and multiomics research presents the opportunity to resolve these gaps. By means of massive radiogenomic datasets, AI-driven algorithms are giving insights on tumor behavior and allowing real-time adaptation of treatment strategies [58] (Table 12.1).

12.4 Technological Innovations in Imaging for Radiotherapy

Advanced imaging modalities that provide morphological, functional, and molecular insights have considerably improved brain tumor radiation (RT) accuracy. Advanced MRI techniques have revolutionized treatment planning by enabling the comprehensive characterization of tumor areas and adjacent critical structures. By means of blood oxygen level-dependent (BOLD) contrasts, functional MRI (fMRI) detects eloquent areas including motor, sensory, and language regions assuring their preservation during high-dose RT [66].

Research including one by Angom et al. (2023) claims that fMRI helps to prevent the speech and movement problems brought on by glioblastoma radiation. The investigation of water diffusion anisotropy in the identification of white matter pathways is a critical component of diffusion tensor imaging (DTI). It mitigates cognitive and motor deficits in the latter phases by maintaining critical networks such as the arcuate fasciculus and corticospinal tract [67]. According to the 2023 study by Martucci et al., DTI-guided RT planning has been beneficial for patients with high-grade gliomas, leading to enhanced functional outcomes post-treatment. Imaging techniques such as PET and SPECT, which detect metabolic and molecular activity, play a crucial role in pinpointing regions of tumors that are actively driving biological changes and in differentiating between radiation-induced alterations and tumor recurrence. The effectiveness of glioblastoma treatment is enhanced by amino acid tracers, such as FET-PET, while fluorodeoxyglucose (FDG)-PET is used to identify hypermetabolic regions [68]. Dong et al. (2023) conducted a recent meta-analysis that emphasized the superiority of FET-PET in detecting recurrent gliomas in terms of diagnostic accuracy compared to conventional MRI. In order to characterize hypoxic regions that may require dose escalation due to their radiation resistance, SPECT, which makes use of tracers such as 99mTc-tetrofosmin, is an extremely useful technique when used. Hybrid imaging modalities such as PET-MRI integrate the metabolic precision of PET with the high-resolution anatomical and functional data of MRI, allowing simultaneous acquisition of complementary datasets [69].

Table 12.1 Radiogenomic biomarkers and their implications in personalized radiotherapy for brain tumors

Biomarker/ genomic profile	Tumor association	Impact on radiotherapy sensitivity	Clinical implications in personalized RT	References
MGMT promoter methylation	Glioblastoma	Increased sensitivity to RT and concurrent temozolomide	Facilitates dose de-escalation in sensitive patients to reduce toxicity. Predicts survival benefit	[59]
IDH1/IDH2 mutations	Gliomas (lower grade, GBM)	Associated with enhanced radiosensitivity and prolonged survival	Enables tailored RT dose for mutant IDH tumors while preserving normal tissue	[60]
TERT promoter mutations	Glioblastomas, oligodendrogliomas	Correlates with treatment resistance and aggressive behavior	Guides intensification of RT or combined therapeutic modalities	[61]
1p/19q codeletion	Oligodendrogliomas	Improved prognosis and favorable RT response	Confirms combined RT and chemotherapy as the preferred treatment approach	[62]
EGFR amplification/ mutations	Glioblastomas (wild-type GBM)	Frequently associated with RT resistance	Suggests adjunctive therapies like radiosensitizers targeting EGFR pathways	[63]
Gene expression signatures	GBM (e.g., proneural, mesenchymal)	Distinct radiosensitivity by transcriptional subtypes	Supports stratified RT protocols, targeting mesenchymal subtypes with more aggressive approaches	[64]
H3K27M mutations	Diffuse midline gliomas	Poor prognosis, limited RT sensitivity	Highlights the necessity for RT dose intensification combined with novel experimental therapies	[65]

PET-MRI is particularly beneficial in re-irradiation cases, as demonstrated in a study by Yan et al. (2022), which reported improved delineation of viable tumor regions and sparing of necrotic tissues. Integrating real-time MRI with a linear accelerator gives MR-LINAC technology adaptive radiation. This method permits real-time tumor monitoring and dosage changes all during treatment [70]. The study conducted by Liu et al. (2023) established the therapeutic applications of MR-LINAC, establishing its efficacy in glioblastoma treatment through improved dosage compliance and reduced irradiation of healthy tissues. Image-guided radiotherapy (IGRT) further enhances precision by addressing intra- and interfractional tumor motion. Able to provide high-resolution pictures at the treatment site, cone-beam computed tomography (CBCT) guarantees perfect alignment by means of positional corrections [71]. For instance, in brain metastases treated with stereotactic radiosurgery (SRS), CBCT has been shown to reduce marginal misses significantly, as reported by Ladbury et al. (2024).

Four-dimensional imaging (4D-CT) incorporates temporal motion data, which is particularly relevant in cranial tumors affected by cerebrospinal fluid (CSF) dynamics or respiratory motion in extracranial metastases [22]. Advanced algorithms enable the integration of this motion data into treatment planning for more accurate dose delivery. Emerging techniques, like oxygen-enhanced MRI, hold promise in identifying hypoxic tumor regions—a known factor in radiation resistance—offering opportunities for targeted dose escalation. Its potential for glioblastoma dose modulation is demonstrated by early-phase clinical investigations, such as those conducted by Telarovic et al. (2021) [72]. Another frontier is the utilization of radiomics, which isolates quantitative imaging biomarkers to forecast treatment outcomes. Machine learning models applied to multiparametric imaging datasets have shown promise in predicting glioblastoma recurrence and therapy response, as outlined in studies by Dorfner et al. (2025) [73]. Radiotherapy for brain tumors is now more precise and personalized because of these technological advancements, which have improved patient outcomes while decreasing side effects. Nevertheless, despite obstacles like high costs, the need for protocol standardization, and the difficulty of data interpretation, continuous advancements in imaging modalities and their integration with AI-driven analytics are pushing the boundaries of effective brain tumor management.

12.5 Advances in Radiotherapy for Treating Recurrent or Metastatic Brain Tumors

The application of new therapeutic approaches and precision-orientated treatments that reduce side effects while improving results has greatly improved the treatment of recurrent and metastatic brain tumors (Table 12.2). To address recurrent brain tumors, two significant methods exist: stereotactic radiosurgery (SRS) and stereotactic body radiotherapy (SBRT). This is a result of recent advancements in radiation therapy. This has led to progression-free survival and manageable toxicity profiles. SBRT, which extends similar precision to extracranial metastases, enables dose fractionation over fewer sessions, thereby reducing the treatment burden for patients with metastatic disease [74]. When it comes to re-irradiation, when tissue tolerance is limited, advanced dose-sparing techniques like IMRT and VMAT work wonders in optimizing dose distribution and lowering the risks of neurocognitive impairment and radiation necrosis [75].

Table 12.2 Summarizing key techniques, their applications, and supporting studies

Technique	Description	Applications	Supporting studies
Stereotactic radiosurgery (SRS)	Delivers high-dose radiation precisely to tumor sites, minimizing exposure to surrounding healthy tissue	Effective for treating small- to medium-sized brain metastases and recurrent tumors	[81]
Stereotactic body radiotherapy (SBRT)	Similar to SRS but used for larger body tumors; in the brain, it treats larger or multiple metastases	Suitable for patients with multiple brain metastases or larger lesions not amenable to SRS	[82]
Intensity-modulated radiotherapy (IMRT)	Uses advanced technology to modulate radiation beams, conforming to the tumor shape	Reduces radiation dose to normal brain tissue, lowering the risk of neurocognitive decline	[83]
Integration with immunotherapy	Combines radiotherapy with immune checkpoint inhibitors to enhance antitumor immune response	Shows promise in improving outcomes for recurrent glioblastomas	[84]
Tumor treating fields (TTFs)	Utilizes alternating electric fields to disrupt cancer cell division	Serves as an adjunctive therapy for recurrent glioblastomas	[85]
Hippocampal avoidance whole-brain radiotherapy (HA-WBRT)	Spares the hippocampus during whole-brain radiotherapy to preserve cognitive function	Beneficial for patients requiring whole-brain radiotherapy	[86]
Management of radiation necrosis	Involves the use of agents like bevacizumab to reduce radiation-induced brain injury	Helps manage side effects in patients undergoing re-irradiation	[87]

Progressive strategies for the management of recurring and metastatic brain tumors are increasingly integrating radiotherapy with systemic medicines to leverage synergistic benefits. The combination of immunotherapy and radiotherapy, including immune checkpoint inhibitors that target PD-1/PD-L1 or CTLA-4 pathways, amplifies antitumor immune responses by using the radiation-induced release of tumor-associated antigens [76]. CheckMate 143 and other clinical trials have demonstrated the potential of combining radiotherapy with nivolumab, a PD-1 inhibitor, to enhance the prognosis of recurrent glioblastoma. Cancer treatment fields (TTFs), a non-invasive technique that modifies mitotic processes in cancer cells, have demonstrated synergistic potential with radiotherapy for the management of both newly diagnosed and recurrent glioblastomas [77]. Recent studies—especially the EF-14 study—have confirmed TTFs as an adjuvant treatment in recurrent situations, thus indicating longer survival periods [78]. Radiation necrosis, a typical consequence of re-irradiation or intensive radiotherapy for metastatic brain lesions, has been improved with the use of neuroprotective medicines. The evidence for bevacizumab, a monoclonal antibody that targets vascular endothelial growth factor (VEGF), comes from observational research and trials such as NCI-014, which demonstrated substantial clinical improvement in persons with refractory cases. It is a successful treatment for both peritumoral edema and symptomatic radiation necrosis. By means of dose fractionation approaches and advanced imaging modalities such MR-LINAC and PET-MRI, the refinement of radiotherapy techniques promises that subsequent radiation exposure concentrates only on viable tumor areas, thus reducing risks to normal brain tissue and expanding the safety profile of re-irradiation [79]. Additionally, decision-making for subsequent radiation treatments is guided by sophisticated biomarkers extracted from liquid biopsy procedures, such as extracellular vesicles and circulating tumor DNA (ctDNA), which are used to track tumor recurrence. Integrating genetic profiles such as MGMT promoter methylation status and IDH mutations to optimize treatment planning for recurrent or metastatic contexts, this precision-driven approach is augmented by advancements in radiogenomics [80]. These developments emphasize the evolving role of radiotherapy as a critical technique in the treatment strategy for recurrent and metastatic brain tumors. In spite of long-term toxicities, tumor heterogeneity, and treatment resistance, ongoing clinical trials and interdisciplinary approaches that integrate radiation with systemic and adjunctive therapies can enhance the quality of life and survival of high-risk patients.

These advancements underscore the evolving role of radiotherapy in effectively treating recurrent and metastatic brain tumors, aiming to improve patient outcomes while minimizing adverse effects.

12.6 Future Directions and Challenges

Recent developments in radiation therapy for brain cancer will necessitate the incorporation of cutting-edge technology, the resolution of important challenges, and the narrowing of the gap between laboratory research and clinical application if we are to develop a truly personalized treatment paradigm. It is anticipated that the analysis of intricate datasets will enable the development of patient-specific radiation plans and biomarkers of therapeutic response, thus revolutionizing treatment planning and outcome prediction. Modern instruments including artificial intelligence (AI) and machine learning will help us to reach this [88]. AI-driven workflows promise greater precision in dose delivery while optimizing fractionation schedules and enabling real-time adaptations to evolving tumor dynamics. Another horizon is provided by nanotechnology, which presents possibilities for tailored drug delivery systems that complement radiation by delivering radiosensitizing chemicals straight to tumor cells, therefore sparing healthy tissue and improving therapeutic efficacy and lowering toxicity [89].

Nevertheless, the path ahead is laden with obstacles that require unified endeavor. Brain tumors are especially challenging to treat due to their characteristic tumor plasticity, which allows cancers to resist targeted therapies and evade diagnosis. Innovations like dynamic radiogenomic profiling and adaptive radiotherapy are required to counteract such resistance by monitoring genetic and phenotypic changes in real time [90]. Equal solutions are necessary to ensure universal applicability due to the high prices of sophisticated radiation modalities and the limited access to these technologies in low-resource regions. Addressing the paucity of standardized protocols for radiogenomic integration remains crucial to realizing the full potential of personalized radiotherapy. Rigorous validation studies and international collaborations are required to harmonize guidelines for biomarker-based treatment plans. Translational research plays a pivotal role in realizing these advancements by bridging discoveries at the bench with clinical applications. In order to bridge the gap between laboratory invention and practical deployment, it is imperative to conduct robust clinical trials and capture real-world data. Collaboration among oncologists, radiologists, geneticists, and bioinformaticians is essential for achieving therapeutically significant and scalable advances. Initiatives to streamline data sharing, improve bioinformatics infrastructure, and develop cost-effective assays are essential steps toward achieving these goals.

12.7 Conclusions

Recent advancements in radiotherapy for brain tumors represent a paradigm shift from traditional treatment methods to precision-driven, patient-specific approaches, significantly enhancing therapeutic outcomes. Medical practitioners can tailor treatment plans to the distinct biology of tumors by utilizing emerging technologies like radiogenomics and advanced imaging techniques, which increases therapeutic efficacy and minimizes side effects. Advancements in adaptive radiation, stereotactic methods, and systemic combinations—including immune checkpoint inhibitors and tumor-treating fields—have substantially enhanced outcomes, particularly for recurring or metastatic brain tumors, where treatment options were previously restricted. Despite these developments, significant challenges still exist, including the high expense of cutting-edge treatments, treatment resistance, and tumor plasticity, especially in low-resource environments. It is crucial to carry out continuous translational research to fill the gap between laboratory findings and practical applications. The primary goals of this endeavor should be to improve biomarker validation, increase access to new technologies, and develop methods that align with the integration of genetic data into clinical practice. Modern technologies like artificial intelligence and machine learning will also greatly boost the accuracy and potency of radiation, therefore enabling improved prognostic predictions and real-time therapeutic optimization. Multidisciplinary interaction across oncology, radiology, genetics, and data science resolving current problems and supporting innovative ideas would help the area to grow. Radiation treatment for brain tumors will depend on its ability to provide customized, efficient, and easily available treatment that will inspire patients once more by means of higher survival rates, better quality of life, and equitable worldwide access to innovative treatments.

References

1. Ostrom QT, Fahmideh MA, Cote DJ, Muskens IS, Schraw JM, Scheurer ME, Bondy ML. Risk factors for childhood and adult primary brain tumors. Neuro-Oncology. 2019;21:1357–75.
2. Aldape K, Brindle KM, Chesler L, et al. Challenges to curing primary brain tumours. Nat Rev Clin Oncol. 2019;16:509–20.
3. Weiss T, Weller M, Roth P. Immunological effects of chemotherapy and radiotherapy against brain tumors. Expert Rev Anticancer Ther. 2016;16:1087–94.
4. Wicki A, Witzigmann D, Balasubramanian V, Huwyler J. Nanomedicine in cancer therapy: challenges, opportunities, and clinical applications. J Control Release. 2015;200:138–57.
5. Oden JT, Lima EABF, Almeida RC, et al. Toward predictive multiscale modeling of vascular tumor growth. Arch Comput Methods Eng. 2016;23:735–79.
6. Hall WA, Small C, Paulson E, et al. Magnetic resonance guided radiation therapy for pancreatic adenocarcinoma, advantages, challenges, current approaches, and future directions. Front Oncol. 2021;11 https://doi.org/10.3389/fonc.2021.628155.
7. De Pietro R, Zaccaro L, Marampon F, Tini P, De Felice F, Minniti G. The evolving role of reirradiation in the management of recurrent brain tumors. J Neuro-Oncol. 2023;164:271–86.
8. Parker AL, Benguigui M, Fornetti J, Goddard E, Lucotti S, Insua-Rodríguez J, Wiegmans AP. Current challenges in metastasis research and future innovation for clinical translation. Clin Exp Metastasis. 2022;39:263–77.
9. Wang J, Zeng Z, Li Z, Liu G, Zhang S, Luo C, Hu S, Wan S, Zhao L. The clinical application of artificial intelligence in cancer precision treatment. J Transl Med. 2025;23:120.
10. Wang K, Tepper JE. Radiation therapy-associated toxicity: etiology, management, and prevention. CA Cancer J Clin. 2021;71:437–54.
11. Taylor A, Powell MEB. Intensity-modulated radiotherapy—what is it? Cancer Imaging. 2004;4:68.
12. Hunte SO, Clark CH, Zyuzikov N, Nisbet A. Volumetric modulated arc therapy (VMAT): a review of clinical outcomes—what is the clinical evidence for the most effective implementation? Br J Radiol. 2022;95 https://doi.org/10.1259/bjr.20201289.
13. Pourparvar Z, Shahbazi-Gahrouei D, Najafizade N, Saeb M, Khaniabadi BM, Khaniabadi PM. Evaluation and comparison of the dose received by the mandible, maxilla, and teeth in two methods of three-dimensional conformal radiation therapy and helical tomotherapy. J Med Signals Sens. 2024;14:26. https://doi.org/10.4103/jmss.jmss_42_23.
14. Scaringi C, Agolli L, Minniti G. Technical advances in radiation therapy for brain tumors. Anticancer Res. 2018;38:6041–5.
15. Xiao Y, Benedict S, Cui Y, et al. Embracing the future of clinical trials in radiotherapy: an NRG oncology CIRO technology retreat whitepaper on pioneering technologies and AI-driven solutions. Int J Radiat Oncol Biol Phys. 2025;122:443. https://doi.org/10.1016/j.ijrobp.2025.01.006.
16. Teh BS, Woo SY, Butler EB. Intensity modulated radiation therapy (IMRT): a new promising technology in radiation oncology. Oncologist. 1999;4:433–42.
17. MacDonald SM, Ahmad S, Kachris S, Vogds BJ, DeRouen M, Gitttleman AE, DeWyngaert K, Vlachaki MT. Intensity modulated radiation therapy versus three-dimensional conformal radiation therapy for the treatment of high grade glioma: a dosimetric comparison. J Appl Clin Med Phys. 2007;8:47.
18. Polkinghorn WR, Dunkel IJ, Souweidane MM, Khakoo Y, Lyden DC, Gilheeney SW, Becher OJ, Budnick AS, Wolden SL. Disease control and ototoxicity using intensity-modulated radiation therapy tumor-bed boost for medulloblastoma. Int J Radiat Oncol Biol Phys. 2011;81:e15. https://doi.org/10.1016/J.IJROBP.2010.11.081.
19. Wu W, Klockow JL, Zhang M, Lafortune F, Chang E, Jin L, Wu Y, Daldrup-Link HE. Glioblastoma multiforme (GBM): an overview of current therapies and mechanisms of resistance. Pharmacol Res. 2021;171:105780.

20. Lawless AK, Fangupo N, Estoesta RP, Perez G, Alzaidi S, Heller G, Hill R, Hong AM. Dosimetric comparison of volumetric modulated arc therapy and intensity modulated radiation therapy for soft tissue sarcoma of the extremities. Pract Radiat Oncol. 2022;12:e306–11.
21. Sheu T, Briere TM, Olanrewaju AM, McAleer MF. Intensity modulated radiation therapy versus volumetric arc radiation therapy in the treatment of glioblastoma—does clinical benefit follow dosimetric advantage? Adv Radiat Oncol. 2018;4:50.
22. Ladbury C, Pennock M, Yilmaz T, et al. Stereotactic radiosurgery in the management of brain metastases: a case-based radiosurgery society practice guideline. Adv Radiat Oncol. 2024;9:101402.
23. Nakano T, Aoyama H, Onodera S, et al. Reduced-dose WBRT combined with SRS for 1–4 brain metastases aiming at minimizing neurocognitive function deterioration without compromising brain tumor control. Clin Transl Radiat Oncol. 2022;37:116–29.
24. Vogelbaum MA, Brown PD, Messersmith H, et al. Treatment for brain metastases: ASCO-SNO-ASTRO guideline. J Clin Oncol. 2022;40:492–516.
25. Topkan E, Kucuk A, Senyurek S, Sezen D, Durankus NK, Akdemir EY, Bolukbasi Y, Pehlivan B, Selek U. Preoperative radiosurgical management of brain metastases: evidence and challenges. In: Advancements in cancer research. Exon Publications; 2023. p. 75–91.
26. Kim GJ, Buckley ED, Herndon JE, et al. Outcomes in patients with 4 to 10 brain metastases treated with dose-adapted single-isocenter multitarget stereotactic radiosurgery: a prospective study. Adv Radiat Oncol. 2021;6:100760.
27. Yamamoto M, Kawabe T, Sato Y, Higuchi Y, Nariai T, Watanabe S, Kasuya H. Stereotactic radiosurgery for patients with multiple brain metastases: a case-matched study comparing treatment results for patients with 2–9 versus 10 or more tumors: clinical article. J Neurosurg. 2014;121:16–25.
28. Saqib M, Zahoor A, Rahib A, Shamim A, Mumtaz H. Clinical and translational advances in primary brain tumor therapy with a focus on glioblastoma-a comprehensive review of the literature. World Neurosurg X. 2024;24:100399.
29. Matsui JK, Perlow HK, Baiyee C, et al. Quality of life and cognitive function evaluations and interventions for patients with brain metastases in the radiation oncology clinic. Cancers (Basel). 2022;14:4301.
30. Curran WJ, Scott CB, Horton J, et al. Recursive partitioning analysis of prognostic factors in three Radiation Therapy Oncology Group malignant glioma trials. J Natl Cancer Inst. 1993;85:704–10.
31. Hwang U-J, Min BJ, Kim M, Kim K-H. Image guided radiation therapy. Prog Med Phys. 2022;33:37–52.
32. De Los SJ, Popple R, Agazaryan N, et al. Image guided radiation therapy (IGRT) technologies for radiation therapy localization and delivery. Int J Radiat Oncol Biol Phys. 2013;87:33–45.
33. Nien H-H, Wang L-Y, Liao L-J, Lin P-Y, Wu C-Y, Shueng P-W, Chung C-S, Lo W-C, Lin S-C, Hsieh C-H. Advances in image-guided radiotherapy in the treatment of oral cavity cancer. Cancers (Basel). 2022;14:4630.
34. Gambacorta MA, Romano A, Caravatta L, Macchia G, Chiloiro G, Galofaro E, Valvo F, Vitolo V, Alterio D, Mantello G. Radiotherapy & total neoadjuvant therapy for recurrent rectal cancer in previously irradiated patients, (RETRY): a multicenter prospective observational study. Radiat Oncol. 2024;19:174.
35. Donato V, Caruso C, Bressi C, Pressello MC, Salvati M, Delitala A, Delfini R. Evaluation of helical tomotherapy in the treatment of high-grade gliomas near critical structures. Tumori. 2012;98:636–42.
36. Zhou G-Q, Lu J-D. The process of the execution of intensity-modulated radiation therapy for nasopharyngeal carcinoma. Springer; 2024. https://doi.org/10.1007/174_2024_476.
37. Nguyen NP, Nguyen ML, Vock J, et al. Potential applications of imaging and image-guided radiotherapy for brain metastases and glioblastoma to improve patient quality of life. Front Oncol. 2013;3:284.

38. Weber DC, Lim PS, Tran S, Walser M, Bolsi A, Kliebsch U, Beer J, Bachtiary B, Lomax T, Pica A. Proton therapy for brain tumours in the area of evidence-based medicine. Br J Radiol. 2020;93 https://doi.org/10.1259/bjr.20190237.
39. Adeberg S, Harrabi SB, Bougatf N, Verma V, Windisch P, Bernhardt D, Combs SE, Herfarth K, Debus J, Rieken S. Dosimetric comparison of proton radiation therapy, volumetric modulated arc therapy, and three-dimensional conformal radiotherapy based on intracranial tumor location. Cancers (Basel). 2018;10 https://doi.org/10.3390/CANCERS10110401.
40. Sampaio F, Langegård U, de Alva PM, et al. Cost-effectiveness of proton beam therapy vs. conventional radiotherapy for patients with brain tumors in Sweden: results from a non-randomized prospective multicenter study. Cost Eff Resour Alloc. 2024;22:66.
41. Petruccelli M, Parent A, Holwell M, Dama H, Tsui G, Liu ZA, Tsang DS. Estimating potential benefits to neurocognition with proton therapy in adults with brain tumors. Int J Part Ther. 2023;9:261–8.
42. Kotecha R, La Rosa A, Mehta MP. How proton therapy fits into the management of adult intracranial tumors. Neuro-Oncology. 2024;26:S26–45.
43. Tseng CL, Zeng KL, Mellon EA, et al. Evolving concepts in margin strategies and adaptive radiotherapy for glioblastoma: a new future is on the horizon. Neuro-Oncology. 2024;26:S3–S16.
44. van Timmeren JE, Chamberlain M, Krayenbuehl J, et al. Treatment plan quality during online adaptive re-planning. Radiat Oncol. 2020;15:203.
45. Winkel D, Bol GH, Kroon PS, et al. Adaptive radiotherapy: the Elekta Unity MR-linac concept. Clin Transl Radiat Oncol. 2019;18:54–9.
46. Gaito S, Burnet NG, Aznar MC, et al. Proton beam therapy in the reirradiation setting of brain and base of skull tumour recurrences. Clin Oncol. 2023;35:673–81.
47. Tanzhu G, Chen L, Ning J, Xue W, Wang C, Xiao G, Yang J, Zhou R. Metastatic brain tumors: from development to cutting-edge treatment. MedComm (Beijing). 2025;6:e70020. https://doi.org/10.1002/mco2.70020.
48. Shah S, Nag A, Sachithanandam SV, Lucke-Wold B. Predictive and prognostic significance of molecular biomarkers in glioblastoma. Biomedicines. 2024;12:2664.
49. He W, Huang W, Zhang L, Wu X, Zhang S, Zhang B. Radiogenomics: bridging the gap between imaging and genomics for precision oncology. MedComm (Beijing). 2024;5:e722. https://doi.org/10.1002/mco2.722.
50. Chen Y, Qu W, Tu J, Qi H. Implications of advances in studies of O6-methylguanine-DNA-methyltransferase for tumor prognosis and treatment. Front Biosci (Landmark). 2023; https://doi.org/10.31083/j.fbl2809197.
51. Álvarez-González E, Sierra LM. Tricarboxylic acid cycle relationships with non-metabolic processes: a short story with DNA repair and its consequences on cancer therapy resistance. Int J Mol Sci. 2024;25:9054.
52. Papageorgakopoulou MA, Bania A, Lagogianni I-A, Birmpas K, Assimakopoulou M. The role of glia telomere dysfunction in the pathogenesis of central nervous system diseases. Mol Neurobiol. 2024;61:5868–81.
53. Harshavardhan BV, Jolly MK. Proneural-mesenchymal antagonism dominates the patterns of phenotypic heterogeneity in glioblastoma. iScience. 2024;27:109184.
54. Al-Arafat T, Mao A, Katsube T, Wang B. Exploring the role of p53 in radiosensitivity: a key player in cancer therapy. Radiat Ther. 2024;4:309–24.
55. Hall WA, Paulson E, Li XA, et al. Magnetic resonance linear accelerator technology and adaptive radiation therapy: an overview for clinicians. CA Cancer J Clin. 2022;72:34–56.
56. Eom K-Y. Hypofractionated radiation therapy for head and neck cancers in the era of intensity-modulated radiation therapy. Radiat Oncol J. 2024;42:1–3.
57. Tomasik B, Garbicz F, Braun M, Bieńkowski M, Jassem J. Heterogeneity in precision oncology. Camb Prisms Precis Med. 2024;2:e2.
58. Ahmed Z, Wan S, Zhang F, Zhong W. Artificial intelligence for omics data analysis. BMC Methods. 2024;1:4.

59. Wick W, Dettmer S, Berberich A, et al. N2M2 (NOA-20) phase I/II trial of molecularly matched targeted therapies plus radiotherapy in patients with newly diagnosed non-MGMT hypermethylated glioblastoma. Neuro-Oncology. 2019;21:95–105.
60. Choate KA, Pratt EPS, Jennings MJ, Winn RJ, Mann PB. IDH mutations in glioma: molecular, cellular, diagnostic, and clinical implications. Biology (Basel). 2024;13:885.
61. Hasanau T, Pisarev E, Kisil O, Nonoguchi N, Le Calvez-Kelm F, Zvereva M. Detection of TERT promoter mutations as a prognostic biomarker in gliomas: methodology, prospects, and advances. Biomedicines. 2022;10:728.
62. Allwohn L, Wolfgang J, Onken J, Wasilewski D, Roohani S, Zips D, Ehret F, Kaul D. Treating oligodendroglioma—an analysis of a homogeneous 1p/19q-codeleted and isocitrate dehydrogenase-mutant patient cohort. Clin Transl Radiat Oncol. 2023;41:100626.
63. Rodriguez SMB, Kamel A, Ciubotaru GV, Onose G, Sevastre A-S, Sfredel V, Danoiu S, Dricu A, Tataranu LG. An overview of EGFR mechanisms and their implications in targeted therapies for glioblastoma. Int J Mol Sci. 2023;24:11110.
64. Steponaitis G, Tamasauskas A. Mesenchymal and proneural subtypes of glioblastoma disclose branching based on GSC associated signature. Int J Mol Sci. 2021;22:4964.
65. Noon A, Galban S. Therapeutic avenues for targeting treatment challenges of diffuse midline gliomas. Neoplasia. 2023;40:100899.
66. McGee KP, Hwang K, Sullivan DC, et al. Magnetic resonance biomarkers in radiation oncology: the report of AAPM Task Group 294. Med Phys. 2021;48:e697. https://doi.org/10.1002/mp.14884.
67. Angom RS, Nakka NMR, Bhattacharya S. Advances in glioblastoma therapy: an update on current approaches. Brain Sci. 2023;13:1536.
68. Martucci M, Russo R, Schimperna F, et al. Magnetic resonance imaging of primary adult brain tumors: state of the art and future perspectives. Biomedicines. 2023;11:364.
69. Dong W, Wang N, Qi Z. Advances in the application of neuroinflammatory molecular imaging in brain malignancies. Front Immunol. 2023;14 https://doi.org/10.3389/fimmu.2023.1211900.
70. Yan Q, Yan X, Yang X, Li S, Song J. The use of PET/MRI in radiotherapy. Insights Imaging. 2024;15:63.
71. Liu X, Li Z, Yin Y. Clinical application of MR-Linac in tumor radiotherapy: a systematic review. Radiat Oncol. 2023;18:52.
72. Telarovic I, Wenger RH, Pruschy M. Interfering with tumor hypoxia for radiotherapy optimization. J Exp Clin Cancer Res. 2021;40:197.
73. Dorfner FJ, Patel JB, Kalpathy-Cramer J, Gerstner ER, Bridge CP. A review of deep learning for brain tumor analysis in MRI. NPJ Precis Oncol. 2025;9:2.
74. Bijlani A, Aguzzi G, Schaal DW, Romanelli P. Stereotactic radiosurgery and stereotactic body radiation therapy cost-effectiveness results. Front Oncol. 2013;3 https://doi.org/10.3389/fonc.2013.00077.
75. Ciammella P, Cozzi S, Botti A, et al. Safety of inhomogeneous dose distribution IMRT for high-grade glioma reirradiation: a prospective phase I/II trial (GLIORAD TRIAL). Cancers (Basel). 2022;14:4604.
76. Voronova V, Vislobokova A, Mutig K, Samsonov M, Peskov K, Sekacheva M, Materenchuk M, Bunyatyan N, Lebedeva S. Combination of immune checkpoint inhibitors with radiation therapy in cancer: a hammer breaking the wall of resistance. Front Oncol. 2022;12 https://doi.org/10.3389/fonc.2022.1035884.
77. Omuro A, Reardon DA, Sampson JH, Baehring J, Sahebjam S, Cloughesy TF, Chalamandaris A-G, Potter V, Butowski N, Lim M. Nivolumab plus radiotherapy with or without temozolomide in newly diagnosed glioblastoma: results from exploratory phase I cohorts of CheckMate 143. Neurooncol Adv. 2022;4 https://doi.org/10.1093/noajnl/vdac025.
78. Khagi S, Kotecha R, Gatson NTN, Jeyapalan S, Abdullah HI, Avgeropoulos NG, Batzianouli ET, Giladi M, Lustgarten L, Goldlust SA. Recent advances in Tumor Treating Fields (TTFields) therapy for glioblastoma. Oncologist. 2024;30 https://doi.org/10.1093/oncolo/oyae227.

79. Khan M, Zhao Z, Arooj S, Liao G. Bevacizumab for radiation necrosis following radiotherapy of brain metastatic disease: a systematic review & meta-analysis. BMC Cancer. 2021;21:167.
80. Soffietti R, Bettegowda C, Mellinghoff IK, et al. Liquid biopsy in gliomas: a RANO review and proposals for clinical applications. Neuro-Oncology. 2022;24:855–71.
81. Soliman H, Das S, Larson DA, Sahgal A. Stereotactic radiosurgery (SRS) in the modern management of patients with brain metastases. Oncotarget. 2016;7:12318–30.
82. Matsui JK, Perlow HK, Upadhyay R, McCalla A, Raval RR, Thomas EM, Blakaj DM, Beyer SJ, Palmer JD. Advances in radiotherapy for brain metastases. Surg Oncol Clin N Am. 2023;32:569–86.
83. Gondi V, Bauman G, Bradfield L, et al. Radiation therapy for brain metastases: an ASTRO clinical practice guideline. Pract Radiat Oncol. 2022;12:265–82.
84. Bausart M, Préat V, Malfanti A. Immunotherapy for glioblastoma: the promise of combination strategies. J Exp Clin Cancer Res. 2022;41:35.
85. Wong ET, Lok E, Swanson KD. An evidence-based review of alternating electric fields therapy for malignant gliomas. Curr Treat Options in Oncol. 2015;16:40.
86. Deepsikha G, Maji T, Lahiri D, Roy S, Bhanja S, Rangineni S, Ray DK, Mohanta BK, Misra DK, De P. Hippocampal avoidance whole brain radiotherapy in brain metastasis using volumetric modulated arc therapy: experience from a Regional Cancer Centre of Eastern India. Rep Pract Oncol Radiother. 2023;28:468–77.
87. Panagiotou E, Charpidou A, Fyta E, Nikolaidou V, Stournara L, Syrigos A, Gkiozos I. High-dose bevacizumab for radiation-induced brain necrosis: a case report. CNS Oncol. 2023;12:CNS98. https://doi.org/10.2217/cns-2023-0002.
88. Barki C, Alsufyani SJ, Softah A, Labidi S, Rahmouni HB. Advancing radiation therapy safety in cancer-care: leveraging AI for adverse reaction prediction. J Radiat Res Appl Sci. 2024;17:101141.
89. Arif M, Nawaz AF, Ullah Khan S, Mueen H, Rashid F, Hemeg HA, Rauf A. Nanotechnology-based radiation therapy to cure cancer and the challenges in its clinical applications. Heliyon. 2023;9:e17252.
90. Rubini D, Gagliardi F, Menditti VS, et al. Genetic profiling in radiotherapy: a comprehensive review. Front Oncol. 2024;14 https://doi.org/10.3389/fonc.2024.1337815.

Chapter 13
Advances in Head and Neck Cancer

Nam P. Nguyen, Olena Gorobets, Vincent Vinh-Hung, Sergio Calleja Blanco, and Brandi R. Page

Abstract Molecular biology of head and neck cancer enables clinicians to personalize treatment based on biomarkers. Oropharyngeal cancer induced by human papillomavirus 16 (HPV-16) typically has a favorable prognosis due to its radiosensitivity and chemosensitivity. However, randomized trials aimed at de-escalating radiation doses or replacing chemotherapy with less toxic agents have not been successful. Discoveries regarding the role of programmed death ligand 1 (PD-L1) on head and neck cancer cells have led to successful treatment with immune checkpoint inhibitors (ICIs) for recurrent disease. Despite this, the optimal sequencing of ICI with radiotherapy remains unclear for the treatment of locally advanced head and neck cancer. Advances in radiation therapy technology and surgical techniques have minimized treatment-related toxicity and may improve patients' quality of life. The introduction of liquid biopsy enables early detection of cancer recurrence and allows for adaptive drug therapy when the cancer develops resistance to medication. Gene therapy holds great promise, pending resolution of regulatory challenges. Additionally, artificial intelligence is poised to revolutionize head and neck cancer treatment by analyzing large datasets and providing rapid solutions to complex issues.

N. P. Nguyen (✉)
Department of Radiation Oncology, Howard University, Washington, DC, USA

O. Gorobets
Department of Oral Surgery, Cancer Tech Care Association, Perpignan, France

V. Vinh-Hung
Department of Radiation Oncology, Centre Hospitalier Public du Contentin, Cherbourg-en-Cotentin, France

S. C. Blanco
Department of Oral Surgery, Howard University, Washington, DC, USA

B. R. Page
Department of Radiation Oncology and Molecular Radiation Sciences, Johns Hopkins University, Baltimore, MD, USA

F. Taghizadeh-Hesary et al. (eds.), *From Radiobiology to Radiation Oncology*,
https://doi.org/10.1007/978-3-031-95724-6_13

Keywords Advances · Molecular biology · Immunotherapy · Radiation technology · Liquid biopsy · Gene therapy · Artificial intelligence

13.1 Overview on Head and Neck Cancer

Every year, there are about 900,000 head and neck cancer patients in the world, resulting in 400,000 deaths. Men are more affected than women likely due to a higher rate of smoking and drinking. Squamous cell carcinoma is the most prevalent histology which arises from the epithelial tissue of the oral cavity, larynx, and pharynx. The cancer is very heterogeneous due to its complex pathogenesis [1]. Multiple etiologic factors produce genetic alterations leading to mutations, and ultimately cancer formation: alcohol, tobacco, poor oral hygiene, betel nuts chewing, Epstein-Barr virus (EBV), and human papillomavirus 16 (HPV 16) [2]. The multistep carcinogenesis is very complex and involves multiple genes. Even though the exact sequence leading to cancer transformation is still debatable, oncogene activation and tumor suppressor gene inactivation play a prominent role. Overexpression of myc, erbB-2, Egfr, and cyclin D and inactivation of p16 and TP53 are frequently observed [3]. The frequency of genes mutation differs depending on the tumor HPV expression. PTEN and PIK3CA mutations are frequently observed among HPV-positive tumors while TP53 and CDKN2A/B alterations are more common in HPV-negative tumors [4]. Additionally, gene mutations frequency is also dependent on anatomic sites, with the laryngeal carcinoma having the highest mutation loads and nasopharyngeal carcinoma the least [5]. Some mutations like HRAS also increase in specific geographic locations like India but remain rare in the Western world [6].

Those genetic alterations lead to significant changes in the tumor cell surface membranes and at the subcellular level. For example, HRAS mutations lead to the activation of the Raf/MEK/ERK and the PI3K pathway which are involved with cell proliferation, differentiation, and survival [7]. The epidermal growth factor receptor (EGFR) is also overexpressed in over 90% of head and neck cancer patients [8]. Increase in cyclin D1 promotes cell cycle progression in the G1-S transition and increases cellular proliferation [9]. Inactivation of TP 53 and Notch-1 also leads to cellular proliferation as they normally inhibit cell cycle progression [10, 11]. Thus, understanding the pathogenesis of head and neck cancer cells may lead to specific intervention targeting a specific pathway such as cetuximab for EGFR overexpression.

13.2 Advances in Molecular Biology of Head and Neck Cancer and Clinical Implications

13.2.1 Correlation Between Oral Sex, P16, and Oropharyngeal Cancer

Across the world, there was an epidemic of oropharyngeal cancer affecting younger patients in recent decades [12]. The increased prevalence of oropharyngeal cancer is linked to the practice of oral sex among young individuals with multiple sex partners [13]. An increase in genito-oral contact among individuals who are infected with human papillomavirus 16 (HPV-16) but may be asymptomatic leads to viral transmission during the sexual act, as saliva is a favorable viral milieu. The virus has a special predilection for the tonsillar crypts. Its DNA becomes integrated into the host cells, producing tonsillitis [14]. Even though the carcinogenesis mechanism of HPV-16 is not fully elucidated, it is postulated to occur through viral proteins E6 and E7, which promote tumor growth through the inactivation of two tumor suppressor proteins, retinoblastoma pRb and tumor protein P53, and epigenetic dysregulation [15]. Adolescents are particularly at risk for viral infection, as the virus was found in children who underwent tonsillectomy for benign diseases [16]. Human papillomavirus 16-induced oropharyngeal cancer is characterized by a high expression of P16 protein in the tumor specimen and an excellent prognosis despite the locally advanced stage at diagnosis, in contrast to HPV-16-negative tumors that are associated with smoking and drinking [17, 18]. P16-positive tumors are both chemosensitive and radiosensitive due to increased apoptosis, impaired DNA repair capacity of cancer cells, and led to a change in the American Joint Committee on Cancer (AJCC) staging for oropharyngeal cancer in the eighth edition [19, 20]. However, the AJCC current staging is misleading, as it does not take into consideration the smoking history, which confers a poor prognosis in P16-positive tumors [21, 22]. Thus, physicians should be vigilant and cautious in the interpretation of the current data. Additionally, there was a call by the oncology community to de-escalate the radiation dose for HPV-induced oropharyngeal cancer, with the belief that reducing the radiation dose may improve the quality of life in those radiosensitive tumors [23, 24]. As a result, NRG-HN005, a phase II-III clinical trial comparing a lower dose of radiotherapy (60 Gy) combined with either cisplatin or nivolumab with standard dose radiation (70 Gy) and cisplatin, was launched for HPV-16-positive oropharyngeal cancer. Patients selected were T1-2N1M0 or T3N0-1 M0 based on AJCC eighth edition, nonsmokers or with a 10 or fewer pack-years smoking history, thus with a low risk of recurrence. However, the survival results were disappointing, leading to an earlier discontinuation of the trial for futility. The 2-year progression-free survival (PFS) was 98% and 88.6% for the standard dose and reduced dose, respectively [25]. Furthermore, attempts to replace cisplatin with cetuximab, which may have a better toxicity profile for low-risk HPV-16-positive oropharyngeal cancer, were also unsuccessful. Standard-dose radiotherapy combined with cisplatin was reported to have a better PFS compared to cetuximab with the same radiation dose. The 3-year PFS was 93% and 80% for cisplatin and cetuximab, respectively [26]. There was no difference in toxicity between the two arms. Thus, cisplatin combined with 70 Gy for HPV-16-positive oropharyngeal cancer remains the standard of care. It is likely that, due to tumor heterogeneity, the emergence of clones of cells that are radio- or chemoresistant may lead to treatment failures.

13.2.2 Programmed Death Ligand 1 (PD-L1) and Clinical Application for Immunotherapy

Programmed death ligand 1 (PD-L1) is a transmembrane glycoprotein that is present on all normal cells to prevent autoimmunity and also on tumor cells to escape destruction by CD8 T cells. Binding of PD-L1 to programmed death 1 (PD-1) receptors on T cells induces inhibition of apoptosis in those cells. Thus, tumors that have a high expression of PD-L1 tend to be biologically aggressive and are associated with a high rate of distant metastases and poor survival [27]. Conversely, those tumors also respond very well to immune checkpoint inhibitor (ICI) immunotherapy, which prevents the binding of PD-L1 to T cells and reestablishes their killing function [28]. Programmed death ligand 1 is highly expressed in head and neck cancer. Compared to other tumors such as bladder, renal cell carcinoma, lung cancer, and melanoma, there is a 60-fold increase in PD-L1 expression in head and neck cancer [29]. For example, nasopharyngeal cancer has the third highest expression of PD-L1 after thymic cancer and diffuse large cell lymphoma [30]. Thus, head and neck cancer provides an ideal target for immunotherapy. High PD-L1 expression, defined as 50% or more or a high tumor proportion score (TPS) or combined positive score (CPS), frequently indicates a high response to ICI. A high TPS has been correlated with an excellent survival independent of tumor histology following immunotherapy [31, 32]. Additionally, compared to systemic chemotherapy, there was a significant reduction in grade 3–5 toxicity, which makes immunotherapy a potential treatment of choice in frail patients such as older patients with locally advanced head and neck cancer [33]. The efficacy of ICI has been investigated in patients with recurrent head and neck cancer. Traditionally, recurrent head and neck cancer carries a poor prognosis, with median survival ranging from 1 to 13 months after salvage chemotherapy [34]. In a randomized study, pembrolizumab alone ($n = 247$) was randomized against investigators' choice of methotrexate, docetaxel, or cetuximab ($n = 248$) for salvage therapy of platinum-resistant head and neck cancer patients [35]. Median survival was 8.4 months and 6.9 months for pembrolizumab and standard–of-care groups, respectively. Grade 3–5 toxicity was also significantly less for pembrolizumab (13%) compared to the standard of care group (18%). Later on, pembrolizumab alone ($n = 301$) was proven superior to the combination of cetuximab and chemotherapy ($n = 300$) for recurrent or metastatic head and neck cancer patients [36]. Pembrolizumab's survival benefit was greatest among patients with a high TPS. Median survival was 14.9 months and 10.7 months for the pembrolizumab alone and cetuximab with chemotherapy group, respectively. Grade 3–5 toxicity was 55% and 83% for pembrolizumab alone and cetuximab with chemotherapy, respectively. As a result, the Food and Drug Administration (FDA) approved pembrolizumab for recurrent or metastatic head and neck cancer with a high TPS. Based on these studies, investigators have also explored durvalumab, another ICI agent combined with radiotherapy as a potential replacement for patients with locally advanced head and neck cancer who are not eligible for cisplatin-based chemotherapy due to contraindications. Patients were randomized between durvalumab ($n = 123$) and cetuximab ($n = 63$) concurrently with standard-dose radiotherapy (70 Gy). However, the trial was terminated prematurely due to poorer PFS in the durvalumab arm [37]. The 2-year PFS was 50.6% and 63.7% for durvalumab and cetuximab, respectively. It is unclear why durvalumab was less effective than cetuximab. One potential hypothesis is that the schedule of concurrent immunotherapy and radiotherapy may produce an immunosuppressive environment compared to sequential radiotherapy followed by immunotherapy. Radiotherapy is known to induce PD-L1 formation in tumors [38]. Thus, it would theoretically be better to sequence radiotherapy first followed by immunotherapy. Another possible hypothesis is the inclusion of PD-L1-negative patients in the trial (19%). Subgroup analysis showed a better survival for patients with a CPS score of 20 or above, compared to those with a CPS score less than 20. However, clinical trials should be performed to confirm this hypothesis.

13.3 Advances in Radiotherapy Techniques

13.3.1 Image-Guided Radiotherapy (IGRT)

Advancements in radiotherapy techniques have aimed at reducing treatment toxicity, and this has led to the evolution from 2-dimensional (2-D) and 3-dimensional (3-D) conformal radiotherapy to intensity-modulated radiotherapy (IMRT) and now image-guided radiotherapy (IGRT). Daily imaging using techniques like cone beam computed tomography (CBCT) or magnetic resonance imaging (MRI) before treatment enables precise visualization of the tumor and patient positioning, reducing the risk of marginal miss and minimizing damage to nearby organs at risk (OARs) [39]. This is particularly important in head and neck cancers, where OARs such as the cochlea, parotids, larynx, pharyngeal muscles, mandible, and spinal cord are sensitive to radiation [40]. Severe complications like hearing loss, xerostomia, hoarseness, dysphagia, aspiration, and osteoradionecrosis can significantly impact patients' quality of life (QOL) following treatment [41]. Aspiration during treatment can lead to fatal complications, including aspiration pneumonia [42, 43]. However, IGRT significantly reduces radiation exposure to these sensitive structures, thereby decreasing complication rates and improving QOL outcomes [44,

47]. For example, hoarseness is significantly reduced through laryngeal sparing, and up to 70% of patients report excellent QOL 1 year posttreatment [48, 49]. Moreover, adaptive therapy, which involves replanning radiotherapy mid-course to account for tumor shrinkage or changes in body contour, can further reduce treatment toxicity. One such approach—decreasing radiation dose to the pharyngeal constrictors—has shown promise in reducing dysphagia and improving QOL [50, 51].

13.3.2 Proton Therapy

Proton therapy is emerging as an optimal radiotherapy option for head and neck cancer due to its ability to spare normal tissues while delivering high doses to tumors. Proton therapy's Bragg peak effect, which limits radiation penetration beyond the tumor, allows high-dose radiation to the target while minimizing damage to surrounding OARs [52, 53]. Studies comparing intensity-modulated proton therapy (IMPT) to IMRT have demonstrated significantly reduced radiation doses to critical structures like the spinal cord, cochlea, and brainstem [54]. Proton therapy is particularly beneficial for tumors located near sensitive areas, such as the sinonasal region. It allows for dose escalation to the tumor, improving local control while reducing the likelihood of treatment-related toxicity [55–57]. For example, a study of 143 patients with sinonasal cancers treated with proton therapy reported an 80% local control rate and 22% grade 3 toxicity [56]. Another study of 64 sinonasal cancer patients showed excellent local control (88%) and no grade 3 toxicity [57]. Proton therapy also improves QOL by reducing complications like gastrostomy tube dependence and opioid use for pain management compared to IMRT [58]. Additionally, proton therapy is a valuable option for re-irradiating patients who have previously received photon therapy. Re-irradiation of head and neck cancer patients often carries a high risk of severe toxicity, but proton therapy's precision offers a promising solution, providing local control rates of 66–75% but death from bleeding is still a threat and ranges from 2% to 4.9% [59–62]. In pediatric patients, proton therapy is preferred to minimize growth-related damage to normal tissues [63]. A study in children reported a 92% local control rate with 32% grade 3 toxicity. Despite its dosimetric advantages, proton therapy remains limited by its high cost and the limited availability of proton therapy centers, making it less accessible for the general head and neck cancer population [64]. Therefore, patient selection is crucial to ensure appropriate referral to specialized proton centers. These advanced radiotherapy techniques—IGRT and proton therapy—represent significant steps forward in enhancing the precision of head and neck cancer treatment while reducing toxicity, ultimately improving patient outcomes and QOL.

13.4 Advances in Surgical Technique: Transoral Robotic Surgery (TORS)

Since receiving FDA approval in 2009 for early-stage oropharyngeal cancer, transoral robotic surgery (TORS) has expanded its indications to include the hypopharynx, supraglottic larynx, parapharyngeal space, and unknown primary tumors [65, 66]. TORS has proven to be effective in conjunction with concurrent neck dissection, and patients may proceed to postoperative adjuvant therapy when necessary [67]. In patients with suitable endoscopic access and small tumors, TORS provides excellent local control, high survival rates, and reduced morbidity. For example, a study involving 410 head and neck cancer patients undergoing TORS reported a 2-year loco-regional control rate of 91.8% and a survival rate of 91% [65]. Although no randomized trials compare TORS with open surgery, TORS is associated with fewer complications such as tracheostomy, pharyngotomy, and flap reconstruction, which are often required in open surgery. Open surgery can involve mandible splitting to access the tumor, leading to higher morbidity and longer recovery times [67]. A meta-analysis comparing TORS to open surgery found that TORS had superior local control, fewer positive resection margins, and shorter postoperative feeding tube dependency [68]. Another study corroborated these findings, showing better survival outcomes for patients undergoing TORS over open surgery. Among 4071 patients with HPV-positive early-stage oropharyngeal cancer, the 5-year survival rates for TORS and open surgery were 84.8% and 80.3%, respectively [69]. Furthermore, a longitudinal study showed good quality of life for oropharyngeal cancer patients post-TORS, though those requiring postoperative adjuvant radiotherapy experienced prolonged dysphagia, which negatively impacted their quality of life [70]. Both TORS and concurrent chemoradiation provided equal survival for patients with oropharyngeal cancer but the toxicity pattern was different with more hearing loss in the radiotherapy group and worse dysphagia and pain in the surgical group [71, 72]. A significant application of TORS is its potential to reduce radiation dose and treatment volume for HPV-16-positive oropharyngeal cancer patients needing adjuvant radiotherapy. In a study involving 150 patients, radiotherapy dose and volume were stratified based on recurrence risk, and a 4-year local control rate of 97.7% was achieved, suggesting the possibility of using this protocol to enhance patient quality of life [73].

13.5 Advances in Liquid Biopsy: Real-Time Monitoring of Disease and Treatment

13.5.1 Outcomes

Liquid biopsy offers a noninvasive method for early cancer diagnosis, recurrence detection, and identification of treatment resistance, in contrast to traditional diagnostic and monitoring methods like imaging [74]. This technique is especially useful for head and neck cancers, which are often heterogeneous, as biopsy of the tumor may not capture all genomic and transcriptomic abnormalities. Liquid biopsy provides complementary information to tumor biopsy, aiding clinicians in patient management. Simple collection of saliva and blood samples allows the detection of circulating tumor biomarkers, such as circulating tumor cells (CTC), circulating tumor DNA (ctDNA), microRNA, mRNA, and exosomes, to monitor disease presence or absence and treatment responses. Preclinical models highlight the potential of liquid biopsy for assessing tumor burden and treatment responses. In a study with rabbits inoculated with papillomavirus-induced VX2 tumor cells, ctDNA and CTC levels were monitored before, during, and after irradiation. A correlation between tumor size and CTC levels was observed, which progressively decreased and became undetectable during and after radiotherapy, demonstrating that liquid biopsy provides real-time information on tumor growth and treatment response [75]. In regions like Southeast Asia, Epstein-Barr virus (EBV)-associated nasopharyngeal cancer can be monitored using plasma EBV DNA levels. These levels correlate positively with tumor load and prognosis, aiding in screening, prognostication, and surveillance for recurrence [76, 77]. The high sensitivity and specificity of EBV DNA make it a reliable marker for early diagnosis, even in asymptomatic individuals, and for monitoring disease recurrence after treatment [78, 79]. In oropharyngeal cancer caused by HPV-16, viral load through circulating HPV-16 DNA has shown high sensitivity and specificity for early cancer diagnosis. A study of 70 patients with biopsy-confirmed HPV-16 oropharyngeal cancer demonstrated 98.4% sensitivity and 98.6% specificity in diagnosing the cancer using circulating HPV-16 DNA [80]. Furthermore, liquid biopsy has shown potential in detecting early recurrence after treatment [81, 82]. In a study of 45 patients with HPV-16-positive oropharyngeal cancer undergoing chemoradiation, sequential liquid biopsy revealed early recurrence an average of 6.6 months before clinical detection [83]. A similar study with 1076 patients found that the predictive value of a positive liquid biopsy for posttreatment recurrence was 95%, supporting its potential as an early diagnostic tool for recurrence [84]. Other studies also corroborated the predictive value of liquid biopsy not only to diagnose early recurrence but to portend a poor prognosis [85–88]. Among patients with HPV-16-negative head and neck cancer, liquid biopsy also predicts early recurrence and a poor prognosis and may complement PET-CT when the result is equivocal after treatment [89–93]. Liquid biopsy is also valuable for identifying gene mutations that can affect treatment responses [94]. For instance, in patients with head and neck cancer undergoing immunotherapy with ICI, liquid biopsy detected epidermal growth factor receptor (EGFR) amplification in two patients, enabling effective treatment with cetuximab [95]. Similarly, a study involving 46 patients receiving cetuximab treatment showed that RAS mutations were associated with disease progression, indicating that liquid biopsy could identify treatment resistance and guide adjustments in therapy [96]. Thus, liquid biopsy offers significant potential for predicting early recurrence, monitoring disease progression, and informing treatment decisions in head and neck cancer.

13.6 Advances in Gene Therapy

The tumor suppressor protein p53 plays a crucial role in the repair of DNA damage. If DNA damage cannot be repaired, p53 halts cell division and induces apoptosis, preventing the damaged cell from proliferating and potentially forming tumors. Alterations in the p53 gene can lead to its dysfunction, which contributes to tumorigenesis. Up to 85% of head and neck cancers exhibit p53 gene mutations, often due to a history of smoking and alcohol consumption [97]. Restoring p53 function theoretically could lead to cancer cell death. In vivo experiments support the role of p53 in halting cancer cell growth. Injection of wild-type p53 via a recombinant adenovector (Ad-p53) into squamous cell tumors in nude mice resulted in significant tumor shrinkage, providing a scientific rationale for gene therapy in head and neck cancers [98]. Injection of Ad-p53 as an adjuvant treatment following surgery in 15 patients with recurrent head and neck carcinoma induced remission in 4 patients with minimal complications [99]. Among 70 patients with recurrent head and neck cancer who received intratumoral injection of Ad-p53, increases in interferon (IFN) and CD8 T cell counts were observed, which were proportional to the dose of viral particles injected [100]. Tumor response was significantly better with higher viral loads, with a response rate of 31% for viral doses greater than 7×10^{10}, compared to 0% for doses less than this threshold. Thus, Ad-p53 injection produced a strong immune response that contributed to tumor shrinkage. In a study of 30 patients with nasopharyngeal cancer treated with two cycles of Ad-p53 combined with chemoradiation ($n = 22$) or chemotherapy ($n = 8$), a 76.7% response rate was observed, with 40% achieving a complete response (CR) and 36.7% a partial response (PR) [101]. Treatment toxicity was acceptable, with 36.7% of patients experiencing grade 3–4 leucopenia and 3.3% experiencing grade 3–4 thrombocytopenia. Several phase II randomized studies conducted in China reported improved survival rates when Ad-p53 was combined with chemotherapy or radiotherapy for patients with head and neck cancer [102–105]. Liu et al. [104] reported on 107 patients with oral cavity cancer who received postoperative radiotherapy alone ($n = 50$) or radiotherapy combined with Ad-p53 ($n = 57$). The 3-year recurrence rate was 32% in the radiotherapy-alone group and 7% in the combined gene therapy group. Toxicity was minimal, consisting of grade 2 fever and flu-like symptoms. Other studies corroborated the safety and efficacy of Ad-p53 when combined with chemotherapy or radiotherapy. A meta-analysis of 566 patients with nasopharyngeal cancer who received chemoradiation or radiotherapy alone ($n = 268$) or combined with Ad-p53 ($n = 298$) reported superior survival in those who received gene therapy [106]. However, no randomized trials on gene therapy have been conducted in the United States, as the Food and Drug Administration (FDA) has not approved the two commercially available Ad-p53 drugs, Advexin and Gendicine, for therapy in the United States. Despite its potential as a promising salvage therapy for recurrent disease, gene therapy's inclusion in the therapeutic arsenal for head and neck cancer patients remains uncertain. Nevertheless, ongoing trials, particularly in China, continue to explore its potential in combination with other therapies.

13.7 Advances in Artificial Intelligence in Head and Neck Cancer Management

Artificial intelligence (AI) is beginning to revolutionize the management of head and neck cancer, particularly through deep learning and natural language processing (NLP). Deep learning, which uses artificial neural networks (ANNs), allows for the processing of vast amounts of medical data and helps uncover complex patterns that were previously challenging to identify. NLP, on the other hand, enables machines to understand, interpret, and generate human language, facilitating active interaction between clinicians and computers. This has profound implications for extracting useful insights from unstructured data sources, such as clinical notes, scientific publications, and electronic health records. With AI's ability to rapidly summarize findings from such data, clinicians and researchers can make more informed decisions regarding patient management and future research directions. Although much of the data for head and neck cancer management remains in the early stages, AI has already begun to make a significant impact, especially in the areas of surgery and radiotherapy. In surgery, the key to success lies in the complete removal of the tumor along with a healthy margin of tissue to prevent recurrence. However, current methods for assessing surgical margins during operations can be slow and sometimes inaccurate. To address this challenge, technologies like hyperspectral imaging (HSI) have been developed, which can detect wavelengths beyond the visible spectrum. HSI works by analyzing the absorption peaks of tumor cells in comparison to normal tissue, providing surgeons with real-time feedback on tumor margins [107]. One study involving deep learning demonstrated an algorithm capable of accurately identifying cancer cells in surgical margins from 102 head and neck cancer patients using HIS [108]. The ability to scan the entire surgical specimen in a single run makes this technology both fast and reliable, potentially reducing operating times and improving surgical precision [109]. While promising, these AI-driven applications of HIS need to be further validated in prospective studies to confirm their benefits in routine clinical practice [110–112]. Preliminary suggests that incorporating HSI and AI in the operating room would reduce the time to assess the surgical margins to 3 min or less [112]. AI is also making waves in the field of radiotherapy, a treatment that is often time-consuming due to the complex planning required to outline tumors and organs at risk (OARs). This process involves the painstaking task of defining these areas on planning CT scans, followed by the creation of a treatment plan that ensures optimal tumor coverage while minimizing radiation exposure to surrounding healthy tissue. Automation of these steps could speed up treatment delivery and reduce costs. Given the close proximity of tumors to sensitive organs in head and neck cancer, this is particularly challenging. AI-based autosegmentation algorithms for autocontouring OARs have shown promise in improving radiotherapy workflows. In a study of 589 patients with head and neck cancer, deep learning contouring (DLC) was found to generate OAR contours that were acceptable to clinicians and outperformed traditional atlas-based methods in terms of precision [113]. This technology not only reduces the workload for physicians but also shortens treatment planning times significantly, with one study showing a reduction in physician workload by 71% [114–117]. The AI-generated contours also demonstrated greater consistency, making them more suitable for clinical studies, as evidenced by a multi-institutional study [118]. Another area where AI is poised to make a significant impact is in the planning of intensity-modulated radiotherapy (IMRT). Traditionally, the planning process for IMRT involves a trial-and-error approach, with dosimetrists manually adjusting constraints within an optimization algorithm until an acceptable plan is achieved. This process is time-consuming and highly dependent on the experience of the planner. However, with AI's advancement, a virtual treatment planner (VTP) can now be used to automate this process [119]. Preliminary studies have shown that plans generated by VTP are comparable to or even outperform those created by human planners [119–123]. For instance, in a study involving 30 head and neck cancer patients treated with IMRT, VTP-generated plans not only provided better sparing of organs at risk but also reduced planning time by almost half, from 64 min for human-generated plans to just 32 min [121]. Further advancements are expected to reduce this time even further to between 10 and 15 min [124]. The implications of AI for radiotherapy are profound, particularly in enhancing efficiency and reducing delays. Autocontouring and autoplanning are two AI applications that are likely to streamline workflows in radiotherapy departments. This becomes especially critical in low-income countries, where there is often a shortage of qualified staff. The integration of AI could help bridge this gap and allow these countries to more effectively participate in international research trials, ultimately improving access to cutting-edge treatments for patients with head and neck cancer [125]. As AI continues to evolve, its role in head and neck cancer management will only grow. From improving surgical precision and reducing operating times to automating complex radiotherapy planning tasks, the potential for AI to enhance the efficiency, accuracy, and accessibility of care is immense. However, further research and prospective trials are needed to fully realize and validate these applications in clinical practice.

13.8 Radiology: Future Prospect in Head and Neck Cancer

PET-CT scan with ^{18}F-fluorodeoxyglucose (FDG) is currently standard in the diagnostic workup for squamous cell head and neck cancer due to its accuracy to diagnose distant metastases and second primaries and for radiotherapy planning. It is also cost-effective for the follow-up of patients who had cervical nodes and received chemoradiation for their head and neck cancer. Those patients frequently required neck dissection if there is residual neck disease after treatment. A randomized study of 564 patients with N2 or N3 disease reported that PET-CT scan was effective for monitoring patients after chemoradiation. Survival and QOL were similar between the group who had surveillance with PET-CT compared to the ones scheduled to have a planned neck dissection after treatment [126].

There were significantly less patients undergoing neck dissection in the PET-CT scan surveillance group which resulted in a saving of 2190 dollars per person. Additionally, in another randomized study, FDG PET-CT scan allowed for radiation dose escalation with no added toxicity among patients who underwent chemoradiation for head and neck cancer [127]. In another study using a different isotope to identify hypoxia within the tumor, ^{18}F-fluoromisonidazole (FMISO), radiation dose escalation to hypoxic areas was also successful without added toxicity [128]. Even though those two studies are only preliminary, the ability of PET-CT scan to identify areas at risk for recurrence could pave the way for future radiation dose escalation studies.

13.9 Conclusion

Both gene therapy and AI are showing remarkable potential in advancing the treatment and management of head and neck cancers. Head and neck cancer patients who suffer from recurrence after treatment have a poor prognosis and poor QOL due to pain and bleeding. Gene therapy targeting the p53 gene mutation is showing promising early results, with minimal toxicity and could be an effective salvage option that should be tested in future randomized trials. AI technologies are improving surgical precision, radiotherapy planning, and treatment delivery. Implementation of AI in low-income countries could be cost-effective and improve the quality of care for cancer patients due to the lack of trained professionals, and also needs to be tested in future studies.

References

1. Miranda-Galvis M, Loveless R, Kowalski LP, Teng Y. Impact of environment factors on head and neck cancer pathogenesis and progression. Cells. 2021;10:389.

2. Constantin M, Chifiriuc MC, Bleotu C, Vrancianu CO, Cristian R, Bertesteanu SV, et al. Molecular pathways and targeted therapies in head and neck cancer pathogenesis. Front Oncol. 2024;14:1373821.
3. Nagai MA. Genetic alterations in head and neck squamous cell carcinomas. Braz J Med Biol Res. 1999;32:897–904.
4. Chung CH, Guthrie VB, Masica DL, Tokheim C, Kang H, Richmon J, et al. Genomic alterations in head and neck squamous cell carcinoma determined by cancer gene-targeted sequencing. Ann Oncol. 2025;26:1216–23.
5. Ju Y, Wu X, Wang H, Li B, Long Q, Zhang D, et al. Genomic landscape of head and neck squamous cell carcinoma across different anatomic sites in Chinese population. Front Genetics. 2021;12:680699.
6. Muthusamy M, Ramani P, Arumugam P. Effect of Harvey ras sarcoma virus mutation on oral squamous carcinoma and its influence on different patient population. Cureus. 2023;15:e45505.
7. Fisher M, Swick A, Nickel KP. Radiosensitization of HRAS mutated head and neck cancer. Int J Radiat Oncol Biol Phys. 2017;99:ES89–90.
8. Rehmani HS, Issaeva N. EGFR in head and neck cancer squamous cell carcinoma: exploring possibilities of novel drug combinations. Ann Translat Med. 2020;8:813.
9. Nakashima T, Kuratomi Y, Yasumatsu R, Masuda M, Koike K, Umezaki T, et al. The effect of cyclin D1 overexpression in human head and neck cancer cells. Eur Arch Otolaryngol. 2005;262:379–83.
10. Zhou G, Liu Z, Myers JN. TP53 mutations in head and neck squamous cell carcinoma and their impact on disease progression and treatment response. J Cell Bioch. 2016;117:2682–92.
11. Fukusumi T, Califano JA. The NOTCH pathway in head and neck: squamous cell carcinoma. J Dental Res. 2018;97:643–53.
12. Zumsteg ZS, Luu M, Rosenberg PS, Elrod JK, Bray F, Vaccarellla S, et al. Global epidemiologic patterns or oropharyngeal cancer incidence trends. JNCI. 2023;115:1544–54.
13. Nguyen NP, Nguyen LM, Sroka T, Hong-Ly B, Chi A, Vos P, et al. Oral sex and oropharyngeal cancer: the role of the primary care physician. Medicine. 2016;95:e4228.
14. Chen R, Sehr P, Waterboer T, Leivo I, Pawlita M, Vaheri A, et al. Presence of DNA of human papillomavirus 16 but no other types in tumor-free tonsillar tissue. J Clin Microbiol. 2005;43:1408–10.
15. Santacroce L, Di Cosola M, Bottalico L, Topi S, Charistos IA, Ballini A, et al. Focus on HPV infection and the molecular mechanisms of oral carcinogenesis. Viruses. 2021;13:559.
16. Duray A, Descamps G, Bettonville M, Sirtaine N, Ernoux-Neufcoeur P, Guenin S, et al. High prevalence of high risk human papillomavirus in palatine tonsils from healthy children and adults. Otolaryngology-Head and Neck Surg. 2011;145:230–5.
17. Nguyen NP, Hong-Ly B, Betz M, Vinh-Hung V. Importance of age as a prognostic factor for tonsillar carcinoma. Ann Surg Oncol. 2010;17:2570–7.
18. Wendt M, Hammarstedt-Nordenvall L, Zupancic M, Friesland S, Landi D, Munck-Wikland E, et al. Long-term survival and recurrence in oropharyngeal squamous cell carcinoma in relation to subsites, HPV, and p16 status. Cancers. 2021;13:2553.
19. Spiotto MT, Taniguchi CM, Klopp AH, Colbert LE, Lin SH, Wang L, et al. Biology of the radio- and chemo-responsiveness in HPV malignancies. Sem Radiat Oncol. 2021;31:274–85.
20. Yip HT, Chopra R, Chakrabarti R, Veena MS, Ramamurthy B, Sritvasan ES, Wang MB. Cisplatin-induced growth arrest of head and neck cancer cells correlates with increased expression of p16 and p53. JAMA Otolaryngol. 2006;132:317–26.
21. Lassen P, Lacas B, Pignon J, Trotti A, Zackrisson B, Zang Q, et al. Prognostic impact of HPV-associated P16-expression and smoking status on outcomes following radiotherapy for oropharyngeal cancer: the MARCH-HPV Project. Radiother Oncol. 2018;126:107–15.
22. Bouland C, Dequanter D, Lechien JR, Hanssens C, Aubain NDS, Digonnet A, et al. Prognostic significance of a scoring system combining P16, smoking, and drinking status in a series of 131 patients with oropharyngeal cancers. Int J Otolaryngol. 2021;8020826:1–6.

23. Wu C, Kuzmin P, Julian R. De-escalation strategies in HPV-associated oropharyngeal cancer: a historical perspective with future direction. Cancers. 2024;16:2733.
24. Rosenberg AJ, Vokes EE. Optimizing treatment de-escalation in head and neck cancer: current and future perspectives. Oncologist. 2021;26:40–8.
25. Yom SS, Harris J, Caudell JJ, Geiger JL, Waldron J, Gillison M, et al. Interim futility of results of NRG-HN005, a randomized, phase II/III non-inferiority trial for non-smoking p16+ oropharyngeal cancer patients. Int J Radiat Oncol Biol Phys. 2024;120:S2–3.
26. Rischin D, King M, Kenny L, Poerceddu S, Wratten C, Macann B, et al. Randomized trial of radiation therapy with weekly cisplatin or cetuximab in low risk HPV-16 associated oropharyngeal cancer (TROG 12.01). Int J Radiat Oncol Biol Phys. 2021;111:876–86.
27. Lin Y, Sung W, Hsieh M, Tsai S, Lai H, Yang S, et al. High PD-L1 expression correlates with metastases and poor prognosis in oral squamous cell carcinoma. PLoS One. 2015;10:e0142656.
28. Patel JJ, Levy DA, Nguyen SA, Knochelman HM, Day TA. Impact of PD-L1 expression and human papillomavirus status in antiPD1/PDL1 immunotherapy for head and neck squamous cell carcinoma-systemic review and meta-analysis. Head Neck. 2020;42:774–86.
29. Heineman TE, Widman A, Kuan EC, St JM. The genetic landscape of program death ligand-1 (PD-L1) alterations in head and neck cancer. Laryngoscope Invest Otolaryngol. 2017;2:99–103.
30. Yarchoan M, et al. PD-L1 expression and tumor mutation burden are independent biomarkers in most cancers. JCI Insight. 2019;4:e126908.
31. Hersbt RS, Baas P, Kim D, Felip E, Perez-Gracia E, Han J, et al. Pembrolizumab versus docetaxel for previously treated PD-L1-positive, advanced non-small cell lung cancer (KEYNOTE-010): a randomized controlled trial. Lancet. 2016;387:1540–50.
32. Colombo N, Dubot C, Lorusso D, Caceres V, Haswgawa K, Shapira-Frommer T, et al. Pembrolizumab for persistent, recurrent, or metastatic cervical cancer. N Engl J Med. 2021;385:1856–67.
33. Mok TSK, Wu Y, Yudaba I, Kowalski DM, Cho BC, Turna HZ, et al. Pembrolizumab versus chemotherapy for previously untreated, PD-L1 expressing locally advanced or metastatic non-small cell lung cancer (KEYNOTE-042): a randomized, open-label, controlled, phase 3 trial. Lancet. 2019;393:1819–30.
34. Magnes T, Melchardt T, Weiss L, Mittermair C, Neureiter D, Kleiser E, et al. Prognostic score in patients with recurrent or metastatic carcinoma of the head and neck treated with cetuximab and chemotherapy. PLoS One. 2017;12:e0180995.
35. Cohen EEW, Souliere D, Le Tourneau C, Dinis J, Licitra L, Ahn M, et al. Pembrolizumab versus methotrexate, docetaxel, or cetuximab for recurrent or head and neck squamous cell carcinoma (KEYNOTE-040): a randomized, open-label phase 3 study. Lancet. 2019;393:156–67.
36. Burtness B, Harrington KJ, Greil R, Soulieres D, Tahara M, De Castro G, et al. Pembrolizumab alone or with chemotherapy versus cetuximab plus chemotherapy for recurrent or metastatic squamous cell carcinoma of the head and neck (KEYNOTE-048): a randomised, open label, phase 3 study. Lancet. 2019;394:1915–28.
37. Mell LK, Torres-Saavedra P, Wong S, Chang S, Kish JA, Minn AJ, et al. Radiotherapy with duvalumab versus cetuximab in patients with locally advanced head and neck cancer and a contraindication to cisplatin (NRG-HN004): an open-label, multicentre, parallel-group, randomised, phase 2–3 trial. Lancet Oncol. 2024;25:1576–88.
38. Mao L, Zhou J, Xiao Y, Yang Q, Wang S, Wu Z, et al. Immunogenic hypofractionated radiotherapy sensitizing head and neck squamous cell carcinoma to anti-PD-L1 therapy in MDSC-dependent manner. Br J Cancer. 2023;128:2126–39.
39. Pollard JM, Wen Z, Sadagopan R, Wang J, Ibbott G. The future of image-guided radiotherapy will be MRI-guided. Br J Radiol. 2017:9020160667.
40. Yu Y, Michaud AL, Sreeraman R, Liu T, Purdy JA, Chen AM. Comparison of daily versus nondaily image-guided radiotherapy protocols for patients treated with intensity-modulated head and neck cancer. Head Neck. 2014;36:992–7.

41. Nguyen NP, Sallah S, Karlsson U. Combined chemotherapy and radiotherapy for head and neck malignancies. Cancer. 2002;94:1131–41.
42. Nguyen NP, Moltz CC, Frank C, Vos P, Smith HJ, Karlsson U, et al. Dysphagia following chemoradiation for locally advanced head and neck cancer. Ann Oncol. 2004;15:383–8.
43. Nguyen NP, Frank C, Moltz CC, Vos P, Smith HJ, Karlsson U, et al. Impact of dysphagia on quality of life after treatment of head and neck cancer. Int J Radiat Oncol Phys Biol. 2005;61:772–8.
44. Nguyen NP, Smith-Raymond L, Vinh-Hung V, Davis R, Desai A, Abraham D, et al. Feasibility of tomotherapy to spare the cochlea from excessive radiation in head and neck cancer. Oral Oncol. 2011;47:414–9.
45. Nguyen NP, Vos P, et al. Feasibility of image-guided radiotherapy based on helical tomotherapy to reduce contralateral parotid dose in head and neck cancer. BMC Cancer. 2012;12:175.
46. Nguyen NP, Ceizyk M, Vos P, Vinh-Hung V, Davis R, Desai A, et al. Effectiveness of image-guided radiotherapy for laryngeal sparing in head and neck cancer. Oral Oncol. 2010;46:283–6.
47. Nguyen NP, Vock J, Chi A, Ewell L, Vos P, Mills M, et al. Effectiveness of intensity-modulated and image-guided radiotherapy to spare the mandible from excessive radiation. Oral Oncol. 2012;48:653–7.
48. Nguyen NP, Abraham D, Desai A, Betz M, Davis R, Sroka T, et al. Impact of image-guided radiotherapy to reduce laryngeal edema following treatment for non-laryngeal non-hypopharyngeal head and neck cancer. Oral Oncol. 2011;47:900.
49. Chen AM, Hsu S, Lamb J, Agazaryan N, Steinberg ML, Low DA, et al. MRI-guided radiotherapy for head and neck cancer: initial clinical experience. Clin Transl Oncol. 2018;20:160–8.
50. Chen AM, Yoshizaki T, Hsu S, Mikaeilian A, Cao M. Image-guided adaptive radiotherapy improves acute toxicity during intensity-modulated radiotherapy for head and neck cancer. J Radiat Oncol. 2018;7:139–45.
51. Weppler S, Quon H, Schinkel C, Yarshenko A, Barbera L, Harjai N, et al. Patient-reported-outcomes-guided adaptive radiation therapy for head and neck cancer. Front Oncol. 2021;11:759–24.
52. Thariat J, Sio T, Blanchart P, Krengli M, Giap H, Nguyen NP. Using proton beam therapy in the elderly population: a snapshot of current perception and practice. Int J Radiat Oncol Biol Phys. 2017;98:840–2.
53. Mohamed N, Lee A, Lee NY. Proton beam radiation therapy treatment for head and neck cancer. Prec Radiat Oncol. 2022;6:59–68.
54. Nguyen ML, Cantrell N, Ahmah S, Henson C. Intensity-modulated proton therapy (IMPT) versus intensity-modulated radiation therapy (IMRT) for the treatment of head and neck cancer. Med Dosim. 2021;46:259–63.
55. Dagan R, Bryant C, Li Z, Yeung D, Justice J, Dzeglewiski P, et al. Outcomes of sinonasal cancer treated with proton therapy. Int J Radiat Oncol Biol Phys. 2016;95:377–85.
56. Dagan R, Bryant C, Li Z, Yeung D, Justice J, Dzeglewiski P, et al. Long-term outcomes from proton therapy for sinonasal cancers. Int J Particle Ther. 2021;8:200–12.
57. Pasalic D, Ludmir EB, Allen PK, Thaker NG, Chapman BV, Hanna EY, et al. Patient-reported outcomes, physician-reported toxicities, and treatment outcomes in a modern cohort of patients with sinonasal cancer treated using proton beam therapy. Radiother Oncol. 2020;148:258–66.
58. McDonald MW, Liu Y, Moore MG, Johnstone PAS. Acute toxicity in comprehensive head and neck radiation for nasopharynx and paranasal sinus cancers: cohort comparison of 3D conformal proton therapy and intensity modulated radiation therapy. Radiat Oncol. 2016;11:32.
59. Romesser PB, Cahlon O, Scher ED, Hug EB, Sine K, DeSelm C, et al. Proton beam reirradiation for recurrent head and neck cancer: multi-institutional report on feasibility and early outcomes. Int J Radiat Oncol Biol Phys. 2016;95:386–96.

60. Krc R, Mendes W, Molitoris J, Ferris M, Song Y, Shetty A, et al. Proton therapy re-irradiation outcomes and genomic landscape of patients with recurrent head and neck cancer. Oral Oncol. 2024;154:106875.
61. Lee A, Woods R, Mahfouz A, Kirpanit S, Cartano O, Mohamed N, et al. Evaluation of proton therapy reirradiation for patients with recurrent head and neck carcinoma. JAMA Network Open. 2023;6:e2250607.
62. McDonald WM, Zoladi-Meybodi M, Lehnert SJ, Estabrook NC, Liu Y, et al. Reirradiation of recurrent and second primary head and neck cancer with proton therapy. Int J Radiat Oncol Biol Phys. 2016;96:808–19.
63. Vogel J, Both S, Kirk M, Chao H, Bagatell R, Li Y, et al. Proton therapy for pediatric head and neck malignancies. Ped Blood Cancer. 2018;65:e26858.
64. Scher DJ, Tishler RB, Pham N, Punglia RS. Cost-effectiveness analysis of intensity modulated radiation therapy versus proton therapy for oropharyngeal squamous cell carcinoma. Int J Radiat Oncol Biol Phys. 2018;101:875–82.
65. de Almeida LR, Magnuson S, Smith RV, Moore E, Lawson G, Remacle M, et al. Oncologic outcomes after transrobotic Surgery. A multi-institutional study. JAMA Otolaryngol Head Neck Surg. 2015;141:1043–51.
66. Patel S, Magnuson JS, Holsinger H, Karni RJ, Richmond JD, Gross ND, et al. Robotic Surgery for primary head and neck squamous cell carcinoma of unknown site. JAMA Otolaryngol Head Neck Surg. 2013;139:1203–11.
67. Rubek N, Channir HI, Charabi BW, Lajer CB, Kiss K, Nielsen HU, et al. Primary transoral robotic surgery with concurrent neck dissection for early stage oropharyngeal squamous cell carcinoma implemented at a Danish head and neck cancer center: a phase II trial on feasibility and tumor margin status. Eur Arch Otorhinolaryngol. 2017;274:2229–37.
68. Liu H, Wang Y, Wu C, Sun X, Li L, Li C, et al. Robotic compared with open operations for cancers of the head and neck: a systematic review and meta-analysis. Br J Oral Maxillofac Surg. 2029;57:967–76.
69. Nguyen AT, Luu M, Mallen-St CJ, Mita AC, Scher KS, Lu DJ, et al. Comparison of survival after transoral robotic surgery vs nonrobotic Surgery in patients with early-stage oropharyngeal squamous cell carcinoma. JAMA Oncol. 2020;6:1555–62.
70. Achim V, Bologne RK, Palmer AD, Graville DJ, Light TJ, Li R, et al. Long-term functional and quality of life outcomes after transoral robotic surgery in patients with oropharyngeal cancer. JAMA Otolaryngol Head Neck Surg. 2018;144:18–27.
71. Nichols C, Theurer J, Prisman E, Read N, Betherlet EB, Tran E, et al. Radiotherapy versus transoral robotic surgery and neck dissection for oropharyngeal squamous cell carcinoma (ORATOR): an open-label, phase 2 randomised trial. Lancet Oncol. 2019;20:1349–59.
72. Nichols C, Theurer J, Prisman E, Read N, Betherlet EB, Tran E, et al. Randomized trial of radiotherapy versus transoral robotic Surgery for oropharyngeal squamous cell carcinoma: Long-term results of the ORATOR trial. J Clin Oncol. 2024;42:4023–8.
73. Lukens JN, MacDuffie E, Siegal T, Poirier K, Weinstein G, Newman JG, et al. Phase II study of radiation dose-de-intensification following TORS and neck dissection for p16+ oropharyngeal squamous cell carcinoma-long-term results of expanded cohort. Int J Radiat Oncol Biol Phys. 2024;25:312.
74. Ghiyasimoghadam N, Shayan N, Mirkatuli HA, Baghbani M, Ameli N, Ashari Z, et al. Does circulating tumor DNA apply as a reliable biomarker for the diagnosis and prognosis of head and neck squamous cell carcinoma. Discover Oncol. 2024;15:427.
75. Muhanna N, Eu D, Chan HHL, Douglas C, Towson JL, Di Grappa MA, et al. Cell-free DNA and circulating tumor cell kinetics in a pre-clinical head and neck cancer model undergoing radiation therapy. BMC Cancer. 2021;21:1075.
76. Lam WKJ, Chan KCA, Lo WMD. Plasma Epstein-Barr virus DNA as an archetipal circulating tumor DNA. J Pathol. 2019;247:641–9.
77. Zhang J, Shu C, Song Y, Li Q, Huang J, Ma X. Epstein-Barr virus level as a novel prognostic factor in nasopharyngeal carcinoma: a meta-analysis. Medicine. 2016;95:e5130.

78. Chan KCA, Woo JKS, King A, Zee BCY, Lam WKJ, Chan SL, et al. Analysis of plasma Epstein-Barr virus DNA to screen for nasopharyngeal carcinoma. N Engl J Med. 2017;377:513–22.
79. Li W, Chen J, Liang B, Li Z, Li J, Yuan X, et al. Long-term monitoring of dynamic changes in plasma EBV DNA for improved prognosis prediction of nasopharyngeal carcinoma. Cancer Med. 2020;10:883–94.
80. Siravegna C, O'Boyle CJ, Varmeh S, Queenan N, Michel A, Stein J, et al. Cell free HPV DNA provides an accurate and rapid diagnosis of HPV associated head and neck cancer. Clin Cancer Res. 2022;28:719–27.
81. Damerla RR, Lee NY, You D, Soni R, Shah R, Reyngold M, et al. Detection of early human papillomavirus-associated cancers by liquid biopsy. JCO Prec Oncol. 2019;3:1–17.
82. Ferrandino RM, Chen S, Kappauf C, Barlow J, Gold BS, Berger MH, et al. Performance of liquid biopsy for diagnosis and surveillance of human papillomavirus-associated oropharyngeal cancer. JAMA Otolaryngol Head Neck Surg. 2023;149:971–7.
83. Chera BS, Adibay D, Lewis E, Pang J, Asarkar AA, Nathan CO. Plasma circulating tumor DNA for the surveillance of cancer recurrence in HPV-associated oropharyngeal cancer. J Clin Oncol. 2020;38:1050–8.
84. Berger BM, Hanna GJ, Posner MR, Genden EM, Lauterszain J, Naber SP, et al. Detection of occult recurrence using circulating tumor tissue modified viral HPV DNA among patients treated for HPV-driven oropharyngeal carcinoma. Clin Cancer Res. 2022;28:4292–301.
85. Lele SJ, Adibay D, Lewis E, Pang J, Asarkar AA, Nathan CO, et al. ctDNA as an adjunct to posttreatment PET for head and neck cancer recurrence risk assessment. Otolaryngol Head Neck Surg. 2024;171:439–44.
86. Califano J, Yousef A, Mostafa H, Valsamakis A, Zhang X, Batis N, et al. Lead time to recurrence after posttreatment plasma and salivary HPV-DNA testing in patients with low-risk HPV oropharyngeal cancer. JAMA Otolaryngol Head Neck Surg. 2023;149:812–9.
87. Rutkowski TW, Mazurek AM, Snietura M, Hejduk B, Jedrzejewska M, Bobek-Billewicz B, et al. Circulating HPV16 DNA may complement imaging assessment of early treatment efficacy in patients with HPV-positive oropharyngeal cancer. J Trans Med. 2020;18:167.
88. Stuheit-Zhao E, Penny L, Zhao Z, Zheng Y, Zou J, Huong SH, et al. Ultrasensitive liquid biopsy in patients with p16-positive oropharyngeal cancer using HPV-seq. Int J Radiat Oncol Biol Phys. 2024;118:e5–6.
89. Flach S, Howarth K, Hackinger S, Pipinikas C, Ellis P, McLay K, et al. Liquid biopsy for minimum residual detection in head and neck squamous cell carcinoma (LIONESS)- a personalized circulating tumor DNA analysis in head and neck squamous cell carcinoma. Br J Cancer. 2022;126:1186–95.
90. Lele SJ, Adilbay D, Lewis E, Pang J, Asarkar AA, Nathan CAO. ctDNA as an adjunct to posttreatment PET for head and neck cancer risk assessment. Otolaryngol Head Neck Surg. 2024;171:439–44.
91. Hilke FJ, Muyas F, Admar J, Kootz B, Nann D, Welz S, et al. Dynamics of cell-free tumor DNA correlated with treatment response of head and neck cancer patients receiving chemoradiotherapy. Radiother Oncol. 2020;151:182–9.
92. Hanna GJ, Dennis MJ, Scarfo N, Mullin MS, Sethi RKV, Seghal K, et al. Personalized ctDNA for monitoring disease status in head and neck squamous cell carcinoma. Clin Cancer Res. 2024;30:3329–36.
93. Kaorey N, Dickinson K, Agnihotram VR, Zeitouni A, Sadeghi N, Burnier JV. The role of ctDNA from liquid biopsy in predicting survival outcomes in HPV-negative head and neck cancer patients. Oral Oncol. 2025;161:107148.
94. Galot R, van Marcke C, Helaers R, Mendola A, Goebbels R, Caignet X, et al. Liquid biopsy for mutational profiling of locoregional recurrent and or/ metastatic head and neck squamous cell carcinoma. Oral Oncol. 2020;104:104631.

95. Kano Y, Noji R, Naito T, Fujiwara S, Ikeda S, Mori T, Harada H, et al. Prospective monitoring for immunotherapy by liquid biopsy and signal pathway alterations in head and neck cancer. J Clin Oncol. 2024;42:16.
96. Braig F, Voiglaender M, Schieferdecker A, Busch C, Laban S, Grob T, et al. Liquid biopsy monitoring uncovers acquired RAS-mediated resistance to cetuximab in a substantial proportion of patients with head and neck squamous cell carcinoma. Oncotarget. 2016;742988–42995.
97. Li M, Sun D, Song N, Chen X, Zhang X, Zheng W, Yu Y, et al. Mutant p53 in head and neck squamous cell carcinoma: Molecular mechanism of gain-of-function and targeting therapy. Oncol Rep. 2023;50:162.
98. Liu T, Zhang W, Taylor DL, Roth JA, Goepfert H, Clayman GL. Growth suppression in human head and neck cancer cells by the introduction of a wild-type p53 gene via a recombinant virus. Cancer Res. 1994;54:3662–7.
99. Clayman GL, Frank DK, Bruso PA, Goepfert H. Adenovirus-mediated wild-type p53 gene transfer as a surgical adjuvant in advanced head and neck cancers. Clin Cancer Res. 1999;3:1715–22.
100. Sobol RE, Menander KB, Chada S, Wiederhold D, Sellman B, Talbott M, et al. Analysis of adenoviral p53 gene therapy in clinical trials in recurrent head and neck squamous cell carcinoma. Front Oncol. 2021;11:645745.
101. He X, Xu J, Guo W, Kong C. Therapeutic p53 gene agent in the treatment of NPC cases: preliminary results. J Clin Oncol. 2015;33:15.
102. Yang M, Li L. Multicenter, open-end, randomized controlled study using p53 gene recombinant adenovirus injection (rAd-p53) combined with chemotherapy for oropharyngeal carcinoma: phase IV clinical trial-progress report and conclusion. J Clin Oncol. 2013;31:15.
103. Li Y, Wang L, Zhang Z, Gao N, Liang C, Huang Y, Han B. Selective intra-arterial infusion of rAd-p53 with chemotherapy for advanced oral cancer: a randomized clinical trial. BMC Med. 2014;12:16.
104. Liu S, Chen P, Hu M, Tao Y, Chen L, Liu H, Wang J, et al. Randomized, controlled phase II study of post-Surgery radiotherapy combined with recombinant adenoviral human p53 gene therapy in treatment of oral cancer. Cancer Gene Ther. 2013;20:375–8.
105. Zhang SW, Xiao SW, Liu CQ, Sun Y, Su X, Li DM, et al. Recombinant adenovirus-p53 gene therapy combined with radiotherapy for head and neck squamous cell carcinoma. Chinese J Oncol. 2005;27:426–8.
106. Yuan C, Su X, Chen Z. Recombinant human adenovirus-p53 therapy for the treatment of nasopharyngeal carcinoma: a meta-analysis. Springer Plus. 2016;5:1885.
107. Pertzborn D, Nguyen H, Huttmann K, Prengel J, Ernst G, Guntinas-Lichius O, et al. Intraoperative assessment of tumor margins in tissue sections with hyperspectral imaging and machine learning. Cancers. 2022;15:213.
108. Halicek M, Dormer JD, Little JV, Chen AY, Myers L, Sumer BD, et al. Hyperspectral imaging of head and neck squamous cell carcinoma for cancer margin detection in surgical specimens from 102 patients using deep learning. Cancers. 2019;11:1367.
109. de Koning SGB, Weijtmans P, Karakullukcu MB, Shan C, Baltussen EJM, Smit LA, et al. Toward assessment of resection margins using hyperspectral diffuse reflection imaging (400–1700 nm) during tongue cancer surgery. Lasers in Surg Med. 2020;52:496–502.
110. Halicek M, Little JV, Wang X, Chen AY, Fei B. Optical biopsy of head and neck cancer using hyperspectral imaging and convolutional neural networks. J Biomed Optics. 2019;24:036007.
111. Trajanovski S, Shan C, Weijtmans PJC, de Koning SGB, Ruers TJM. Tongue tumor detection in hyperspectral images using deep learning semantic segmentation. IEEE Trans Biomed Eng. 2021;68:1330–40.
112. Bali A, Bitter T, Mafra M, Ballmaier J, Kouka M, Schneider G, et al. Endoscopic in vivo hyperspectral imaging for head and neck tumor surgeries using a medically approved CE-certified camera with rapid visualization during surgery. Cancers. 2024;16:3785.

113. van Dijk LV, van den Bosch L, Aljabar P, Pressutti D, Both S, Steenbakkers RJHM, et al. Improving automatic delineation for head and neck organs at risk by deep learning contouring. Radiother Oncol. 2020;142:115–23.
114. Temple SWP, Rowbottom CJ. Gross failure rates and failure modes for a commercial AI-based auto-segmentation algorithm in head and neck cancer patients. J Appl Clin Med Phys. 2024;25:e14273.
115. Lucido JJ, DeWees TA, Leavitt TR, Anand A, Beltran CJ, Brokke MD, et al. Validation of clinical acceptability of Deep-learning-based automated segmentation of organs-at-risk for head and neck radiotherapy treatment planning. Front Oncol. 2023;13:1137803.
116. Shen G, Jin X, Sun C, Li Q. Artificial intelligence radiotherapy planning: automatic segmentation of human organs in CT images based on a modified convolutional network. Front Public Health. 2022;10:813135.
117. Sarria GR, Kugel F, Roehner F, Layer J, Dejonckheere C, Scafa D, et al. Artificial intelligence-based autosegmentation: advantages in delineation, absorbed dose-distribution, and logistics. Adv Radiat Oncol. 2024;9(101):394.
118. Nielsen CP, Lorenzen EL, Jensen K, Sarup N, Brink C, Smulders B, Holm AIS, et al. Consistency in contouring of organs at risk by artificial intelligence vs oncologists in head and neck cancer patients. Acta Oncol. 2023;62:1418–25.
119. Hansen CR, Bertelsen A, Hazell I, Zukauskaite R, Gyldenkerne N, Johansen J, Eriksen JG, et al. Automatic treatment planning improves the clinical quality of head and neck cancer treatment plans. Clin Transl Radiat Oncol. 2016;1:2–8.
120. Gao Y, Park YK, Jia X. Human-like intelligent automatic treatment planning of head and neck cancer radiation therapy. Phys Med Biol. 2024;69:115049.
121. Hansen CR, Bertelsen A, Hazell I, Zukauskaite R, Gydenkerne N, Johansen J, et al. Automatic treatment planning improves the clinical quality of head and neck cancer treatment plans. Clin Transl Radiat Oncol. 2016;1:2–8.
122. McIntosh C, Welch M, McNiven A, Jaffray DA, Purdie TG. Fully automated treatment planning for head and neck radiotherapy using a voxel-based dose prediction and dose mimicking method. Phys Med Biol. 2017;62:5926–44.
123. Biston M, Costea M, Gassa F, Serre A, Voet P, Larson R, et al. Evaluation of fully automated a priori MCO treatment planning in VMAT for head and neck cancer. Phys Med. 2021;87:31–8.
124. Li X, Sheng Y, Wu QJ, Ge Y, Brizek DM, Mowery YM, et al. Clinical commissioning and introduction of an in-house artificial intelligence (AI) platform for automated head and neck intensity modulated radiation therapy treatment planning. J Appl Clin Med Phys. 2025;16:e14558.
125. Court LE, Aggarwal A, Jhingran A, Naidoo K, Netherton T, Olanrewaju A, et al. Artificial intelligence-based radiotherapy contouring and planning to improve global access to cancer care. JCO Glob Oncol. 2024;10:e2300376.
126. Mehanna H, Wong W, McConkey CC, Rahman JK, Robinson M, Hartley AGJ, et al. PET-CT surveillance versus neck dissection in advanced head and neck cancer. N Engl J Med. 2016;374:1444–54.
127. de Leeuw AL, Giralt J, Tao Y, Benavente S, Nguyen TF, Hoebers FJP, et al. A multicentric randomized controlled phase III trial of adaptive and 18F-FDG-PET-guided dose-redistribution in locally advanced head and neck squamous cell carcinoma (ARTFORCE). Radiother Oncol. 2024;196:110281.
128. Welz S, Paulsen F, Pfannenberg C, Reynold M, Reischl G, Nikolaou K, et al. Dose escalation to hypoxic subvolumes in head and neck cancer: a randomized phase II study using dynamic [^{18}F]FMISO PET/CT. Radiother Oncol. 2022;171:30–6.

Chapter 14
Application of Radiotherapy in Oligometastatic Disease

Kasra Kolahdouzan, Romina Abyaneh, Reza Ghalehtaki, and Felipe Couñago

Abstract Oligometastatic disease (OMD) represents an intermediate state between localized and widespread metastatic cancer, characterized by limited metastatic foci. This chapter explores the application of radiotherapy, particularly stereotactic ablative radiotherapy (SABR), in managing OMD. SABR delivers high and precise radiation doses to metastatic sites, offering potential survival benefits, improved progression-free survival (PFS), and delayed need for initiating or changing systemic therapies. The chapter reviews the rationale for radiotherapy in OMD, emphasizing advancements in imaging, treatment planning, and delivery techniques, such as adaptive radiotherapy (ART) and MR-guided radiotherapy (MRgRT), which enhance precision and minimize toxicity. Patient selection is critical, with factors such as metastasis number, location, and tumor biology guiding treatment decisions. Clinical applications of radiotherapy in non-small cell lung cancer, prostate cancer, and breast cancer are discussed, highlighting the role of SABR in improving outcomes. However, challenges remain, including the lack of standardized definitions and protocols for OMD. The chapter underscores the need for further research, particularly prospective randomized trials and biomarker studies, to refine patient

K. Kolahdouzan · R. Ghalehtaki (✉)
Department of Radiation Oncology, Cancer Institute, IKHC, Tehran University of Medical Sciences, Tehran, Iran

Radiation Oncology Research Center, Cancer Research Institute, Tehran University of Medical Sciences, Tehran, Iran
e-mail: rghalehtaki@tums.ac.ir

R. Abyaneh
Radiation Oncology Research Center, Cancer Research Institute, Tehran University of Medical Sciences, Tehran, Iran

F. Couñago
Department of Medicine, Faculty of Medicine, Health and Sport Sciences, Universidad Europea de Madrid, Madrid, Spain

GenesisCare, Madrid, Spain

Hospital Universitario San Francisco de Asís, Madrid, Spain

Hospital Universitario La Milagrosa, Madrid, Spain

F. Taghizadeh-Hesary et al. (eds.), *From Radiobiology to Radiation Oncology*,
https://doi.org/10.1007/978-3-031-95724-6_14

selection and treatment strategies, ultimately improving the therapeutic ratio and quality of life for patients with OMD.

Keywords Oligometastatic disease · Radiotherapy · Stereotactic ablative radiotherapy · Adaptive radiotherapy

14.1 Introduction

14.1.1 Definition of Oligometastatic Disease

According to the famous cancer theory, which believes that cancer is a spectrum of diseases from a localized to a widely spread metastatic disease, Weichselbaum and Hellman introduced the term "oligometastasis" in 1995 [1]. Some metastatic patients have limited foci of metastases, making them biologically different from widely disseminated metastatic cases. Oligometastatic disease (OMD) has been proposed as an intermediate state between localized disease and systemic metastatic disease. Dingemans et al., in 2019, defined OMD as the presence of five or fewer metastatic foci in three or fewer organs, independent of the primary tumor and the location of the lesions [2]. However, this limit is not well-established, nor is the maximum lesion size that defines OMD. Some OMD cases are diagnosed de novo. However, some are converted to OMD after a period of systemic treatments for a previously widespread disease. The latter situation is more accurately defined as "induced oligometastatic" or "oligopersistent" disease [3].

According to the 2020 European Society for Radiotherapy and Oncology/ European Organization for Research and Treatment of Cancer (ESTRO/EORTC) consensus, the ability to deliver safe and clinically meaningful radiotherapy with curative intent to all metastatic sites is the minimum requirement for defining OMD in the context of radiotherapy. In 2020, the ESTRO and EORTC published a classification of OMD, with 9 subtypes, to guide the prognostic differences between them [4].

Subclassifications of OMD:

- Genuine OMD
  - De novo OMD (first-time diagnosis of OMD)
    - Synchronous OMD
    - Metachronous OMD
      - Metachronous oligorecurrence
      - Metachronous oligoprogression
  - Repeat OMD (previous history of OMD)
    - Repeat oligorrecurrence

    - Repeat OMD.
        - Repeat oligopersistence
        - Repeat oligoprogression
- Induced OMD
    - Induced oligorrecurrence
    - Induced OMD
        - Induced oligopersistence
        - Induced oligoprogression

Studies have explored the prognostic significance of OMD. Chen HJ et al. conducted a retrospective study on lung OMD patients treated with radical radiotherapy, finding that the "induced oligoprogression" subtype had a statistically worse prognosis in both univariate ($p = 0.02$) and multivariate ($p = 0.01$) analyses [5].

Another retrospective study of patients with bone-only oligometastatic prostate cancer treated with stereotactic body radiotherapy (SBRT), with patients classified into oligometastatic states based on the ESTRO/EORTC classification system, showed that patients in a metachronous oligometastatic state were 4.50 times more likely to experience metastatic progression compared with those with synchronous oligometastases (p = 0,03), and 6.69 times more likely to experience any failure (p = 0,04). Further validation in prospective series is needed to establish this classification as a prognostic tool [6].

Also, another study in which patients from the phase II SABR-5 trial were stratified based on ESTRO/EORTC criteria found that OMD groups differed significantly in progression-free survival (PFS) ($P < 0.001$) but not overall survival (OS) (p-value = 0.069). The OMD classification was an independent predictor of both PFS (p-value = 0.005) and OS (p-value = 0.002). Of the five classification factors, only chronicity (synchronous, HR: 0.52; p-value = 0.027) and oligoprogression (HR: 2.05; p-value = 0.004) were independently prognostic factors for OS, concluding that this classification could be used to identify specific patient groups for clinical trials [7].

14.1.2 Current Treatment Landscapes

There are a number of treatments for OMD, from only local therapies to the affected sites to systemic treatments alone. Among local therapies, surgery, radiotherapy, and thermal ablative treatments are the most evidence-based therapies. Most patients with OMD receive systemic therapies, with or without local treatments to the affected sites. The selection between these modalities depends on patients' performance, age, comorbidities, and disease characteristics, including size, location, and number of lesions, affected organs, etc. [8]. There is no head-to-head prospective comparison between these modalities; however, due to the inherent complications of surgery, it remains an option for young and well-going patients with synchronous metastases in the brain, lung, or liver who are expected to have a longer survival [9]. The main differentiating factor in choosing between thermal ablation and radiotherapy is the location and size of the lesions [10]. For example, lesions smaller than 3 cm in liver metastasis and located at least 3 mm from the main vessels are potential candidates for thermal ablation.

14.1.3 Rationale for Radiotherapy in Oligometastatic Disease

Due to recent advancements in the planning and delivery of radiotherapy, including image guidance and adaptive planning, patient immobilization, and organ motion tracking, a safe and effective dose of radiation can be delivered to the affected sites of metastases [11]. All of these advancements are collectively applied in the stereotactic ablative body radiotherapy (SABR) or stereotactic radiotherapy (SRT) techniques for extracranial and intracranial metastases, respectively. By delivering tumoricidal doses using these techniques, the burden of the disease can be reduced, and consequently, the efficacy of systemic therapy may be enhanced. As a result, the outcomes such as PFS or even OS could be improved [12]. For example, in the landmark SABR-COMET trial, about one-fifth of patients survived more than 5 years without any new recurrences after receiving radiotherapy to the oligometastatic sites, a number that has been considered "magical" for metastatic patients in previous decades [13]. Other benefits of adding radiotherapy to the standard of care in OMD patients include prolonged disease control within the current systemic therapy line and a reduced need to pursue next-line systemic treatment.

14.1.4 Patient Selection

Appropriate patient selection among OMD cases for radiotherapy involves evaluating patient-, tumor-, and technique-related factors. Key patient factors include age, performance status, comorbidities, competing risks, and patient preference. Tumor-related considerations include metastases sequence (synchronous vs. metachronous), oligometastatic state (de novo vs. induced), primary tumor status, response to systemic therapy, need for other local therapies, and tumor radiosensitivity. For instance, the CURB study showed better PFS in lung cancer patients compared to breast cancer patients, attributed to non-small cell lung cancer (NSCLC) resistance to systemic therapy and higher radiosensitivity [14, 15]. Another important tumor-related factor that needs to be further explored is the genomic identifiers of response to SABR [16].

From a technical standpoint, aside from the requirement for image-guided and accurate treatment planning and delivery systems, good candidates for SABR are patients with lesions suitable for receiving a biologically effective dose (BED) of at least 100 Gy [12]. Additionally, the ability to target all metastatic sites may be a marker of better outcomes, as shown in the ORIOLE study [17]. However, the inability to deliver higher doses does not necessarily preclude patients from receiving SABR, as lower doses may still be effective when combined with potent systemic therapies, including immunotherapy, which may have synergistic interactions with SABR.

14.2 Radiotherapy Techniques for Oligometastatic Disease

The advent of new technologies has revolutionized the treatment landscape of metastatic cancer patients, particularly those with OMD. More traditional treatment techniques, such as conventional radiotherapy and three-dimensional conformal radiotherapy (3DCRT), have been largely replaced by modern techniques, including intensity-modulated radiotherapy (IMRT), volumetric arc therapy (VMAT), and stereotactic radiotherapy using photon or proton beams. These methods rely on precise imaging for diagnosis, lesion detection, and accurate treatment delivery, ensuring optimal tumor coverage while sparing healthy tissues.

14.2.1 Imaging

The first step in delivering effective radiotherapy is the use of appropriate imaging, which is essential for both detecting OMD and assessing treatment response. A number of imaging modalities are used, including computed tomography (CT), magnetic resonance imaging (MRI), and functional or metabolic imaging like positron-emission tomography (PET) [18]. However, several challenges in this regard need to be addressed. The choice of imaging modality depends on the tumor type, the timing of oligometastasis, prior treatments, and, most importantly, the location of metastases. Conventional imaging, such as CT scans, is the most commonly used modality for ruling out metastases. It can detect pulmonary and lymph node lesions with acceptable accuracy, but its sensitivity is low for other areas, such as bones and soft tissues [19]. For soft tissue, MRI, especially when combined with diffusion-weighted imaging (DWI), can better delineate cancer lesions from surrounding normal tissues. DWI can also be useful in assessing tumor response [20]. These characteristics make MRI the best option for detecting liver, bone, and central nervous system metastasis. Due to these excellent capabilities, there are now commercialized MRI-guided linear accelerators, called MRgRT, that can be ideal for SABR, with real time, non-ionizing image guidance for accurately delineating the target and monitoring it during the delivery of radiation [21].

For the metastasis of prostate cancer, choline and prostate-specific membrane antigen (PSMA) PET-CT provide more sensitive and specific imaging compared to CT and MRI. PET-CT is not only a valuable imaging technique for staging and detecting oligometastasis but also an appropriate method for response monitoring due to its unique capability of reporting standardized uptake values (SUVs). However, it has some limitations in the OMD state that depend on tumor size and biology. One important limitation of PET-CT is the uptake of radiotracers caused by inflammation, which can confound the evaluation of response in the irradiated volumes [22].

14.2.2 SBRT

SABR, also known as SBRT, delivers high radiation doses to small, well-defined tumors in 1–5 fractions, making it ideal for OMD with limited metastatic sites (typically 1–5). It can delay disease progression and potentially improve survival [3, 23]. SABR uses advanced imaging (CT, MRI, PET-CT) for precise tumor localization and planning [24]. Patients are immobilized with personalized devices, and image-guided radiotherapy (IGRT) ensures accuracy by verifying tumor position and adjusting for movement. High doses are delivered with sub-millimeter precision, sparing healthy tissues and minimizing side effects [25].

For OMD, SABR is typically used to target metastases in the lung, liver, spine, or other sites. The goal is to achieve local control, potentially delaying the need for systemic therapies like chemotherapy or extending the time period on the current line of systemic therapy, thus delaying the need to proceed with next-line systemic treatment (NEST). Multidisciplinary collaboration is needed, mainly between radiation oncologists, surgeons, medical oncologists, and radiologists, to ensure patient selection is appropriate and that the benefits outweigh the risks. Standard practice includes precise contouring of the tumor and nearby dose-limiting critical structures, optimizing dose distribution, and adhering to strict dose constraints to avoid toxicity.

SABR is usually delivered in an outpatient setting, with each session lasting 15–60 minutes delivered once a day or every other day. Common side effects are site-specific and generally mild, such as asthenia or localized discomfort. Rigorous follow-up is essential to monitor treatment response in the index lesion and to detect new metastases early [13].

14.3 Clinical Applications of Radiotherapy

With the development of more effective local and systemic therapies, including robotic-assisted or laparoscopic surgical techniques, SABR, targeted therapies, immunotherapy, and the development of more accurate functional and non-functional diagnostic imaging, a new paradigm shift has started in the treatment of metastatic cancers, especially in patients with a low burden of metastatic disease.

Physicians are more frequently referring these patients for local therapies, a cornerstone of which is advanced radiotherapy techniques like SABR.

SABR is available in many centers and is widely used for treating brain, lung, bone, liver, and other metastases. The investigators of the SABR-COMET phase 2 trial showed that SABR to oligometastatic sites in multiple cancer subtypes (including breast, lung, prostate, colorectal, etc.) improves the 5-year OS rates from 17.7% up to 42.3% (p-value: 0.006) and the 5-year PFS rates from 3.2% to 17.3% (p-value: 0.001) [23].

Oligoprogression is another phenomenon for which SABR has clinical implications. It is defined as the progression of up to five metastatic sites in patients under otherwise effective systemic treatment who maintain stable metastasis control in other metastatic sites. It has been shown that SABR in this setting could improve PFS [14], prolong lesional control [26], and delay the need to start second-line systemic therapies.

Effective management of oligometastatic patients requires a multidisciplinary team to address disease and patient-specific factors for optimal care. The following sections summarize current evidence on the role of radiotherapy in treating various primary oligometastatic cancer sites.

14.3.1 Non-Small Cell Lung Cancer

Patient Selection

For biopsy-proven NSCLC with up to 5 metastatic sites, comprehensive staging with whole-body FDG/PET-CT and imaging is recommended. After confirming oligometastases, a multidisciplinary team evaluates each case to decide between standard systemic treatment alone or combined with definitive local therapy, if feasible and safe. Due to limited phase 3 trials and uncertainties in treatment sequencing, symptomatic lesions are prioritized for palliation. In asymptomatic synchronous oligometastatic NSCLC, at least 3 months of systemic therapy is advised. If the disease remains stable upon re-staging, definitive local treatment for all metastatic and locoregional sites can be pursued [27].

Clinical Outcomes

Most recently, the NRG-LU002 phase 2/3 study reported its results. In this study, patients with NSCLC with up to 3 extracranial metastases staying at least stable after 4 cycles of systemic therapy were randomized 1:2 to receive maintenance therapy alone or combined with local consolidative therapy (radiotherapy and/or surgery). The 2-year PFS rate was 36% (95% CI: 24.8, 47.2) in the control group versus 40% (95% CI: 31.5, 48.6) in the intervention group (HR: 0.93 (95% CI: 0.66,

1.31), p-value: 0.66). This study closed early after failing to achieve the predefined efficacy parameters [28].

Other than the SABR-COMET trial, which showed significant OS benefit (18% with lung cancer) with the addition of SABR to oligometastatic sites, two other small phase 2 randomized controlled trials (RCTs) also showed PFS advantage with SABR (and/or surgery) to oligometastatic sites in NSCLC [29, 30]. In the era of targetable mutations like epidermal growth factor receptor (EGFR), which justify using highly efficient specific tyrosine kinase inhibitors (TKIs), using SABR in oligometastatic NSCLC has also been explored. A recent study by Wang et al. showed that in EGFR-mutated synchronous oligometastatic NSCLC, other than the TKIs, the addition of SABR to both the primary tumor and all oligometastatic sites also leads to an improvement in PFS (median PFS 12.5 vs. 20.2 months, p-value <0.001) and OS (median OS 17.4 vs. 25.5 months, p-value <0.001) [31].

14.3.2 Prostate Cancer

Patient Selection

As with other malignancies, there is significant discrepancy regarding a clear definition of low burden or OMD in prostate cancer, and further research is needed to establish a precise definition. These definitions are often elaborated in the context of clinical trials, such as the CHAARTED and LATITUDE trials, which provide frameworks for categorizing disease burden.

In the CHAARTED trial [32], high-volume disease was defined as the presence of visceral metastases and/or four or more bone metastases, with at least one outside the vertebral column and pelvis. In contrast, the LATITUDE trial [33] focused on high-risk metastatic castration-sensitive prostate cancer (mCSPC) and defined high-risk disease based on the presence of at least two of three factors: a Gleason score of 8 or higher, three or more bone lesions, or measurable visceral metastases (excluding lymph nodes). These differing definitions highlight the need for a standardized approach to defining disease burden in prostate cancer, which could help guide treatment decisions and improve patient outcomes.

Currently, at least metastatic low-volume de novo prostate cancer patients must receive definitive doses of radiotherapy to the prostate. Also, new evidence gathers around locally treating the sites of metastases by either radiotherapy or surgery in oligometastatic prostate cancer patients, with routine use of PSMA PET as the imaging method of choice for staging and re-staging [34].

Clinical Outcomes

The CHAARTED trial [32], which showed an OS benefit for the addition of docetaxel to androgen deprivation therapy (ADT) in high-volume disease, initially defined high burden of disease in prostate cancer as: "presence of 4 or more bone metastases with at least one of them located in sites other than the vertebral bodies or pelvis and/or presence of visceral metastases." This definition might not be representative of true OMD, as patients with multiple bone metastases confined only to the axial skeleton will be categorized as "low burden," contrasting the more commonly accepted definition of less than five metastases in other sites of malignancy. Despite this, many large RCTs, some of which are discussed next, have stuck to the original definition presented by the CHAARTED to make comparisons possible between their reported results.

The STAMPEDE (arm H) [35] clinical trial demonstrated a survival benefit for treatment of the primary tumor in the setting of low-burden (or oligometastatic) prostate cancer. The subgroup of low-burden metastatic patients in the STAMPEDE achieved a 3-year OS rate of 63% with the addition of radiotherapy of the primary site to the standard of care versus 58% without the addition of radiotherapy (HR:0.68, 95%CI: 0.52–0.90, *p*-value: 0.007).

The HORRAD trial [36] also explored the efficacy of radiotherapy in prostate cancer patients with documented bone metastases when added to anti-androgen therapy. This study showed a significant prostate-specific antigen (PSA) recurrence-free survival benefit with the addition of radiotherapy (p-value: 0.02), but no significant difference was found in OS ($p = 0.4$). The interaction of hazard ratios of radiotherapy efficacy regarding OS in the subgroup of patients with less than five bone metastases (HR:0.68, 95%CI: 0.42–1.10) versus ≥5 bone metastases (HR: 1.06, 95%CI: 0.80–1.39) showed a reduction of radiotherapy efficacy in the latter group.

More frequent use of second-generation anti-androgen drugs like abiraterone acetate in the treatment of metastatic prostate cancer has significantly improved oncologic outcomes in these patients. After significant OS benefits of abiraterone acetate addition to ADT shown by the LATITUDE trial [33], recently, the PEACE-1 trial delved further into this issue of whether the addition of radiotherapy to the newest standard of care of ADT (with or without docetaxel) plus abiraterone would also improve survival outcomes or not [37]. This study demonstrated improved radiographic PFS with the addition of radiotherapy in low-burden metastatic prostate cancer (HR:0.65, 99.9%CI: 0.36–1.19, *p*-value: 0.019) as well as improved castration resistance-free survival regardless of the disease burden (HR: 0.74, (95%CI: 0.60–0.92), *p*-value: 0.0069) and also a reduction of serious genitourinary adverse events with the implementation of radiotherapy regardless of disease burden and systemic treatment used (22.3% versus 12.2%, *p*-value <0.05), but did not improve OS ($p = 0{,}21$).

Other than radiotherapy to the primary disease site, SABR (either single fraction or fractionated) of the metastatic sites in oligometastatic prostate cancer has also been explored in several studies, most of them conducted retrospectively. The maximum number of metastases accepted as oligometastasis definition in these studies has been between 2 and 5, and radiotherapy has resulted in a 22% to 83% 2-year PFS rate in these patients [38].

In a phase 2 basket RCT (the EXTEND trial), 87 patients with oligometastatic prostate cancer (defined as up to 5 metastases) who had received at least 2 months of hormone therapy were randomized to continue hormone therapy with or without definitive radiation to all sites of metastatic disease. The combination therapy resulted in significant prolongation of median PFS (15.8 months vs. not reached, HR:0.25, 95%CI: 0.12–0.55, *p*-value <0.001) [39]. Similar results were achieved with SABR versus observation in the phase 2 ORIOLE RCT of prostate cancer patients with up to 3 metastases randomized 2:1 to receive SBRT vs. observation. At 6 months, progression occurred in 19% of patients in the SBRT arm versus 61% in the control arm (P = 0.005), with an improved PFS in the SBRT arm (HR: 0.30; 95% CI, 0.11–0.81, p-value: 0.002) as well as a systemic immune response elicited by the SABR in favor of the abscopal effects of high-dose radiotherapy [17]. The STOMP trial included 62 metachronous oligometastatic patients with ≤3 extracranial lesions. With a median follow-up of 3 years, a median ADT-free survival of 13 months was found in the control group versus 21 months in the metastasis-directed therapy group with no evidence of ≥2 toxicities [16].

The ARTO phase 2 trial randomized 157 patients with oligometastatic castration-resistant prostate cancer with up to 3 bone or nodal metastases to first-line abiraterone acetate and prednisone (AAP) or AAP plus SBRT to all metastatic sites and showed a significant improvement in 6-month biochemical response (92% versus 68.3%, OR: 5.34, 95% CI: 2.05–13.88, *p*-value: 0.001), complete biochemical response in 38.8%, and PFS improvement (HR: 0.35, 95% CI: 0.21 to 0.57, *p*-value <0.001) [40]. These studies show that metastasis-directed therapy in oligometastatic prostate cancer prolongs the time for the development of new metastases and the need to change systemic therapy.

Despite all the mentioned results, issues like the difference between synchronous and metachronous OMD, the use of ADT concurrently with metastasis-directed therapy, the role of modern imaging technologies in metastasis detection (e.g., PET/PSMA scan), and the use of genomic/molecular biomarkers for treatment stratification remain unresolved.

14.3.3 Breast Cancer

Patient Selection

Oligometastasis has been defined as 1–5 metastases in the majority of breast cancer trials, and according to the ESTRO-ASTRO consensus document [41]. Although it is long known that there are different patterns of metastasis and natural history regarding the site and disease burden based on the hormone receptor status of the tumor, these factors have not been introduced into a more detailed definition of oligometastasis in breast cancer. In a retrospective study on results of SABR for the treatment of liver metastases, it was shown that breast cancer metastases were more radiosensitive than other histologies, and tumor control probability was significantly higher regardless of other confounding factors when breast cancer liver metastases were treated with SABR [42].

Clinical Outcomes

There has been much debate regarding the role of local treatment of metastatic breast cancer patients, with most studies but one [43] showing no OS benefit [44–47]. Other than that, metastasis-directed therapy has also been explored.

Metastatectomy of isolated lung metastases with a breast cancer origin has shown a 5-year OS rate of 46% (95%CI: 43–49%) according to a systematic review of 1937 breast cancer patients [48].

Moreover, as mentioned previously, the SABR-COMET randomized controlled phase 2 trial showed PFS and OS benefits from the use of SABR in oligometastatic cancers. Of note, 18% of patients in that trial had primary breast cancer [23]. Other retrospective studies have also supported favorable PFS and OS rates with SABR used in oligometastatic breast cancer patients [49–51].

Despite favorable rates of OS, PFS, and local control (LC), and low rates of complication for SABR of oligoprogressive metastases in retrospective studies [52], a subgroup analysis in a phase 2 RCT that used SABR to target the oligoprogressive sites did not show a PFS benefit in patients with breast cancer in contrast to patients with NSCLC [14].

14.4 Emerging Technologies

14.4.1 Adaptive Radiotherapy

Adaptive radiotherapy (ART) represents a transformative advancement in radiation oncology, leveraging advanced imaging and artificial intelligence to dynamically adjust treatment plans in response to anatomical changes during therapy. Unlike a conventional static treatment plan, ART refines radiation delivery based on real-time variations in tumor size, shape, position, and surrounding organs at risk (OAR) and minimizes radiation exposure to healthy tissues, improving both efficacy and safety [53].

Three major imaging modalities currently support online ART: MRI, cone-beam CT (CBCT), and PET-based systems. CBCT, widely used for online ART, provides in-room imaging before each fraction, but it is limited by lower resolution and artifacts and cannot adapt to intra-fraction treatment. MRI-based ART offers superior soft-tissue contrast and avoids ionizing radiation, enabling precise tumor delineation, allowing the adaptation of the treatment volume before each fraction, and enabling real-time motion tracking during treatment. PET-based ART, which integrates metabolic data for treatment adaptation, is still awaiting FDA approval for clinical use but holds significant promise for future applications [53, 54].

Recent studies highlight the clinical benefits of ART in OMD. A retrospective study demonstrated that image-guided SBRT targeting both the primary tumor and all oligometastases improved OS, PFS, and polyprogression-free survival (PPFS) compared to SBRT targeting the primary tumor alone in metastatic pancreatic cancer patients [55]. Additionally, a prospective in silico study evaluated CBCT-guided stereotactic adaptive radiotherapy (CT-STAR) for abdominal oligometastases. The study found that CT-STAR resolved violations of OAR constraints in 75% of fractions and improved V100 and D95 tumor coverage in over 63% and 50% of cases, with a median adaptation time per fraction of 22.59 minutes. These findings support CT-STAR as a viable strategy for abdominal oligometastases, with clinical trials currently in development [56].

MR-Guided Radiotherapy

MR-guided radiotherapy (MRgRT) integrates real-time MRI with radiation delivery systems, offering enhanced soft-tissue contrast for precise tumor targeting. This technology is particularly advantageous for tumors near critical organs or in motion-prone areas, such as abdominopelvic tumors. MRgRT enables real-time adjustments to treatment plans during each session, reducing margins, enhancing precision, and minimizing toxicity [57–59].

One of the most promising developments in MRgRT is stereotactic MRI-guided adaptive radiotherapy (SMART). The SOFT trial, a multicenter phase II study, evaluated MR-guided adaptive SABR for infra-diaphragmatic metastases in oligometastatic patients. The study reported a 1-year LC rate of 89% and a median PFS of 7.1 months, with 50% of patients remaining off systemic therapy during follow-up. These results suggest that MR-guided SABR can extend the time to start systemic treatments [60]. Further evidence supporting MRgRT comes from a study by Yang et al., which demonstrated that online adaptation improved coverage or organ sparing in 86.6% of fractions for abdominopelvic oligometastases. The 12-month LC rate was 93%, with minimal grade 3 toxicities. The study concluded that SMART is feasible, safe, and effective for treating abdominopelvic oligometastases [61].

Building on these findings, recent studies have explored the dosimetric feasibility of fractionated and single-fraction treatments using MRgRT and real-time reoptimization to improve target coverage and minimize OAR constraint violations.

A dosimetric study by Lee et al. has also explored the feasibility of single-fraction MRgRT and demonstrated that reoptimized MRI-guided SABR plans for abdominal and thoracic oligometastases improved target coverage and avoided OAR dose constraint violations, highlighting the potential of single-fraction MRgRT for OMD [62]. Furthermore, a phase I trial by Henke et al. in recurrent ovarian cancer patients treated 17 oligometastases with SMART (35Gy/5 fractions). All treatment fractions were successfully delivered, with online-adaptive plans used in 58% of cases. Results showed 94% LC at 3 months, a median PFS of 10.9 months, and a median systemic-therapy-free survival of 11.5 months. SMART demonstrated safety, minimal toxicity, high LC, and improved quality of life (QoL), supporting its viability for high-dose radiation in ovarian oligometastases [63]. Lastly, Henke et al. assessed MR-guided SMART for oligometastatic or unresectable abdominal malignancies in a phase I trial, delivering 50 Gy/5 fractions (BED 100 Gy) with strict OAR constraints. Daily online MR-guided plans were often needed to maintain OAR limits or boost PTV dose, improving PTV coverage in many fractions. No serious acute toxicities occurred within 6 months. The study confirmed SMART as a feasible, safe approach, enhancing PTV dose coverage while preserving or improving OAR sparing [64].

A recent retrospective study explored MRI-guided repeated SBRT (MRg-reSBRT) as salvage therapy for recurrent liver or abdominal oligometastases. Real-time tumor visualization and adaptive re-irradiation improved targeting precision, achieving a 1-year OS of 73% with no significant toxicities. These results underscore MRg-reSBRT as a promising strategy for disease control, supporting its integration into multi-modality treatments for oligoprogressive or recurrent OMD [65].

14.4.2 Clinical Trials and Research Needs

Prospective Randomized Trials

Prospective randomized trials are essential to establish the efficacy and safety of emerging radiotherapy technologies in OMD. While retrospective and single-arm studies, such as the SOFT trial [60], have provided valuable insights. Randomized trials are needed to compare these advanced techniques with standard treatments and define their role in clinical practice.

Biomarkers for Patient Selection

Biomarkers play a critical role in identifying patients who are most likely to benefit from aggressive local therapy. A retrospective cohort study evaluated pretreatment FDG-PET parameters in oligometastatic NSCLC patients undergoing high-dose IGRT. Higher metabolic tumor volume (MTV) and total lesion glycolysis (TLG) were associated with shorter OS, suggesting that these metrics could serve as prognostic biomarkers for patient selection [66].

Similarly, a multi-institutional pooled analysis identified key prognostic factors, including primary tumor type, disease-free interval, and number of metastases, that influence long-term survival in extracranial oligometastatic patients. This study proposed a risk stratification framework to optimize patient selection for metastasis-directed radiotherapy, highlighting the importance of personalized treatment strategies [67]. Further studies should be conducted to establish biomarkers as a definitive decision-making tool.

Optimization of Treatment Planning and Delivery

Optimizing treatment planning and delivery is crucial for maximizing the therapeutic ratio in OMD. Previous studies have demonstrated the benefits of online-adaptive MR-guided SBRT in improving target coverage and reducing OAR constraint violations.

Henke et al. compared nonadaptive SBRT plans, based on initial CT anatomy, with adaptive plans adjusted in real time using daily MRI imaging. In 19/30 cases, applying the original plan to daily anatomy led to OAR constraint violations, which were corrected through adaptive planning. Additionally, target coverage improved in 21/30 cases while maintaining or enhancing OAR sparing. These results suggest that MR-guided adaptive SBRT enables tumor dose escalation while reducing toxicity, potentially enhancing the therapeutic ratio for treating central thoracic and abdominal malignancies [68].

Additionally, a study comparing MR-linac-guided SBRT with conventional CBCT-linac treatment for lymph node oligometastases found that MR-linac significantly improved adherence to dose constraints, particularly in patients with multiple metastases [69]. These findings underscore the potential of advanced imaging and adaptive planning to enhance precision and treatment outcomes.

The ONE STOP trial (NCT06236516) is recruiting to address logistical challenges in SBRT for oligometastatic and primary lung tumors. It introduces a simulation-free workflow using HyperSight CBCT and adaptive radiotherapy to deliver single-fraction SBRT in 2–3 h. This approach reduces treatment time while maintaining precision through real-time anatomical adaptations, potentially expanding access for remote patients. If successful, the trial could transform SBRT delivery, showcasing the role of innovation in enhancing radiotherapy outcomes for OMD.

Another innovative ongoing phase I/II trial (NCT05501665) is investigating the safety and efficacy of split-course adaptive radiotherapy (SCART) in combination with immunotherapy for patients with IVA-B (M1b or M1c) or locally advanced lung cancer. SCART is a novel approach that modifies radiation dosing and field size based on tumor response, aiming to maximize treatment effectiveness while reducing toxicity. This strategy is particularly relevant in oligometastatic NSCLC, where aggressive local therapy may improve disease control and prolong survival. By integrating SCART with immune checkpoint inhibitors, the trial explores whether radiation can enhance systemic immune activation (abscopal effect), potentially improving overall treatment response.

14.5 Conclusion

Radiotherapy, particularly SABR, is a cornerstone in treating OMD, offering precise, non-invasive tumor control with minimal toxicity and potential survival/QoL benefits. Advances in imaging, treatment planning and delivery, and insights into tumor biology enable high-dose delivery while sparing healthy tissues. However, patient selection remains critical, guided by factors like metastasis number, location, and tumor biology. Emerging biomarkers and advanced imaging is enhancing patient stratification, and combining radiotherapy with systemic therapies (e.g., immunotherapy) shows considerable promise. Key challenges persist, including OMD definition heterogeneity and lack of standardized protocols. Further RCTs and biomarker-driven research are needed to refine treatment strategies.

14.5.1 Summary of Key Points

- OMD is defined as ≤5 metastatic foci in ≤3 organs, representing an intermediate state between localized and widespread metastatic cancer. It can be de novo (diagnosed initially) or induced (after systemic therapy).
- SABR delivers high and precise radiation doses in 1–5 fractions, targeting metastases while sparing healthy tissues. Proven to enhance systemic therapy efficacy and survival and delay the need for next-line treatments; however, advanced imaging (CT, MRI, PET-CT) and multidisciplinary collaboration are needed for optimal patient selection and planning.
- ART adjusts treatment plans using real-time imaging (MRI, CBCT, PET), improving precision, reducing toxicity, and enhancing outcomes in OMD. MRgRT combines real-time MRI with radiation delivery, offering superior soft-tissue contrast and precise tumor targeting, with studies showing high LC and minimal toxicity.
- Biomarkers, such as FDG-PET parameters (e.g., metabolic tumor volume, total lesion glycolysis), are critical for identifying oligometastatic patients likely to benefit from local therapies like SABR/SBRT or metastasis-directed radiotherapy. However, RCTs are essential to establish the efficacy, safety, and comparative benefits of these advanced techniques, as current evidence relies heavily on retrospective and single-arm studies.

Conflict of Interest None to declare.

References

1. Hellman S, Weichselbaum RR. Oligometastases. J Clin Oncol. 1995;13(1):8–10.

2. Dingemans A-MC, Hendriks LE, Berghmans T, Levy A, Hasan B, Faivre-Finn C, et al. Definition of synchronous oligometastatic non–small cell lung cancer—a consensus report. J Thorac Oncol. 2019;14(12):2109–19.
3. Guckenberger M, Lievens Y, Bouma AB, Collette L, Dekker A, Nandita M, et al. Characterisation and classification of oligometastatic disease: a European Society for Radiotherapy and Oncology and European Organisation for Research and Treatment of Cancer consensus recommendation. Lancet Oncol. 2020;21(1):e18–28.
4. Guckenberger M, Lievens Y, Bouma AB, Collette L, Dekker A, de Souza NM, et al. Characterisation and classification of oligometastatic disease: a European Society for Radiotherapy and Oncology and European organisation for research and treatment of Cancer consensus recommendation. Lancet Oncol. 2020;21(1):e18–28.
5. Chen H-J, Tu C-Y, Hsia T-C, Fang H-Y, Chen C-H, Chen S-C, et al. Prognostic significance of oligometastatic disease classification by the ESTRO/EORTC of cancer for patients with lung cancer treated with definitive radical radiotherapy. Anticancer Res. 2020;40(10):5895–9.
6. Nicholls L, Chapman E, Khoo V, Suh Y-e, Tunariu N, Wang Y, van As N. Metastasis-directed therapy in prostate cancer: prognostic significance of the ESTRO/EORTC classification in oligometastatic bone disease. Clin Oncol. 2022;34(1):63–9.
7. Baker S, Mou B, Jiang W, Liu M, Bergman A, Schellenberg D, et al. Validation of the prognostic utility of ESTRO/EORTC oligometastatic disease classification: a secondary analysis from the population-based phase II SABR-5 trial. Int J Radiat Oncol Biol Phys. 2022;114(5):849–55.
8. Garde-Noguera J, Martín-Martín M, Obeso A, López-Mata M, Crespo IR, Pelari-Mici L, et al. Current treatment landscape for oligometastatic non-small cell lung cancer. World J Clin Oncol. 2022;13(6):485.
9. Mitchell KG, Farooqi A, Ludmir EB, Corsini EM, Sepesi B, Gomez DR, et al. Pulmonary resection is associated with long-term survival and should remain a therapeutic option in oligometastatic lung cancer. J Thorac Cardiovasc Surg. 2021;161(4):1497–504. e2
10. De Baere T, Tselikas L, Delpla A, Roux C, Varin E, Kobe A, et al. Thermal ablation in the management of oligometastatic colorectal cancer. Int J Hyperth. 2022;39(1):627–32.
11. Nugent K, Good J. The oligometastatic paradigm and the role of radiotherapy. Clin Med. 2023;23(1):61–4.
12. Pacifico P, Colciago RR, De Felice F, Boldrini L, Salvestrini V, Nardone V, et al. A critical review on oligometastatic disease: a radiation oncologist's perspective. Med Oncol. 2022;39(12):181.
13. Harrow S, Palma DA, Olson R, Gaede S, Louie AV, Haasbeek C, et al. Stereotactic radiation for the comprehensive treatment of oligometastases (SABR-COMET): extended long-term outcomes. Int J Radiat Oncol Biol Phys. 2022;114(4):611–6.
14. Tsai CJ, Yang JT, Shaverdian N, Patel J, Shepherd AF, Eng J, et al. Standard-of-care systemic therapy with or without stereotactic body radiotherapy in patients with oligoprogressive breast cancer or non-small-cell lung cancer (Consolidative Use of Radiotherapy to Block [CURB] oligoprogression): an open-label, randomised, controlled, phase 2 study. Lancet. 2024;403(10422):171–82.
15. Hanna GG, McDonald F. SBRT for oligoprogressive disease: using the evidence to maximise the benefits. Lancet. 2024;403(10422):122–4.
16. Deek MP, Van der Eecken K, Sutera P, Deek RA, Fonteyne V, Mendes AA, et al. Long-term outcomes and genetic predictors of response to metastasis-directed therapy versus observation in oligometastatic prostate cancer: analysis of STOMP and ORIOLE trials. J Clin Oncol. 2022;40(29):3377–82.
17. Phillips R, Shi WY, Deek M, Radwan N, Lim SJ, Antonarakis ES, et al. Outcomes of observation vs stereotactic ablative radiation for oligometastatic prostate cancer: the ORIOLE phase 2 randomized clinical trial. JAMA Oncol. 2020;6(5):650–9.
18. de Souza NM, Liu Y, Chiti A, Oprea-Lager D, Gebhart G, Van Beers BE, et al. Strategies and technical challenges for imaging oligometastatic disease: recommendations from the

European Organisation for Research and Treatment of Cancer imaging group. Eur J Cancer. 2018;91:153–63.
19. Hövels A, Heesakkers R, Adang E, Jager G, Strum S, Hoogeveen Y, et al. The diagnostic accuracy of CT and MRI in the staging of pelvic lymph nodes in patients with prostate cancer: a meta-analysis. Clin Radiol. 2008;63(4):387–95.
20. Padhani AR, Liu G, Mu-Koh D, Chenevert TL, Thoeny HC, Takahara T, et al. Diffusion-weighted magnetic resonance imaging as a cancer biomarker: consensus and recommendations. Neoplasia. 2009;11(2):102–25.
21. Chetty IJ, Doemer AJ, Dolan JL, Kim JP, Cunningham JM, Dragovic J, et al. MRI-guided radiotherapy (MRgRT) for treatment of oligometastases: review of clinical applications and challenges. Int J Radiat Oncol Biol Phys. 2022;114(5):950–67.
22. Pirasteh A, Lovrec P, Pedrosa I. Imaging and its impact on defining the oligometastatic state. Semin Radiat Oncol. 2021; https://doi.org/10.1016/j.semradonc.2021.03.006. Elsevier
23. Palma DA, Olson R, Harrow S, Gaede S, Louie AV, Haasbeek C, et al. Stereotactic ablative radiotherapy for the comprehensive treatment of oligometastatic cancers: long-term results of the SABR-COMET phase II randomized trial. J Clin Oncol. 2020;38(25):2830–8.
24. Bauckneht M, Lanfranchi F, Albano D, Triggiani L, Linguanti F, Urso L, et al. Diverse imaging methods may influence long-term oncologic outcomes in Oligorecurrent prostate cancer patients treated with metastasis-directed therapy (the PRECISE-MDT study). J Nucl Med. 2024;65:1202.
25. Kinj R, Muggeo E, Schiappacasse L, Bourhis J, Herrera FG. Stereotactic body radiation therapy in patients with oligometastatic disease: clinical state of the art and perspectives. Cancers. 2022;14(5):1152.
26. Schellenberg D, Gabos Z, Duimering A, Debenham B, Fairchild A, Huang F, et al. Stereotactic ablative radiation for Oligoprogressive cancers: results of the randomized phase 2 STOP trial. Int J Radiat Oncol Biol Phys. 2025;121(1):28–38.
27. Iyengar P, All S, Berry MF, Boike TP, Bradfield L, Dingemans AC, et al. Treatment of Oligometastatic non-small cell lung cancer: an ASTRO/ESTRO clinical practice guideline. Pract Radiat Oncol. 2023;13(5):393–412.
28. Iyengar P, Hu C, Gomez DR, Timmerman RD, Simone CB, Robinson CG, et al. NRG-LU002: randomized phase II/III trial of maintenance systemic therapy versus local consolidative therapy (LCT) plus maintenance systemic therapy for limited metastatic non-small cell lung cancer (NSCLC). American Society of Clinical Oncology; 2024.
29. Gomez DR, Tang C, Zhang J, Blumenschein GR Jr, Hernandez M, Lee JJ, et al. Local consolidative therapy vs. maintenance therapy or observation for patients with Oligometastatic non-small-cell lung cancer: long-term results of a multi-institutional, phase II, randomized study. J Clin Oncol. 2019;37(18):1558–65.
30. Iyengar P, Wardak Z, Gerber DE, Tumati V, Ahn C, Hughes RS, et al. Consolidative radiotherapy for limited metastatic non-small-cell lung cancer: a phase 2 randomized clinical trial. JAMA Oncol. 2018;4(1):e173501.
31. Wang XS, Bai YF, Verma V, Yu RL, Tian W, Ao R, et al. Randomized trial of first-line tyrosine kinase inhibitor with or without radiotherapy for synchronous Oligometastatic EGFR-mutated non-small cell lung cancer. J Natl Cancer Inst. 2023;115(6):742–8.
32. Kyriakopoulos CE, Chen YH, Carducci MA, Liu G, Jarrard DF, Hahn NM, et al. Chemohormonal therapy in metastatic hormone-sensitive prostate cancer: long-term survival analysis of the randomized phase III E3805 CHAARTED trial. J Clin Oncol. 2018;36(11):1080–7.
33. Fizazi K, Tran N, Fein L, Matsubara N, Rodriguez-Antolin A, Alekseev BY, et al. Abiraterone acetate plus prednisone in patients with newly diagnosed high-risk metastatic castration-sensitive prostate cancer (LATITUDE): final overall survival analysis of a randomised, double-blind, phase 3 trial. Lancet Oncol. 2019;20(5):686–700.
34. Zilli T, Achard V, Dal Pra A, Schmidt-Hegemann N, Jereczek-Fossa BA, Lancia A, et al. Recommendations for radiation therapy in oligometastatic prostate cancer: an ESTRO-ACROP Delphi consensus. Radiother Oncol. 2022;176:199–207.

35. Parker CC, James ND, Brawley CD, Clarke NW, Hoyle AP, Ali A, et al. Radiotherapy to the primary tumour for newly diagnosed, metastatic prostate cancer (STAMPEDE): a randomised controlled phase 3 trial. Lancet. 2018;392(10162):2353–66.
36. Boeve LMS, Hulshof M, Vis AN, Zwinderman AH, Twisk JWR, Witjes WPJ, et al. Effect on survival of androgen deprivation therapy alone compared to androgen deprivation therapy combined with concurrent radiation therapy to the prostate in patients with primary bone metastatic prostate cancer in a prospective randomised clinical trial: data from the HORRAD trial. Eur Urol. 2019;75(3):410–8.
37. Bossi A, Foulon S, Maldonado X, Sargos P, MacDermott R, Kelly P, et al. Efficacy and safety of prostate radiotherapy in de novo metastatic castration-sensitive prostate cancer (PEACE-1): a multicentre, open-label, randomised, phase 3 study with a 2 × 2 factorial design. Lancet. 2024;404(10467):2065–76.
38. Rogowski P, Roach M 3rd, Schmidt-Hegemann NS, Trapp C, von Bestenbostel R, Shi R, et al. Radiotherapy of oligometastatic prostate cancer: a systematic review. Radiat Oncol. 2021;16(1):50.
39. Tang C, Sherry AD, Haymaker C, Bathala T, Liu S, Fellman B, et al. Addition of metastasis-directed therapy to intermittent hormone therapy for Oligometastatic prostate cancer: the EXTEND phase 2 randomized clinical trial. JAMA Oncol. 2023;9(6):825–34.
40. Francolini G, Allegra AG, Detti B, Di Cataldo V, Caini S, Bruni A, et al. Stereotactic body radiation therapy and Abiraterone acetate for patients affected by Oligometastatic castrate-resistant prostate cancer: a randomized phase II trial (ARTO). J Clin Oncol. 2023;41(36):5561–8.
41. Lievens Y, Guckenberger M, Gomez D, Hoyer M, Iyengar P, Kindts I, et al. Defining oligometastatic disease from a radiation oncology perspective: an ESTRO-ASTRO consensus document. Radiother Oncol. 2020;148:157–66.
42. Klement RJ, Guckenberger M, Alheid H, Allgauer M, Becker G, Blanck O, et al. Stereotactic body radiotherapy for oligo-metastatic liver disease – influence of pre-treatment chemotherapy and histology on local tumor control. Radiother Oncol. 2017;123(2):227–33.
43. Soran A, Ozmen V, Ozbas S, Karanlik H, Muslumanoglu M, Igci A, et al. Randomized trial comparing resection of primary tumor with no surgery in stage IV breast cancer at presentation: protocol MF07-01. Ann Surg Oncol. 2018;25(11):3141–9.
44. Badwe R, Hawaldar R, Nair N, Kaushik R, Parmar V, Siddique S, et al. Locoregional treatment versus no treatment of the primary tumour in metastatic breast cancer: an open-label randomised controlled trial. Lancet Oncol. 2015;16(13):1380–8.
45. Fitzal F, Bjelic-Radisic V, Knauer M, Steger G, Hubalek M, Balic M, et al. Impact of breast surgery in primary metastasized breast cancer: outcomes of the prospective randomized phase III ABCSG-28 POSYTIVE trial. Ann Surg. 2019;269(6):1163–9.
46. Khan SA, Zhao F, Goldstein LJ, Cella D, Basik M, Golshan M, et al. Early local therapy for the primary site in de novo stage IV breast cancer: results of a randomized clinical trial (E2108). J Clin Oncol. 2022;40(9):978–87.
47. Shien T, Nakamura K, Shibata T, Kinoshita T, Aogi K, Fujisawa T, et al. A randomized controlled trial comparing primary tumour resection plus systemic therapy with systemic therapy alone in metastatic breast cancer (PRIM-BC): Japan Clinical Oncology Group Study JCOG1017. Jpn J Clin Oncol. 2012;42(10):970–3.
48. Fan J, Chen D, Du H, Shen C, Che G. Prognostic factors for resection of isolated pulmonary metastases in breast cancer patients: a systematic review and meta-analysis. J Thorac Dis. 2015;7(8):1441–51.
49. Cha C, Ahn SG, Yoo TK, Kim KM, Bae SJ, Yoon C, et al. Local treatment in addition to endocrine therapy in hormone receptor-positive and HER2-negative Oligometastatic breast cancer patients: a retrospective multicenter analysis. Breast Care (Basel). 2020;15(4):408–14.
50. Trovo M, Furlan C, Polesel J, Fiorica F, Arcangeli S, Giaj-Levra N, et al. Radical radiation therapy for oligometastatic breast cancer: results of a prospective phase II trial. Radiother Oncol. 2018;126(1):177–80.

51. Milano MT, Katz AW, Zhang H, Okunieff P. Oligometastases treated with stereotactic body radiotherapy: long-term follow-up of prospective study. Int J Radiat Oncol Biol Phys. 2012;83(3):878–86.
52. Yan B, Ramadan S, Jerzak KJ, Louie AV, Donovan E. The effectiveness and safety of stereotactic body radiation therapy (SBRT) in the treatment of Oligoprogressive breast cancer: a systematic review. Curr Oncol. 2023;30(7):6976–85.
53. Dona Lemus OM, Cao M, Cai B, Cummings M, Zheng D. Adaptive radiotherapy: next-generation radiotherapy. Cancers. 2024;16(6):1206.
54. Lavrova E, Garrett MD, Wang Y-F, Chin C, Elliston C, Savacool M, et al. Adaptive radiation therapy: a review of CT-based techniques. Radiol Imaging Cancer. 2023;5(4):e230011.
55. Jiang L, Ye Y, Feng Z, Liu W, Cao Y, Zhao X, et al. Stereotactic body radiation therapy for the primary tumor and oligometastases versus the primary tumor alone in patients with metastatic pancreatic cancer. Radiat Oncol. 2024;19(1):111.
56. Schiff JP, Stowe HB, Price A, Laugeman E, Hatscher C, Hugo GD, et al. In Silico trial of computed tomography-guided stereotactic adaptive radiation therapy (CT-STAR) for the treatment of abdominal Oligometastases. Int J Radiat Oncol Biol Phys. 2022;114(5):1022–31.
57. Da Silva Mendes V, Nierer L, Li M, Corradini S, Reiner M, Kamp F, et al. Dosimetric comparison of MR-linac-based IMRT and conventional VMAT treatment plans for prostate cancer. Radiat Oncol. 2021;16(1):133.
58. Fink CA, Buchele C, Baumann L, Liermann J, Hoegen P, Ristau J, et al. Dosimetric benefit of online treatment plan adaptation in stereotactic ultrahypofractionated MR-guided radiotherapy for localized prostate cancer. Front Oncol. 2024;14:1308406.
59. Fast MF, Cao M, Parikh P, Sonke J-J. Intrafraction motion management with MR-guided radiation therapy. Semin Radiat Oncol. 2024; https://doi.org/10.1016/j.semradonc.2023.10.008. Elsevier
60. van Overeem Felter M, Møller PK, Josipovic M, Bekke SN, Bernchou U, Serup-Hansen E, et al. 1-year efficacy results after MR-guided risk-adapted stereotactic radiotherapy of infra-diaphragmatic oligometastases in a multicenter phase II trial. Radiother Oncol. 2025;110748:110748.
61. Yang DD, Brennan VS, Huynh E, Williams CL, Han Z, Ampofo N, et al. Stereotactic magnetic resonance-guided adaptive radiation therapy (SMART) for abdominopelvic Oligometastases. Int J Radiat Oncol Biol Phys. 2022;114(5):941–9.
62. Lee S, Yadav P, van der Kogel AJ, Bayouth J, Bassetti MF. In Silico single-fraction stereotactic ablative radiation therapy for the treatment of thoracic and abdominal Oligometastatic disease with online adaptive magnetic resonance guidance. Adv Radiat Oncol. 2021;6(3):100652.
63. Henke LE, Stanley JA, Robinson C, Srivastava A, Contreras JA, Curcuru A, et al. Phase I trial of stereotactic MRI-guided online adaptive radiation therapy (SMART) for the treatment of Oligometastatic ovarian cancer. Int J Radiat Oncol Biol Phys. 2022;112(2):379–89.
64. Henke L, Kashani R, Robinson C, Curcuru A, DeWees T, Bradley J, et al. Phase I trial of stereotactic MR-guided online adaptive radiation therapy (SMART) for the treatment of oligometastatic or unresectable primary malignancies of the abdomen. Radiother Oncol. 2018;126(3):519–26.
65. Chiloiro G, Panza G, Boldrini L, Romano A, Placidi L, Nardini M, et al. REPeated mAgnetic resonance Image-guided stereotactic body Radiotherapy (MRIg-reSBRT) for oligometastatic patients: REPAIR, a mono-institutional retrospective study. Radiat Oncol. 2024;19(1):52.
66. Chin AL, Kumar KA, Guo HH, Maxim PG, Wakelee H, Neal JW, et al. Prognostic value of pretreatment FDG-PET parameters in high-dose image-guided radiotherapy for oligometastatic non-small-cell lung cancer. Clin Lung Cancer. 2018;19(5):e581–e8.
67. Hong JC, Ayala-Peacock DN, Lee J, Blackstock AW, Okunieff P, Sung MW, et al. Classification for long-term survival in oligometastatic patients treated with ablative radiotherapy: a multi-institutional pooled analysis. PLoS One. 2018;13(4):e0195149.
68. Henke L, Kashani R, Yang D, Zhao T, Green O, Olsen L, et al. Simulated online adaptive magnetic resonance–guided stereotactic body radiation therapy for the treatment of oligometa-

static disease of the abdomen and central thorax: characterization of potential advantages. Int J Radiat Oncol Biol Phys. 2016;96(5):1078–86.
69. Winkel D, Bol GH, Werensteijn-Honingh AM, Intven MPW, Eppinga WSC, Hes J, et al. Target coverage and dose criteria based evaluation of the first clinical 1.5T MR-linac SBRT treatments of lymph node oligometastases compared with conventional CBCT-linac treatment. Radiother Oncol. 2020;146:118–25.

Chapter 15
Emerging Role of Flash Radiotherapy

Bedanta Bhattacharjee, K. Sandhanam, Shriyeta Biswas, Ram Kumar Sahu, and Jiyauddin Khan

Abstract Nearly ten million people die from cancer each year, making it one of the top causes of mortality worldwide in 2024. Delivering ultra-high-dose radiation in a split second makes FLASH radiotherapy a novel method in cancer treatment. This method departs greatly from traditional radiotherapy by making use of the distinct biological reactions that high-dose radiation, given quickly, can trigger. Electron FLASH therapy, proton FLASH therapy, and photon FLASH therapy utilize electron beams, photon beams, proton beams, and heavy-ion beams, employing linear electron accelerators (LINACs) to deliver radiation at rates exceeding 40 Gy/s. These therapies are effective in targeting cancerous cells with high specificity and reduced cytotoxicity to healthy cells.. FLASH therapy has demonstrated promise in reducing radiation-induced toxicity and enhancing tumor control compared to conventional methods. Recent research indicates that FLASH radiation therapy improves the overall therapeutic window and minimizes acute adverse effects. Mechanistically, the rapid delivery of radiation appears to activate different cellular and molecular pathways, including those involved in DNA damage repair, immune modulation, and oxidative stress responses. Preclinical models have demonstrated promising results, indicating that FLASH radiotherapy can be effective across various tumor types, including those resistant to traditional radiation therapies.

B. Bhattacharjee · S. Biswas
School of Pharmaceutical Sciences, Girijananda Chowdhury University-Tezpur Campus, Tezpur, Assam, India

K. Sandhanam
Department of Pharmacology, SRM College of Pharmacy, SRM Institute of Science and Technology, Chengalpattu, Tamil Nadu, India

R. K. Sahu
Department of Pharmaceutical Sciences, HemvatiNandanBahugunaGarhwal University (A Central University), Chauras Campus, TehriGarhwal, Uttarakhand, India

J. Khan (✉)
School of Pharmacy, Management and Science University, Selangor Darul Ehsan, Malaysia
e-mail: jiyauddin_khan@msu.edu.my

F. Taghizadeh-Hesary et al. (eds.), *From Radiobiology to Radiation Oncology*,
https://doi.org/10.1007/978-3-031-95724-6_15

Moreover, advancements in accelerator technology are facilitating the application of FLASH radiation in clinical settings, paving the way for professional trials that will further elucidate its efficacy and safety profile.

Keywords FLASH radiotherapy · Ultra-high-dose radiation · Tumor control · DNA damage repair · Photon beam

15.1 Introduction

Radiotherapy (RT) remains a fundamental part of cancer treatment, considering that more than 65% of cancer cases require it eventually during their care. Since the introduction of ionizing radiation with the invention of X-rays in 1895, it has provided significant benefits to millions of patients around the globe [1]. The field of RT has experienced remarkable advancements, transitioning from traditional photon-based techniques to the targeted use of fragments in therapies involving protons and heavy ions [2, 3]. The introduction of FLASH radiotherapy (FLASH-RT) heralds a new era in the treatment of cancer. As a state-of-the-art, third-generation radiotherapy technology, FLASH-RT has the capacity to transform oncological care and modify our approach to fighting cancer [4]. FLASH-RT represents a groundbreaking approach in cancer treatment, delivering radiation at ultra-high and excessive dosages that are orders of magnitude higher than those that are now employed in standard therapeutic treatment. This innovative technique aims to achieve effective tumor eradication while minimizing harm to healthy tissues brought on by radiation. In traditional radiation therapy, maintaining a balance between killing cancer cells and sparing healthy tissue is a significant challenge [5]. RT, a noninvasive treatment designed to damage and eliminate tumor cells, is a key component in managing 30–50% of cancer cases. However, one of its major challenges is the unintentional damage it causes to the healthy tissues surrounding the tumor. The effectiveness of radiotherapy relies on delivering a sufficient dose to eradicate tumor cells while reducing harm to adjacent normal tissues. FLASH-RT is an innovative approach that addresses this limitation by using ultra-high dose rates [6]. This cutting-edge technique not only matches the tumor-killing efficacy of conventional dose-rate radiotherapy (CON-RT) but also significantly reduces damage to healthy tissues, offering a safer and potentially more effective treatment option. FLASH-RT overcomes this limitation by delivering radiation at dosage rates of 40 Gy/s or more. Studies in multiple animal models have demonstrated that FLASH-RT achieves comparable or superior tumor control compared to conventional RT while causing significantly less injury to normal cells and tissues [7]. The unique FLASH-RT protective effect on healthy tissues is thought to stem from its influence on toxic oxygen-reactive species. Unlike conventional RT, which generates high levels of reactive oxygen species (ROS) that can damage normal tissues, FLASH-RT produces lower ROS levels, thereby reducing oxidative stress and preserving normal tissue integrity. This dual benefit of enhanced tumor control and reduced normal

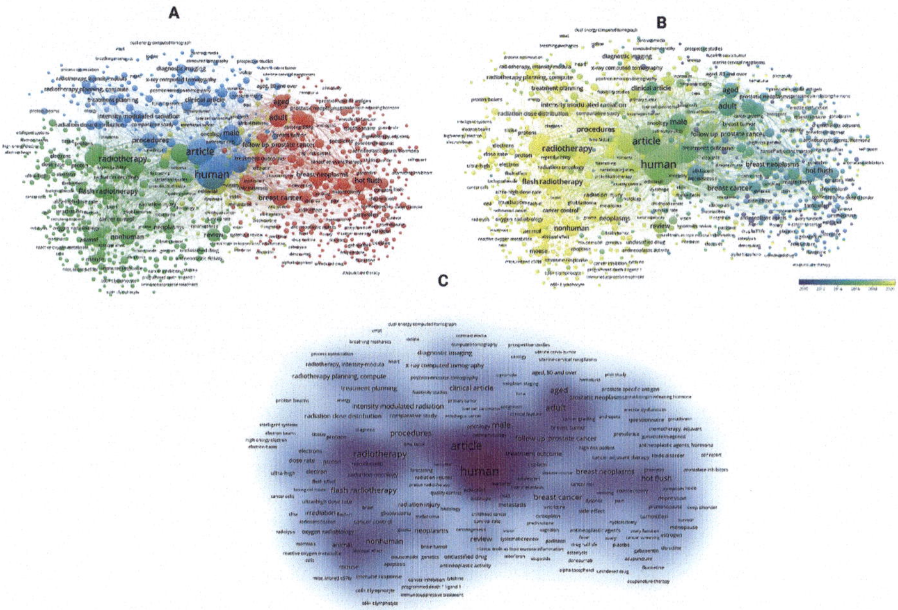

Fig. 15.1 Scientific research bibliometric network map on radiotherapy as a FLASH therapeutic approach for cancer studies. The visualization of the network highlights the four colored clusters (**a**), indicating scientific journals' relatedness, while the visualization of an overlay (**b**) illustrates the occurrence of the keyword's period from 2010 (blue) to 2020 (yellow). The figure was generated using VOSviewer, with data collected from the Web of Science Core Collection and Scopus databases, utilizing the keywords "Flash" and "radiotherapy," and "cancer" within the timeframe 2010–2020, (**c**) Density visualization

tissue toxicity positions FLASH-RT as a promising frontier in radiotherapy, with the potential to revolutionize cancer treatment and improve patient outcomes [8]. In recent years, researchers have increasingly focused on FLASH-RT as a promising approach for cancer treatment, as depicted in the bibliometric map shown in Fig. 15.1. The timeline in the figure illustrates the progression of research on this innovative technique, highlighting its evolution and growing significance in oncology.

FLASH-RT presents significant advantages over traditional radiotherapy (CON-RT). A primary benefit is its capacity to minimize lymphocyte depletion, which may enhance the effectiveness of subsequent systemic therapies, including chemotherapy, immunotherapy, and targeted treatments. While CON-RT paired with immune checkpoint inhibitors has shown improved tumor control, the effectiveness and safety of combining FLASH-RT with chemotherapy or immunotherapy are still under active investigation [9]. Recent studies reveal promising findings regarding FLASH-RT potential. For instance, research indicates that abdominopelvic FLASH irradiation can continue to enhance the invasion of CD8 T-cells in tumors, potentially enhancing anti-PD-1 therapy's effectiveness in treating ovarian

Table 15.1 Comparison of FLASH-RT and Conventional RT, classification, and their applications

Modality	Year	Mean dose rate (Gy/s)	Delivery time	Usage of cancer types	Clinical applicability	References
Conventional	Various	0.02–0.05	10–30 min	Most solid tumors	Standard treatment for various cancers, including breast and prostate.	[14]
FLASH	2022	40	<1 s	Ovarian cancer, abdominal tumors	Potentially enhances tumor control while preserving normal tissues.	[15]
Electron FLASH therapy	2022	40	<1 s	Various cancers, particularly skin tumors	Effective in delivering high doses with minimal surrounding tissue damage.	[16]
Photon FLASH therapy	2022	40	<1 s	Ovarian cancer, abdominal tumors	Improves immune response and tumor infiltration by T-cells.	[17]
Proton FLASH therapy	2022	20	<1 s	Pediatric cancers, brain tumors	Reduces radiation exposure to surrounding healthy tissues.	[18]
Role of linear electron accelerators	Various	0.02–0.05	Varies (minutes to hours)	Broad range of cancers	Versatile in delivering various forms of radiation therapy; foundational technology for many modalities.	[19]

cancer [10]. Similarly, the research found that FLASH X-rays in conjunction with anti-PD-L1 therapy were equally as efficient as CON-RT in managing abdominal tumors, as depicted in Table 15.1. These results imply that FLASH-RT could expand the therapeutic opportunities for oncological radio-immunotherapy, leading to improved treatment outcomes and better quality of life for patients [11]. Despite these advancements, clinical and preclinical investigations into the combination of FLASH-RT with chemotherapy or immunotherapy remain limited. Comprehensive evidence from future studies is essential to substantiate its incorporation into combination treatment protocols. In addition to its protective effects on healthy tissues, FLASH-RT offers the potential for dose escalation strategies [12]. By allowing higher radiation doses while reducing damage to adjacent normal tissues, FLASH-RT could further enhance tumor control. This innovative technique has the potential to refine conventional radiotherapy practices and establish itself as a transformative approach to cancer treatment, ultimately enhancing the patient's quality of life and survival rates [13].

FLASH-RT was first utilized in a human therapeutic environment. The inaugural patient to receive treatment using FLASH-RT was a 75-year-old man, who was identified as having multidrug-resistant CD30+ skin lymphoma T-cell, having already undergone over treatments of 110 local radiations for skin lesions. The progressive tumor lobules were treated with an overall 15 Gy dosage, administered in 10 pulses of 1 microsecond. It took roughly 90 milliseconds to complete, achieving a mean dosage rate of approximately 167 Gy/s. Following the treatment, only mild grade 1 dermatitis and temporary grade 1 edema were noted in the soft tissue around the tumor [20]. The initial clinical experiments produced promising results. While it is premature to draw conclusive findings from a single experiment, these outcomes highlight the potential of clinical use of FLASH technology. Ongoing research and additional clinical trials are being conducted to further explore its mechanisms and broaden its use in oncology [14]. This chapter uncovers the FLASH-RT mechanism, classification, and clinical properties related to cancer.

15.2 Background and History of FLASH Radiotherapy

In the world of radiation, FLASH-RT is a relatively recent idea. Its origins can be found in the early 1900s, when scientists noticed certain significant consequences of high radiation doses. For more than a century, radiotherapy has been a pillar of cancer therapy, using ionizing radiation in order to break cancer cells' DNA. Despite its effectiveness, conventional radiation (CRT) frequently causes severe harm to healthy tissues, resulting in both immediate and long-term adverse effects that restrict the therapeutic dosage [14]. In the late 1950s, one of the first FLASH-RT-related observations was made. Currently referred to as FLASH-RT, the phenomenon is called the FLASH effect [21]. They experimented with irradiating *Serratia marcescens* with megavoltage X-rays at an extremely high dosage rate. According to the study, the bacteria showed reduced radio-sensitivity in a nitrogen–oxygen mixture when subjected to ultra-high radiation rates (10–20 krads/μs) as opposed to being exposed to standard dose rates (1 krads/min) in 100% nitrogen. The response usually seen in anaerobic environments was in line with this decreased radio-sensitivity. These findings demonstrated that *Serratia marcescens* was affected by both the oxygen concentration and the radiation dose rate [21]. The outcomes of the mammalian cells exposed to radiation at a UHDR of 3.5×10^6 krad/s were interesting. A single pulse had a superior survival rate than two pulses when the cells were exposed to a dose of up to 1 krads [22]. The findings were corroborated by comparable results in their studies employing UHDR irradiation of 1 krads for a 15-nanosecond pulse on hamster and HeLa cells [23]. Results from studies on small animals showed that the presence of oxygen during irradiation decreased the radiation's ability to cause death [24]. According to their research, the radiation's capacity to induce skin reactions significantly decreased at a rate of 0.5 milligrams per minute [25]. This provided more evidence that oxygen consumption is directly related to the FLASH effect, which is characterized by decreased effectiveness. Together, these preclinical and cell studies demonstrated how oxygen's presence during radiation and the FLASH effect are related. The findings underscored the significance of oxygen-related parameters in FLASH-RT and the effect of oxygen consumption on the observed decrease in radiation effectiveness [25]. It was observed in 2014 that when mice were exposed to the same dose of FLASH radiation, none of them experienced fibrosis or pneumonia. Mice exposed to FLASH radiation began to develop fibrosis and pneumonia when the dosage was increased to 30 Gy [26]. Additionally, it was found that flash irradiation at 17 Gy stopped acute apoptosis in blood vessels and bronchi and inhibited the activation of transforming growth factor-β (TGF-β). Six cats with T2/T3N0M0 nasal plane squamous cell carcinoma that had spread locally, and who also had fibrinoid necrosis and hair loss, participated in a research trial for larger animals. The trial was a Phase I escalation research in which a single dosage between 25 and 41 Gy was given [27]. The findings showed that FLASH-RT had a "protective effect," causing less harm to healthy tissue than traditional radiation. Over a 16-month period, the experimental group's progression-free survival rate was 84%. The findings supported potential advantages of FLASH-RT and created the foundation for future studies on its effects on people. A research study including the initial human patient to undergo FLASH-RT was carried out in 2019. The patient was recommended a dose of 15 Gy for CD30+ T-cell cutaneous lymphoma with multiple resistance. A 5.6 MeV linear accelerator was used to administer the treatment in an incredibly brief 90 milliseconds. Three weeks after the treatment, a clinical examination showed only edema and grade-I epithelitis in the soft tissues surrounding the tumors [20]. The follow-up assessment ultimately determined that FLASH therapy was both practical and safe, with a positive outcome for the patient.

15.3 Overview of FLASH Radiotherapy

The FLASH-RT technology involves the delivery of ultra-high-dose-rate radiation to the targeted cells. Without sacrificing the treatment's anti-cancer effects, FLASH has been demonstrated to lessen radiation-induced damage in normal tissues [28]. Highly conformal radiation beams may be precisely delivered to the target, thanks to modern technology, reducing damage to surrounding healthy tissues.

15.3.1 Types of FLASH-RT

Electron FLASH Therapy

Using electron beams to administer radiation at rates higher than 40 Gy/s, electron-based flash irradiation (eFLASH) is a prospective UHDR technology. Because electrons have a limited depth of penetration, this approach works especially well for treating superficial or shallow tumors. Technical problems with the acceleration of electrons in a medical apparatus of a convenient size (i.e. one that is neither too large nor too complex) restrict the use of electrons with extremely high energies (between 50 and 250 MeV) that can reach deeper depths. Nonetheless, efforts are being made to construct appropriate accelerators [29]. Another benefit was that compared to dose distributions obtained with protons, those produced by very high-energy electrons appear to be less dependent on body inhomogeneities [26]. Controlling dosage distribution, adapting to different tumor forms, delivering treatment quickly, and having a wide range of applications are what set electron beam therapy apart. For individuals with tumors, especially those suffering from superficial tissue or skin tumors, including skin cancer, lymphoma, and cutaneous melanoma, it was a safe and efficient therapy option. Nowadays, a large amount of research shows that electron beam FLASH-RT is better than traditional dose-rate radiation at protecting healthy tissues. Biological responses to the ultra-high-dose radiation lead to the quick loss of oxygen in exposed tissues because radiation-induced ROS use more molecular oxygen [30]. The temporary hypoxia provides less protection to tumor cells, which are more reliant on ROS-mediated DNA damage. Normal tissues are protected against excessive damage in low-oxygen environments because they rely more on direct DNA breakage than on ROS for damage. Distinct cellular reactions are because of their increased rates of proliferation and genomic instability; tumor cells are less able to adjust to the stress.

Normal cells are better able to heal because they have stronger DNA repair systems. The generation capability of ROS in normal tissues is overpowered by FLASH dose rates, which affect normal tissues, causing less oxidative stress, which lowers the level of lipid peroxidation, protein damage, and DNA double-strand breaks [30].

Proton FLASH Therapy

Proton FLASH radiation is a cutting-edge technique that harnesses the therapeutic advantages of the FLASH effect by administering protons at extremely high rates of dosage (≥40 Gy/s). The majority of tumors can receive energy at therapeutic depth from carbon ions or protons with energies of 200-300 MeV/n, since they can travel 15–20 cm through the body. Higher dosage rates given in nanoseconds longer than what is advised to create FLASH effects are involved in proton radiation [31]. For CON-RT, 100 milliseconds of radiation times are still insufficient to produce the FLASH effect. To assess the impact on normal tissues and possible lung-sparing effects, conventional radiation was compared to FLASH-RT using proton beams [32]. It was discovered that due to a better immunological response, proton-based FLASH-RT may save normal tissues (both acute and late). Biological responses to ultra-high-dose radiation are strongly related to the FLASH effect, which keeps tumors under control while protecting normal tissues from radiation harm. Reactive oxygen species are produced when protons cause a sudden explosion of ionizing events [33]. As a result, the irradiated tissue temporarily loses molecular oxygen. Because tumor cells depend on oxygen to maintain ROS-mediated DNA damage, the hypoxic environment is less effective in these cells. Normal tissues are therefore protected from severe injury since they have effective repair systems and lower oxygen requirements [33]. The proton flash causes direct damage to DNA, including breaks in the single and double strands, at a rapid rate, resulting in DNA damage responses and repair kinetics. Normal cells recover well because they have a higher capacity for repair. Tumor cells cannot adequately recover because their repair mechanisms are faulty or overloaded, improving the therapeutic ratio by guaranteeing tumor cell killing while protecting healthy tissue. High-dose-rate proton administration decreases the accumulation of ROS in healthy tissues because oxygen is used quickly and is temporarily unavailable. By utilizing the Bragg peak phenomenon, proton radiation ensures accurate energy deposition inside tumor tissues while causing the least amount of harm to nearby normal tissues. This characteristic minimizes negative effects on surrounding tissues while permitting dose escalation within the tumor, improving therapeutic safety and efficacy [34]. Using the pair of proton beams for FLASH-RT at different dosage rates and CON-RT, the mice's right posterior legs received 35 Gy of uniform physical dosages (toxicity study) or 15 Gy (tumor control study). The results demonstrated that the FLASH group's leg contractures, cutaneous toxicity, and plasma and cutaneous TGF-β1 levels were all considerably lower than those of the conventional group. Proton radiation was beneficial for tumors close to delicate tissues or important organs because of its precision in tumor localization and energy delivery, as well as its capacity to maximize tissue protection and minimize adverse effects [17]. Further highlighting the possible superiority of proton radiotherapy in FLASH mode, comparative studies have indicated that proton irradiation is more effective than photon radiotherapy.

Photon FLASH Therapy

Photon flash radiation makes use of the FLASH effect by utilizing extremely high dose rates of high-energy X-rays, or photons. Because of their adaptability and capacity to enter deep tissues, photons can be used to treat a variety of tumors. Due to the nature of the interaction between photons and tissue, traditional radiotherapy usually uses 15 millivolt photon beams, which is adequate to achieve good dose coverage for all tumors. However, to attain ultra-high dose rates for photons, the conversion efficiency of electron beams to photon beams must be low [35]. Only a small fraction of the electron energy fluence was transferred to photons; the bulk was squandered through other processes, including heat. This suggests that a source for a FLASH photon accelerator must be able to create a lot more electrons (by a factor of 1000) than what the devices that are now available on the market can produce [35]. In addition, there are issues with converting the energy of so many electrons into photons by speeding them up. Because of their remarkable tissue-penetrating qualities, photon beams can effectively enter tumor tissues that are deeply seated. There is a lot of space to grow the use of FLASH-RT photons in animal models, especially in investigating how well it works to treat deep lesions, including those in the thorax, abdomen, pelvis, and head and neck [35].

15.3.2 Radiobiological Mechanism of FLASH-RT

The radiobiological FLASH-RT mechanism, in which the consequences of ultra-high rate dosage were reported, proving that hypoxia was brought on by high dose-rate irradiation [30]. Since there are fewer DNA damage sites in FLASH-RT than after normal rate dosage irradiation, it has been demonstrated that ultra-high rate dosage irradiation can protect mammalian cells, including cancer cells, through temporary hypoxia. According to several publications, FLASH-RT results in a reduced number of dicentric chromosomes compared to traditional dose-rate radiation. There were differences between G2 cell cycle arrest after FLASH-RT and traditional dose-rate irradiation. When it comes to distinguishing the biological effects of FLASH-RT from those of conventional dose-rate irradiation, myosin light chain activation is essential [30]. The four main hypotheses related to the mechanism of FLASH-RT are discussed below:

Oxygen Depletion Hypothesis

The oxygen depletion theory suggests that FLASH radiation treatment, which provides radiation at extremely high dose rates, produces a transient decrease in oxygen consumption inside tissues, which leads to minimal radiation toxicity to healthy cells [36]. Chinese hamster ovarian cells (CHO-K1) were exposed to carbon-ion FLASH-RT and conventional dose rates in order to examine cell viability at 0%, 0.5%, 4%, and 21% O_2 oxygenation levels. They were given FLASH or a standard dose rate of 7.5 Gy. There were no discernible variations in FLASH cell survival and conventional radiation in experiments conducted under normoxic (21% O_2) and hypoxic (0% O_2) settings. Rather, the FLASH protective impact was amplified in hypoxic conditions with 0.5% O_2, indicating that the FLASH protective effect is oxygenation-dependent and more pronounced in hypoxic environments [37]. One potential biological explanation for the FLASH effect is oxygen depletion. Preclinical studies indicate that while the brief irradiation duration inhibits extracellular oxygen from migrating into cells, high intracellular oxygen is quickly depleted by high-dose-rate radiation, resulting in a hypoxic condition. Tissue resistance to radiation is increased by this brief hypoxia [38]. It was hypothesized that tumor tissues, which are already hypoxic, remain sensitive to radiation therapy because they do not develop further radiation resistance as a result of oxygen deprivation. To completely investigate the biological impacts of FLASH, future studies should use tumor models that are both hypoxic and well-oxygenated. The hypoxia hypothesis is probably not the only mechanism underlying the current data, as other processes also contribute to the protective impact of FLASH on healthy tissues and organs.

Oxidative Stress Response Hypothesis

Many scientists have proposed in recent years that the interactions between free radicals created by RT could be the source of the distinct biological effects caused by FLASH radiation. Atoms or molecules having unpaired electrons that are extremely reactive, known as free radicals, have the ability to set off a chain of chemical events inside cells that can cause damage or even death. The fundamental tenet of the idea of free radical interaction process is that immediate high concentrations of free radicals react with one another due to the exceptionally fast dosage rate of FLASH radiation [39]. Free radicals combine to create stable molecules, and their recombination shortens the cycle of reactions, lowering lipid peroxidation, protein peroxidation, and DNA damage, among other cell-damaging effects. ROS are created when other free radicals combine with oxygen. Cell damage can result from ROS spilling over and interacting with biological targets, such as lipids and DNA [40]. Normal cells have decreased quantities of redox and active iron compared to cancer cells, indicating notable metabolic changes between the two types of cells. By increasing the lowering intracellular metal ion activity and reducing the action of antioxidant enzymes, normal cells can efficiently and quickly remove ROS, lowering cellular toxicity. On the other hand, tumor cells have an overabundance of active metal ions, including Fe^{2+} and active iron, which might intensify the peroxidation chain reaction [41].

DNA Damage Repair Hypothesis

Numerous factors influence DNA damage from ultra-high dose rate radiation, but little is known about how much of an impact each factor has. It has been proposed that differences in DNA repair and breaking may be connected to changes in cells exposed to varying amounts of radiation. While the irradiation method had no effect on cancer cells, normal tissue cells' DNA was damaged less at ultra-high dose rates than at standard dose rates when the TRP53 binding protein 1 (53BP1) was employed as a biomarker [26]. Hypoxia and depletion of oxygen effects of FLASH-RT lead to a brief hypoxic condition in the irradiated area due to the rapid oxygen consumption during radiation delivery. The generation of ROS, which causes a large portion of indirect DNA damage, is decreased by hypoxia. Normal cells benefit from this brief hypoxia, while tumor cells are less protected since they depend more on oxygen to worsen radiation-induced DNA damage.

Modulation of DNA repair includes FLASH-RT, which affects important DNA repair mechanisms, making tumor cells more vulnerable to FLASH-RT-induced DNA damage if their homologous recombination (HR) pathways are compromised [42]. The severity of the DNA damages brought on by FLASH-RT may exceed the main repair pathway in tumor cells, resulting in repair errors and cell death.

Immune Response Hypothesis

In animal models exposed to radiation, fibrotic lesions and inflammation were reduced when radiation was administered at an extremely high dose rate, suggesting a distinct immune response to FLASH radiation [43]. Regulatory T-cells (Tregs) and myeloid-derived suppressor cells (MDSCs) were two examples of immunosuppressive cells whose recruitment and activity may be reduced by FLASH-RT. Cytotoxic T-cells (CD8+), which were essential for anti-tumor immunity, can infiltrate more readily. Long-term negative effects can result from CON-RT, such as causing chronic inflammation in healthy tissues. FLASH-RT considerably lessens this by lowering the release of pro-inflammatory cytokines such as interleukin-6 (IL-6) and tumor necrosis factor-alpha (TNF-α), and reducing inflammatory reactions in healthy tissues caused by macrophages. FLASH-RT is a promising adjunct to immunological treatments, such as immune checkpoint inhibitors (anti-CTLA-4, anti-PD-1/PD-L1, etc., antibodies), due to its capacity to influence the immune system while sparing healthy tissues. FLASH-RT may improve the effectiveness of these treatments by improving the presentation of tumor antigens and decreasing immunosuppressive signals [43].

15.4 Therapeutic Benefit of FLASH Radiotherapy

15.4.1 Normal Tissue-Preventive Effect

The primary advantage of FLASH-RT is that it protects the normal tissue. In 1971, a study was conducted on the healthy mouse having gastrointestinal problems demonstrated that FLASH-RT treatment can produce hypoxia in exposed tissues, reducing radiation responsiveness and sustaining healthy tissues at dose rates of >60 Gy/min [44]. After a couple of years, trials in rat skin revealed that FLASH exposure might attenuate the negative effects at radiation dose levels, i.e., 5000 Gy/min [45]. The results from the two examinations suggested that under a hypoxic condition, the irradiated tissues' radiation sensitivity remained unchanged. In a study conducted by Favaudon et al. on thoracic irradiation of healthy mice, contrast irradiation was carried out using FLASH-RT (≥40 Gy/s) or CON-RT (≤0.03Gy/s). The treatment of mice with 17-Gy of CON-RT for a period of 8 weeks exhibited significant fibrosis at the irradiation area. Moreover, mice exposed to 17 Gy of FLASH-RT did not exhibit any fibrosis. However, increasing the dose of FLASH-RT from 16 to 30 Gy resulted in the development of modest pigmentation in the irradiated region [26]. Fouillade et al. reported that FLASH-RT exposure led to faster tissue recovery and decreased pulmonary fibrosis in mice [43]. Additionally, FLASH-RT substantially lowered hematopoietic stem cell destruction following entire body radiation therapy for acute lymphoblastic leukemia when compared with CON-RT [46]. Alaghband et al. explored the neuropreventive activity of CON-RT and FLASH-RT on a young medulloblastoma mouse model with minimal radiation sensitivity. They reported that a single dose administration of FLASH-RT at 8 Gy resulted in no cognitive dysfunction within 2–4 months. These findings establish a hypothetical foundation for FLASH-RT in clinical settings, including glioblastoma and pediatric medulloblastoma [47]. Vozenin et al. used six cats and one mini-pig with nasal squamous cell carcinoma to compare the effects of CON-RT at a dosage of 0.083 Gy/s and FLASH-RT at 300 Gy/s. Following radiation exposure to several mini-pig skin areas, they found that the FLASH-RT area exhibited modest depilation after 3 weeks' irradiation but no significant toxicity. On the other hand, severe fibrosis and non-regeneration of the hair follicles and irradiated skin were seen in the CON-RT area. The study on mini pigs and cats produced comparable findings, with minimal toxicity following FLASH-RT. Additionally, FLASH-RT was shown to improve the overall survival rate of unhealthy cats [27]. The practical use of FLASH-RT has been supported by the results of this higher mammalian study. In all trials, FLASH-RT was found to be more effective in conserving healthy tissue as compared to CON-RT. Although FLASH-RT has been demonstrated to have beneficial effects, there have also been contentious findings. Venkatesulu et al. reported that FLASH-RT at a dose rate of 35 Gy/s caused lymphocyte apoptosis and greater gastrointestinal toxicity in abdominal irradiation mice compared to CON-RT at a dose of 0.1 Gy/s. This finding implies that tissue heterogeneity or other unidentified biological variables may restrict the utility of FLASH-RT against normal tissue protecting function [48]. Consequently, more investigation into FLASH-RT limits is required.

15.4.2 Enhanced Tumor Control

FLASH-RT has a significant protective impact on healthy tissues, but its primary benefit is its anti-tumor activity. In 2014, Favaudon et al. investigated the effect of ultra-high doses on normal and tumor tissues in C57BL/6 J mice [26]. The head and neck tumor cell line (Hep-2) and human breast cancer cell line (HBCx-12A) derived from xenografts received similar dosages of CON-RT and FLASH-RT. The study found that both xenograft models were effective in reducing tumor development [26]. In an allograft tumor mice model, CON-RT or FLASH-RT with a cumulative dosage of 15 Gy showed similar anti-tumor efficacy. But when the dosage rate was increased by upto 28 Gy, these mice exposed to FLASH experienced a better lifespan period than those exposed to CON-RT. The measurement of their toxicity level supports the notion that FLASH-RT greatly increases the chance of survivability compared to CON-RT. Loo et al. demonstrated that FLASH-RT induced toxicity within healthy tissues at a dosage of 70–210 Gy/s, which is noticeably higher compared to the toxicity-causing 0.05 Gy/s dose of CON-RT [49].

15.5 In Vitro- and In Vivo-Based Evidence with FLASH Radiotherapy

15.5.1 In Vitro Outcomes

A recent study used a linear electron accelerator (Kinetron) operating at 4.5 MeV to reveal human lung fibroblast cell lines (MRC5, IMR-90), mouse lungs (C57BL/6 J wild type and Terc−/− mice), as well as a human lung epithelial cancer cell line (A-549) to radiation to ascertain the mechanism underlying FLASH protective effect on normal tissues. A clear organ-sparing effect was shown in the study when FLASH was compared to CON-RT. The DNA damage response was analyzed by immunofluorescence of 53BP1 and H2AX foci, which showed that the frequency of replicative senescence with FLASH was decreased, lung progenitor cells were protected from radiation-induced damage, and normal lung cells had less DNA damage in-vitro [43]. In an in-vitro study, Buonanno et al. exposed normal human lung fibroblasts to extremely high dosage rates of proton irradiation up to 1000 Gy/s to assess the effects of FLASH-RT. Both long-term and acute tissue consequences were investigated, including senescence induction, pro-inflammatory marker TGF expression, H2AX foci formation, and clonogenic survival. Ultra-high dose rates significantly impacted long-term effects, as evidenced by decreased senescence induction and pro-inflammatory marker expression, which suggested delayed unfavorable ramifications, whereas the FLASH dosage rate had little effect on acute reactions on tumor cells [32]. Beddok et al. examined three lung cell lines, i.e. two non-transformed human lung fibroblasts, MRC-5 and IMR-90, and one human lung cancer cell line, which were isolated using the same LINAC (4.5 MeV electrons) after FLASH and CON-RT. Either FLASH (>40 Gy) or conventional (0.03 Gy/s) irradiation was used to expose cells to 5 Gy. The MTT assay was used to determine cell viability, and the DNA damage response was assessed using immunofluorescence. The findings showed that there was no appreciable difference between the two modalities in terms of DNA damage or cell viability. For instance, the cancer cell line's average number of H2AX foci after both treatments was about the same (30.9 for CON and 29.10 for FLASH, $p = 0.6$), while MRC-5 cells showed no significant differences between the two therapies. These results contradicted the in-vivo findings of other investigations, suggesting that the FLASH effect may be influenced by the cellular environment or immune response, necessitating further in-vivo research [50]. HeLa cells were exposed to proton beams supplied using laser acceleration for pulsed irradiation at extremely high dose rates (109 Gy/s) in order to assess their impact on cell cycle arrest, colony formation, and apoptosis. The experiment compared pulsed and continuous irradiation modes, with a 3 Gy dose delivered using a 20 MeV proton beam [34]. Immunofluorescence analysis at 10 h post-irradiation revealed a significant reduction in the G2 fraction after pulsed irradiation compared to continuous, likely due to variations in the complexity of the damage or a prolonged G2 arrest time with pulsed radiation. Other endpoints showed significant similar results between the two irradiation modes. This arrangement was subsequently examined in tumor-bearing animals to compare pulsed and continuous proton beams with conventional photon irradiation. Notably, the ultra-high dose rates (109 Gy/s) used in this study may have effects different from traditional FLASH, which functions at radiation rates far lower than 40 Gy/s [51].The in-vitro report on FLASH suggested positive approach for cancer treatment effects in lung cancer fibroblasts and liver cancer epithelial cell lines. It was proposed that the FLASH effect may be influenced by the immune system and/or the cellular microenvironment. Further, in-vivo findings were discussed in the following section.

15.5.2 In Vivo Outcomes

Researchers used FLASH radiation (at >40 Gy/s) to test the effects on healthy tissues and the tumor's reaction to high dose rates in a study that looked at immunocompetent mice with orthotropic lung tumors and naked mice with human lung tumor xenografts. Lesser lung fibrogenesis and lower apoptosis in normal epithelial cells and smooth muscle were observed using FLASH irradiation as compared to traditional, prolonged unit-dose radiation (15 Gy at <0.03 Gy/s). Following this, researchers looked at the effects of increasing the dosage and found that 30 Gy of FLASH was needed to achieve the identical degree of fibrosis as 17 Gy of traditional elucidation; nevertheless, they found no major side effects from levels below 23 Gy of FLASH. As far as tumor control goes, FLASH was just as effective as CON-RT, which raises the possibility that ultra-high dose rate radiation could eliminate cancers with less collateral damage to healthy tissues [52]. A study by Liljedahl et al. used a 10 Gy of electrons flashed across a mouse brain at FLASH dose rates, and the animal's cognitive recall abilities were significantly preserved, particularly at average dosage rates between 30 Gy/s and 100 Gy/s. On the other hand, traditional irradiation caused permanent memory loss. While using X-rays at a dosage rate of 37 Gy/s, a comparable sparing effect was noted. After 14 Gy, the FLASH protective effect on nude mice's neurocognitive functions began to wane. The recognition memory function of the mice remained similar to that of the non-irradiated controls even after hypo-fractionation schemes of 3 × 10 Gy and 2 × 7 Gy. One possible explanation for the persistence of cell division in the hippocampus following FLASH irradiation is that the area responsible for memory continues to generate new neurons and glial cells. In addition, five out of ten pro-inflammatory cytokines were found to be elevated in the hippocampus after CON-RT, but only three cytokines were elevated by FLASH-RT, suggesting a milder immunological response to radiation. It is worth mentioning that when immunocompetent mice were given 28 Gy of radiation to subcutaneously injected glioblastoma cells, the anti-tumor effects persisted for up to 100 days after both CON-RT and FLASH irradiation [53]. The effects of FLASH and CON-RT on the maintenance of brain function in young mice after whole-brain irradiation were examined in a recent study by Alaghband and colleagues. The 8 Gy dose utilized in the study was administered at a rate of 4.4×10^6 Gy/s for FLASH and 0.077 Gy/s for CON-RT radiation. The findings demonstrated that, in contrast to traditional radiation, FLASH therapy maintained the neurogenic niche, hippocampal neurogenesis, and healthy amounts of growth hormone. Furthermore, 4 months after treatment, FLASH therapy produced normal or very normal results on tests of socializing, memory, and learning, whereas traditional radiation therapy resulted in notable deficiencies in these domains. According to these results, FLASH therapy may be a viable strategy to reduce the long-term negative effects of brain radiation when treating children's brain tumors [54]. In contrast to CON-RT at 4 Gy (<0.07 Gy/s), Chabi et al. reported that mice with normal hematopoiesis and lymphoblastic leukemia were subjected to 4 Gy of FLASH radiation (200 Gy/s). When FLASH was tested over an extended period in two xenografts produced from patients (PDX), it was found to reduce leukemic cells more than CON. In mice administered with FLASH, tumor control persisted for up to 7 weeks. Additionally, FLASH showed a more potent inhibitory effect on the development of T-cell acute leukemia cells. FLASH, however, slowed the growth of the tumor in two additional PDXs, whereas CON produced total remission. This implies that the response to FLASH in T-cell acute leukemia may be influenced by intrinsic variables. The FLASH-sensitive individuals shared comparable genetic defects, which may have contributed to their sensitivity, according to genomic studies. Furthermore, in three out of four cases, FLASH therapy prevented tumor growth and retained hematopoietic stem/progenitor cells, while CON-treated cells were overtaken by leukemia infiltration [46]. Levy et al. examined the comparison of FLASH (216 Gy/s) and CON-RT (0.079 Gy/s) for complete abdominal irradiation in mice with ovarian cancer to determine the impact of radiation-induced GI damage in the treatment of metastatic ovarian cancer and normal mice. Compared to CONV RT, they discovered that FLASH induced less gastrointestinal syndrome mortality. In contrast to the CON-treated mice, 90% of the FLASH-treated mice survived and recovered, even though both classifications of mice exposed to 16 Gy decreased their body weight by over 25%. The intestinal mucosa of FLASH-treated animals was similar to that of non-irradiated mice, while both groups of normal mice displayed mucosal injury. Furthermore, crypt base cells exposed to a sub-lethal dosage of 14 Gy FLASH saw decreased apoptosis and DNA damage, which improved intestinal function and epithelial integrity preservation when compared to CON-RT [55]. In contrast to CON-RT, Ruan et al. examined the effects of various dosage rates and temporal pulses. FLASH dosage rates ranging from 7.5 to 12.5 Gy successfully preserved gastrointestinal function in their mouse models. They discovered that, comparable to several pulses with a 216 Gy/s average rate, one pulse at the maximum dosage rate offered the intestinal crypt the most protection. Furthermore, the effects of radiation on the gut microbiota varied; CON-RT resulted in a lower microbial richness, which was associated with more intestinal damage [56].

15.6 Clinical Applications of FLASH Radiotherapy

FLASH-RT is now being tested in a small number of experiments. The first trial included a 75-year-old man who had skin lesions that were unpleasant and poorly tolerated. He had previously been treated with CON-RT for skin lymphoma that was CD30+ T-cell and resistant to many drugs. Switching to FLASH-RT resulted in similar tumor control while also reducing toxicity. It is worth mentioning that the assessments conducted after treatment showed positive results, with no noticeable decrease in skin thickness when compared to conventional radiation. These detections provide initial validation that FLASH-RT is both feasible and safe, and they highlight its potential to preserve normal tissue integrity while improving tumor control [20]. The findings of numerous prior preclinical investigations have demonstrated that FLASH-RT is accompanied by the development of the FLASH effect; therefore, in theory, FLASH-RT is the best option for treating patients with advanced, incurable tumors since it can manage them with minimal harm. In recent years, many studies have shown that FLASH-RT has the potential to be beneficial in the treatment of various types of solid tumors [9]. At present, the most harmful adverse effect of radiation therapy for brain tumors is the development of irreparable traumatic brain injury, which has a significant impact on the quality of life for patients who survive. When comapred with FLASH-RT, it has shown promising results, particularly in time of minimizing neurocognitive adverse effects, lengthening memory time and decreasing the release of pro-inflammatory cytokines that promote inflammation [57]. It is extremely important to safeguard the way patients' brains work and enhance the quality of life of those with brain tumors. Radioactive pulmonary fibrosis is a severely harmful adverse effect of lung cancer radiation therapy, and there is no effective medication available to treat it. It has been demonstrated that FLASH-RT can stop the typical radiotoxicity of the lungs from occurring by preventing the initiation of the Smad/TGF-β cascade [58]. Additionally, FLASH-RT can be used to treat gastrointestinal cancers, including colorectal cancer. Following comprehensive abdominal radiation, researchers assessed intestinal damage and discovered that FLASH-RT might markedly reduce the depletion of cells that are growing in the intestinal crypts, hence mitigating the acute intestinal reaction. Research utilizing animal models has demonstrated that by inhibiting macrophage growth, FLASH-RT lowers radiation enteritis. FLASH-RT has been used to treat superficial tumors, such as subcutaneous cancers and skin cancers [16]. FLASH-RT mitigates skin toxicity by reducing the synthesis of cytokines that promote inflammation, including TGF-β1. These findings underscore the FLASH-RT's significant therapeutic potential for many solid malignancies. FLASH-RT is expected to be an effective prospect for enhancing cancer therapy while reducing detrimental effects on normal tissues [59]. Nevertheless, further clinical investigations and experiments are necessary to thoroughly ascertain its effectiveness and safety across all cancer types.

15.7 Conclusion and Future Directions

The emergence of FLASH-RT marks a transformative advancement in cancer treatment, offering a rapid and effective alternative to conventional methods. By delivering ultra-high doses of radiation in fractions of a second, FLASH therapy efficiently targets tumors while minimizing harm to healthy cells. This innovative approach has shown a unique biological response, enhancing tumor control and reducing acute side effects, particularly for tumors resistant to traditional treatments. Current preclinical and early clinical studies suggest that FLASH-RT can enhance patient outcomes by providing a less harmful and more efficient treatment option. The future of this technique looks promising, with ongoing advancements in accelerator technology and a growing body of research supporting its efficacy and safety. Continued clinical trials are essential to validate the benefits observed in preclinical models and establish standardized protocols for its application across various tumor types. Understanding the mechanisms behind the FLASH effect, particularly its impact on DNA repair and immune modulation, may lead to its integration with other therapeutic approaches, including immunotherapy and chemotherapy. Ultimately, FLASH-RT not only holds the potential to reshape cancer care by improving efficacy and reducing toxicity but also represents a significant leap forward in the quest for more effective cancer treatments. Continued investment in research and experimental trials will be crucial for reaching its maximum potential, making FLASH therapy a mainstay in oncology practice.

References

1. Sahu RK, Ruhi S, Jeppu AK, Al-Goshae HA, Syed A, Nagdev S, et al. Malignant mesothelioma tumors: molecular pathogenesis, diagnosis, and therapies accompanying clinical studies. Front Oncol. 2023;13:1204722.
2. Mee T, Kirkby NF, Defourny NN, Kirkby KJ, Burnet NG. The use of radiotherapy, surgery and chemotherapy in the curative treatment of cancer: results from the FORTY (Favourable Outcomes from RadioTherapY) project. Br J Radiol. 2023;96(1152):20230334.
3. Bhattacharjee B, Kuppusamy S, Ghose Bishwas NR, Rynjah D, Sahu RK. Advanced smart nanomaterial-based system for treatment of cervical cancer: a systematic review. J Drug Deliv Sci Technol [Internet]. 2024 [cited 2025 Jan 6];102:106327. Available from: https://linkinghub.elsevier.com/retrieve/pii/S1773224724009961
4. Lv Y, Lv Y, Wang Z, Lan T, Feng X, Chen H, et al. FLASH radiotherapy: a promising new method for radiotherapy (review). Oncol Lett. 2022;24(6):419.
5. Kim JS, Kim HJ. FLASH radiotherapy: bridging revolutionary mechanisms and clinical frontiers in cancer treatment—a narrative review. Ewha Med J. 2024;47(4):e54.
6. Carlos-Reyes A, Muñiz-Lino MA, Romero-Garcia S, López-Camarillo C, Hernández-de La Cruz ON. Biological adaptations of tumor cells to radiation therapy. Front Oncol. 2021;11:718636.
7. Alhaddad L, Osipov AN, Leonov S. FLASH radiotherapy: benefits, mechanisms, and obstacles to its clinical application. IJMS. 2024;25(23):12506.
8. Geirnaert F, Kerkhove L, Montay-Gruel P, Gevaert T, Dufait I, De Ridder M. Exploring the metabolic impact of FLASH radiotherapy. Cancers. 2025;17(1):133.

9. Yan O, Wang S, Wang Q, Wang X. FLASH radiotherapy: mechanisms of biological effects and the therapeutic potential in cancer. Biomol Ther. 2024;14(7):754.
10. Eggold JT, Chow S, Melemenidis S, Wang J, Natarajan S, Loo PE, et al. Abdominopelvic FLASH irradiation improves PD-1 immune checkpoint inhibition in preclinical models of ovarian cancer. Mol Cancer Ther. 2022;21(2):371–81.
11. Wang Y, Qi SN, Bi N, Li YX. FLASH radiotherapy combined with immunotherapy: from biological mechanisms to blockbuster therapeutics. Transl Oncol. 2025;51:102183.
12. Ahire V, Ahmadi Bidakhvidi N, Boterberg T, Chaudhary P, Chevalier F, Daems N, et al. Radiobiology of combining radiotherapy with other cancer treatment modalities. In: Baatout S, editor. Radiobiology textbook [internet]. Cham: Springer; 2023 [cited 2025 Jan 20]. p. 311–86. Available from: https://link.springer.com/10.1007/978-3-031-18810-7_6
13. Schulte R, Johnstone C, Boucher S, Esarey E, Geddes CGR, Kravchenko M, et al. Transformative technology for FLASH radiation therapy. Appl Sci. 2023;13(8):5021.
14. Taylor PA, Moran JM, Jaffray DA, Buchsbaum JC. A roadmap to clinical trials for FLASH. Med Phys. 2022;49(6):4099–108.
15. Limoli CL, Vozenin MC. Reinventing radiobiology in the light of FLASH radiotherapy. Ann Rev Cancer Biol. 2023;7(1):1–21.
16. Giannini N, Gadducci G, Fuentes T, Gonnelli A, Di Martino F, Puccini P, et al. Electron FLASH radiotherapy in vivo studies. A systematic review. Front Oncol. 2024;14:1373453.
17. Shukla S, Saha T, Rama N, Acharya A, Le T, Bian F, et al. Ultra-high dose-rate proton FLASH improves tumor control. Radiother Oncol. 2023;186:109741.
18. Alrasheed AS, Aleid AM, Alharbi RA, Alhodibi MH, Alhussain AA, Alessa AA, et al. Brainstem toxicity following proton beam radiation therapy in pediatric brain tumors: a systematic review and meta-analysis. Cancers. 2024;16(21):3655.
19. Hao J, Magnelli A, Godley A, Yu JS. Use of a linear accelerator for conducting in vitro radiobiology experiments. JoVE. 2019;147:59514.
20. Bourhis J, Sozzi WJ, Jorge PG, Gaide O, Bailat C, Duclos F, et al. Treatment of a first patient with FLASH-radiotherapy. Radiother Oncol. 2019;139:18–22.
21. Taylor E, Létourneau D. How quickly does FLASH need to be delivered? A theoretical study of radiolytic oxygen depletion kinetics in tissues. Phys Med Biol [Internet]. 2024 [cited 2025 Jan 29];69(11):115008. Available from: https://iopscience.iop.org/article/10.1088/1361-6560/ad3e5e
22. Ito A, Kitabatake S, Furuichi W, Takase N, Nakahara T, Akiyama T, et al. LET dependence of 8-hydroxy-2′-deoxyguanosine (8-OHdG) generation in mammalian cells under air-saturated and hypoxic conditions: a possible experimental approach to the mechanism of the decreasing oxygen effect in the high-LET region. Radiat Res [Internet]. 2024 [cited 2025 Jan 29];201(3):189. Available from: https://bioone.org/journals/radiation-research/volume-201/issue-3/RADE-23-00046.1/LET-Dependence-of-8-Hydroxy-2-deoxyguanosine-8-OHdG-Generation/10.1667/RADE-23-00046.1.full
23. Berry RJ, Hall EJ, Forster DW, Storr TH, Goodman MJ. Survival of mammalian cells exposed to x rays at ultra-high dose-rates. Br J Radiol. 1969;42(494):102–7.
24. Zhang Q, Gerweck LE, Cascio E, Yang Q, Huang P, Niemierko A, et al. Proton FLASH effects on mouse skin at different oxygen tensions. Phys Med Biol. 2023;68(5):055010.
25. Hageman E, Che PP, Dahele M, Slotman BJ, Sminia P. Radiobiological aspects of FLASH radiotherapy. Biomolecules. 2022;12(10):1376.
26. Favaudon V, Caplier L, Monceau V, Pouzoulet F, Sayarath M, Fouillade C, et al. Ultrahigh dose-rate FLASH irradiation increases the differential response between normal and tumor tissue in mice. Sci Transl Med. 2014;6(245):245ra93.
27. Vozenin MC, De Fornel P, Petersson K, Favaudon V, Jaccard M, Germond JF, et al. The advantage of FLASH radiotherapy confirmed in mini-pig and cat-cancer patients. Clin Cancer Res. 2019;25(1):35–42.

28. Matuszak N, Suchorska WM, Milecki P, Kruszyna-Mochalska M, Misiarz A, Pracz J, et al. FLASH radiotherapy: an emerging approach in radiation therapy. Rep Pract Oncol Radiother. 2022;27(2):344–51.
29. Lagzda A, Angal-Kalinin D, Jones J, Aitkenhead A, Kirkby KJ, MacKay R, et al. Influence of heterogeneous media on Very High Energy Electron (VHEE) dose penetration and a Monte Carlo-based comparison with existing radiotherapy modalities. Nucl Instrum Methods Phys Res B [Internet]. 2020 [cited 2025 Jan 29];482:70–81. Available from: https://linkinghub.elsevier.com/retrieve/pii/S0168583X20304043
30. Montay-Gruel P, Bouchet A, Jaccard M, Patin D, Serduc R, Aim W, et al. X-rays can trigger the FLASH effect: ultra-high dose-rate synchrotron light source prevents normal brain injury after whole brain irradiation in mice. Radiother Oncol. 2018;129(3):582–8.
31. Beyreuther E, Brand M, Hans S, Hideghéty K, Karsch L, Leßmann E, et al. Feasibility of proton FLASH effect tested by zebrafish embryo irradiation. Radiother Oncol. 2019;139:46–50.
32. Buonanno M, Grilj V, Brenner DJ. Biological effects in normal cells exposed to FLASH dose rate protons. Radiother Oncol. 2019;139:51–5.
33. Schmid TE, Dollinger G, Hable V, Greubel C, Zlobinskaya O, Michalski D, et al. Relative biological effectiveness of pulsed and continuous 20 MeV protons for micronucleus induction in 3D human reconstructed skin tissue. Radiother Oncol. 2010;95(1):66–72.
34. Zlobinskaya O, Siebenwirth C, Greubel C, Hable V, Hertenberger R, Humble N, et al. The effects of ultra-high dose rate proton irradiation on growth delay in the treatment of human tumor xenografts in nude mice. Radiat Res. 2014;181(2):177–83.
35. Liu F, Shi J, Zha H, Li G, Li A, Gu W, et al. Development of a compact linear accelerator to generate ultrahigh dose rate high-energy X-rays for FLASH radiotherapy applications. Med Phys. 2023;50(3):1680–98.
36. Cao X, Zhang R, Esipova TV, Allu SR, Ashraf R, Rahman M, et al. Quantification of oxygen depletion during FLASH irradiation in vitro and in vivo. Int J Radiat Oncol Biol Phys. 2021;111(1):240–8.
37. Scarmelotto A, Delprat V, Michiels C, Lucas S, Heuskin AC. The oxygen puzzle in FLASH radiotherapy: a comprehensive review and experimental outlook. Clin Transl Radiat Oncol [Internet]. 2024 Nov [cited 2025 Jan 29];49:100860. Available from: https://linkinghub.elsevier.com/retrieve/pii/S240563082400137X
38. Sunnerberg JP, Tavakkoli AD, Petusseau AF, Daniel NJ, Sloop AM, Schreiber WA, et al. Oxygen consumption in vivo by ultra-high dose rate electron irradiation depends upon baseline tissue oxygenation. Int J Radiat Oncol Biol Phys [Internet]. 2024 [cited 2025 Jan 29];S0360301624035107. Available from: https://linkinghub.elsevier.com/retrieve/pii/S0360301624035107
39. Wardman P. Radiotherapy using high-intensity pulsed radiation beams (FLASH): a radiation-chemical perspective. Radiat Res. 2020;194(6):607–17.
40. Jansen J, Beyreuther E, García-Calderón D, Karsch L, Knoll J, Pawelke J, et al. Changes in Radical Levels as a Cause for the FLASH effect: Impact of beam structure parameters at ultra-high dose rates on oxygen depletion in water. Radiother Oncol [Internet]. 2022 [cited 2025 Jan 29];175:193–6. Available from: https://linkinghub.elsevier.com/retrieve/pii/S0167814022042451
41. Friedl AA, Prise KM, Butterworth KT, Montay-Gruel P, Favaudon V. Radiobiology of the FLASH effect. Med Phys. 2022;49(3):1993–2013.
42. Zakaria AM, Colangelo NW, Meesungnoen J, Azzam EI, Plourde MÉ, Jay-Gerin JP. Ultra-high dose-rate, pulsed (FLASH) radiotherapy with carbon ions: generation of early, transient, highly oxygenated conditions in the tumor environment. Radiat Res. 2020;194(6):587–93.
43. Fouillade C, Curras-Alonso S, Giuranno L, Quelennec E, Heinrich S, Bonnet-Boissinot S, et al. FLASH irradiation spares lung progenitor cells and limits the incidence of radio-induced senescence. Clin Cancer Res. 2020;26(6):1497–506.
44. Hornsey S, Bewley DK. Hypoxia in mouse intestine induced by electron irradiation at high dose-rates. Int J Radiat Biol Relat Stud Phys Chem Med. 1971;19(5):479–83.

45. Field SB, Bewley DK. Effects of dose-rate on the radiation response of rat skin. Int J Radiat Biol Relat Stud Phys Chem Med. 1974;26(3):259–67.
46. Chabi S, To THV, Leavitt R, Poglio S, Jorge PG, Jaccard M, et al. Ultra-high-dose-rate FLASH and conventional-dose-rate irradiation differentially affect human acute lymphoblastic leukemia and normal hematopoiesis. Int J Radiat Oncol Biol Phys. 2021;109(3):819–29.
47. Alaghband Y, Cheeks SN, Allen BD, Montay-Gruel P, Doan NL, Petit B, et al. Neuroprotection of radiosensitive juvenile mice by ultra-high dose rate FLASH irradiation. Cancers (Basel). 2020;12(6):1671.
48. Venkatesulu BP, Sharma A, Pollard-Larkin JM, Sadagopan R, Symons J, Neri S, et al. Ultra high dose rate (35 Gy/sec) radiation does not spare the normal tissue in cardiac and splenic models of lymphopenia and gastrointestinal syndrome. Sci Rep. 2019;9(1):17180.
49. Loo BW, Schuler E, Lartey FM, Rafat M, King GJ, Trovati S, et al. (P003) delivery of ultra-rapid flash radiation therapy and demonstration of normal tissue sparing after abdominal irradiation of mice. Int J Radiat Oncol Biol Phys [Internet]. 2017 [cited 2025 Jan 29];98(2):E16. Available from: https://linkinghub.elsevier.com/retrieve/pii/S0360301617305047
50. Beddok A, Fouillade C, Quelennec E, Favaudon V. OC-0030: in vitro study of FLASH vs. conventional dose-rate irradiation: cell viability and DNA damage repair. Radiother Oncol. 2017;123:S9–10.
51. Auer S, Hable V, Greubel C, Drexler GA, Schmid TE, Belka C, et al. Survival of tumor cells after proton irradiation with ultra-high dose rates. Radiat Oncol. 2011;6(1):139.
52. Favaudon V, Fouillade C, Vozenin MC. Radiothérapie «flash» à très haut débit de dose: un moyen d'augmenter l'indice thérapeutique par minimisation des dommages aux tissus sains? Cancer/Radiothérapie. 2015;19(6–7):526–31.
53. Liljedahl E, Konradsson E, Gustafsson E, Jonsson KF, Olofsson JK, Ceberg C, et al. Long-term anti-tumor effects following both conventional radiotherapy and FLASH in fully immunocompetent animals with glioblastoma. Sci Rep. 2022;12(1):12285.
54. Alaghband Y, Cheeks SN, Allen BD, Montay-Gruel P, Doan NL, Petit B, et al. Neuroprotection of radiosensitive juvenile mice by ultra-high dose rate FLASH irradiation. Cancers. 2020;12(6):1671.
55. Levy K, Natarajan S, Wang J, Chow S, Eggold JT, Loo PE, et al. Abdominal FLASH irradiation reduces radiation-induced gastrointestinal toxicity for the treatment of ovarian cancer in mice. Sci Rep. 2020;10(1):21600.
56. Ruan JL, Lee C, Wouters S, Tullis IDC, Verslegers M, Mysara M, et al. Irradiation at ultra-high (FLASH) dose rates reduces acute normal tissue toxicity in the mouse gastrointestinal system. Int J Radiat Oncol Biol Phys. 2021;111(5):1250–61.
57. Montay-Gruel P, Acharya MM, Petersson K, Alikhani L, Yakkala C, Allen BD, et al. Long-term neurocognitive benefits of FLASH radiotherapy driven by reduced reactive oxygen species. Proc Natl Acad Sci USA. 2019;116(22):10943–51.
58. Ramia P, Bodgi L, Mahmoud D, Mohammad MA, Youssef B, Kopek N, et al. Radiation-induced fibrosis in patients with head and neck cancer: a review of pathogenesis and clinical outcomes. Clin Med Insights Oncol. 2022;16:11795549211036898.
59. Chow JCL, Ruda HE. Mechanisms of action in FLASH radiotherapy: a comprehensive review of physicochemical and biological processes on cancerous and normal cells. Cells. 2024;13(10):835.

Chapter 16
Novel Approaches to Counteract Radiotherapy Toxicities: Focusing on Radioprotectors and Radiation Mitigators

Seyed Mohammad Javad Mortazavi, Reza Fardid, James S. Welsh, Joseph J. Bevelacqua, and Fateme Zaker

Abstract Radiotherapy plays a crucial role in treating cancer, but it often comes with the unfortunate side effect of damaging healthy tissues. This damage is largely due to the production of reactive oxygen species (ROS), which can cause cellular injury. In this chapter, we dive into the world of radioprotectors and mitigators—substances that aim to protect healthy cells while still allowing radiation to effectively target cancer cells. One of the most well-known radioprotectors is amifostine, the only FDA-approved option. It works by neutralizing free radicals and reducing

---

Authors S. M. J. Mortazavi, Reza Fardid, and James S Welsh have equally contributed to this work.

S. M. J. Mortazavi (✉)
Ionizing and Non-Ionizing Radiation Protection Research Center (INIRPRC), Shiraz University of Medical Sciences, Shiraz, Iran

Medical Physics and Engineering Department, School of Medicine, Shiraz University of Medical Sciences, Shiraz, Iran

R. Fardid
Ionizing and Non-Ionizing Radiation Protection Research Center (INIRPRC), Shiraz University of Medical Sciences, Shiraz, Iran

Radiology and Radiobiology Department, School of Paramedical Sciences, Shiraz University of Medical Sciences, Shiraz, Iran

J. S. Welsh
Department of Radiation Oncology, dward Hines Jr VA Hospital, Hines, IL, USA

Department of Radiation Oncology, Stritch School of Medicine, Loyola University, Chicago, IL, USA

J. J. Bevelacqua
Bevelacqua Resources, Richland, Washington, DC, USA

F. Zaker
Ionizing and Non-Ionizing Radiation Protection Research Center (INIRPRC), Shiraz University of Medical Sciences, Shiraz, Iran

F. Taghizadeh-Hesary et al. (eds.), *From Radiobiology to Radiation Oncology*,
https://doi.org/10.1007/978-3-031-95724-6_16

toxicity, but its use is limited due to side effects and the need for intravenous administration. Vitamin C, renowned for its antioxidant properties, has shown promise in protecting organs from radiation damage, though finding the right dosage is still a challenge. Another fascinating compound is curcumin, derived from turmeric. It serves a dual purpose: protecting healthy cells while making cancer cells more vulnerable to radiation. However, its clinical use is hindered by its poor bioavailability, and much research is underway to improve how the body absorbs it. Similarly, hesperidin, a flavonoid found in citrus fruits, has shown potential in protecting the lungs and liver from radiation injury. One of the more interesting compounds is melatonin, a hormone best known for regulating sleep cycles. Beyond that, it has anti-inflammatory and antioxidant effects, which could contribute to its radioprotective properties. An ideal radioprotector would effectively shield normal cells from damage without hindering the radiation's ability to kill cancer cells. Achieving this delicate balance depends on factors like how the substances are taken up by cells, how they're activated, and the unique environments in normal versus cancerous tissues. While animal studies have shown these substances hold promise, more rigorous clinical trials are needed to fully establish their safety, effectiveness, and real-world potential. This chapter highlights the latest advances in radioprotective strategies and the hope they bring for improving cancer treatments. If research continues to progress, we could see major breakthroughs in cancer care, helping reduce side effects and improve the overall quality of life for patients undergoing radiotherapy.

Keywords Radioprotectors · Radiotherapy · Reactive oxygen species (ROS) · Antioxidants · Cancer treatment

16.1 New Biological Ways to Reduce Radiotherapy Side Effects

Humans are continuously exposed to ionizing radiation (IR) from both natural sources, such as cosmic rays, and artificial ones, including medical imaging and radiotherapy [1]. Advances in radiation technology have significantly expanded its applications in medicine, particularly in diagnostic and therapeutic procedures [2, 3]. Despite its indispensable role in modern medicine, IR poses risks to healthy tissues [4]. A major concern is the generation of reactive oxygen species (ROS) during radiation exposure, which induces oxidative damage in cellular macromolecules [5, 6]. This oxidative stress is also implicated in cancer development. As radiation interacts with living tissues, it generates reactive oxygen species (ROS), such as superoxide (O_2^{-}•), hydrogen peroxide (H_2O_2), hydroxyl radicals (OH•), and nitrogen dioxide radicals (NO_2•). These reactive molecules can damage DNA, leading to cellular dysfunction or death [7].

To mitigate radiation-induced damage, radioprotectors with free radical-scavenging properties have been investigated [8, 9]. When IR interacts with cells, it induces ROS formation, which compromises essential biomolecules, including DNA, proteins, and membrane lipids [10]. Free radicals also trigger lipid peroxidation, leading to the accumulation of harmful byproducts such as malondialdehyde (MDA) and lipid peroxides [11]. MDA is particularly concerning due to its cytotoxic effects and its role in suppressing protective enzymes, acting as a co-carcinogen [12].

Radiotherapy is a cornerstone of cancer treatment, with studies suggesting that up to 70% of cancer patients require it. It delivers high doses of ionizing radiation (typically 50–80 Gy) over multiple sessions to control or eradicate tumors while minimizing damage to adjacent healthy tissues. However, achieving this balance remains a challenge, especially in highly proliferative organs susceptible to radiation toxicity [13–15]. Unintended exposure of normal tissues can lead to significant functional impairment [16]. The potential of certain compounds to reduce radiation-induced damage was first recognized in 1949 [17]. Research has since identified DNA damage, primarily mediated by free radicals, as the key mechanism underlying radiation-induced cellular injury and apoptosis. This has led to the development of radioprotective agents, including free radical scavengers, which may be used prophylactically or in emergency settings, such as during radiotherapy, for radiation workers, astronauts, or in cases of accidental exposure [8, 9, 18, 19].

Radiotherapy remains essential for both curative and palliative cancer treatments. However, IR-induced ROS formation can lead to oxidative damage of proteins, lipids, and nucleic acids. Among these, lipid components of cell membranes are particularly vulnerable due to lipid peroxidation, which affects membrane integrity, enzyme activity, and cell viability. Radiation exposure also depletes the body's total antioxidant capacity, exacerbating oxidative stress. While mammals possess intrinsic antioxidant defenses, they are often insufficient, highlighting the importance of dietary antioxidants. Several naturally derived antioxidants have demonstrated the ability to reduce free radical damage and protect biomolecules from radiation-induced oxidative stress [7, 20, 21].

Despite its efficacy, radiotherapy is often limited by collateral damage to healthy tissues. IR can directly disrupt atomic structures within cells, causing chemical and biological alterations. Indirectly, it induces water radiolysis, generating ROS that damage DNA, proteins, and lipids. The resulting oxidative stress has been linked to various degenerative conditions, including cardiovascular diseases, cancer, arthritis, atherosclerosis, asthma, and renal and hepatic dysfunction. Both direct and indirect radiation effects trigger a cascade of biochemical and molecular events that determine whether cells undergo repair or permanent damage leading to apoptosis [22–24]. Radioactive iodine therapy (RAIT), commonly used for thyroid conditions, also has associated complications. These include nasolacrimal duct obstruction [25], salivary gland inflammation (sialadenitis) [26], bone marrow suppression [27], and, in some cases, secondary malignancies [28, 29]. Studies suggest that oxidative stress in non-thyroidal tissues expressing the iodide transporter contributes to these adverse effects [30].

The damaging effects of low-linear energy transfer (LET) radiation are primarily driven by the generation of highly reactive free radicals with extremely short lifespans [31]. In oxygen-rich environments, these radicals lead to ROS formation, resulting in DNA strand breaks, protein oxidation, and membrane disruption. Oxidative DNA damage is a key driver of mutations and carcinogenesis [32]. Furthermore, transient free radicals play a crucial role in mediating radiation-induced apoptosis [33].

Certain tissues, including the gastrointestinal system, bone marrow, testis, tongue, lung, and microvascular structures, exhibit high radiosensitivity. When tumors are located within or near these organs, the risk of radiation-induced toxicity is substantial, significantly impacting patient quality of life. This is particularly evident in gastrointestinal cancers, where radiation exposure to the digestive tract leads to severe complications. Among gastrointestinal organs, the small intestine is the most radiosensitive, with the duodenum being especially vulnerable during treatments targeting nearby organs such as the liver, stomach, and pancreas. Pathological assessments have revealed that radiotherapy for gastric lymphoma can cause persistent damage to duodenal blood vessels and villi, leading to chronic inflammation, ulceration, and prolonged bleeding [34–39].

Total-body exposure to proton irradiation, whether accidental or therapeutic, can induce severe pathophysiological effects resembling acute radiation syndrome (ARS). ARS symptoms include anorexia, nausea, vomiting, diarrhea, leukopenia, and immune suppression. Damage to the intestinal mucosa, compromised barrier function, and reduced antioxidant capacity further increase susceptibility to infections and toxicities, particularly in patients receiving concurrent chemotherapy or immunotherapy. At the cellular level, IR-induced DNA double-strand breaks result in genomic instability, ROS overproduction, and activation of molecular pathways leading to tissue injury and cell death. Given the association between compromised intestinal barrier function and inflammatory diseases, targeting stress-induced intestinal damage could offer therapeutic benefits. However, the toxicity of many synthetic radioprotectors has limited their clinical utility, driving interest in natural herbal products as safer, effective alternatives for radioprotection [40–44] (Fig. 16.1).

16.1.1 Amifostine

Amifostine (WR-2721) is the first FDA-approved broad-spectrum cytoprotective agent in the United States, widely used in clinical settings to protect healthy tissues during radiotherapy and chemotherapy. Initially developed for radiation protection in nuclear warfare, amifostine requires intravenous administration within 30 min before radiation exposure. However, its use is limited by side effects, including reversible hypotension, nausea, and vomiting, which reduce its practicality, particularly in military settings. The development of more convenient, effective, and safer forms of amifostine remains a priority [45–51]. Oral administration is preferable for

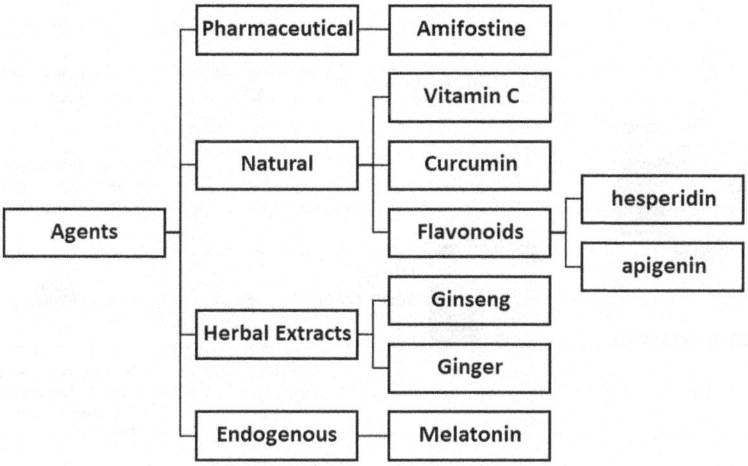

Fig. 16.1 Some of the most important radiation protectors discussed in this chapter

patient convenience, but amifostine lacks oral bioavailability. Methacrylic acid copolymer (MAC) L100-55, a pH-responsive polymer, has been used to improve drug delivery to the gastrointestinal tract by protecting compounds from degradation in stomach acid. Gula et al. developed enteric-coated amifostine microcapsules using Eudragit L100-55, demonstrating that oral administration of these microcapsules provided effective radioprotection compared to the uncoated drug. This approach enhances amifostine bioavailability by ensuring its release in the intestine [46, 52–57].

Amifostine, a prodrug initially developed under the U.S. Army's Anti-Radiation Drug Development Program, is an inorganic thiophosphate rapidly converted into its active form, WR-1065, by alkaline phosphatase in vascular endothelial cells. Studies indicate that amifostine selectively protects normal tissues during radiotherapy without conferring protection to tumors (Fig. 16.2). However, its effectiveness depends on precise timing relative to radiation exposure, as metabolism is required for activation. Additionally, significant adverse effects may limit its clinical utility in certain cases [58–61].

Unlike other cytoprotective agents that target specific organs, amifostine provides broad-spectrum protection across multiple tissues. Research has shown that it mitigates radiation-induced damage, enhances bone formation, and improves bone-implant stability in irradiated sites by reducing inflammation and preserving bone mineral integrity [62–67]. A comparative study evaluating radioprotective agents revealed that while both amifostine and NAC reduced apoptosis and provided tissue protection, melatonin demonstrated superior efficacy. In the melatonin-treated group, morphological damage was nearly fully repaired, with apoptotic index levels approaching those of the control group. The study did not identify an optimal administration method for amifostine, suggesting melatonin as a potential alternative due to fewer constraints and greater efficacy in certain scenarios [68].

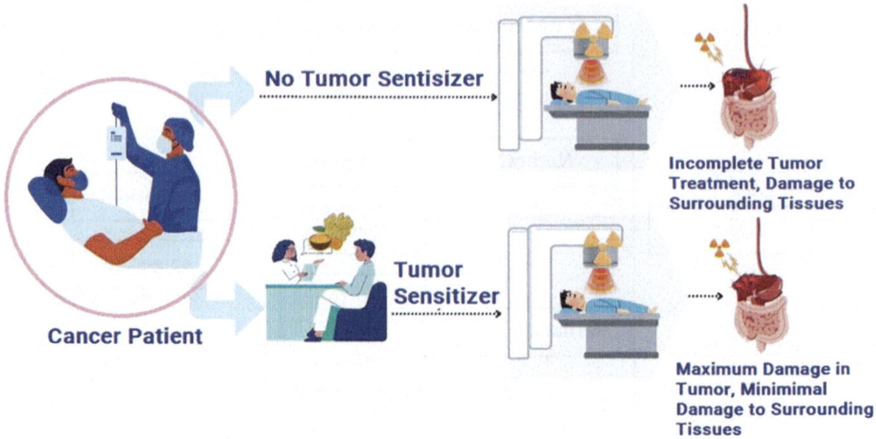

Fig. 16.2 It is essential that a tumor sensitizer like amifostine (WR2721) selectively shields normal tissues during radiotherapy without extending any protection to tumor cells

Histopathological analysis of prostate tissue showed that amifostine influenced MUC1 secretion, with expression levels varying based on radiation dose and fractionation. While MUC1 secretion was elevated in the AMF, AMF + RT, and RT groups compared to controls, further research is required to elucidate the implications of this effect [69].

16.1.2 Vitamin C

Vitamin C (ascorbic acid, $AscH_2$; ascorbate, $AscH^-$) is a water-soluble antioxidant with two ionizable hydroxyl groups [70]. Unlike most animals, humans, other primates, guinea pigs, and some bat species lack the ability to synthesize vitamin C and must obtain it from dietary sources [70, 71]. Extensive research highlights vitamin C's ability to protect DNA from ionizing radiation-induced damage [31]. Studies have demonstrated that pre-treatment with vitamin C significantly reduces radiation-induced gastrointestinal damage in animal models [72]. As a potent antioxidant and free radical scavenger, vitamin C enhances the antioxidant defense systems of the liver and kidneys, mitigating oxidative stress caused by radiation exposure [73, 74]. Its radioprotective properties have been confirmed in various studies, demonstrating its potential to reduce radiation-induced damage through multiple mechanisms [75–79]. The impact of vitamin C on radiotherapy outcomes appears to depend on dosage and cancer subtype. Lower doses of vitamin C, whether used alone or in combination with radiotherapy, influence cell proliferation in a cell-line-specific manner. In breast cancer patients undergoing radiotherapy, vitamin C intake was associated with a lower histological grade and body mass index (BMI) without increasing cancer recurrence risk. However, in some cases, low-dose vitamin C supplementation led to increased cancer cell proliferation and reduced radiosensitivity, suggesting that its effects on radiotherapy efficacy vary depending on tumor characteristics and vitamin C concentration [80]. A single dose of vitamin C administered within 24 h post-radiation exposure can mitigate high-dose ionizing radiation damage, such as that caused by solar particle events [81]. Combining vitamin C with adipose tissue-derived stem cell (AdSC) transplantation has shown promise in improving survival rates following lethal irradiation. This combined approach may help counteract tissue damage from chemotherapy and toxic agents. Mortazavi et al. advocate for the combined use of AdSCs and vitamin C as an effective treatment strategy for managing radiation-induced injuries [82]. This strategy is particularly relevant in space missions, where astronauts may encounter sudden solar particle events. NASA's success in cultivating vitamin C-rich vegetables aboard the International Space Station through the Veggie system provides a potential source of fresh antioxidants for astronauts. Further research is needed to optimize vitamin C supplementation for space travel, and AI-driven nutritional guidance may assist astronauts in selecting and growing the most suitable crops for optimal health [80]. Combining vitamin C with radiotherapy has shown minimal impact on colony formation, cell cycle arrest, and protein expression compared to radiotherapy alone [83]. However, high doses of vitamin C have demonstrated strong radioprotective effects, making it a promising candidate for developing non-toxic, affordable, and widely accessible radioprotectors for scenarios involving lethal radiation exposure. Comparative studies suggest that vitamin C provides greater radioprotection than vitamins A and E [84]. Vitamin C has been investigated for its ability to protect against radiation-induced liver damage. As a potent antioxidant, it neutralizes free radicals and supports enzymatic processes by reducing essential metal ions such as iron and copper. It also regenerates other antioxidants like vitamin E and modulates epigenetic factors such as TET enzymes. By reducing oxidative stress and enhancing hepatic antioxidant defenses, vitamin C may prevent radiation-induced liver fibrosis and other radiation-related injuries [85–87].

Radiation exposure increases lipid peroxidation, resulting in oxidative stress-related damage to DNA and cell membranes. Pre-treatment with vitamin C has been shown to reduce lipid peroxidation and improve wound healing in animal models of whole-body gamma radiation exposure. Additionally, vitamin C enhances antioxidant enzyme activity, including superoxide dismutase (SOD), catalase, and glutathione peroxidase (GPx), thereby reducing oxidative stress-related tissue damage [88–90].

Vitamin C's hepatoprotective effects are well-documented, as it restores normal liver enzyme levels (AST, ALT, and malondialdehyde [MDA]) and counteracts the toxic effects of radiation. Additionally, it prevents liver fibrosis, cirrhosis, and inflammation by enhancing antioxidant defenses and reducing lipid peroxidation [86, 91–102]. A study on radiation-induced liver injury found that fractional electron irradiation caused severe hepatic damage, including weight loss, reduced liver mass, elevated liver enzyme levels (ALT, AST, ALP, and bilirubin), increased inflammatory markers (IL-1β, IL-6, and TNF-α), and oxidative stress indicators (MDA). Histological analysis revealed hepatocyte degeneration, fibrosis, and impaired microvascular circulation. Vitamin C pre-treatment significantly mitigated these effects by reducing oxidative stress and inflammation while restoring the balance between cell proliferation and apoptosis. These findings suggest that vitamin C has potential for preventing radiation-induced liver fibrosis and improving radiotherapy outcomes, though further research is needed to validate its clinical application [103]. Radiation exposure also leads to kidney damage, as evidenced by increased uric acid, urea, creatinine, and thiobarbituric acid reactive substances (TBARS) levels, along with decreased antioxidant activity (SOD, CAT, and GPx). Histological and ultrastructural analyses confirmed radiation-induced renal damage. Treatment with daidzein and/or vitamin C reduced kidney damage by lowering oxidative stress and restoring antioxidant enzyme activity, with the combination proving more effective than either compound alone [104]. Similarly, UVC radiation exposure significantly elevated hepatic inflammatory markers (TNF-α, IL-1β, iNOS, and IDO-1), activated caspase-3, and induced DNA fragmentation. Histological and ultrastructural analyses confirmed these biochemical changes. Co-treatment with vitamins C and B12 mitigated radiation-induced liver damage, with vitamin C demonstrating superior radioprotection by reducing oxidative stress, inflammation, and DNA damage. These findings support the potential use of vitamin C and B12 as protective agents for workers exposed to UVC radiation [105].

16.1.3 Curcumin

Curcumin, a phenolic compound derived from the roots of Curcuma longa (turmeric), exhibits dual functionality as a radiosensitizer for tumors and a radioprotector for normal tissues [106]. It interacts with various molecular targets, influencing gene regulation and apoptosis [107]. Lycopene, a red, insoluble carotenoid found in many yellow and red fruits and vegetables, is well-documented for its potent free radical scavenging and anti-inflammatory properties [108]. Fardid et al. investigated the radioprotective effects of curcumin and lycopene at radiation doses of 2 and 6 Gy using a micronucleus assay. Their findings suggest that the combination of curcumin and lycopene provides significant radioprotection, particularly at lower radiation doses [109]. Curcumin possesses strong antioxidant, anti-carcinogenic, anti-mutagenic, and anti-inflammatory properties, contributing to its notable anticancer potential. It has been shown to inhibit tumor cell proliferation, regulate transcription factors, and enhance wound healing following radiation exposure. Recent research by Mosieniak et al. demonstrated that curcumin induces double-strand DNA breaks, leading to genetic instability. This process triggers signaling pathways that push tumor cells past cell cycle arrest, leading to mitotic catastrophe and subsequent cell death [110–114]. Curcumin has also demonstrated protective effects against genotoxic damage from environmental toxins such as arsenic and fluoride, reducing DNA damage in human peripheral blood lymphocytes. Additionally, it shields hepatic tissues from carbon tetrachloride (CCl_2) toxicity. However, its low bioavailability—due to poor intestinal absorption and rapid metabolism—limits its therapeutic efficacy. Studies have shown that intraperitoneal curcumin administration effectively activates the glutathione (GSH) antioxidant system, providing hepatic protection in vivo. Despite these limitations, clinical trials indicate that curcumin is well-tolerated in humans, with no dose-limiting toxicity observed even at doses as high as 10 g/day [115–117]. Curcumin's broad therapeutic benefits include applications in arthritis, asthma, inflammatory bowel disease, nephrotoxicity, psoriasis, diabetes, Alzheimer's disease, multiple sclerosis, neurodegenerative disorders, cardiovascular diseases, and cancer. It inhibits reactive oxygen species (ROS) formation, neutralizes free radicals, and enhances the activity of key antioxidant enzymes such as superoxide dismutase (SOD), catalase, and glutathione peroxidase (GSH-Px), thereby exerting significant protective effects against oxidative stress [113, 118–121]. Historically used in traditional medicine for treating inflammation, thrombosis, and liver disorders, curcumin has shown promise as a chemopreventive agent in Phase 1 clinical trials, demonstrating efficacy without significant adverse effects [122, 123]. Studies in mice have confirmed curcumin's radioprotective benefits, showing reduced chromosomal aberrations and oxidative stress markers following γ-irradiation. Ionizing radiation increases levels of TBARS, hydroperoxide (HP), xanthine oxidase (XO), DNA fragmentation, and caspase-3 activation, while depleting GSH and antioxidant enzymes (SOD, catalase, GPx). Curcumin administered before or after radiation exposure mitigates liver toxicity by lowering oxidative stress markers and restoring antioxidant levels, underscoring its potential in preventing γ-ray-induced cellular damage [124]. Özgen SÇ reported that 100% of irradiated subjects developed cataracts, whereas the cataract rate dropped to 40% in the curcumin-treated group, with only mild cataract formation (grades 1 and 2). The irradiation group exhibited decreased antioxidant enzyme activity and elevated malondialdehyde (MDA) levels, while curcumin treatment restored antioxidant enzyme levels and significantly reduced oxidative damage. These findings suggest that curcumin could be an effective radioprotective agent against radiation-induced cataract formation [125]. Radiotherapy-induced oxidative damage to the skin is characterized by increased MDA levels and reduced antioxidant enzyme activity (CAT, SOD, GSH-Px). However, pre- and post-treatment with curcumin significantly counteracted this damage, enhancing enzymatic antioxidant activity and preserving skin integrity. Further research is needed to validate curcumin's clinical efficacy in radiotherapy-induced skin toxicity [126].

In a study examining the effects of iodine-131 (I-131) exposure, the number of DNA double-strand breaks (DSBs) increased by 102.9% compared to controls. However, curcumin and trehalose supplementation significantly reduced DSB levels, with curcumin demonstrating superior radioprotective effects compared to trehalose [127]. In irradiated rats, body weight reduction and severe hepatic cell damage were observed. The radiation group exhibited hepatocyte edema, inflammatory infiltration, and vacuolar degeneration, whereas the curcumin-treated group showed significantly fewer pathological changes. Curcumin administration decreased serum levels of ALT, AST, ALP, LDH, and MDA while increasing total protein (TP), SOD, CAD, and GSH levels. It also reduced apoptosis and inflammation by inhibiting the NF-κB signaling pathway, suggesting its potential to mitigate radiation-induced liver damage [128].

16.1.4 Hesperidin

Hesperidin (HES) is a flavonoid widely found in citrus fruits, particularly in their peels and membranes. Unripe oranges contain more than 14% hesperidin, making citrus waste a valuable extraction source [129–134]. Hesperidin has diverse biological activities, including antimicrobial [135], antifungal [136], anticancer [137], anti-inflammatory [138], and anti-allergic properties [139]. It also exhibits lipid-lowering [140] and antioxidant effects [141], making it a promising natural compound for overall health. Animal studies indicate that oral hesperidin administration for 7 days before radiotherapy can reduce troponin-1 levels, a marker of cardiac injury, thereby protecting the heart from radiation-induced damage [142]. In rats exposed to 2 Gy and 8 Gy gamma radiations, hesperidin treatment effectively reduced inflammatory cell infiltration and prevented radiation-induced vascular damage, demonstrating its potential as a radioprotector for radiotherapy patients and radiation workers [143]. Fardid et al. found that hesperidin at doses of 50 and 100 mg/kg induced apoptosis by modulating the expression of Bax, Bcl-2, and their ratio, which regulate cell death [144]. Additionally, oral hesperidin administration protected against gamma radiation-induced lung damage and oxidative stress in rats, likely due to its ability to neutralize free radicals and stabilize cell membranes. Further studies confirmed hesperidin's role in reducing lung tissue inflammation, fibrosis, and mast cell activity, thereby mitigating radiation pneumonitis and fibrosis [144, 145]. A combination of diosmin and hesperidin provided significant radioprotective effects in albino rats, restoring antioxidant enzyme levels (SOD, GPx, GSH), reducing lipid peroxidation, DNA damage, and apoptosis. A dose of diosmin (200 mg/kg) and hesperidin (100 mg/kg) normalized antioxidant levels and mitigated radiation-induced liver damage. Histopathological analysis confirmed that pretreated livers were shielded from radiation-induced harm, and no metabolic disturbances or DNA damage were observed, highlighting the safety and efficacy of this combination [146]. Hesperidin at a concentration of 16.38 μM demonstrated optimal radioprotective effects, reducing genetic damage (micronuclei formation, chromosome aberrations, DNA fragmentation) and oxidative stress in lymphocytes exposed to gamma radiation. Pre-treatment significantly reduced radiation-induced lipid peroxidation and preserved antioxidant enzyme levels, further supporting its potential as a radioprotector [147]. Oral hesperidin administration (50–100 mg/kg for 7 days) significantly mitigated gamma radiation-induced tissue damage, as evidenced by improved biochemical and histological markers. Periodic acid–Schiff staining revealed reduced necrotic tissue damage, with a dose-dependent protective effect in hesperidin-treated animals [148].

Radiation exposure significantly increased MDA levels and decreased GSH levels in lens tissue, indicating oxidative stress. However, hesperidin co-administration markedly reduced MDA levels and restored GSH levels, suggesting its potential to prevent radiation-induced cataracts by reducing oxidative stress in lens tissue [149].

16.1.5 Apigenin

Apigenin, a flavonoid compound from the flavone family, is found in various fruits and vegetables, including bell peppers, broccoli, celery, onions, tomatoes, garlic, and guava, as well as in plant-based beverages like tea and wine. It is known for its potent antioxidant, anti-inflammatory, and anti-tumor properties [150–157]. Apigenin has demonstrated strong protein-inhibitory activity, with binding energies of −8.21 kcal/mol for p38 mitogen-activated protein kinases and −5.34 kcal/mol for cyclin-dependent kinase 4. It has also been shown to outperform established inhibitors of xanthine oxidase and DNA methyltransferase. Notably, apigenin and related flavonoids can inhibit oncogenic KRAS mutations at codon 12 (G12C, G12D, G12V), which are frequently implicated in solid tumors. Their high binding affinity to these mutated proteins provides a potential advantage over therapies targeting specific KRAS mutations, while their generally favorable safety profile further supports their potential clinical utility [158]. Apigenin has also demonstrated radioprotective effects in both in vitro and in vivo models. It reduces chromosomal damage and cytogenetic abnormalities in human lymphocytes exposed to radiation. Research by Begum et al. confirmed that apigenin protects against oxidative stress and hematopoietic damage caused by γ-radiation in mice. Additionally, it activates the Nrf2 pathway by promoting nuclear translocation of Nrf2, thereby enhancing the expression of genes involved in antioxidant defense [159–167].

In a study by Rithidech, apigenin-treated human lymphocyte cultures (2.5, 5.0, and 10 μg/mL) showed no significant increase in micronuclei (MN) frequency, whereas a higher concentration (25 μg/mL) resulted in an increase. However, in irradiated cultures, apigenin significantly reduced MN formation in a dose-dependent manner, suggesting a protective effect against radiation-induced genotoxicity [163].

Exposure to gamma radiation increases reactive oxygen species (ROS) production, apoptosis, and mitochondrial dysfunction in human peripheral blood lymphocytes. Pre-treatment with apigenin significantly mitigated these effects by reducing ROS levels, decreasing apoptosis, and preserving mitochondrial membrane potential (ΔΨm). Furthermore, apigenin preserved the antioxidant defense system by maintaining the activity of superoxide dismutase (SOD), catalase, glutathione peroxidase (GPx), and reduced glutathione (GSH) while inhibiting lipid peroxidation and micronuclei formation [168]. Histological analysis of mice exposed to 7 Gy of radiation revealed severe crypt-villus damage in the intestines. However, pre-treatment with apigenin significantly restored crypt-villus architecture and reduced intestinal injury. Apigenin modulated the expression of apoptotic markers (p53, p21, Bax, caspase-3, and caspase-9) and inhibited NF-κB activation in gastrointestinal tissue, suggesting that its radioprotective effects are mediated via NF-κB signaling modulation [169]. In C57 mice subjected to lethal irradiation, apigenin treatment enhanced survival, accelerated crypt-villus regeneration, and promoted epithelial cell differentiation. Apigenin also increased intestinal crypt cell proliferation while reducing apoptosis. It upregulated Nrf2 and HO-1 expression, reducing oxidative stress in irradiated tissues. These findings highlight apigenin's potential as a radioprotective agent in cancer treatment and nuclear exposure scenarios [170].

Administering apigenin at 15 mg/kg increased the radiation LD50 (lethal dose for 50% survival) from 8.2 to 10 Gy, with a dose-modifying factor of 1.21. In Swiss albino mice exposed to 7 Gy radiation, lipid peroxidation levels increased while antioxidant enzyme activity and hematological parameters (hemoglobin, white blood cell, and red blood cell counts) declined. Apigenin pre-treatment mitigated these changes, reducing lipid peroxidation, restoring antioxidant levels, and improving hematological status, further supporting its radioprotective potential [166].

16.1.6 Ginseng

Panax ginseng (PG), commonly known as Asian or Korean ginseng, is a traditional herbal remedy used to combat fatigue, enhance immunity, and improve metabolic health. Research has demonstrated that oral PG administration improves blood glucose control in diabetes [171, 172], enhances immune defense against infections [173], and reduces cancer risk [174]. Animal studies have further highlighted its ability to counteract oxidative stress in diabetic models [175]. Ginseng, derived from the steamed root of Panax ginseng Meyer, undergoes chemical modifications during processing that enhance the bioavailability of its active compounds, including ginsenosides, polysaccharides, peptides, polyacetylene alcohols, and fatty acids. Korean red ginseng is well-documented for its antibacterial, antiviral, antioxidant, anti-tumor, and immunomodulatory effects [176–178]. In 2012, China's Ministry of Health officially recognized ginseng as a food product, and its popularity has since expanded to countries such as the U.S., Canada, and Japan. The bioactive compounds in ginseng contribute to its wide-ranging biological effects, including anticancer, antioxidant, radioprotective, anti-diabetic, and immune-modulating properties [44, 179–182]. In the context of radioactive iodine therapy (RAIT), which is known to increase micronucleus (MN) frequency—a marker of chromosomal damage—ginseng has shown significant protective effects. Patients who received PG supplementation exhibited a reduction in MN frequency compared to control and placebo groups. Additionally, PG had no adverse impact on liver or kidney toxicity markers, suggesting its potential as a protective agent in RAIT for differentiated thyroid cancer patients [183]. Lee et al. demonstrated that both red and fermented red ginseng reduced intestinal inflammation and improved gut barrier integrity, while Seong et al. found that fermented wild ginseng alleviated colitis severity in a dextran sodium sulfate-induced model. These findings indicate that ginseng compounds possess anti-inflammatory and gut-protective properties. Although whole ginseng root offers health benefits, its absorption in the intestine is limited. In contrast, ginseng oligopeptides, which have higher bioavailability, may provide greater physiological benefits [184]. Animal studies suggest that Korean red ginseng improves survival rates, reduces mucositis severity, and minimizes histological damage to oral mucosa and submandibular glands. Its protective effects are attributed to inhibition of the caspase-3 apoptotic signaling pathway, preventing radiation-induced apoptosis in oral tissues [185]. Mice treated with ginseng oligopeptides exhibited reduced plasma LPS and DAO levels, markers of inflammation and oxidative stress, and improved intestinal barrier function. A 15-day pre-treatment with ginseng oligopeptides demonstrated radioprotective effects, while an additional 15-day post-treatment accelerated recovery from radiation-induced damage. These findings highlight ginseng oligopeptides as a promising therapeutic approach for radiation-induced hematopoietic, gastrointestinal, and oxidative damage in cancer patients [186]. Irradiation increased malondialdehyde (MDA), 8-hydroxy-2′-deoxyguanosine (8-OHdG), and caspase-3 levels while depleting GSH in kidney tissue. Severe renal damage, including tubular necrosis and vacuolization, was observed. However, ginseng-treated groups exhibited higher GSH levels and reduced oxidative stress markers, suggesting that ginseng may serve as a protective antioxidant against radiation-induced kidney damage [187]. X-irradiation exposure significantly increased MDA levels and decreased GSH activity in the intestines. Treatment with Korean red ginseng reduced oxidative damage and caspase-3 activation while enhancing GSH levels, demonstrating its potential to protect intestinal tissue from radiation-induced injury [188].

16.1.7 Ginger

Ginger (Zingiber officinale) is a widely used spice with a long history of medicinal applications, particularly in Ayurveda, the traditional Indian system of medicine. It contains numerous bioactive compounds, including terpenes (e.g., ar-curcumin, farnesene, zingiberene) and oxygenated monoterpenes (e.g., linalool, geraniol, borneol), which contribute to its antioxidant, anti-inflammatory, and free radical-scavenging properties [189–196]. Ginger's anti-inflammatory effects are primarily mediated by its ability to inhibit prostaglandin and leukotriene biosynthesis via suppression of 5-lipoxygenase synthetase activity. Additionally, ginger modulates NF-κB signaling by reducing inducible nitric oxide synthase (iNOS) expression and inhibiting mitogen-activated protein kinase (MAPK) phosphorylation, particularly ERK1/2 [197–200]. In a study evaluating radiation-induced kidney damage, daily administration of ginger extract for 10 days before radiation exposure (2, 4, or 8 Gy) significantly improved histological and biochemical markers of renal function. These findings suggest that ginger mitigates radiation-induced oxidative DNA damage and inflammation through its antioxidant and anti-inflammatory properties

[201]. Pre-treatment with Zingiber officinale extract (ZOE) significantly reduced radiation sickness severity and mortality in mice across all radiation doses. ZOE provided protection against both gastrointestinal (GI) and bone marrow syndromes, with a dose reduction factor of 1.15. The optimal protective dose was 10 mg/kg, which is 1/50 of the lethal dose (LD50) of 500 mg/kg. Radiation exposure increased lipid peroxidation and depleted glutathione (GSH) levels, but these effects were significantly reduced by ZOE pre-treatment. Additionally, ZOE exhibited dose-dependent antimicrobial activity against Pseudomonas aeruginosa, Salmonella typhimurium, Escherichia coli, and Candida albicans [189]. In a study on radiation-induced intestinal injury, gamma radiation exposure resulted in structural damage to jejunal tissue, including villus distortion, crypt cell proliferation, and inflammatory cell infiltration. Additionally, there was a marked reduction in goblet cells and an increase in collagen deposition. However, ginger-treated animals exhibited notable restoration of jejunal mucosal structure, suggesting its potential as a radioprotective agent for intestinal health [202].

16.1.8 Melatonin

Melatonin (N-acetyl-5-methoxytryptamine), the primary hormone secreted by the pineal gland, is a potent antioxidant and immune modulator. It functions as both a direct free radical scavenger and an indirect antioxidant by enhancing the body's endogenous antioxidant defense mechanisms [203–205]. Melatonin has been shown to effectively neutralize hydroxyl radicals (OH•), making it a promising candidate for radiation protection [206]. Melatonin is widely recognized for its role in sleep regulation and is FDA-approved for various clinical applications. It has strong antioxidant and anti-inflammatory properties with minimal adverse effects on normal cells and tissues. Specifically, melatonin has demonstrated efficacy in blocking radiation-induced pathways linked to oral mucositis [204, 206–213]. In addition to being synthesized in the pineal gland, melatonin is naturally present in various foods, including cereals, olives, walnuts, tomatoes, pineapples, ginger, and legumes. Its radioprotective effects stem from its ability to neutralize free radicals, enhance antioxidant enzyme activity, and inhibit apoptosis and inflammation. Studies suggest that melatonin protects multiple organs—including the eyes, brain, liver, spleen, skin, and lungs—from radiation-induced damage [214–222]. Melatonin's radioprotective mechanisms include its ability to directly scavenge hydroxyl radicals, thereby preventing oxidative damage to cellular macromolecules. Due to its high lipid solubility, melatonin readily crosses cell membranes and biological barriers, inhibiting lipid peroxidation and oxidative stress. Additionally, melatonin enhances the activity of superoxide dismutase (SOD) and glutathione peroxidase (GSH), while its metabolites—N1-acetyl-N2-formyl-5-methoxykynuramine (AFMK) and N-acetyl-5-methoxykynuramine (AMK)—also function as free radical scavengers [203, 223–231].

Studies on radiation-induced liver damage indicate that radiation exposure significantly increases malondialdehyde (MDA) levels while depleting glutathione (GSH). However, melatonin pre-treatment effectively reduced MDA levels and restored GSH levels, suggesting its protective role against radiation-induced oxidative stress in hepatic tissue [232]. Radiation exposure also upregulated TNF-α gene expression nearly five-fold in irradiated rats compared to controls. Melatonin pre-treatment significantly downregulated TNF-α expression, in addition to reducing MDA levels. These findings suggest that melatonin may help prevent radiation-induced spinal cord injury by inhibiting inflammatory signaling pathways [233]. In a study by Fardid et al., melatonin was found to mitigate the radiation-induced bystander effect, which can cause oxidative DNA damage and upregulation of key enzymes such as COX-2 and iNOS in non-targeted lung tissues. Melatonin reduced DNA damage in non-targeted cells, offering a promising strategy for protecting healthy tissues from radiation exposure [234]. Melatonin has also been shown to protect against radiation-induced gastrointestinal toxicity. In irradiated rats, symptoms of mucositis—including inflammation, cell death, Brunner's gland damage, and goblet cell loss—were observed. Melatonin treatment significantly reduced these effects, and when combined with metformin, further alleviated some radiation-induced toxicity. However, melatonin alone was more effective in preventing duodenal damage than the melatonin-metformin combination [235]. Radiotherapy-induced gut toxicity is associated with mitochondrial oxidative stress, impaired energy metabolism, and activation of the NLRP3 inflammasome. Oral melatonin gel treatment demonstrated protective effects in the small intestine by maintaining mitochondrial function and reducing inflammation. Melatonin-treated rats exhibited reduced NF-κB and NLRP3 inflammasome activation, improved mucosal integrity, and enhanced intestinal healing. These findings suggest that melatonin gel could be a valuable preventive therapy for gut toxicity in cancer patients undergoing radiotherapy [236]. Post-irradiation treatment with melatonin significantly reduced MDA levels while increasing antioxidant enzyme activities (SOD and catalase) in intestinal tissue. Pre-treatment with varying melatonin doses also provided significant tissue protection, with a 100 mg/kg dose being more effective than 50 mg/kg. These results highlight melatonin's ability to protect against radiation-induced intestinal damage by reducing oxidative stress and enhancing antioxidant enzyme activity [237]. In studies on radiation-induced lung injury, pre-treatment with melatonin—either via intraperitoneal injection or oral administration—significantly reduced MDA levels in red blood cells (RBCs), plasma, and lung tissue. Early-stage lung injury was associated with increased infiltration of lymphocytes and macrophages, but melatonin treatment reduced lung injury scores. These findings confirm melatonin's radioprotective effects, reducing lipid peroxidation and mitigating early- and late-stage radiation-induced lung damage [238] (Tables 16.1, 16.2, 16.3, 16.4, 16.5, 16.6, 16.7, 16.8, and 16.9).

Table 16.1 Overview of radioprotective agents

Agent	Mechanism of action	Strengths	Challenges
amifostine	Scavenges free radicals	FDA-approved, strong protection	Side effects, IV-only administration
Vitamin C	Boosts antioxidant defenses	Safe, widely available	Requires precise dosing for efficacy
Curcumin	Anti-inflammatory, antioxidant	Reduces fibrosis and cell death	Poor bioavailability
Hesperidin	Antioxidant, anti-inflammatory	Protects tissues from radiation damage	Limited human studies
Apigenin	Antioxidant	Potential radioprotector	Mechanism and clinical efficacy need further study
Ginseng	Antioxidant	Broad health benefits	Standardization of active components needed
Ginger	Antioxidant, anti-inflammatory, free radical scavenger	Protects multiple organ systems	Optimal dosing and long-term effects require further research
Melatonin	Scavenges radicals, boosts enzymatic activity	Effective across multiple organs	Optimal dosing is unclear

16.2 Shielding Normal Tissues While Sparing Tumor Cells

Radiotherapy is a cornerstone of cancer treatment, yet its collateral damage to normal tissues can limit therapeutic doses and impair patient quality of life. To overcome this challenge, researchers have investigated agents that, like amifostine (WR2721), selectively protect normal tissues during radiotherapy without extending protection to tumor cells. The following agents represent some of the promising alternatives:

16.2.1 Tempol

Tempol is a nitroxide compound known for its potent free radical-scavenging activity. Preclinical studies have shown that Tempol effectively reduces radiation-induced oxidative damage in normal tissues [241]. Its antioxidant properties appear to be selectively active in healthy cells, thereby preserving the radiosensitivity of tumor cells and maintaining the overall efficacy of radiotherapy.

Table 16.2 Studies on amifostine

Author(s), Year	Study design	Key findings
Nevra Aydemir Celep, 2023 [67]	35 Wistar rats divided into control and four experimental groups. Radiation exposure (15 Gy in 3 fractions) followed by titanium implant placement. Amifostine (200 mg/kg) administered before radiation in two groups. Rats were euthanized after 28 days, and tibial bone samples analyzed	Amifostine accelerated bone formation, improved implant healing, reduced inflammation, and increased bone-implant contact
Arzu Gezer, 2022 [239]	Adult male rats divided into four groups: control, amifostine, RT-saline, and RT + amifostine. Histological analysis performed 4 weeks post-treatment, apoptosis assessed via TUNEL assay	Amifostine administration 30 min before RT protected testicular germ cells against apoptosis
U. Seker, 2020 [68]	35 female Sprague-Dawley rats divided into five groups (Control, RT, Melatonin, amifostine, NAC). Histopathological and apoptosis analysis performed 48 h post-irradiation	Amifostine and NAC groups showed tissue protection and reduction in apoptosis index
S. Cakina, 2020 [240]	50 female Wistar rats divided into five groups (Control, RT, RT + amifostine, RT + Melatonin, RT + amifostine + Melatonin). A single 8 Gy dose was administered	Amifostine increased total oxidant status (TOS). Combined administration with melatonin reduced radiation-induced renal oxidative stress
Sevda Eliş Yildiz, 2023 [69]	32 male Sprague-Dawley rats divided into four groups (Control, amifostine-only, RT-only, RT + amifostine). Prostate tissue analyzed for MUC1 expression	Immunohistochemical analysis showed increased MUC1 secretion in RT and amifostine groups compared to controls

Table 16.3 Studies on vitamin C

Author(s), Year	Study design	Key findings
S.M.J. Mortazavi, 2015 [81]	120 male Wistar rats divided into control and radiation-exposed groups. Single vitamin C doses were administered 1, 12, or 24 h post-exposure	Vitamin C treatment post-radiation did not improve survival rates
S.M.J. Mortazavi, 2024 [82]	45 rats divided into three groups: one received oral vitamin C (400 mg/kg), another received adipose-derived stem cells (AdSCs), and a third received both. All groups were exposed to 10 Gy radiation	The combination of AdSCs and vitamin C resulted in a significant survival increase (73.3%) compared to AdSCs alone (60%) and vitamin C alone (13.3%)
Grigory Demyashkin, 2024 [103]	120 male Wistar rats divided into four groups: control, radiation-only (30 Gy), vitamin C pre-treatment before radiation, and vitamin C alone. Liver tissue analyzed at multiple time points	Vitamin C reduced oxidative stress, apoptosis, and inflammation, preventing progression of radiation-induced liver fibrosis

Table 16.4 Studies on Curcumin

Author(s), Year	Study Design	Key Findings
Reza Fardid, 2021 [109]	Lycopene and curcumin (5 μg/mL each) were tested at 2 and 6 Gy radiation doses. Micronucleus assay used to assess genotoxicity	Pre-treatment with lycopene and curcumin reduced micronucleus formation at 2 Gy
Sameh Tawfik, 2013 [124]	Male mice received curcumin (400 μmol/kg) before, after, or both before and after 3 Gy γ-radiation	Curcumin exhibited both radioprotective and radio-recovery effects
Seher Çimen Özgen, 2012 [125]	15 Gy radiation exposures in rats, curcumin (100 mg/kg) administered via intragastric intubation for 28 days	Cataract incidence dropped from 100% in the irradiation group to 40% in the curcumin group. Curcumin increased antioxidant enzyme activity and reduced MDA levels

Table 16.5 Studies on hesperidin

Author(s), Year	Study design	Key findings
Hazem K.A. Sarhan, 2024 [146]	Rats were pretreated with a diosmin-hesperidin combination for 30 days before exposure to 10 Gy of gamma radiation. Various cellular and biochemical markers were assessed	Pre-treatment improved antioxidant indices, reduced DNA damage, and decreased apoptosis
Somayeh Sajadi, 2021 [142]	Sixty male rats divided into four groups (Sham, Hesperidin-only, RT-only, RT + Hesperidin). Hesperidin (100 mg/kg) administered for 7 days before radiation	Hesperidin significantly decreased radiation-induced cardiac injury markers

Table 16.6 Studies on Apigenin

Author(s), Year	Study design	Key findings
Naziya Begum, 2022 [169]	Swiss albino mice received apigenin (15 mg/kg) for 6 days before exposure to 7 Gy whole-body irradiation	Apigenin pre-treatment restored intestinal crypt and villus structure post-radiation
Tanat Peanlikhit, 2024 [158]	Molecular docking study using AutoDock software compared binding energies of apigenin and other flavonoids to 24 cancer-related proteins	Apigenin showed strong binding affinity to mutated KRAS proteins, suggesting its potential as an anticancer agent

16.2.2 MnSOD Gene Therapy

Manganese superoxide dismutase (MnSOD) is a critical mitochondrial enzyme that neutralizes superoxide radicals generated during irradiation. Gene therapy strategies aimed at upregulating MnSOD in normal tissues have demonstrated promising radioprotective effects in experimental models. By bolstering the intrinsic

Table 16.7 Studies on Ginseng

Author(s), Year	Study design	Key findings
Vida Omrani, 2024 [183]	40 patients undergoing 131I treatment divided into five groups. Micronuclei frequency and serum markers were measured	Ginseng treatment significantly reduced radiation-induced micronuclei formation
JW Chang, 2014 [185]	Rats exposed to 20 Gy irradiation, divided into four groups (Control, Ginseng-only, RT-only, RT + Ginseng)	Ginseng improved survival rates, oral intake, and reduced severity of mucositis

Table 16.8 Studies on Ginger

Author(s), Year	Study design	Key findings
Ganesh Chandra Jagetia, 2003 [189]	Mice pretreated with ginger extract (10 mg/kg) before exposure to 6–12 Gy radiation	Ginger reduced radiation sickness severity and mortality

antioxidant defenses of healthy cells, MnSOD gene therapy can mitigate radiation-induced injury without inadvertently protecting tumor cells [242, 243].

16.2.3 Ex-Rad (ON 01210.Na)

Ex-Rad is a small-molecule radioprotector that has been evaluated in various preclinical studies. It appears to modulate cellular stress response pathways, promoting repair and recovery in normal tissues following radiation exposure. Importantly, Ex-Rad does not interfere with the cytotoxic effects of radiation on tumor cells, making it a compelling candidate for clinical use as a normal tissue radioprotector [244].

16.2.4 Palifermin

Palifermin, a recombinant human keratinocyte growth factor, is already approved for reducing severe oral mucositis in patients undergoing chemoradiotherapy. By stimulating the growth and repair of epithelial cells, palifermin selectively enhances the resilience of normal mucosal tissues against radiation-induced damage. Its targeted action helps protect healthy tissue without compromising the radiosensitivity of malignant cells [245].

Table 16.9 Studies on Melatonin

Author(s), Year	Study design	Key findings
Shirazi A.R., 2012 [232]	42 rats divided into six groups: control, melatonin-only, radiation-only, and four melatonin pre-treatment groups (10, 100, 200, 400 μg/kg) exposed to 6 Gy whole-body gamma irradiation	Melatonin pre-treatment significantly reduced MDA levels and increased GSH levels in irradiated rats
Gholam Hassan Haddadi, 2015 [233]	Rats divided into four groups: control, radiation-only, melatonin (100 mg/kg) + radiation, and melatonin-only. Melatonin was administered orally before and after radiation	TNF-α gene expression was significantly lower in the melatonin pre-treatment group, while MDA levels were reduced compared to the radiation-only group
Reza Fardid, 2017 [234]	Male Sprague-Dawley rats received 3 Gy γ-radiation (1.25 MeV) to a 2 × 2 cm^2 pelvic area. Melatonin (100 mg/kg) was administered orally and via intraperitoneal injection. Biomarkers of oxidative stress were analyzed in lung tissue outside the radiation field	Melatonin reduced the radiation-induced bystander effect, lowering 8-OHdG, COX-2, and iNOS levels in non-targeted lung tissues
Masoud Najaf, 2019 [235]	30 male rats divided into six groups: control, melatonin-only, melatonin + metformin, radiation-only (10 Gy), radiation + melatonin, and radiation + melatonin + metformin. Duodenal tissues analyzed histologically	Melatonin reduced inflammation, necrosis, Brunner's gland damage, and goblet cell loss. Melatonin and metformin together alleviated some radiation-induced toxicity, though melatonin alone was more effective
Beatriz Fernández-Gil, 2017 [236]	Male Wistar rats received 7.5 Gy X-ray irradiation to the tongue daily for 5 days. Post-radiation, melatonin gel (45 mg/day) or vehicle was applied locally for 21 days	Melatonin gel demonstrated protective effects on the small intestine
Ahmed Eleojo Musa, 2019 [237]	30 male Wistar rats divided into six groups: control, radiation-only (8 Gy), melatonin-only (50 mg/kg or 100 mg/kg), and radiation + melatonin (50 mg/kg or 100 mg/kg). Biochemical analysis of intestinal tissues was conducted	Melatonin post-treatment significantly reduced MDA levels while increasing SOD and catalase (CAT) activity in irradiated intestinal tissues
Raziyeh Tahamtan, 2015 [238]	62 rats divided into five groups: control, oral melatonin, oral melatonin + 18 Gy thoracic radiation, intraperitoneal melatonin + 18 Gy thoracic radiation, and radiation-only. Early-phase lung injury assessed at 48 h, late-phase injury evaluated at 8 weeks	Melatonin pre-treatment significantly reduced MDA levels in RBCs and mitigated early and late-stage radiation-induced lung damage

16.2.5 *Other Emerging Agents*

Additional compounds, such as certain angiotensin-converting enzyme (ACE) inhibitors (e.g., captopril) and antioxidant agents like melatonin, have also shown potential in reducing radiation-induced normal tissue injury. While these agents are under active investigation, further research is necessary to fully establish their selectivity and clinical utility in safeguarding normal tissues during radiotherapy. In summary, finding alternative radioprotective agents that offer promise in mitigating the adverse effects of radiotherapy on normal tissues, thereby potentially allowing higher radiation doses to be used safely in the treatment of cancer, is of crucial importance. Each agent works through different mechanisms, ranging from free radical scavenging to the enhancement of cellular repair pathways, contributing to a multi-faceted approach to radioprotection.

16.3 Conclusion

Radioprotectors and mitigators are essential for reducing the adverse effects of radiotherapy. Amifostine remains the most well-established radioprotector, but natural compounds such as curcumin, vitamin C, hesperidin, melatonin, ginger, and ginseng offer promising alternatives with fewer side effects. These agents exhibit antioxidant, anti-inflammatory, and free radical-scavenging properties, which contribute to their protective effects against radiation-induced damage. Moving forward, further research should focus on optimizing these compounds for greater bioavailability, improved delivery methods, and enhanced clinical efficacy. More large-scale clinical trials are necessary to confirm their benefits and establish standardized protocols for their use in radiotherapy patients. Exploring combination therapies may also provide synergistic protective effects, ultimately improving patient outcomes and minimizing radiation-related toxicities.

References

1. Meister M, Karam P. The health effects of low level radiation. New York: American Council on Science and Health; 2005. p. 2–4.
2. Krishna G, Fiedler R, Theiss J. Simultaneous evaluation of clastogenicity, aneugenicity and toxicity in the mouse micronucleus assay using immunofluorescence. Mutat Res Lett. 1992;282(3):159–67.
3. Hall E, Giaccia A. Radiobiology for the radiologist. 6th ed. Lippincott Williams & Wilkins, Philadelphia[Google Scholar]; 2006.
4. Soltani B, Ghaemi N, Sadeghizadeh M, Najafi F. Curcumin confers protection to irradiated THP-1 cells while its nanoformulation sensitizes these cells via apoptosis induction. Cell Biol Toxicol. 2016;32:543–61.

5. Miller DL, Weinstock MA. Nonmelanoma skin cancer in the United States: incidence. J Am Acad Dermatol. 1994;30(5):774–8.
6. Hejazi J, Rastmanesh R, Taleban F-A, Molana S-H, Hejazi E, Ehtejab G, et al. Effect of curcumin supplementation during radiotherapy on oxidative status of patients with prostate cancer: a double blinded, randomized, placebo-controlled study. Nutr Cancer. 2016;68(1):77–85.
7. Riley P. Free radicals in biology: oxidative stress and the effects of ionizing radiation. Int J Radiat Biol. 1994;65(1):27–33.
8. Bhosle SM, Huilgol NG, Mishra KP. Enhancement of radiation-induced oxidative stress and cytotoxicity in tumor cells by ellagic acid. Clin Chim Acta. 2005;359(1–2):89–100.
9. Weiss JF, Landauer MR. Radioprotection by antioxidants A. Ann N Y Acad Sci. 2000;899(1):44–60.
10. Mansour HH, Hafez HF, Fahmy NM, Hanafi N. Protective effect of N-acetylcysteine against radiation induced DNA damage and hepatic toxicity in rats. Biochem Pharmacol. 2008;75(3):773–80.
11. Chevion S, Or R, Berry EM. The antioxidant status of patients subjected to today body irradiation. IUBMB Life. 1999;47(6):1019–27.
12. Taysi S, Uslu C, Akcay F, Sutbeyaz MY. Malondialdehyde and nitric oxide levels in the plasma of patients with advanced laryngeal cancer. Surg Today. 2003;33:651–4.
13. Mofid B, Razzaghdoust A, Kashi ASY, Mirzaei HR. A comparative study of cisplatin-based definitive chemo-radiation in non-metastatic squamous cell carcinoma of the esophagus. Electron Physician. 2016;8(10):3069.
14. Saadipoor A, Razzaghdoust A, Simforoosh N, Mahdavi A, Bakhshandeh M, Moghadam M, et al. Randomized, double-blind, placebo-controlled phase II trial of nanocurcumin in prostate cancer patients undergoing radiotherapy. Phytother Res. 2019;33(2):370–8.
15. Razzaghdoust A, Mozdarani H, Mofid B. Famotidine as a radioprotector for rectal mucosa in prostate cancer patients treated with radiotherapy. Strahlenther Onkol. 2014;190(8):739.
16. Topol EJ, Leya F, Pinkerton CA, Whitlow PL, Hofling B, Simonton CA, et al. A comparison of directional atherectomy with coronary angioplasty in patients with coronary artery disease. N Engl J Med. 1993;329(4):221–7.
17. Dale WM, Gray LH, Meredith W. The inactivation of an enzyme (carboxypeptidase) by X-and α-radiation. Philos Trans R Soc Lond Ser A Math Phys Sci. 1949;242(840):33–62.
18. Karran P. DNA double strand break repair in mammalian cells. Curr Opin Genet Dev. 2000;10(2):144–50.
19. Pipalová I, Viklická Š. Noradrenaline reduces cardiovascular effects of the combined dipyridamole and AMP administration but preserves radioprotective effects of these drugs on hematopoiesis in mice. Physiol Res. 1993;42:333–40.
20. Akpolat M, Tarladaçalışır YT, Dökmeci D, Turan FN, Uzal MC. Protective effects of ibuprofen and L-carnitine against whole body gamma irradiation-induced duodenal mucosal injury. Balkan Med J. 2011;2011(1):80–5.
21. Varma SD, Kovtun S, Hegde KR. Role of ultraviolet irradiation and oxidative stress in cataract formation—medical prevention by nutritional antioxidants and metabolic agonists. Eye Contact Lens. 2011;37(4):233–45.
22. El-Bahr SM. Biochemistry of free radicals and oxidative stress. Biochemistry. 2013;1(5567/5cjiŋt):11–111.
23. Shedid SM, Abdel-Magied N, Saada HN. Role of betaine in liver injury induced by the exposure to ionizing radiation. Environ Toxicol. 2019;34(2):123–30.
24. Azzam EI, Jay-Gerin J-P, Pain D. Ionizing radiation-induced metabolic oxidative stress and prolonged cell injury. Cancer Lett. 2012;327(1–2):48–60.
25. Singer MC, Marchal F, Angelos P, Bernet V, Boucai L, Buchholzer S, et al. Salivary and lacrimal dysfunction after radioactive iodine for differentiated thyroid cancer: American head and neck society endocrine surgery section and salivary gland section joint multidisciplinary clinical consensus statement of otolaryngology, ophthalmology, nuclear medicine and endocrinology. Head Neck. 2020;42(11):3446–59.

26. Reza S, Sultana S, Begum F, Perveen R, Jabin Z, Nahar N, et al. Delayed citrus stimulation and prevention of salivary glands damage in patients of thyroid carcinoma treated by radioiodine therapy. Bangladesh J. 2018;21(2):92.
27. Duskin-Bitan H, Leibner A, Amitai O, Diker-Cohen T, Hirsch D, Benbassat C, et al. Bone-marrow suppression in elderly patients following empiric radioiodine therapy: real-life data. Thyroid. 2019;29(5):683–91.
28. Molenaar R, Pleyer C, Radivoyevitch T, Sidana S, Godley A, Advani A, et al. Risk of developing chronic myeloid neoplasms in well-differentiated thyroid cancer patients treated with radioactive iodine. Leukemia. 2018;32(4):952–9.
29. Verburg FA, Hoffmann M, Iakovou I, Konijnenberg MW, Mihailovic J, Gabina PM, et al. Errare humanum Est, sed in errare perseverare diabolicum: methodological errors in the assessment of the relationship between I-131 therapy and possible increases in the incidence of malignancies. Springer; 2020. p. 519–22, 47
30. Sengoz T, Kilic-Toprak E, Yaylali O, Kilic-Erkek O, Ozdemir Y, Oymak B, et al. Hemorheology and oxidative stress in patients with differentiated thyroid cancer following I-131 ablation/metastasis treatment. Clin Hemorheol Microcirc. 2020;74(2):209–21.
31. Cai L, Koropatnick J, Cherian MG. Roles of vitamin C in radiation-induced DNA damage in presence and absence of copper. Chem Biol Interact. 2001;137(1):75–88.
32. Feig DI, Reid TM, Loeb LA. Reactive oxygen species in tumorigenesis. Cancer Res. 1994;54(7_Supplement):1890s–4s.
33. Roots R, Okada S. Estimation of life times and diffusion distances of radicals involved in X-ray-induced DNA strand breaks or killing of mammalian cells. Radiat Res. 1975;64(2):306–20.
34. Mofid B, Razzaghdoust A, Peyghambarlou P. Identifying predictive factors for severe anemia in cancer patients treated with cytotoxic chemotherapy. Ann Oncol. 2018;29:ix133.
35. Elhammali A, Patel M, Weinberg B, Verma V, Liu J, Olsen JR, et al. Late gastrointestinal tissue effects after hypofractionated radiation therapy of the pancreas. Radiat Oncol. 2015;10:1–8.
36. Kim H, Lim DH, Paik SW, Yoo BC, Koh KG, Lee JH, et al. Predictive factors of gastroduodenal toxicity in cirrhotic patients after three-dimensional conformal radiotherapy for hepatocellular carcinoma. Radiother Oncol. 2009;93(2):302–6.
37. Prior P, Tai A, Erickson B, Li XA. Consolidating duodenal and small bowel toxicity data via isoeffective dose calculations based on compiled clinical data. Pract Radiat Oncol. 2014;4(2):e125–e31.
38. Iwamuro M, Tanaka T, Kanzaki H, Kawano S, Kawahara Y, Iwasaki Y, et al. Deterioration of duodenal lymphangiectasia after radiotherapy for gastric MALT lymphoma. ecancermedicalscience. 2017;11:11.
39. Holyoake DL, Aznar M, Mukherjee S, Partridge M, Hawkins MA. Modelling duodenum radiotherapy toxicity using cohort dose-volume-histogram data. Radiother Oncol. 2017;123(3):431–7.
40. Danjoux C, Rider W, Fitzpatrick P. The acute radiation syndrome: a memorial to William Michael court-Brown. Clin Radiol. 1979;30(5):581–4.
41. Paganetti H, Yock T, Tarbell NJ, Trofimov A. Optimization of radiotherapy treatment delivery technology to minimize radiation injury. Human radiation injury. Philadelphia: Lippincott Williams & Wilkins; 2011. p. 114–25.
42. Soto-Pantoja DR, Ridnour LA, Wink DA, Roberts DD. Blockade of CD47 increases survival of mice exposed to lethal total body irradiation. Sci Rep. 2013;3(1):1038.
43. Groschwitz KR, Hogan SP. Intestinal barrier function: molecular regulation and disease pathogenesis. J Allergy Clin Immunol. 2009;124(1):3–20.
44. Lee T-K, Johnke RM, Allison RR, O'Brien KF, Dobbs LJ Jr. Radioprotective potential of ginseng. Mutagenesis. 2005;20(4):237–43.
45. Cook JA, Naz S, Anver MR, Sowers AL, Fabre K, Krishna MC, et al. Cancer incidence in C3H mice protected from lethal total-body radiation after amifostine. Radiat Res. 2018;189(5):490–6.

46. Gula A, Ren L, Zhou Z, Lu D, Wang S. Design and evaluation of biodegradable enteric microcapsules of amifostine for oral delivery. Int J Pharm. 2013;453(2):441–7.
47. Gurney JG, Bass JK, Onar-Thomas A, Huang J, Chintagumpala M, Bouffet E, et al. Evaluation of amifostine for protection against cisplatin-induced serious hearing loss in children treated for average-risk or high-risk medulloblastoma. Neuro-Oncology. 2014;16(6):848–55.
48. Levi M, DeRemer SJ, Dou C, Ensminger WD, Smith DE. Disposition of WR-1065 in the liver of tumor-bearing rats following regional vs systemic administration of amifostine. Biopharm Drug Dispos. 2004;25(1):27–35.
49. Pamujula S, Kishore V, Rider B, Agrawal KC, Mandal TK. Radioprotection in mice following oral administration of WR-1065/PLGA nanoparticles. Int J Radiat Biol. 2008;84(11):900–8.
50. Ranganathan K, Simon E, Lynn J, Snider A, Zhang Y, Nelson N, et al. Novel formulation strategy to improve the feasibility of amifostine administration. Pharm Res. 2018;35:1–8.
51. Trajković S, Dobrić S, Jaćević V, Dragojević-Simić V, Milovanović Z, Đorđević A. Tissue-protective effects of fullerenol C60 (OH) 24 and amifostine in irradiated rats. Colloids Surf B Biointerfaces. 2007;58(1):39–43.
52. Molkentine JM, Fujimoto TN, Horvath TD, Grossberg AJ, Garcia CJG, Deorukhkar A, et al. Enteral activation of WR-2721 mediates radioprotection and improved survival from lethal fractionated radiation. Sci Rep. 2019;9(1):1949.
53. Van der Vijgh W, Korst A. Amifostine (Ethyol®): pharmacokinetic and pharmacodynamic effects in vivo. Eur J Cancer. 1996;32:S26–30.
54. Korst A, Eeltink C, Vermorken J, Van der Vijgh W. Pharmacokinetics of amifostine and its metabolites in patients. Eur J Cancer. 1997;33(9):1425–9.
55. Franco P, De Marco I. Eudragit: a novel carrier for controlled drug delivery in supercritical antisolvent coprecipitation. Polymers. 2020;12(1):234.
56. Prusty A, Mishra AK, Gupta BK. Development and evaluation of matrix tablet by taking new chemicals combination of chitosan and Eudragit-l 100. J Young Pharm. 2016;8(3):168.
57. Vlachou M, Kikionis S, Siamidi A, Kyriakou S, Tsotinis A, Ioannou E, et al. Development and characterization of Eudragit®-based electrospun nanofibrous mats and their formulation into nanofiber tablets for the modified release of furosemide. Pharmaceutics. 2019;11(9):480.
58. Smoluk GD, Fahey RC, Calabro-Jones PM, Aguilera JA, Ward JF. Radioprotection of cells in culture by WR-2721 and derivatives: form of the drug responsible for protection. Cancer Res. 1988;48(13):3641–7.
59. Vesna J, Danica J, Kamil K, Viktorija D-S, Silva D, Sanja T, et al. Effects of fullerenol nanoparticles and amifostine on radiation-induced tissue damages: histopathological analysis. J Appl Biomed. 2016;14(4):285–97.
60. Koukourakis MI, editor. Amifostine: is there evidence of tumor protection? Seminars in oncology. Elsevier; 2003.
61. Brizel DM, Wasserman TH, Henke M, Strnad V, Rudat V, Monnier A, et al. Phase III randomized trial of amifostine as a radioprotector in head and neck cancer. J Clin Oncol. 2000;18(19):3339–45.
62. Rainaldi G, Ferrante A, Indovina PL, Santini MT. Induction of apoptosis or necrosis by ionizing radiation is dose-dependent in MG-63 osteosarcoma multicellular spheroids. Anticancer Res. 2003;23(3B):2505–18.
63. Barnett GC, Kerns SL, Noble DJ, Dunning AM, West CM, Burnet NG. Incorporating genetic biomarkers into predictive models of normal tissue toxicity. Clin Oncol. 2015;27(10):579–87.
64. Wang J, Biju MP, Wang M-H, Haase VH, Dong Z. Cytoprotective effects of hypoxia against cisplatin-induced tubular cell apoptosis: involvement of mitochondrial inhibition and p53 suppression. J Am Soc Nephrol. 2006;17(7):1875–85.
65. Kaur P, Bansal M. Influence of selenium induced oxidative stress on spermatogenesis and lactate dehydrogenase-X in mice testis. Asian J Androl. 2004;6(3):227–32.
66. Rao M, Narayana K, Benjamin S, Bairy K. L-ascorbic acid ameliorates postnatal endosulfan induced testicular damage in rats. Indian J Physiol Pharmacol. 2005;49(3):331.

67. Celep NA, Kara H, Erbaş E, Doğan E. Radioprotective role of amifostine on osteointegration of titanium implants in the tibia of rats. J Vet Sci. 2023;24(3)
68. Seker U, Aktas A, Nergiz Y, Zincircioglu S, Ketani M. Investigation of the protective effects of melatonin, amifostine (WR-2721), and N-acetylcysteine on radiotherapy-induced uterine tissue injury in rats. Int J Radiat Res. 2020;18(4):791–8.
69. Yildiz S, Gezer A, Karadağ Sari E. Effects of Amifostine application on mucin 1 secretion in prostate tissue of radiotherapy-treated rats. Turk Onkoloji Dergisi-Turkish J Oncol. 2023;38
70. Du J, Cullen JJ, Buettner GR. Ascorbic acid: chemistry, biology and the treatment of cancer. Biochim Biophys Acta Rev Cancer. 2012;1826(2):443–57.
71. Nishikimi M, Fukuyama R, Minoshima S, Shimizu N, Yagi K. Cloning and chromosomal mapping of the human nonfunctional gene for L-gulono-gamma-lactone oxidase, the enzyme for L-ascorbic acid biosynthesis missing in man. J Biol Chem. 1994;269(18):13685–8.
72. Yamamoto T, Kinoshita M, Shinomiya N, Hiroi S, Sugasawa H, Matsushita Y, et al. Pretreatment with ascorbic acid prevents lethal gastrointestinal syndrome in mice receiving a massive amount of radiation. J Radiat Res. 2010;51(2):145–56.
73. Mathew D, Nair CKK, Jacob JA, Biswas N, Mukherjee T, Kapoor S, et al. Ascorbic acid monoglucoside as antioxidant and radioprotector. J Radiat Res. 2007;48(5):369–76.
74. Adaramoye OA, Okiti OO, Farombi EO. Dried fruit extract from Xylopia aethiopica (Annonaceae) protects Wistar albino rats from adverse effects of whole body radiation. Exp Toxicol Pathol. 2011;63(7–8):635–43.
75. Fujii Y, Kato TA, Ueno A, Kubota N, Fujimori A, Okayasu R. Ascorbic acid gives different protective effects in human cells exposed to X-rays and heavy ions. Mutat Res/Genetic Toxicol Environ Mutagenesis. 2010;699(1–2):58–61.
76. Kanter M, Akpolat M. Vitamin C protects against ionizing radiation damage to goblet cells of the ileum in rats. Acta Histochem. 2008;110(6):481–90.
77. Mozdarani H, Ghoraeian P. Modulation of gamma-ray-induced apoptosis in human peripheral blood leukocytes by famotidine and vitamin C. Mutat Res/Genetic Toxicol Environ Mutagenesis. 2008;649(1–2):71–8.
78. Mozdarani H, Nazari E. Frequency of micronuclei in 4–8 cell mouse embryos generated after maternal gamma-irradiation in the presence and in the absence of vitamin C. Radiat Environ Biophys. 2007;46:417–22.
79. Mozdarani H, Nazari E. Cytogenetic damage in preimplantation mouse embryos generated after paternal and parental gamma-irradiation and the influence of vitamin C. Reproduction. 2009;137(1):35–43.
80. Mortazavi A, Yarbaksh H, Zarandi BFBB, Yarbakhsh R, Ghadimi-Moghaddam F, Mortazavi SMJ, et al. Cultivation of vitamin C-rich vegetables for space-radiation mitigation. Radiat Ther. 2024;4(1):101–14.
81. Mortazavi S, Foadi M, Mozdarani H, Haghani M, Mosleh SM, Abolghasemi P, et al. Future role of vitamin C in radiation mitigation and its possible applications in manned deep space missions: survival study and the measurement of cell viability. Physica Medica. 2015;30:e97.
82. Mortazavi SMJ, Kadivar F, Zare S, Kanani A, Mosleh-Shirazi MA, Masoudi S, et al. The role of adipose tissue-derived stem cells together with vitamin C on survival of rats with acute radiation syndrome. J Biomed Phys Eng. 2024;
83. Khazaei S, Nilsson L, Adrian G, Tryggvadottir H, Konradsson E, Borgquist S, et al. Impact of combining vitamin C with radiation therapy in human breast cancer: does it matter? Oncotarget. 2022;13:439.
84. Mortazavi S, Rahimi S, Mosleh-Shirazi M, Arjomandi M, Soleimani A, Hossein-Abadi OK, et al. A comparative study on the life-saving radioprotective effects of vitamins A, E, C and over-the-counter multivitamins. J Biomed Phys Eng. 2015;5(2):59.
85. Yue X, Rao A. TET family dioxygenases and the TET activator vitamin C in immune responses and cancer. Blood J Am Soc Hematol. 2020;136(12):1394–401.
86. Kaźmierczak-Barańska J, Boguszewska K, Adamus-Grabicka A, Karwowski BT. Two faces of vitamin C—antioxidative and pro-oxidative agent. Nutrients. 2020;12(5):1501.

87. Spoelstra-de Man AM, Elbers PW, Oudemans-Van Straaten HM. Vitamin C: should we supplement? Curr Opin Crit Care. 2018;24(4):248–55.
88. Suhail N, Bilal N, Khan H, Hasan S, Sharma S, Khan F, et al. Effect of vitamins C and E on antioxidant status of breast-cancer patients undergoing chemotherapy. J Clin Pharm Ther. 2012;37(1):22–6.
89. Szeto YT, To TL, Pak SC, Kalle W. A study of DNA protective effect of orange juice supplementation. Appl Physiol Nutr Metab. 2013;38(5):533–6.
90. Mallikarjun Rao KVN. Ascorbic acid treatment enhances the wound healing in mice exposed to radiation. Indian J Radiat Res. 2004;1:24.
91. Akbari A. An overview of the characteristics and function of vitamin C in various tissues: relying on its antioxidant function. In Press; 2016.
92. Didier AJ, Stiene J, Fang L, Watkins D, Dworkin LD, Creeden JF. Antioxidant and anti-tumor effects of dietary vitamins A, C, and E. Antioxidants. 2023;12(3):632.
93. Gęgotek A, Skrzydlewska E. Antioxidative and anti-inflammatory activity of ascorbic acid. Antioxidants. 2022;11(10):1993.
94. Kurutas EB. The importance of antioxidants which play the role in cellular response against oxidative/nitrosative stress: current state. Nutr J. 2015;15:1–22.
95. Padayatty SJ, Levine M. Vitamin C: the known and the unknown and goldilocks. Oral Dis. 2016;22(6):463–93.
96. Chambial S, Dwivedi S, Shukla KK, John PJ, Sharma P. Vitamin C in disease prevention and cure: an overview. Indian J Clin Biochem. 2013;28:314–28.
97. Lutsenko EA, Cárcamo JM, Golde DW. Vitamin C prevents DNA mutation induced by oxidative stress. J Biol Chem. 2002;277(19):16895–9.
98. Tan BL, Norhaizan ME, Liew W-P-P, Sulaiman Rahman H. Antioxidant and oxidative stress: a mutual interplay in age-related diseases. Front Pharmacol. 2018;9:1162.
99. Abdulrazzaq AM, Badr M, Gammoh O, Abu Khalil AA, Ghanim BY, Alhussainy TM, et al. Hepatoprotective actions of ascorbic acid, alpha lipoic acid and silymarin or their combination against acetaminophen-induced hepatotoxicity in rats. Medicina. 2019;55(5):181.
100. Autifi MAH, Mohamed WY, Abdul Haye WM, Elbaz KR. The possible protective role of vitamin C against toxicity induced by lead acetate in liver and spleen of adult albino rats (light and electron microscopic study). Egypt J Hosp Med. 2018;73(10):7650–8.
101. Hassnin K, Hashem KS. The Hepatoprotective effects of vitamin C and micronized vitamin C against paracetamol induced hepatotoxicity in rats: a comparative study. J Vet Med Res. 2013;22(1):46–52.
102. Yu SJ, Bae S, Kang JS, Yoon J-H, Cho EJ, Lee J-H, et al. Hepatoprotective effect of vitamin C on lithocholic acid-induced cholestatic liver injury in Gulo (−/−) mice. Eur J Pharmacol. 2015;762:247–55.
103. Demyashkin G, Parshenkov M, Koryakin S, Skovorodko P, Shchekin V, Yakimenko V, et al. Targeting oxidative stress: the potential of vitamin C in protecting against liver damage after electron beam therapy. Biomedicines. 2024;12(10):2195.
104. Salah M, Abdul-Hamid M, Soliman SM, Osman H. Histopathological and ultrastructural studies on the protective effect of Daidzein and vitamin C against gamma irradiation-induced kidney alterations in albino rats. Egy J Histol. 2020;43(2):427–40.
105. Attia AA, Hamad HA, Fawzy MA, Saleh SR. The prophylactic effect of vitamin C and vitamin B12 against ultraviolet-C-induced hepatotoxicity in male rats. Molecules. 2023;28(11):4302.
106. Goel A, Aggarwal BB. Curcumin, the golden spice from Indian saffron, is a chemosensitizer and radiosensitizer for tumors and chemoprotector and radioprotector for normal organs. Nutr Cancer. 2010;62(7):919–30.
107. Liu X, Zhang R, Shi H, Li X, Li Y, Taha A, et al. Protective effect of curcumin against ultraviolet a irradiation-induced photoaging in human dermal fibroblasts. Mol Med Rep. 2018;17(5):7227–37.

108. Islamian JP, Mehrali H. Lycopene as a carotenoid provides radioprotectant and antioxidant effects by quenching radiation-induced free radical singlet oxygen: an overview. Cell J (Yakhteh). 2015;16(4):386.
109. Fardid R, Fazilatpour H, Ebrahimi N, Sadeghi A, Zare T. Evaluation of the effects of curcumin and lycopene treatment on human lymphocytes before 2 and 6 Gy of 6 MV X-ray irradiation. Iran J Med Phys. 2021;18(2):117–22.
110. Chen G, Chen Y, Yang N, Zhu X, Sun L, Li G. Interaction between curcumin and mimetic biomembrane. Sci China Life Sci. 2012;55:527–32.
111. Qiao Q, Jiang Y, Li G. Curcumin improves the antitumor effect of X-ray irradiation by blocking the NF-κB pathway: an in-vitro study of lymphoma. Anti-Cancer Drugs. 2012;23(6):597–605.
112. Plengsuriyakarn T, Viyanant V, Eursitthichai V, Picha P, Kupradinun P, Itharat A, et al. Anticancer activities against cholangiocarcinoma, toxicity and pharmacological activities of Thai medicinal plants in animal models. BMC Complement Altern Med. 2012;12:1–19.
113. Jagetia GC, Rajanikant GK. Acceleration of wound repair by curcumin in the excision wound of mice exposed to different doses of fractionated γ radiation. Int Wound J. 2012;9(1):76–92.
114. Mosieniak G, Sliwinska M, Piwocka K, Sikora E. Curcumin abolishes apoptosis resistance of calcitriol-differentiated HL-60 cells. FEBS Lett. 2006;580(19):4653–60.
115. Leong PK, Chiu PY, Ko KM. Prooxidant-induced glutathione antioxidant response in vitro and in vivo: a comparative study between schisandrin B and curcumin. Biol Pharm Bull. 2012;35(4):464–72.
116. Tiwari H, Rao MV. Curcumin supplementation protects from genotoxic effects of arsenic and fluoride. Food Chem Toxicol. 2010;48(5):1234–8.
117. Bar-Sela G, Epelbaum R, Schaffer M. Curcumin as an anti-cancer agent: review of the gap between basic and clinical applications. Curr Med Chem. 2010;17(3):190–7.
118. Jagetia G, Rajanikant G. Effect of curcumin on radiation-impaired healing of excisional wounds in mice. J Wound Care. 2004;13(3):107–9.
119. Jagetia GC. Radioprotection and radiosensitization by curcumin. In: The molecular targets and therapeutic uses of curcumin in health and disease. New York: Springer; 2007. p. 301–20.
120. Jagetia GC, Aggarwal BB. "Spicing up" of the immune system by curcumin. J Clin Immunol. 2007;27:19–35.
121. Zhou H, Beevers CS, Huang S. The targets of curcumin. Curr Drug Targets. 2011;12(3):332–47.
122. Yahyapour R, Shabeeb D, Cheki M, Musa AE, Farhood B, Rezaeyan A, et al. Radiation protection and mitigation by natural antioxidants and flavonoids: implications to radiotherapy and radiation disasters. Curr Mol Pharmacol. 2018;11(4):285–304.
123. Cheng AnnLii CA, Hsu ChihHung HC, Lin JenKun LJ, Hsu MowMing HM, Ho YunnFang HY, Shen TzungShiahn ST, et al. Phase I clinical trial of curcumin, a chemopreventive agent, in patients with high-risk or pre-malignant lesions. Anticancer Res. 2001;21:2895.
124. Tawfik SS, Abouelella AM, Shahein YE. Curcumin protection activities against γ-rays-induced molecular and biochemical lesions. BMC Res Notes. 2013;6:1–10.
125. Özgen SÇ, Dökmeci D, Akpolat M, Karadağ ÇH, Gündüz Ö, Erbaş H, et al. The protective effect of curcumin on ionizing radiation-induced cataractogenesis in rats. Balkan Med J. 2012;2012(4):358–63.
126. Shabeeb D, Musa AE, Abd Ali HS, Najafi M. Curcumin protects against radiotherapy-induced oxidative injury to the skin. Drug Des Devel Ther. 2020;14:3159–63.
127. Jafarpour SM, Safaei M, Mohseni M, Salimian M, Aliasgharzadeh A, Fahood B. The radioprotective effects of curcumin and trehalose against genetic damage caused by I-131. Indian J Nucl Med. 2018;33(2):99–104.
128. Li W, Jiang L, Lu X, Liu X, Ling M. Curcumin protects radiation-induced liver damage in rats through the NF-κB signaling pathway. BMC Complementary Med Ther. 2021;21:1–10.
129. Tiwari AK. Imbalance in antioxidant defence and human diseases: multiple approach of natural antioxidants therapy. Curr Sci. 2001:1179–87.

130. Lee Y-S, Huh J-Y, Nam S-H, Moon S-K, Lee S-B. Enzymatic bioconversion of citrus hesperidin by aspergillus sojae naringinase: enhanced solubility of hesperetin-7-O-glucoside with in vitro inhibition of human intestinal maltase, HMG-CoA reductase, and growth of helicobacter pylori. Food Chem. 2012;135(4):2253–9.
131. Emim JADS, Oliveira AB, Lapa AJ. Pharmacological evaluation of the anti-inflammatory activity of a citrus bioflavonoid, hesperidin, and the isoflavonoids, duartin and claussequinone, in rats and mice. J Pharm Pharmacol. 1994;46(2):118–22.
132. Garg A, Garg S, Zaneveld L, Singla A. Chemistry and pharmacology of the citrus bioflavonoid hesperidin. Phytother Res. 2001;15(8):655–69.
133. Middleton E Jr. Effect of plant flavonoids on immune and inflammatory cell function. In: Flavonoids in the living system. New York: Plenum Press; 1998. p. 175–82.
134. Montanari A, Chen J, Widmer W. Citrus flavonoids: a review of past biological activity against disease: discovery of new flavonoids from Dancy tangerine cold pressed peel oil solids and leaves. In: Flavonoids in the living system. New York: Plenum Press; 1998. p. 103–16.
135. Cushnie TT, Lamb AJ. Antimicrobial activity of flavonoids. Int J Antimicrob Agents. 2005;26(5):343–56.
136. Krolicki Z, Lamer-Zarawska E. Investigation of antifungal effect of flavonoids. 1 [hesperidin, naringin, phellodendroside, luteolin-7-glucoside, hipotethin-7-glucoside, quecetin, celastroside, amentoflavone; Botrytis cinerea, Trichoderma glaucum, aspergillus fumigatus]. Herba Polonica (Poland). 1984;30(1)
137. Ahmadi A, Shadboorestan A. Oxidative stress and cancer; the role of hesperidin, a citrus natural bioflavonoid, as a cancer chemoprotective agent. Nutr Cancer. 2016;68(1):29–39.
138. Dalgård C, Nielsen F, Morrow JD, Enghusen-Poulsen H, Jonung T, Hørder M, et al. Supplementation with orange and blackcurrant juice, but not vitamin E, improves inflammatory markers in patients with peripheral arterial disease. Br J Nutr. 2008;101(2):263–9.
139. Lee N-K, Choi S-H, Park S-H, Park E-K, Kim D-H. Antiallergic activity of hesperidin is activated by intestinal microflora. Pharmacology. 2004;71(4):174–80.
140. Monforte MT, Trovato A, Kirjavainen S, Forestieri A, Galati EM, Lo Curto RB. Biological effects of hesperidin, a Citrus flavonoid(note II): hypolipidemic activity on experimental hypercholesterolemia in rat. Farmaco (Societ Ã chimica italiana: 1989). 1995;50(9):595–9.
141. Kuntić V, Brborić J, Holclajtner-Antunović I, Uskoković-Marković S. Evaluating the bioactive effects of flavonoid hesperidin–a new literature data survey. Vojnosanit Pregl. 2014;71(1):60.
142. Sajadi S, Haddadi G, Kadivar F, Fardid R. Hesperidin modulates Troponin-1 serum level and decrease heart tissue injury of irradiated rats. J Biomed Phys Eng. 2021;11(3):377.
143. Zh G, Fardid R, Haddadi G, Derakhshanfar A, Kouhpayeh A, Haddadi Z. Histopathologic evaluation of radio-protective effect of hesperidin on the liver of Sprague Dawely rats. J Biomed Phys Eng. 2020;10(1):7.
144. Fardid R, Ghorbani Z, Haddadi G, Behzad-Behbahani A, Arabsolghar R, Kazemi E, et al. Effects of hesperidin as a radio-protector on apoptosis in rat peripheral blood lymphocytes after gamma radiation. J Biomed Phys Eng. 2016;6(4):217.
145. Rezaeyan A, Fardid R, Haddadi G, Takhshid M, Hosseinzadeh M, Najafi M, et al. Evaluating radioprotective effect of hesperidin on acute radiation damage in the lung tissue of rats. J Biom Phys Eng. 2016;6(3):165.
146. Sarhan HK, Saleh A, Hammam OA, Atta AH, El-Nehrery IM. Protective effect of Diosmin-Hesperidine combination on gamma radiation-induced apoptosis in liver of albino rats. Molecular and Immunohistchemistry studies. Egypt J Chem. 2024;67(11):571–84.
147. Kalpana K, Devipriya N, Srinivasan M, Menon VP. Investigation of the radioprotective efficacy of hesperidin against gamma-radiation induced cellular damage in cultured human peripheral blood lymphocytes. Mutat Res/Genetic Toxicol Environ Mutagenesis. 2009;676(1–2):54–61.

148. Pradeep K, Ko KC, Choi MH, Kang JA, Chung YJ, Park SH. Protective effect of hesperidin, a citrus flavanoglycone, against γ-radiation-induced tissue damage in Sprague–Dawley rats. J Med Food. 2012;15(5):419–27.
149. Karimi N, Monfared AS, Haddadi GH, Soleymani A, Mohammadi E, Hajian-Tilaki K, et al. Radioprotective effect of hesperidin on reducing oxidative stress in the lens tissue of rats. Int J Pharm Invest. 2017;7(3):149.
150. Miean KH, Mohamed S. Flavonoid (myricetin, quercetin, kaempferol, luteolin, and apigenin) content of edible tropical plants. J Agric Food Chem. 2001;49(6):3106–12.
151. F'guyer S, Afaq F, Mukhtar H. Photochemoprevention of skin cancer by botanical agents. Photodermatol Photoimmunol Photomed. 2003;19(2):56–72.
152. Wei H, Tye L, Bresnick E, Birt DF. Inhibitory effect of apigenin, a plant flavonoid, on epidermal ornithine decarboxylase and skin tumor promotion in mice. Cancer Res. 1990;50(3):499–502.
153. Hirano T, Oka K, Akiba M. Antiproliferative effects of synthetic and naturally occurring flavonoids on tumor cells of the human breast carcinoma cell line, ZR-75-1. Res Commun Chem Pathol Pharmacol. 1989;64(1):69–78.
154. Rosenberg RS, Grass L, Jenkins DJ, Kendall CW, Diamandis EP. Modulation of androgen and progesterone receptors by phytochemicals in breast cancer cell lines. Biochem Biophys Res Commun. 1998;248(3):935–9.
155. Kobayashi T, Nakata T, Kuzumaki T. Effect of flavonoids on cell cycle progression in prostate cancer cells. Cancer Lett. 2002;176(1):17–23.
156. Caltagirone S, Rossi C, Poggi A, Ranelletti FO, Natali PG, Brunetti M, et al. Flavonoids apigenin and quercetin inhibit melanoma growth and metastatic potential. Int J Cancer. 2000;87(4):595–600.
157. Yin F, Giuliano AE, Van Herle AJ. Growth inhibitory effects of flavonoids in human thyroid cancer cell lines. Thyroid. 1999;9(4):369–76.
158. Peanlikhit T, Aryal U, Welsh J, Shroyer K, Rithidech K. In silico comparison of apigenin and related compounds with diverse pharmacological activities using AutoDock 4.2. 6. 2024.
159. Tang D, Chen K, Huang L, Li J. Pharmacokinetic properties and drug interactions of apigenin, a natural flavone. Expert Opin Drug Metab Toxicol. 2017;13(3):323–30.
160. Li R-R, Pang L-L, Du Q, Shi Y, Dai W-J, Yin K-S. Apigenin inhibits allergen-induced airway inflammation and switches immune response in a murine model of asthma. Immunopharmacol Immunotoxicol. 2010;32(3):364–70.
161. Patel D, Shukla S, Gupta S. Apigenin and cancer chemoprevention: progress, potential and promise. Int J Oncol. 2007;30(1):233–45.
162. Karamese M, Erol HS, Albayrak M, Findik Guvendi G, Aydin E, Aksak Karamese S. Antioxidant and anti-inflammatory effects of apigenin in a rat model of sepsis: an immunological, biochemical, and histopathological study. Immunopharmacol Immunotoxicol. 2016;38(3):228–37.
163. Rithidech KN, Tungjai M, Whorton EB. Protective effect of apigenin on radiation-induced chromosomal damage in human lymphocytes. Mutat Res/Genetic Toxicol Environ Mutagenesis. 2005;585(1–2):96–104.
164. Begum N, Prasad NR, Kanimozhi G, Hasan AQ. Apigenin ameliorates gamma radiation-induced cytogenetic alterations in cultured human blood lymphocytes. Mutat Res/Genetic Toxicol Environ Mutagenesis. 2012;747(1):71–6.
165. Begum N, Prasad NR, Thayalan K. Apigenin protects gamma-radiation induced oxidative stress, hematological changes and animal survival in whole body irradiated Swiss albino mice. Int J Nutr Pharmacol Neurol Dis. 2012;2(1):45–52.
166. Paredes-Gonzalez X, Fuentes F, Su Z-Y, Kong A-NT. Apigenin reactivates Nrf2 anti-oxidative stress signaling in mouse skin epidermal JB6 P+ cells through epigenetics modifications. AAPS J. 2014;16:727–35.

167. Paredes-Gonzalez X, Fuentes F, Jeffery S, Saw CLL, Shu L, Su ZY, et al. Induction of NRF2-mediated gene expression by dietary phytochemical flavones apigenin and luteolin. Biopharm Drug Dispos. 2015;36(7):440–51.
168. Begum N, Prasad NR. Apigenin, a dietary antioxidant, modulates gamma radiation-induced oxidative damages in human peripheral blood lymphocytes. Biomed Prev Nutr. 2012;2(1):16–24.
169. Begum N, Rajendra Prasad N, Kanimozhi G, Agilan B. Apigenin prevents gamma radiation-induced gastrointestinal damages by modulating inflammatory and apoptotic signalling mediators. Nat Prod Res. 2022;36(6):1631–5.
170. Liu D, Peng R, Chen Z, Yu H, Wang S, Dong S, et al. The protective effects of Apigenin against radiation-induced intestinal injury. Dose-Response. 2022;20(3):15593258221113791.
171. Arring NM, Millstine D, Marks LA, Nail LM. Ginseng as a treatment for fatigue: a systematic review. J Altern Complement Med. 2018;24(7):624–33.
172. Jovanovski E, Komishon A, Au-Yeung F, Zurbau A, Jenkins AL, Sung M-K, et al. Vascular effects of combined enriched Korean red ginseng (Panax Ginseng) and American ginseng (Panax Quinquefolius) administration in individuals with hypertension and type 2 diabetes: a randomized controlled trial. Complement Ther Med. 2020;49:102338.
173. Riaz M, Rahman NU, Zia-Ul-Haq M, Jaffar HZ, Manea R. Ginseng: a dietary supplement as immune-modulator in various diseases. Trends Food Sci Technol. 2019;83:12–30.
174. Zhu H, Liu H, Zhu J-H, Wang S-Y, Zhou S-S, Kong M, et al. Efficacy of ginseng and its ingredients as adjuvants to chemotherapy in non-small cell lung cancer. Food Funct. 2021;12(5):2225–41.
175. Kim HY, Kang KS, Yamabe N, Yokozawa T. Comparison of the effects of Korean ginseng and heat-processed Korean ginseng on diabetic oxidative stress. Am J Chin Med. 2008;36(05):989–1004.
176. Park H-M, Kim S-J, Kim J-S, Kang H-S. Reactive oxygen species mediated ginsenoside Rg3-and Rh2-induced apoptosis in hepatoma cells through mitochondrial signaling pathways. Food Chem Toxicol. 2012;50(8):2736–41.
177. Park SE, Park C, Kim SH, Hossain MA, Kim MY, Chung HY, et al. Korean red ginseng extract induces apoptosis and decreases telomerase activity in human leukemia cells. J Ethnopharmacol. 2009;121(2):304–12.
178. Xu X, Ling Q, Wei Q, Wang K, Zhou B, Zhuang L, et al., editors. Korean red ginseng: a new approach for the treatment of graft-versus-host disease after liver transplantation. Transplantation proceedings. Elsevier; 2011.
179. Senchina DS, Shah NB, Doty DM, Sanderson CR, Hallam JE. Herbal supplements and athlete immune function—what's proven, disproven, and unproven. Exerc Immunol Rev. 2009;15(66):106.
180. Shishtar E, Sievenpiper JL, Djedovic V, Cozma AI, Ha V, Jayalath VH, et al. The effect of ginseng (the genus panax) on glycemic control: a systematic review and meta-analysis of randomized controlled clinical trials. PLoS One. 2014;9(9):e107391.
181. Pan H-Y, Qu Y, Zhang J-K, Kang T-G, Dou D-Q. Antioxidant activity of ginseng cultivated under mountainous forest with different growing years. J Ginseng Res. 2013;37(3):355.
182. Wong AS, Che C-M, Leung K-W. Recent advances in ginseng as cancer therapeutics: a functional and mechanistic overview. Nat Prod Rep. 2015;32(2):256–72.
183. Omrani V, Fardid R, Alavi M, Haddadi G, Takhshid MA. Protective effects of Panax Ginseng against 131I-induced genotoxicity in patients with differentiated thyroid cancer. J Cancer Res Ther. 2024;20(1):304–10.
184. Seong MA, Woo JK, Kang J-H, Jang YS, Choi S, Jang YS, et al. Oral administration of fermented wild ginseng ameliorates DSS-induced acute colitis by inhibiting NF-κB signaling and protects intestinal epithelial barrier. BMB Rep. 2015;48(7):419.
185. Chang JW, Choi JW, Lee BH, Park JK, Shin YS, Oh Y-T, et al. Protective effects of Korean red ginseng on radiation-induced oral mucositis in a preclinical rat model. Nutr Cancer. 2014;66(3):400–7.

186. He L-X, Wang J-B, Sun B, Zhao J, Li L, Xu T, et al. Suppression of TNF-α and free radicals reduces systematic inflammatory and metabolic disorders: Radioprotective effects of ginseng oligopeptides on intestinal barrier function and antioxidant defense. J Nutr Biochem. 2017;40:53–61.
187. Yilmaz H, Karakoc Y, Tumkaya L, Mercantepe T, Sevinc H, Yilmaz A, et al. The protective effects of red ginseng and amifostine against renal damage caused by ionizing radiation. Hum Exp Toxicol. 2022;41:09603271221143029.
188. Ozdemir A, Tumkaya L, Mercantepe T, Celik Samanci T, Uyan M, Kalcan S, et al. The protective effects of ginseng on x-irradiation-induced intestinal damage in rats. Radiat Environ Biophys. 2023;62(3):395–402.
189. Jagetia GC, Baliga MS, Venkatesh P, Ulloor JN. Influence of ginger rhizome (Zingiber officinale Rosc) on survival, glutathione and lipid peroxidation in mice after whole-body exposure to gamma radiation. Radiat Res. 2003;160(5):584–92.
190. Nadkarni A. popular press Ltd. Mumbai; 1976.
191. Umadevi K, Vanitha V, Vijayalakshmi K. Antimicrobial activity of three INDIAN medicinal plants- an in vitro study. Bioscan. 2011;6(1):25–8.
192. Rastogi RP, Mehrotra B. Compendium of Indian medicinal plants, vol. 1. CSIR: New Delhi; 1990. p. 118–22.
193. Aeschbach R, Löliger J, Scott B, Murcia A, Butler J, Halliwell B, et al. Antioxidant actions of thymol, carvacrol, 6-gingerol, zingerone and hydroxytyrosol. Food Chem Toxicol. 1994;32(1):31–6.
194. Tanaka T. Cancer chemoprevention by natural-products. Oncol Rep. 1994;1(6):1139–55.
195. Kong A-NT, Yu R, Chen C, Mandlekar S, Primiano T. Signal transduction events elicited by natural products: role of MAPK and caspase pathways in homeostatic response and induction of apoptosis. Arch Pharm Res. 2000;23:1–16.
196. Ahmed RS, Seth V, Banerjee B. Influence of dietary ginger (Zingiber officinales Rosc) on antioxidant defense system in rat: comparison with ascorbic acid. Indian J Exp Biol. 2000;38:443.
197. Aktan F, Henness S, Tran VH, Duke CC, Roufogalis BD, Ammit AJ. Gingerol metabolite and a synthetic analogue Capsarol™ inhibit macrophage NF-κB-mediated iNOS gene expression and enzyme activity. Planta Med. 2006;72(08):727–34.
198. Zick SM, Turgeon DK, Ren J, Ruffin MT, Wright BD, Sen A, et al. Pilot clinical study of the effects of ginger root extract on eicosanoids in colonic mucosa of subjects at increased risk for colorectal cancer. Mol Carcinog. 2015;54(9):908–15.
199. McGaffin KR, Witham WG, Yester KA, Romano LC, O'Doherty RM, McTiernan CF, et al. Cardiac-specific leptin receptor deletion exacerbates ischaemic heart failure in mice. Cardiovasc Res. 2011;89(1):60–71.
200. Jung HW, Yoon C-H, Park KM, Han HS, Park Y-K. Hexane fraction of Zingiberis Rhizoma Crudus extract inhibits the production of nitric oxide and proinflammatory cytokines in LPS-stimulated BV2 microglial cells via the NF-kappaB pathway. Food Chem Toxicol. 2009;47(6):1190–7.
201. Saberi H, Keshavarzi B, Shirpoor A, Gharalari FH, Rasmi Y. Rescue effects of ginger extract on dose dependent radiation-induced histological and biochemical changes in the kidneys of male Wistar rats. Biomed Pharmacother. 2017;94:569–76.
202. Raafat MH, Shoukry Y. The possible protective role of ginger in gamma radiation-induced Jejunal enteropathy in adult male rats. A light and electron microscopic study. Egyptian. J Anat. 2020;43(2):293–314.
203. Rodriguez C, Mayo JC, Sainz RM, Antolín I, Herrera F, Martín V, et al. Regulation of antioxidant enzymes: a significant role for melatonin. J Pineal Res. 2004;36(1):1–9.
204. Shirazi A, Ghobadi G, Ghazi-Khansari M. A radiobiological review on melatonin: a novel radioprotector. J Radiat Res. 2007;48(4):263–72.
205. Tan DX. Melatonin: a potent, endogenous hydroxyl radical scavenger. Endocr J. 1993;1:57.

206. Karbownik M, Reiter RJ. Antioxidative effects of melatonin in protection against cellular damage caused by ionizing radiation. Proc Soc Exp Biol Med Minirev. 2000;225(1):9–22.
207. Reiter RJ, Tan D-X, Herman TS, Thomas CR Jr. Melatonin as a radioprotective agent: a review. Int J Radiat Oncol Biol Phys. 2004;59(3):639–53.
208. Kim J-M, Yoo H, Kim J-Y, Oh SH, Kang JW, Yoo BR, et al. Metformin alleviates radiation-induced skin fibrosis via the downregulation of FOXO3. Cell Physiol Biochem. 2018;48(3):959–70.
209. Kolivand S, Motevaseli E, Cheki M, Mahmoudzadeh A, Shirazi A, Fait V. The anti-apoptotic mechanism of metformin against apoptosis induced by ionizing radiation in human peripheral blood mononuclear cells. Klin Onkol. 2017;30(5):372–9.
210. Xu G, Wu H, Zhang J, Li D, Wang Y, Wang Y, et al. Metformin ameliorates ionizing irradiation-induced long-term hematopoietic stem cell injury in mice. Free Radic Biol Med. 2015;87:15–25.
211. Esposito E, Cuzzocrea S. Antiinflammatory activity of melatonin in central nervous system. Curr Neuropharmacol. 2010;8(3):228–42.
212. Favero G, Franceschetti L, Bonomini F, Rodella LF, Rezzani R. Melatonin as an anti-inflammatory agent modulating inflammasome activation. Int J Endocrinol. 2017;2017(1):1835195.
213. Ortiz F, Acuña-Castroviejo D, Doerrier C, Dayoub JC, López LC, Venegas C, et al. Melatonin blunts the mitochondrial/NLRP 3 connection and protects against radiation-induced oral mucositis. J Pineal Res. 2015;58(1):34–49.
214. Meng X, Li Y, Li S, Zhou Y, Gan R-Y, Xu D-P, et al. Dietary sources and bioactivities of melatonin. Nutrients. 2017;9(4):367.
215. Karslioğlu I, Ertekin MV, Taysi S, Koçer I, Sezen O, Gepdiremen A, et al. Radioprotective effects of melatonin on radiation-induced cataract. J Radiat Res. 2005;46(2):277–82.
216. Erol FS, Topsakal C, Ozveren MF, Kaplan M, Ilhan N, Ozercan IH, et al. Protective effects of melatonin and vitamin E in brain damage due to gamma radiation: an experimental study. Neurosurg Rev. 2004;27:65–9.
217. El-Missiry M, Fayed T, El-Sawy M, El-Sayed A. Ameliorative effect of melatonin against gamma-irradiation-induced oxidative stress and tissue injury. Ecotoxicol Environ Saf. 2007;66(2):278–86.
218. Taysi S, Koc M, Büyükokuroğlu ME, Altınkaynak K, Şahin YN. Melatonin reduces lipid peroxidation and nitric oxide during irradiation-induced oxidative injury in the rat liver. J Pineal Res. 2003;34(3):173–7.
219. Sharma S, Haldar C. Melatonin prevents X-ray irradiation induced oxidative damagein peripheral blood and spleen of the seasonally breeding rodent, Funambulus pennanti during reproductively active phase. Int J Radiat Biol. 2006;82(6):411–9.
220. Shabeeb D, Najafi M, Musa AE, Keshavarz M, Shirazi A, Hassanzadeh G, et al. Biochemical and histopathological evaluation of the radioprotective effects of melatonin against gamma ray-induced skin damage. Curr Radiopharm. 2019;12(1):72–81.
221. Shabeeb D, Najafi M, Keshavarz M, Musa AE, Hassanzadeh G, Hadian MR, et al. Recent finding in repair of the peripheral nerve lesions using pharmacological agents: common methods for evaluating the repair process. Cent Nerv Syst Agents Med Chem. 2018;18(3):161–72.
222. Najafi M, Shirazi A, Motevaseli E, Geraily G, Amini P, Shabeeb D, et al. Evaluating the expression of NOX2 and NOX4 signaling pathways in rats' lung tissues following local chest irradiation; modulatory effect of melatonin. Int J Mol Cell Med. 2018;7(4):220.
223. Citrin D, Cotrim AP, Hyodo F, Baum BJ, Krishna MC, Mitchell JB. Radioprotectors and mitigators of radiation-induced normal tissue injury. Oncologist. 2010;15(4):360–71.
224. Reiter RJ, Paredes SD, Manchester LC, Tan D-X. Reducing oxidative/nitrosative stress: a newly-discovered genre for melatonin. Crit Rev Biochem Mol Biol. 2009;44(4):175–200.
225. Reiter RJ, Tan D-x, Burkhardt S. Reactive oxygen and nitrogen species and cellular and organismal decline: amelioration with melatonin. Mech Ageing Dev. 2002;123(8):1007–19.

226. Reiter RJ, Korkmaz A, Ma S, Rosales-Corral S, Tan D-X. Melatonin protection from chronic, low-level ionizing radiation. Mutat Res/Rev Mutat Res. 2012;751(1):7–14.
227. Reiter RJ, Tan D-x, Manchester LC, Qi W. Biochemical reactivity of melatonin with reactive oxygen and nitrogen species: a review of the evidence. Cell Biochem Biophys. 2001;34:237–56.
228. Reiter RJ. The pineal gland and melatonin in relation to aging: a summary of the theories and of the data. Exp Gerontol. 1995;30(3–4):199–212.
229. Kaya H, Delibas N, Serteser M, Ulukaya E, Özkaya O. The effect of melatonin on lipid peroxidation during radiotherapy in female rats. Strahlenther Onkol. 1999;175(6):285–8.
230. Okatani Y, Wakatsuki A, Shinohara K, Kaneda C, Fukaya T. Melatonin stimulates glutathione peroxidase activity in human chorion. J Pineal Res. 2001;30(4):199–205.
231. Galano A, Tan DX, Reiter RJ. On the free radical scavenging activities of melatonin's metabolites, AFMK and AMK. J Pineal Res. 2013;54(3):245–57.
232. Shirazi A, Fardid R, Mihandoost E. Protective effect of low dose melatonin on radiation-induced damage to rat liver. J Biomed Phys Eng. 2012;2(2)
233. Haddadi GH, Fardid R. Oral administration of melatonin modulates the expression of tumor necrosis factor-alpha (TNF-alpha) gene in irradiated rat cervical spinal cord. Rep Pract Oncol Radiother. 2015;20(2):123–7.
234. Fardid R, Salajegheh A, Mosleh-Shirazi MA, Sharifzadeh S, Okhovat MA, Najafi M, et al. Melatonin ameliorates the production of COX-2, iNOS, and the formation of 8-OHdG in non-targeted lung tissue after pelvic irradiation. Cell Journal (Yakhteh). 2017;19(2):324.
235. Najafi M, Cheki M, Hassanzadeh G, Amini P, Shabeeb D, Musa AE. The radioprotective effect of combination of melatonin and metformin on rat duodenum damage induced by ionizing radiation: a histological study. Adv Biomed Res. 2019;8(1):51.
236. Fernández-Gil B, Moneim AEA, Ortiz F, Shen Y-Q, Soto-Mercado V, Mendivil-Perez M, et al. Melatonin protects rats from radiotherapy-induced small intestine toxicity. PLoS One. 2017;12(4):e0174474.
237. Musa AE, Shabeeb D, Alhilfi HSQ. Protective effect of melatonin against radiotherapy-induced small intestinal oxidative stress: biochemical evaluation. Medicina. 2019;55(6):308.
238. Tahamtan R, Monfared AS, Tahamtani Y, Tavassoli A, Akmali M, Mosleh-Shirazi MA, et al. Radioprotective effect of melatonin on radiation-induced lung injury and lipid peroxidation in rats. Cell J (Yakhteh). 2015;17(1):111.
239. Gezer A, Karadag-Sari E. The role of amifostine in preventing radiotherapy induced testicular tissue damage in rats. Biotech Histochem. 2022;97(3):215–21.
240. Cakina S, Gulyasar T, Özen A, Parlar S, Cukur Z, Kurkcu N, et al. Comparison of protective effects of melatonin and Amifostine on radiation-induced renal oxidative stress in rats. J Hellenic Vet Med Soc. 2020;71(1):2005–10.
241. Spitz DR, Oberley LW, Buettner GR. Tempol as a radioprotector: the potential role of a redox cycling compound. Free Radic Biol Med. 1994;16(4):537–47.
242. Epperly MW, et al. Adenoviral MnSOD gene therapy ameliorates radiation-induced pulmonary injury. Radiat Res. 2001;156(2):146–55.
243. Epperly MW, et al. Intratracheal administration of manganese superoxide dismutase plasmid liposomes mitigates radiation-induced lung injury. Radiat Res. 2002;158(2):186–94.
244. Gupta RC. Emerging radioprotective agents for clinical use: ex-rad. Expert Opin Investig Drugs. 2010;19(6):783–791.244.
245. Bonner JA, et al. Palifermin for oral mucositis after intensive therapy for hematologic cancers. N Engl J Med. 2004;351(25):2590–8.

Chapter 17
Advances in Radiotherapy Using Artificial Intelligence

Shubham Kumar, Bimlesh Kumar, Sumel Ashique, Utpal Bhui, and Sheetu

Abstract Radiation therapy is one of the medical fields where artificial intelligence (AI) has emerged as an influential force with enormous potential to change treatment procedures completely and results. The development of radiation therapy and the vital part AI plays in improving accuracy and effectiveness in healthcare. The fundamental ideas of neural networks and machine learning are described, along with important AI techniques that are now used in radiotherapy. AI's critical contributions to tumor identification, treatment planning, adaptive radiation approaches, outcome prediction, and patient-specific dosimetry are among the applications that are highlighted. The article discusses AI-driven advances that enhance workflow efficiency and treatment homogeneity, including automated contouring, real-time treatment monitoring, and quality assurance. The future of precision and individualized radiation is demonstrated by emerging technologies such as radiomics, federated learning, and AI-assisted advanced therapies. Data privacy, ethical considerations, algorithmic interpretability, and clinical validation are some of the directions in which contemporary AI has been making the technology more efficacious, even though it still maintains constraints. AI strategies should be optimally initiated as a part of a comprehensive, multicentric, and prospective validation process to ensure both their effectiveness in a wide range of clinical scenarios and the quality of their precision. This study reviews AI's potential for individual precision medication, how to decrease differences in the provision of radiotherapy, and how to embrace collaboration in the research community. The implementation of AI in the radiotherapy workflow can be an effective solution for the increasing complexity of modern radiotherapy without failing to address the need for individualized patient care, radiation dose recommendation refinement, and patient outcome improvement.

S. Kumar · B. Kumar (✉) · U. Bhui · Sheetu
School of Pharmaceutical Sciences, Lovely Professional University Phagwara, Phagwara, Punjab, India

S. Ashique
School of Pharmaceutical Sciences, Lovely Professional University Phagwara, Phagwara, Punjab, India

Department of Pharmaceutical Technology, Bharat Technology, Uluberia, West Bengal, India

F. Taghizadeh-Hesary et al. (eds.), *From Radiobiology to Radiation Oncology*,
https://doi.org/10.1007/978-3-031-95724-6_17

Keywords Radiotherapy · Artificial intelligence · Deep learning · Algorithm

17.1 Introduction

17.1.1 Overview of Radiotherapy

Radiotherapy is the use of light waves to kill cancerous cells. However, there are many side effects too. Approximately 50% of all cancer patients get radiotherapy for their disease. Radiotherapy is used to recover from cancer or to manage painful symptoms; that is, it is palliative [1]. The earliest recorded use of irradiation as a treatment was by Emil Grubbé in Chicago in 1896, when it was applied to manage the postoperative recurrence of breast cancer. Early radiotherapy focused on surface X-ray irradiations, but technological advances during World Wars I and II introduced deeper-penetrating sources like cobalt-60, betatrons, and linear accelerators. Radiotherapy is delivered externally (X-rays, γ-rays) or internally (brachytherapy, radioisotopes). Proton therapy shows promise, though its benefits remain debated. A radiotherapy utilization rate exceeding 50% is now a benchmark for effective cancer care delivery in developed countries. In the USA, this translated to more than 600,000 patients receiving radiotherapy treatments annually [2].

17.1.2 Role of Artificial Intelligence in Healthcare

Artificial intelligence is poised to revolutionize healthcare by enhancing various processes and improving patient outcomes. Artificial intelligence (AI) can augment clinicians' work by assisting in the diagnostic process, identifying opportunities for prevention, and providing personalized treatment recommendations. The ability of the AI to understand the patient from the electronic health record is very important since we live in a time when health data is widely available. AI is empowering healthcare to become more transparent in decision models. It is also about ethics in the field of AI, where partiality and the establishment of liability are the main aspects that the research draws attention to. Again, common evaluation metrics used to measure the capability of AI are vital, as they will enable the healthcare practice to be done accurately and securely [3, 4].

17.1.3 Importance of AI in Modern Radiotherapy

Modern technology is AI, the most important aspect that has been contained to be very much used in modern radiotherapy; it brings the ability range to the patient for the better to have a more accurate treatment. Planning optimization of AI in treatment is the main activity of AI. The AI algorithms can recommend the most precise therapy parameters that are specific to each patient by analyzing large datasets from previous cases, as a result of which better treatments are developed. At the same time, AI enhances the accuracy of medical image analysis, in particular using machine learning techniques, which allow the detection and characterization of tumors more precisely. This is crucial to plan radiotherapy effectively [5].

In addition, AI also helps to avoid human mistakes and save time by doing everyday tasks like scrutinizing cancer lumps and pointing out the organs in danger by itself, which in turn ensures a consistent and successful therapy frequency. AI, too, facilitates the monitoring of patient responses to treatments in real time. Besides, they also allow for modifications and practice to ensure maximum efficiency as the patients change. Another field where AI stands out is predictive analytics. It is a module that provides the prognosis of patient outcomes by using a given method of treatment based on existing data. What this means for medical professionals is that they can have their choices backed by data and also give reasonable expectations to the patients. Moreover, AI can easily merge with other innovations such as radiation delivery and imaging items to form a safer and more cohesive treatment process. Therefore, regarding AI, it becomes clear that it is more individualized and efficacious and shows a real new step in cancer care [6].

17.2 Basics of Artificial Intelligence in Radiotherapy

17.2.1 Machine Learning and Deep Learning Basics

AI can make a revolution by redefining the diagnosis and treatment of many cancers, which are now difficult to treat. As a non-robotized AI, when we refer to the application of artificial intelligence to the field of oncology and specifically to image analysis, it is not AI that represents a complicated solution to every problem in this area but rather data-based AI that deeply understands the domain knowledge and can deliver relevant predictions. The newly introduced assumption of AI in diagnosing cancer has concerned some oncologists. The human-in-the-loop AI can be a great tool for catching rare disease examples to learn from and for making rare disease diagnoses using deep learning algorithms. We see some difficulties in turning the system from an AI descriptive diagnostic to a therapeutic model. For this, together with AI, we must redesign our diagnostic and treatment processes by developing AI-support models to facilitate diagnostics, treatments, and patient care [7].

These inventions ensure that the carrying out of the treatment is properly done, precision is increased, and the operation of administration is streamlined. The process of radiotherapy is more precise now, and the waiting time is shorter due to AI integration, which provides a platform that requires less staff time, minimizes the variability of treatment outcomes and the burden on medical staff, but at the same time permits more personalized care. There are several barriers to optimizing AI in the radiation oncology field, such as the lack of standardized data and ad hoc treatment planning software. Despite the problems, the increasing computing power and the newer algorithms are bringing the benefits of AI to a wider number of cases. AI is redefining the therapy landscape by advancing precision, productivity, and the overall cancer care delivery standard using machine learning and deep learning [8, 9].

17.2.2 Key AI Algorithms Used in Radiotherapy

Radiotherapy in the hands of AI is a real game-changer, as it not only shortens patient recovery time but also guarantees the most precise and resource-effective treatment. The technical stuff I am writing is about neural networks, which are the basic part of deep learning that is intended to mimic the computational abilities of the human brain, and these are the main characters in the entire shift. They are specialists in computer vision tasks such as automated detection and object recognition in medical images, identifying different body parts, and classifying these into different structures or categories. Convolutional neural networks (CNNs), an example of a special category, are doing great in this area [10]. Their main tasks are the segmentation of the organ and the determination of the tumor margin for radiation therapy, which is crucial for treatment planning. Another AI technique, reinforcement learning, enhances the therapy dose parameters adjustment by considering such parameters as patient reactions, learning from the performed actions in the past, and optimum setting up of the treatment. Support vector machines (SVMs) for high-dimensional datasets are used for effective differentiation between the two cases and job classification while the imaging data deals with tissue analysis [11]. To this end, random forests, an ensemble learning set, are employed for patient prognosis and diagnosis, dividing the patients into their groups and forecasting based on their clinical and imaging characteristics. Recurrent neural networks (RNNs) and autoencoders are two examples of deep learning networks, among many others, that deal with multimodal data and risk prediction through the integration of imaging and clinical data in radiation therapy selection. Mostly, when forecasting, Gradient Boosting Machines (GBMs) are used. The AI algorithms process the data, thus enabling the therapy to use and optimize them in their clinical workflow, decision-making, and cancer therapy, taking it up for patients to be more precise, effective, and personalized [10, 11].

17.2.3 Integration of AI into Existing Radiotherapy Frameworks

Inserting AI into the radiotherapy sets brings many benefits and the most prominent disadvantages of AI in radiotherapy. Radiation therapy is elaborated, and AI is mentioned as a way to transform it, nearly leading to AI being the main actor.

Consequently, deep learning (DL) technology, which improves the accuracy and speed of image processing, treatment planning, and predicting patient outcomes, is one of the major trends in this field. On the other hand, the COVID-19 pandemic highlighted the urgent need for innovative solutions across various aspects of healthcare. The abrupt nature of these changes underscored the importance of adaptability. In this context, the emergence of artificial intelligence (AI) and digital radiology, where advancements in one domain significantly influence the other, has proven to be particularly impactful. Their integration should be regarded as a vital component of modern healthcare systems [6, 12].

The main issue with radiation is that there are no standards for data gathering and analysis. For AI and applications to be reliable and effective, it is necessary to have organized data storage in one place and cooperation networks. Regulations are an especially important subject matter to deal with for the successful application of AI in clinics; hence, the visualization of medical imaging and planning of treatments have to meet the standards [12]. The problem could only be solved if the research scientists, physicians, and the government would collaborate on diverse issues. These alliances may give a significant boost to the recovery of patients and the elimination of the barriers to progress in therapy through the use of AI in full capacity in radiotherapy. The future of AI in medicine is promising; it is foreseen that it will improve diagnostic skills, automate complex tasks, and individualize treatment regimens. To optimally utilize it, AI for changeable radiation procedures to high-quality ones and the improvement of patients' outcomes ought to be realized through the constant studying and developing processes [13].

17.3 Applications of AI in Radiotherapy

17.3.1 Tumor Detection and Localization

The integration of AI with radiation significantly enhances the localization of tumors and the accuracy of their detection with the added advantage of an improved treatment outcome. AI-based methods enhance target definition to ensure that radiation is delivered precisely to the tumor while minimizing exposure to adjacent normal organs. AI also facilitates multi-modality image registration, which uses image data across multiple modalities [e.g., computerized tomography (CT), magnetic resonance imaging (MRI)] to generate a more comprehensive understanding of the tumor characteristics and behavior [14].

Another important advantage is that it is possible to observe tumors in real time since AI algorithms enable the continued observation of tumor motion while treatment continues. For instance, single-view volumetric imaging, particularly effective for tumors affected by respiration, facilitates observation of the tumor's location and consequent adjustment of treatment. AI-driven image-guided radiotherapy (IGRT) algorithms identify and correct positional errors that could be made during treatment, thus minimizing geometric uncertainties, particularly in 3D volumetric imaging [15]. Yet another significant application is adaptive radiotherapy, which facilitates real time updating of the treatment plan concerning the position of the tumor and the patient's anatomy. In light of possible large anatomical changes during fractionated radiation treatment, this versatility is especially necessary. Non-invasive soft-tissue registration becomes feasible due to AI-based IGRT methods, rendering the technique more practical for use in clinical routine than established fiducial-based techniques requiring invasive interventions [16].

Furthermore, by providing accurate tumor localization and real-time target positioning confirmation, AI enhances treatment confidence by ensuring that the prescribed dose is delivered as planned. This improves the chances of successful tumor control while reducing the risk of undesirable effects on healthy tissues. On balance, the application of AI in radiation enhances accuracy and improves treatment safety and efficacy, which is beneficial to both patients and physicians [17].

17.3.2 Treatment Planning Optimization

AI can mechanize significant processes, enhance accuracy, and personalize approaches for challenging patients without sacrificing essential human governance; it has a fundamental role to play in maximizing treatment planning. It minimizes trial-and-error methods and ensures consistently high-quality treatment plans by enhancing several steps of the treatment planning process, including fluence map generation, DVH projections, and dose distribution estimates. An important ingredient of such advancements is machine learning (ML) models, which are usually classified into two categories. The models predicting DVH, dosage, and fluence map in the first category rely on historic clinical data to predict optimal treatment parameters, translating experience to enhance planning quality. By mimicking human decision-making, the second type of planning strategy model assists optimization engines in arriving at iterative increments that culminate in optimal treatment outcomes [18, 19]. Treatment planning is also expected to become increasingly accurate, efficient, and trustworthy as AI models improve and training data increases. AI also plays a significant role in addressing simple and complex problems. Fluence map prediction models can quickly generate clinically viable treatment plans for less complex patients, whereas dose prediction models and DVHs help to traverse trade-offs between various dose-volume constraints for more complex cases. Future treatment planning with AI will involve blending different models to address different patient anatomy and cancer types, ensuring more personalized and effective care [19, 20]. Human input is still needed in treatment planning to safeguard patient health and ensure clinical decisions are according to best practice, even with these advances. AI amplifies automation and speed, but knowledge from healthcare professionals is still paramount for verifying and enhancing AI-suggested plans. Overall, AI revolutionizes treatment planning by accelerating processes, refining accuracy, and enabling tailored solutions while still demonstrating the worth of human input to clinical decision-making [19, 21].

17.3.3 Adaptive Radiotherapy Techniques

Adaptive radiotherapy (ART) is a state-of-the-art method of treating cancer that dynamically modifies radiation treatment regimens in response to patient anatomical changes throughout treatment. Through its capability to consistently adapt by recognizing the changes in the size, contour, and location of the tumor, ART advancement further enhances the precision of the procedure and ensures the injury to healthy tissues is kept to a minimum and radioactive substances are delivered properly [22, 23].

The two primary approaches for the implementation of ART are offline, where adjustments are done between the sessions after the previous imaging data has been reviewed, and online, where adjustments are done in real time during the treatment session with the help of rapid imaging feedback. ART is a kind of radiation therapy that is based not only on internal organ misplacement and movement but also on the localization of the targeted tumor. It heavily depends on the imaging technology. State-of-the-art technology like CT, MRI, and positron emission tomography (PET) scans is used to exactly measure any organ movement and the reduction in tumor size. One of the crucial tools of ART is gating, which can be particularly useful in lung cancer treatment. This technique is all about synchronizing the radiation with the patient's breathing cycle to focus on the tumor only, whereas the nearby tissues will be protected. Moreover, biologically adaptive radiation therapy designs treatment plans that are focused not only on the anatomy of the organ but also on the biological response and thereby take into consideration how organs and tumors function and grow. The on/off changes in function are included in the plan to optimize radiation delivery [23, 24].

Furthermore, ART's high-precision applications, especially in lung cancer treatment, are exemplified by Stereotactic Body Radiation Therapy (SBRT), which uses daily imaging to adapt radiation plans, ensuring accurate dose delivery to the tumor while protecting organs at risk. ART represents a significant advancement in radiotherapy, improving treatment efficacy, reducing side effects, and improving patient outcomes by continuously adapting to anatomical and biological changes throughout treatment. Emerging technologies like MRI-guided and cone-beam CT (CBCT)-guided ART provide enhanced imaging capabilities, improving tumor targeting while reducing radiation exposure to normal tissues [25].

17.3.4 Prediction of Treatment Outcomes

AI is transforming cancer radiotherapy by improving predictive modeling, incorporating radionics, and maximizing treatment personalization. Machine learning (ML) and deep learning (DL) algorithms process large datasets to forecast patient outcomes, facilitating risk stratification and treatment planning. Radiomics, which harvests quantitative features from medical images, improves AI's capacity to evaluate tumor behavior and response to treatment. AI also enables dynamic imaging analysis, utilizing real-time information from cone-beam computed tomography to track tumor evolution and early signs of toxicity. AI-based feature selection enhances predictive models for patient survival, enabling treatment de-escalation without diminishing efficacy. These models, however, need to be tested by large trials before clinical implementation to ascertain safety and reliability. Eventually, AI-powered innovations in radiotherapy are of tremendous potential in pushing the field of personalized treatment strategies for cancer [12, 26].

17.3.5 Patient-Specific Dosimetry

Patient-specific dosimetry is being greatly improved by the incorporation of sophisticated computational methods into radiotherapy. Automated segmentation is a crucial application in which deep learning (DL) models accurately detect tumors and the tissues around them. This enhances treatment planning by guaranteeing exact radiation targeting while reducing exposure to healthy tissues. Furthermore, predictive models are being created to foresee mechanical mistakes in treatment equipment. By seeing possible problems before they affect treatment results, these models are essential to patient-specific quality assurance (QA) [27, 28].

A significant development is the evaluation of gamma passing rates (GPR) for volumetric-modulated arc treatment (VMAT) and intensity-modulated radiation therapy (IMRT) schemes. Treatment safety is increased by using GPR prediction to make sure the delivered radiation dose closely resembles the intended dose. Furthermore, to maintain high QA standards and guarantee the efficacy and dependability of treatment equipment, computational models are being used to forecast machine beam data [29, 30].

Healthcare workers will be able to concentrate more on patient care as a result of the anticipated simplification of radiation procedures brought about by the incorporation of these technologies into clinical workflows. Going forward, it is anticipated that continued research and development in this area will bring new approaches that improve the accuracy and security of radiation therapy treatments. All of these developments work together to increase clinical efficiency, patient safety, and treatment precision [27].

17.4 AI-Driven Innovations in Radiotherapy Workflow

17.4.1 Automated Contouring and Segmentation

In radiation therapy, automated contouring and segmentation are essential because they precisely identify tumors and the surrounding organs at risk (OARs). By using sophisticated computational techniques, automated contouring reduces the need for doctors to manually enter information by automatically identifying anatomical components in medical imaging. The customization of the radiotherapy to the neoplasms with the help of the procedure and with the tissue of those living beings that do the procedure is improvised [31].

This scientific paper covers the recent developments in the field of diagnosis and treatment of cancer, mainly focusing on medical imaging, which is the major focus of the research. The study shows how the two methods, OncoStudio and Protege AI, which are based on automatic cancer patient segmentation with deep learning techniques in the form of CNNs, work to intensify the accuracy and speed of the segmentation process [32].

Based on a performance test, OncoStudio, which was generated using a dataset mainly consisting of patients, proved to be more accurate in segmentation than Protege AI. This is an example of the importance of population-targeted models to increase the benefit of the treatment plan. In general, the computer-aided technology of automatic contouring and segmentation has been a significant achievement in radiation therapy, thereby improving the accuracy, and hence, the efficiency in the treatment of cancer patients [33].

17.4.2 Real-Time Treatment Monitoring

AI-driven remote patient monitoring (RPM) systems have greatly improved symptom monitoring accuracy, enabling prompt responses that aid in efficiently managing symptom burden. These devices are also predictive; research indicates that problems can be detected as early as 48 h before a clinical diagnosis, allowing for preventative care and lowering hospitalization rates. According to research, patients who use RPM have decreased hospital readmission rates because real-time monitoring facilitates better post-treatment health management. Reports indicate that up to 83.5% of patients adhere to RPM therapies, and when the technology is easy to use and improves symptom tracking, overall patient satisfaction increases [34, 35].

AI-powered RPM is expected to play a significant role in cancer treatment, especially in the post-pandemic healthcare setting where remote monitoring is becoming more and more important. To confirm its long-term effects and investigate its incorporation into standard cancer treatment, more investigation is required [35].

17.4.3 Quality Assurance with AI

The credibility of study findings is ensured by rigorous evaluation techniques in AI-driven remote patient monitoring (RPM) for cancer care. Studies that were included in this were evaluated for validity and replicability using quality evaluation methodologies, such as the Critical Appraisal Skills Programme (CASP) checklists. Bias evaluation, a crucial step in this approach, found and eliminated studies with high potential for bias, including selection and performance bias, to preserve the integrity of the research.

Strict inclusion and exclusion criteria were also followed, choosing only excellent research that was published after the COVID-19 epidemic started and that particularly examined AI uses in RPM for cancer patients. The review emphasizes the necessity of continued research to solve issues with scalability, cost-effectiveness, and integration into conventional healthcare procedures, despite encouraging outcomes. The successful real-world application of AI-driven RPM in cancer care will depend on maintaining ongoing quality assurance [34, 36].

17.4.4 Workflow Efficiency and Error Reduction

AI integration in interventional radiotherapy (IRT) has shown great promise for improving treatment accuracy and workflow efficiency. AI can cut catheter reconstruction time by up to 75% by automating repetitive processes like contouring and treatment planning, freeing up medical staff to concentrate on more intricate patient care facets. Furthermore, by evaluating intricate patient data, AI-driven treatment planning optimizes the distribution of radiation doses, guaranteeing accurate positioning of radioactive sources while reducing exposure to healthy tissues [37]. Additionally, AI improves overall patient safety by identifying up to 57% of patient safety events that could otherwise go undetected, strengthening quality assurance (QA). AI solutions help save time in a variety of clinical workflow stages, freeing up medical professionals to focus more on patient care. Enabling substantial infrastructure improvements by process standardization and shortening the learning curve for IRT, cancer care inefficiencies can be overcome, and the most modern drugs and practices can be made available to under-resourced healthcare facilities. Notwithstanding the above, effective clinical validation is still a must to conduct the software technology AI for bringing healthcare services on board for clinicians and not to replace them. In a nutshell, AI eliminates operational errors, streamlines workflow, and augments patient safety with IRT; still, its pros need to be monitored and certified for errors [38].

17.5 Emerging AI Technologies in Radiotherapy

17.5.1 Federated Learning in Oncology

AI solutions support a variety of clinical workflows, and thus, medical professionals have the liberty to work on patient care tasks, among other things. Besides, when it comes to efficiency, AI with a uniform process and a shorter learning curve for IRT can be used as a universal remedy in global disparities in cancer care and as part of the healthcare settings that have fewer resources, still giving access to innovative treatments. Nevertheless, clinical validation is vitally significant to ensure the working nature of AI systems is safe and beneficial, which, in turn, should help doctors rather than displace them. AI performs better in treatment, reduces the number of errors in IRT, and also facilitates workflow; whereas, still, its application should be observed and verified very closely. The study showed that patients who were identified as low-risk by the model had the highest likelihood of curative treatment, which indicated better survival outcomes. In the future, federated learning will be one of the keys to advancing research by increasing the size of the set of data and making the model more generalizable. Future research should also be focused on the inclusion of systemic therapy and additional patient factors in the decision-support tools in oncology. The data that are leveraged through federated learning, besides ensuring privacy, offer new avenues to treatment and patient care through cancer care [39].

17.5.2 AI-Assisted Proton and Carbon Therapy

By targeting precisely the carbon and proton beams on the tumor while limiting the damage to the healthy tissue of the patient, AI algorithms are improving therapy and minimizing any negative side effects. The treatment process is run through AI planning, a technological process that lets doctors easily make immediate changes based on the patient's response and real-time imaging; therefore, this is more of an individualized and specialized therapy [40]. AI implemented in both monitoring and feedback systems allows for automatic evaluation of treatments, which gives people the right information to help them make the most suitable decisions for their health. Using data analysis, AI identifies trends in large datasets of previous treatments, makes up forecasts of patient reactions, and enables doctors to develop treatment programs that are clinically supported. AI is also in regards to MRI and CT scans, which are also considered to be an imaging technique, increasing the efficiency of delineation of the tumors from the surrounding tissues and the hills. The constant endeavor of the research is about making a thorough examination of the patient's safety and effectiveness with the help of AI-driven novel algorithms and methodologies. On the whole, when faithful localization, real-time monitoring, and data-driven decision-making are merged with the treatment, patient care is significantly affected by AI exclusion in proton and carbon treatment, i.e., cancer management could dramatically benefit [41, 42].

17.5.3 Multi-modality Imaging and AI Integration

The combination of multiple imaging methods, including PET (positron emission tomography), CT (computerized tomography), and MRI (magnetic resonance imaging), provides a complete picture of the tumors in multi-modality imaging. This comprehensive diagnostic approach is a necessity in oncology, for example, for the precise diagnosis, treatment planning, and prognosis, as each imaging modality provides specific information about the tumors. Nonetheless, conventional imaging techniques such as spatial resolution, sensitivity, and functional insights become the problems that can hamper the successful design of individualized treatment plans and the efficient evaluation of treatment outcomes. By examining complex medical images from multiple modalities and the involvement of artificial intelligence (AI), the quality of the analysis is enhanced. The application of AIs, especially CNNs, can enable feature/extraction automation and tumor segmentation in the case of AI-aided methods, significantly increasing the accuracy and reducing the manual labor of the diagnosis [43].

The proposed framework, Multi-Modal Imaging and Enhanced Tumor Characterization (MMIETC), combines multimodal imaging data with advanced machine learning methods. This methodology employs CNNs for feature extraction and segmentation and random forests for predictive biomarker identification to enhance tumor characterization. The system combines information from many imaging modalities to provide a complete view of tumor traits, enabling clinical decision-making with accurate segmentations and customized treatment recommendations. By facilitating dynamic treatment modifications, this integration enhances patient outcomes. The dependence on excellent multimodal datasets, however, presents a problem because they might not always be accessible. The combination of AI and multimodal imaging is a breakthrough in cancer that will eventually improve patient care by enabling more accurate tumor descriptions and tailored treatment plans [44].

17.5.4 Radiomics and AI in Personalized Radiotherapy

Radiomics and AI are driving personalized radiotherapy toward medical precision and success. Radiomic biomarkers are extracted from medical images using a comprehensive analysis method called radiomics to incorporate different phenotypic features of cancers. These biomarkers for imaging trajectory help in precision medicine decision-making by linking imaging features to patient outcomes [45]. The precision medicine decision process is supported by these biomarkers that correlate imaging features with patient outcomes. AI, particularly machine learning, boosts the prognostic powers of radiomics by forming statistical models predicting treatment outcomes, survival, and probable adverse side effects by using extracted radiomic features. Radiomics is popular for stratifying patients, referring patients to their unique radiomic signatures. In addition, the AI interoperability augments the prediction of treatment efficacy and toxicity, which optimizes radiation dose and mitigates side effects. Ongoing studies are meant to determine best practices concerning feature selection methods and machine learning algorithms for the most significant improvement in the prognosis [12, 43].

17.6 Challenges and Limitations of AI in Radiotherapy

17.6.1 Data Privacy and Security Concerns

Data sharing in healthcare must strike a balance between the need to secure protected health information and the advantages of thorough medical histories, making patient privacy protection a crucial concern. It becomes more complicated to decide which data is pertinent to a given doctor-patient contact, which calls for clear guidelines for safe exchange, use, and preservation. Advanced information systems that use strong encryption techniques to stop unwanted access are becoming more and more necessary to avoid breaches of sensitive data. Transparency in AI models is still a worry since it can be challenging to understand how they operate, particularly when data protection laws restrict access to training datasets to organizations that have received ethical approvals [46, 47]. Big data and AI integration in healthcare also present ethical questions because patients might not be aware that their anonymized data is being used to create AI models. These concerns must be addressed by regulatory agencies to guarantee moral and responsible data use. Additionally, as new medical technologies are developed, the exponential increase in healthcare data necessitates a robust IT infrastructure that can safely store and handle massive volumes of data. Concerns about data security and privacy must be addressed because AI has the potential to revolutionize healthcare. This calls for improved transparency in AI, the creation of safe data management systems, and careful evaluation of the moral ramifications of exploiting patient data [48].

17.6.2 Ethical and Regulatory Challenges

Deep Learning (DL) algorithms in healthcare necessitate access to private health data, which raises ethical questions about data privacy and the need for express patient consent. These models are a real issue since the degree of effectiveness of these models rests on the diversity of the training data. If the datasets are not representative of the overall population, the models that are built may further augment existing healthcare inequalities by embedding gaps in treatment outcomes. A second fear is clinician dependence on AI, which may challenge clinicians to autonomously adapt and integrate their clinical judgment, potentially underappreciating the doctor's base of knowledge about their patients and the significance of patient-recorded symptoms [13, 27].

The use of dynamic AI models that update treatment recommendations as new data or applications of the model become available raises concerns about patient autonomy and the continuity of the informed consent process [5]; they amplified the need for transparency about how decisions are made by the AI system. The other critical ethical concerns include equitable access to AI-enabled clinical solutions, as restricted access to such solutions may further widen the disparities of cancer treatment along with the access to other medical interventions across diverse racial and socioeconomic groups. Durable monitoring frameworks, such as clear data governance guidelines and nonstop cooperation between tech inventors, medical experts, and ethicists, are necessary to address these issues and guarantee the ethical and responsible application of AI in healthcare [15, 22].

17.6.3 Interpretability of AI Models

AI models provide transparency into their decision-making processes, a crucial feature in the medical industry where critical decisions are made. However, many challenges remain on the path toward applying radionics approaches in clinical practice. The lack of reproducible performance in a diversity of clinical environments is a big problem preventing such models from being widely accepted and trusted by clinicians. Explainable artificial intelligence (XAI) is needed now to understand how certain AI models generate predictions, to identify possible biases in the training data, and to instill confidence in the model's prediction behavior on new data [27, 32, 35].

XAI defines research directions to build approaches that can demonstrate how deep learning techniques work, both statistically and visually, thus helping model interpretation. These interpretability techniques are diverse and range from visualization techniques to quantitative metrics on the quality of explanations produced by the models, although they are still under exploration to define what exactly a "good" explanation is. The most serious challenge to achieving interpretability, for example, is the complexity of deep learning models such as VGG-16 that contain millions of parameters and make it difficult to draw meaningful insights about their behavior. To make them clinically useful, future studies will have to continue tackling these key aspects, increasing the data harmonization and improving the generalizability of models across different datasets and clinical scenarios [18, 27].

17.6.4 Reliability and Generalizability in Clinical Practice

One of the most prominent roadblocks to the clinical implementation of synthetic CT (sCT) is the lack of consensus over commissioning & QA procedures. Due to this inconsistency, sCT methods are not uniform with hospitals, leading to varying reliability of sCT in different clinical settings. To ensure that sCT can be used reliably in clinical procedures, it is important to develop strong QA systems that include specialized software capable of analyzing sCT images and giving immediate feedback on the reliability of the images. Such software should include tests designed specifically to evaluate sCT without the need for matching planning CT (pCT) [15, 33].

Integration of sCT into clinical practice requires several stages, including sCT generator algorithm development and training, independent cohort validation, and prospective clinical validation. All of these stages, however, are critical for sCT to be gradually integrated into routine clinical practice. Sufficient commercial patient-specific QA (PSQA) solutions have not been developed for widespread adoption. Effective differentiation from simple visual inspection to complex dose recalculations could be relatively easy or difficult based on the treatment configuration, like MR-Linac or IGRT-Linac systems [18].

A significant barrier is the lack of prospective studies evaluating the use of sCT in clinical practice. This lack of knowledge limits the translation of the findings to many clinical settings. Standardized procedures for implementing and evaluating these systems are essential to facilitate generalizability and reliability for both developers and clinical communities. Standardization of these components will ultimately lead to more standardized procedures, which will in turn help improve the integration of sCT into clinical processes [21, 23]. While sCT has great potential to enhance radiotherapy (RT) workflows, its clinical reliability and adoption depend on the availability of efficient commercial QA tools, the standardization of QA protocols, and the enforcement of rigorous validation practices [15].

17.7 Future Directions and Opportunities

17.7.1 AI for Enhancing Precision Medicine

The massive volumes of genomic data generated by next-generation sequencing (NGS) require the use of artificial intelligence (AI) technologies such as natural language processing (NLP) and machine learning. These methods require the capacity to translate big data into therapeutically relevant insight for precision medicine. AI systems investigate genomic profiles to forecast the risk of diseases, expand diagnostics, and produce precision medicines to build personalized treatments tailored to an individual's genetic makeup, lifestyle, and environmental context [28, 29].

AI has also shown incredible accuracy in analyzing medical imaging and genomic data, frequently surpassing skilled pathologists in the diagnosis of some cancers. This feature improves the precision of cancer diagnosis and treatment suggestions. Furthermore, through research on both public and private datasets, AI is crucial in detecting heterogeneity in medication response, which makes it a prime contender for integration into individualized treatment regimens, particularly in the oncology domain [31].

Despite its potential, there remain obstacles to AI adoption in healthcare, such as repeatability, algorithmic openness, and data requirements. For AI to be safely and successfully incorporated into clinical practice, these issues must be resolved. Additionally, healthcare workers must be well trained in AI applications as AI technology becomes more and more common [27, 32].

It is anticipated that AI will continue to propel the development of precision medicine. However, continued efforts are required to overcome current obstacles and maximize AI's integration into healthcare workflows to fully realize its potential in clinical settings. This is transforming precision medicine by improving diagnostic precision, facilitating customized therapies, and effectively handling complicated genomic data. However, further development and improvement will be required to ensure its widespread and successful application [26, 35, 45].

17.7.2 Role of AI in Reducing Global Disparities in Radiotherapy Access

AI is essential to improving radiotherapy because it makes customized treatment plans possible, increases diagnostic precision, streamlines workflows, and increases access to cutting-edge technologies. AI guarantees that patients receive radiation therapy customized to their unique disease types and genetic profiles by evaluating enormous volumes of data. This is especially helpful in areas with little access to specialized physicians [28]. AI-based classifiers lower detection error rates and help pathologists diagnose cancer more reliably and accurately. The lack of adequate medical facilities where timely and accurate diagnosis can ensure prompt treatment of patients and improve patient outcomes. In addition to diagnosis, AI enables a more efficient radiotherapy workflow through improved imaging, treatment planning, and quality assurance [45]. Clinical AI enables healthcare professionals in low-resource areas to provide efficient therapy without overwhelming the workforce. Moreover, cloud-based AI platforms are making cutting-edge radiotherapy techniques more affordable by bridging the treatment availability gap between developed and underdeveloped nations [45, 46].

AI also addresses the challenge of limited radiotherapy expertise by providing training and support to healthcare workers. With real-time guidance and decision support, AI allows local practitioners to provide high-quality care. Moreover, AI's capacity to analyze vast amounts of data aids in the identification of trends and insights that guide public health strategies, enabling optimal resource allocation and targeted interventions in areas with high cancer prevalence and limited treatment availability [28, 29].

17.7.3 Collaborative Efforts in AI and Radiotherapy Research

DL discovery and application radiotherapy are advanced; however, the magnitude and pace of know-how can be challenging for researchers, physicians, and regulatory agencies to navigate across academic frontiers. These partnerships are essential to solve the complex challenges of integrating AI-based technology into clinical practice. The association is necessary to build shared databases and frameworks that will maximize AI's potential in imaging research. This is especially true in Italy, where AI's use in imaging has been rapidly developing. By building strong datasets, experts would be able to aid in the effectiveness, dependability, and accuracy of deep learning algorithms [35, 37, 41].

Artificial intelligence, deep learning, and radionics working together have a profound effect on every step of image-guided radiation therapy, from diagnosis and treatment planning to even predicting the results. Collaborative approaches are greatly needed so that these processes can be more sophisticated and personalized in terms of patient care [20]. There are limitations to the adoption of AI in digital radiology and radiotherapy, for example, compatibility and standardization. These challenges will require coordinated efforts to align progress in the two fields and ensure that advances in one are broadly reciprocated in the other. Continued collaboration will stimulate innovative strategies and solutions that will advance patient care and better define radiation therapeutic modalities. The persuasion may make key steps that enhance the worth and integration of AI-driven technologies in healthcare by cultivating interdisciplinary collaborations and creating common research frameworks [17, 21, 29].

17.8 Conclusions

AI has made a big impact on treatment planning in radiotherapy. It looks at huge sets of data from past cases, which helps create custom therapy limits for each patient. This results in treatments that work better and patients who do better overall. Also, machine learning has made medical image analysis much better. Now, we can spot and describe tumors more, which is crucial to plan radiotherapy just right. AI takes care of many tasks, like outlining tumors and finding organs that might be at risk. This cuts down on human mistakes, saves time, and makes sure therapy is given the same way every time. AI goes beyond mere automation. It monitors patients, allowing doctors to adjust treatments based on individual responses. This adaptability ensures optimal therapy effectiveness as a patient's condition evolves. AI can forecast patient outcomes with specific treatment approaches. This capability helps medical teams make informed decisions and establish achievable goals. AI also integrates with advanced technologies such as radiation delivery and imaging systems. This collaboration enhances the safety and efficacy of radiotherapy. AI goes beyond mere automation. It monitors patients, allowing doctors to adjust treatments based on individual responses. This adaptability ensures optimal therapy effectiveness as a patient's condition evolves. AI can forecast patient outcomes with specific treatment approaches. This capability helps medical teams make informed decisions and establish achievable goals. AI also integrates with advanced technologies such as radiation delivery and imaging systems. This collaboration enhances the safety and efficacy of radiotherapy.

References

1. Siddique S, Chow JC. Artificial intelligence in radiotherapy. Rep Pract Oncol Radiother. 2020;25(4):656–66.
2. Radiation Oncology Market Size, Share & Trends Analysis Report by 2033; retrived from https://www.biospace.com/press-releases/radiation-oncology-market-size-share-trends-analysis-report-by-2033#:~:text=In%20the%20U.S.%2C%20approximately%20600%2C000,of%20North%20America%20(RSNA). Accessed 12 Apr 2025.
3. El-Jerjawi NS, Murad WF, Harazin D, Qaoud AN, Jamala MN, Abunasser BS, et al. The role of artificial intelligence in revolutionizing health: challenges, applications, and future prospects. Int J Acad Appl Res. 2024;8:7–15.
4. Ahmed Z, Mohamed K, Zeeshan S, Dong X. Artificial intelligence with multi-functional machine learning platform development for better healthcare and precision medicine. Database. 2020;2020:baaa010.
5. Huynh E, Hosny A, Guthier C, Bitterman DS, Petit SF, Haas-Kogan DA, et al. Artificial intelligence in radiation oncology. Nat Rev Clin Oncol. 2020;17(12):771–81.
6. Francolini G, Desideri I, Stocchi G, Salvestrini V, Ciccone LP, Garlatti P, et al. Artificial intelligence in radiotherapy: state of the art and future directions. Med Oncol. 2020;37:1–9.
7. Manco L, Maffei N, Strolin S, Vichi S, Bottazzi L, Strigari L. Basic of machine learning and deep learning in imaging for medical physicists. Phys Med. 2021;83:194–205.
8. Jeong C, Goh Y, Kwak J. Challenges and opportunities to integrate artificial intelligence in radiation oncology: a narrative review. Ewha Med J. 2024;47(4)
9. Mugabe KV. Barriers and facilitators to the adoption of artificial intelligence in radiation oncology: a New Zealand study. In: Technical innovations & patient support in radiation oncology, vol. 18. Oxford: Elsevier; 2021. p. 16–21.
10. Huang D, Bai H, Wang L, Hou Y, Li L, Xia Y, et al. The application and development of deep learning in radiotherapy: a systematic review. Technol Cancer Res Treat. 2021;20:15330338211016386.
11. Ono T, Iramina H, Hirashima H, Adachi T, Nakamura M, Mizowaki T. Applications of artificial intelligence for machine-and patient-specific quality assurance in radiation therapy: current status and future directions. J Radiat Res. 2024:rrae033.
12. Sheng K. Artificial intelligence in radiotherapy: a technological review. Front Med. 2020;14(4):431–49.
13. Putz F, Fietkau R. The increasing role of artificial intelligence in radiation oncology: how should we navigate it? Strahlenther Onkol. 2025;201(3):207–9.
14. Bi WL, Hosny A, Schabath MB, Giger ML, Birkbak NJ, Mehrtash A, Allison T, Arnaout O, Abbosh C, Dunn IF, Mak RH. Artificial intelligence in cancer imaging: clinical challenges and applications. CA Cancer J Clin. 2019;69(2):127–57.
15. Sarma G, Kashya H, Medhi PP, Kalita R, Lahkar D. Unravelling the landscape of image-guided radiotherapy: a comprehensive overview. Palliative Med Pract. 2024;
16. Zhao W, Shen L, Islam MT, Qin W, Zhang Z, Liang X, Zhang G, Xu S, Li X. Artificial intelligence in image-guided radiotherapy: a review of treatment target localization. Quant Imaging Med Surg. 2021;11(12):4881.
17. Sun Z, Wang K, Wu C, Chen Y, Kong Z, She L, Song B, Luo N, Wu P, Wang X, Zhang X. Using an artificial intelligence model to detect and localize visible clinically significant prostate cancer in prostate magnetic resonance imaging: a multicenter external validation study. Quant Imaging Med Surg. 2023;14(1):43.
18. Rajendran P, Yang Y, Niedermayr TR, Gensheimer M, Beadle B, Le Q-T, et al. Large language model-augmented learning for auto-delineation of treatment targets in head-and-neck cancer radiotherapy. Radiother Oncol. 2025:110740.
19. Liu S, Pastor-Serrano O, Chen Y, Gopaulchan M, Liang W, Buyyounouski M, et al. Automated radiotherapy treatment planning guided by GPT-4Vision. arXiv preprint arXiv:240615609. 2024.

20. Dong Z, Chen Y, Gay H, Hao Y, Hugo GD, Samson P, et al. Large-language-model empowered dose volume histogram prediction for intensity modulated radiotherapy. arXiv preprint arXiv:240207167. 2024.
21. Alowais SA, Alghamdi SS, Alsuhebany N, Alqahtani T, Alshaya AI, Almohareb SN, et al. Revolutionizing healthcare: the role of artificial intelligence in clinical practice. BMC Med Educ. 2023;23(1):689.
22. Dona Lemus OM, Cao M, Cai B, Cummings M, Zheng D. Adaptive radiotherapy: next-generation radiotherapy. Cancers. 2024;16(6):1206.
23. Glide-Hurst CK, Lee P, Yock AD, Olsen JR, Cao M, Siddiqui F, et al. Adaptive radiation therapy (ART) strategies and technical considerations: a state of the ART review from NRG oncology. Int J Radiat Oncol Biol Phys. 2021;109(4):1054–75.
24. Kitson SL. Modern medical imaging and radiation therapy. Cyber Security| Big Data| AI Open Med Science. 2024.
25. Raman S. Impact of technological advances in treatment planning, image guidance, and treatment delivery on target margin design for prostate cancer radiotherapy: an updated review. Br J Radiol. 2023;97:31.
26. Zwanenburg A, Price G, Löck S. Artificial intelligence for response prediction and personalisation in radiation oncology. Strahlenther Onkol. 2024;1–8:266.
27. Osman AF, Maalej NM. Applications of machine and deep learning to patient-specific IMRT/VMAT quality assurance. J Appl Clin Med Phys. 2021;22(9):20–36.
28. Xue X, Luan S, Ding Y, Li X, Li D, Wang J, et al. Treatment plan complexity quantification for predicting gamma passing rates in patient-specific quality assurance for stereotactic volumetric modulated arc therapy. J Appl Clin Med Phys. 2024;25:e14432.
29. Lam D, Zhang X, Li H, Deshan Y, Schott B, Zhao T, et al. Predicting gamma passing rates for portal dosimetry-based IMRT QA using machine learning. Med Phys. 2019;46(10):4666–75.
30. Fondevila D, Rios P, Peñalva DD, Arbiser S. Predicting gamma passing rates for portal dosimetry-based IMRT QA using deep learning. Int J Radiat Oncol Biol Phys. 2021;111(3):e110–e1.
31. Jin D, Guo D, Ge J, Ye X, Lu L. Towards automated organs at risk and target volumes contouring: defining precision radiation therapy in the modern era. J Natl Cancer Center. 2022;2(4):306–13.
32. Kosmin M, Ledsam J, Romera-Paredes B, Mendes R, Moinuddin S, De Souza D, et al. Rapid advances in auto-segmentation of organs at risk and target volumes in head and neck cancer. Radiother Oncol. 2019;135:130–40.
33. Choi SH, Park JW, Cho Y, Yang G, Yoon HI. Automated organ segmentation for radiation therapy: a comparative analysis of AI-based tools versus manual contouring in Korean cancer patients. Cancers. 2024;16(21):3670.
34. Aziz F, Bianchini D, Olawade DB, Boussios S. The impact of AI-driven remote patient monitoring on cancer care: a systematic review. Anticancer Res. 2025;45(2):407–18.
35. Islam MM. Exploring the Impact of Artificial Intelligence in Healthcare. J Artif Intell General Sci. 2024;2(1):228–38. ISSN: 3006-4023.
36. Shaik T, Tao X, Higgins N, Li L, Gururajan R, Zhou X, et al. Remote patient monitoring using artificial intelligence: current state, applications, and challenges. Wiley Interdiscip Rev Data Mining Knowl Discov. 2023;13(2):e1485.
37. Fionda B, Boldrini L, D'Aviero A, Lancellotta V, Gambacorta M, Kovács G, et al. Artificial intelligence (AI) and interventional radiotherapy (brachytherapy): state of art and future perspectives. J Contemp Brachytherapy. 2020;12(5):497–500.
38. Adeyemi C, Adegoke BO, Odugbose T. The impact of healthcare information technology on reducing medication errors: A review of recent advances. International journal of frontiers in medicine and surgery research [online]. 2024;5(2):20–9.
39. Field M, Vinod S, Delaney GP, Aherne N, Bailey M, Carolan M, et al. Federated learning survival model and potential radiotherapy decision support impact assessment for non–small cell lung cancer using real-world data. Clin Oncol. 2024;36(7):e197–208.

40. Isaksson LJ, Mastroleo F, Vincini MG, Marvaso G, Zaffaroni M, Gola M, et al. The emerging role of artificial intelligence in proton therapy: a review. Crit Rev Oncol Hematol. 2024;204:104485.
41. Feng H, Shan J, Vargas CE, Keole SR, Rwigema J-CM, Ding Y, et al. Online adaptive proton therapy facilitated by artificial intelligence–based autosegmentation in pencil beam scanning proton therapy. Int J Radiat Oncol Biol Phys. 2024;121:822.
42. Xinyang Z, Pengbo H, Yazhou L, Xinguo L, Yuanyuan M, Guosheng S, et al. 5–50 DR-only carbon-ion radiotherapy treatment planning via deep learning. Phys Med. 2022;100:120–8.
43. Soltaninejad M, Zhang L, Lambrou T, Yang G, Allinson N, Ye X. MRI brain tumor segmentation using random forests and fully convolutional networks. arXiv preprint arXiv:190906337. 2019.
44. Sheikh F, Patil S. Integrating multi-modal imaging and computational techniques for enhanced tumor characterization in oncology. PLoS One. 2025;20(5):e0323752.
45. Brocki L, Chung NC. Integration of Radiomics and tumor biomarkers in interpretable machine learning models. Cancers. 2023;15(9):2459.
46. Faridoon A, Kechadi MT, editors. Healthcare data governance, privacy, and security-a conceptual framework, EAI international conference on body area networks. Springer; 2024.
47. Jha D, Durak G, Sharma V, Keles E, Cicek V, Zhang Z, et al. A conceptual framework for applying ethical principles of AI to medical practice. Bioengineering. 2025;12(2):180.
48. Bose S, Marijan D. A survey on privacy of health data lifecycle: a taxonomy, review, and future directions. arXiv preprint arXiv:231105404. 2023.

Chapter 18
Radiogenomics: A Potential Approach for Personalized Radiotherapy

Ali Amini and Babak Behnam

Abstract Radiogenomics, the integration of genomic data with radiation response, is emerging as a transformative approach for personalized radiotherapy. By identifying genetic biomarkers that predict tumor sensitivity to radiation, radiogenomics enables tailored treatment strategies to maximize efficacy and minimize toxicity. Recent studies highlight the role of key genes, such as *ataxia-telangiectasia mutated (ATM)*, *BRCA1/2*, *RAD51*, *CHEK1*, and *HIF-1α*, in modulating radiation response through DNA repair, cell cycle regulation, hypoxia adaptation, and immune modulation. For instance, mutations in DNA repair genes like *ATM* and *BRCA1/2* can impair the ability to fix radiation-induced double-strand breaks, rendering tumors more susceptible to radiation when combined with targeted therapies like PARP inhibitors. Similarly, hypoxia-related genes such as *HIF-1α* and *CA9* promote tumor survival under low oxygen, contributing to radioresistance. Advances in next-generation sequencing and machine learning have refined radiogenomics, allowing for multi-omics integration to predict treatment outcomes. Immune-related biomarkers (*PD-L1, CTLA4*) guide radiotherapy-immunotherapy combinations, leveraging the abscopal effect to enhance systemic responses. Additionally, *TP53* and *MRE11* influence tumor sensitivity by regulating apoptosis and DNA damage response pathways. Despite its potential, radiogenomics faces challenges, including tumor heterogeneity, limited large-scale validation, and ethical concerns regarding genetic data usage. Addressing these barriers through interdisciplinary research and regulatory oversight is essential for fully integrating radiogenomics into clinical practice, ultimately improving patient outcomes.

Keywords Radiogenomics · DNA repair · Radiosensitivity · Hypoxia · Immunotherapy · Precision radiotherapy

A. Amini
American University, Washington, DC, USA

B. Behnam (✉)
American University, Washington, DC, USA

Avicenna Biotech Research, Clarksburg, MD, USA
e-mail: bbehnam@american.edu; babak.behnam@gmail.com

F. Taghizadeh-Hesary et al. (eds.), *From Radiobiology to Radiation Oncology*,
https://doi.org/10.1007/978-3-031-95724-6_18

18.1 Introduction

Radiogenomics integrates radiation oncology with genomics to predict individual responses to radiotherapy by identifying genetic determinants of radiosensitivity, toxicity, and tumor control. By analyzing genomic biomarkers, such as single nucleotide polymorphisms (SNPs) and DNA repair gene mutations, it enables personalized treatment strategies [2]. Advances in artificial intelligence and machine learning have improved predictive modeling, optimizing radiation dosing while minimizing toxicity [3]. Tailored radiation protocols based on genetic profiles enhance treatment efficacy and patient survival [4]. As a key component of precision medicine, radiogenomics bridges the gap between genomics and radiation oncology.

18.1.1 Definition and Significance in Precision Medicine

Radiogenomics is an emerging field that integrates genomics with radiotherapy to predict individual responses to radiation treatment. It focuses on identifying genetic biomarkers that influence radiosensitivity, radiation-induced toxicity, and tumor response, thereby enabling more precise and personalized radiotherapy strategies. By leveraging genomic data, radiogenomics aims to minimize adverse effects and enhance therapeutic efficacy, ensuring that patients receive optimized radiation doses tailored to their genetic profile. The significance of radiogenomics in precision medicine lies in its potential to revolutionize cancer treatment. Genetic variants in DNA repair pathways (e.g., *BRCA1, ATM, TP53*) and SNPs influence patient-specific responses to radiation. Studies have demonstrated that integrating radiogenomic biomarkers with clinical and imaging data enhances treatment personalization, reduces toxicity, and improves overall survival rates [3, 2]. Furthermore, the advent of artificial intelligence and machine learning has accelerated the identification of genomic signatures predictive of radiation response, further refining precision radiotherapy approaches [4]. Radiogenomics is a paradigm shift in oncology, offering a pathway toward truly individualized cancer treatment by aligning radiation therapy with a patient's genetic landscape.

18.1.2 The Relationship Between Genomics and Radiation Response

Genomics plays a transformative role in determining individual responses to radiation therapy by influencing radiosensitivity, DNA repair capacity, and tumor control. Variations in genes involved in DNA damage response, oxidative stress regulation, and cell cycle control significantly impact how normal and cancerous tissues react to radiation exposure. Key genes such as *ATM, TP53, BRCA1, RAD51,*

XRCC1, and *SOD2* regulate pathways essential for DNA double-strand break repair, apoptosis, and reactive oxygen species (ROS) detoxification, directly affecting radiation-induced cytotoxicity and therapeutic efficacy [5]. SNPs and mutations in these genes contribute to interindividual differences in radiation tolerance and treatment outcomes. For instance, ATM-deficient patients exhibit heightened radiosensitivity due to impaired homologous recombination repair, leading to increased normal tissue toxicity and secondary malignancies post-radiotherapy [3]. Additionally, *TP53* mutations, commonly found in radioresistant tumors, alter apoptotic pathways, allowing malignant cells to evade radiation-induced cell death [4]. Epigenetic modifications, including DNA methylation and histone modifications, further shape radiation response by modulating gene expression without altering the DNA sequence. Recent studies indicate that radiation itself induces epigenetic reprogramming, which can lead to adaptive resistance in surviving tumor cells, complicating treatment efficacy [6]. Beyond intrinsic tumor factors, the genomic landscape of the tumor microenvironment (TME) also influences radiotherapy outcomes. Tumor-infiltrating immune cells exhibit distinct radiation responses based on genetic and transcriptional profiles, affecting the inflammatory and immunosuppressive environment post-irradiation. Recent evidence suggests that tumors with defects in STING-mediated immune signaling show reduced immunogenic cell death following radiation, limiting the effectiveness of radiation-induced immune activation [7]. The integration of genomics into radiation oncology has paved the way for precision radiotherapy, where patients' genetic profiles inform individualized treatment plans. Radiogenomic biomarkers now help stratify patients into risk categories for radiation toxicity, allowing for dose adjustments and protective strategies. Overall, genomics serves as a critical determinant of radiation response, shaping the efficacy and toxicity of radiotherapy. As precision medicine evolves, incorporating genomic insights into radiotherapy planning holds immense promise for improving treatment personalization, minimizing adverse effects, and enhancing therapeutic efficacy.

18.2 Genetic Determinants of Radiation Sensitivity

Radiation sensitivity, the varying degree to which individuals or cells respond to ionizing radiation, is influenced by a complex interplay of genetic factors. Recent studies have shed light on the genetic determinants underlying this phenomenon, emphasizing the role of DNA repair mechanisms, cell cycle regulation, and oxidative stress responses. Key genes, including *ATM, TP53, BRCA1, XRCC1, RAD51*, and *SOD2*, play crucial roles in modulating cellular responses to radiation exposure [5]. Mutations in these genes impact the ability of cells to repair DNA double-strand breaks, a critical factor in radiosensitivity. For example, ATM-deficient individuals exhibit extreme radiosensitivity due to impaired homologous recombination repair, leading to increased radiation toxicity and heightened cancer risk [3]. One key determinant is the *ATM* (ataxia telangiectasia mutated) gene, which encodes a critical protein kinase for detecting DNA double-strand breaks (DSBs). Mutations in *ATM* are associated with increased radiation sensitivity, as seen in patients with ataxia-telangiectasia [8]. Similarly, the *BRCA1* and *BRCA2* genes, involved in homologous recombination repair, play a pivotal role in radiation response. Deficiencies in these genes lead to impaired DSB repair and heightened radiosensitivity [9]. Additionally, polymorphisms in genes encoding antioxidant enzymes, such as *SOD2* (superoxide dismutase 2), influence cellular resilience to radiation-induced oxidative stress [10]. Furthermore, the *TP53* gene, a tumor suppressor regulating apoptosis and cell cycle arrest, modulates radiation sensitivity. Mutations in TP53 can lead to defective apoptosis and increased radiation resistance [11]. Emerging research also highlights the role of non-coding RNAs, such as microRNAs, in regulating radiation response pathways [12]. These findings underscore the importance of genetic profiling in predicting radiation sensitivity, paving the way for personalized radiotherapy strategies.

18.2.1 Role of Single Nucleotide Polymorphisms (SNPs) in Radiosensitivity

SNPs contribute to interindividual differences in radiation tolerance and treatment outcomes, and the analyses of SNPs enable personalized treatment strategies [2]. These SNPs genomic biomarkers, involved in DNA repair, oxidative stress response, and cell cycle regulation, are critical genetic variations that influence individual differences in radiosensitivity, impacting both cancer treatment outcomes and risk of radiation-induced toxicity. For instance, SNPs in the *XRCC1* gene, which encodes a protein essential for base excision repair, have been linked to altered radiation sensitivity. A common SNP (Arg399Gln) in *XRCC1* has been associated with increased risk of radiation-induced damage due to reduced DNA repair capacity [13]. Similarly, SNPs in the *ATM* gene, such as *rs1801516*, have been shown to modulate radiation response by affecting the repair of DNA double-strand breaks (DSBs) [2]. Polymorphisms in antioxidant-related genes, such as *SOD2* (*rs4880*), also play a significant role. The *SOD2* Ala16Val SNP affects mitochondrial antioxidant activity, influencing cellular resilience to radiation-induced oxidative stress [10]. Additionally, SNPs in the *TP53* gene, such as *rs1042522*, have been associated with variations in apoptosis and cell cycle arrest, impacting radiosensitivity [11]. An analysis of 20 DSB-related genes in patients with head and neck squamous cell carcinoma (HNSCC) treated with radiotherapy identified 14 SNPs linked to survival outcomes, with 10 SNPs—found in MRE11, XRCC5, RAD51, LIG4, CHEK1, TP53, ATM, XRCC6, and NBN—serving as independent predictors of therapy failure and poor prognosis. Among them, only CHEK1 rs558351 retained statistical significance after multiple-testing correction [14]. Emerging research also emphasizes the role of SNPs in non-coding regions, such as microRNA-binding sites, which can regulate gene expression post-radiation exposure [12]. These findings underscore the potential of SNP profiling in predicting radiosensitivity and personalizing radiotherapy.

18.2.2 DNA Repair Gene Mutations (e.g., BRCA1/2, ATM, TP53*)*

Mutations in DNA repair genes, such as *BRCA1/2, ATM*, and *TP53*, significantly influence individual radiosensitivity, impacting both cancer therapy efficacy and the risk of adverse effects. These genes play critical roles in maintaining genomic stability, and their dysfunction can alter cellular responses to ionizing radiation. The *BRCA1* and *BRCA2* genes, essential for homologous recombination repair (HRR), are pivotal in repairing DNA double-strand breaks (DSBs). Mutations in these genes impair HRR, leading to increased radiosensitivity and genomic instability. For example, *BRCA1/2*-deficient cells exhibit heightened sensitivity to radiation due to their inability to efficiently repair DSBs [15]. This has implications for radiotherapy, as *BRCA1/2* mutation carriers may experience better tumor control but also higher toxicity in normal tissues. The *ATM* gene, encoding a kinase that activates DSB repair pathways, is another key player. *ATM* mutations, as seen in ataxia-telangiectasia patients, result in defective DSB detection and repair, leading to extreme radiosensitivity [8]. Similarly, TP53 mutations disrupt apoptosis and cell cycle regulation, often conferring resistance to radiation in tumor cells but increasing sensitivity in normal tissues [11]. Understanding these mutations enables personalized radiotherapy approaches, optimizing tumor control while minimizing side effects. Recent studies also highlight the potential of targeting DNA repair pathways to enhance radiosensitivity in tumors.

18.2.3 Influence of Genetic Variants on Normal Tissue Toxicity

Genetic variants play a significant role in determining individual susceptibility to normal tissue toxicity following radiotherapy, with recent studies highlighting the impact of polymorphisms in DNA repair, inflammatory response, and oxidative stress pathways. These variants can predict the risk of adverse effects, enabling personalized treatment strategies. For example, the Arg399Gln, one of the SNPs in the *XRCC1* gene, has been associated with increased radiation-induced toxicity due to impaired base excision repair [13]. Similarly, SNPs in the *ATM* gene (including rs1801516) influence DSB repair efficiency, correlating with higher rates of tissue damage in carriers [16]. As mentioned above, the SNPs in the genes regulating oxidative stress, such as *SOD2* (rs4880 and Ala16Val), contribute to normal tissue toxicity and exacerbate radiation-induced oxidative damage [10]. Additionally, *polymorphisms in inflammatory response genes*, such as *TGFB1* (rs1800469), have been linked to fibrosis and other late toxicities [5].

Emerging research also implies non-coding genetic variants, such as microRNA-related SNPs, in modulating radiation response [12]. These findings underscore the importance of genetic profiling to predict and mitigate normal tissue toxicity, improving the safety and efficacy of radiotherapy.

18.3 Molecular Pathways in Radiation Response

Radiation therapy is a cornerstone in cancer treatment, yet its efficacy is influenced by complex molecular pathways that dictate cellular response. Recent sophisticated studies have elucidated key mechanisms underlying radiation-induced DNA damage and repair. The ataxia-telangiectasia mutated (ATM) pathway, for instance, plays a pivotal role in detecting double-strand breaks (DSBs) and initiating repair through homologous recombination (HR) and non-homologous end joining (NHEJ) [17]. Additionally, the p53 tumor suppressor protein regulates cell cycle arrest and apoptosis in response to radiation, with its dysfunction often leading to radioresistance [18]. Emerging research highlights the role of the PI3K/AKT/mTOR pathway, including the molecular pathways of hypoxia and oxidative stress, in promoting cell survival post-irradiation, with inhibitors of this pathway showing promise in sensitizing tumors to radiation [19]. Furthermore, the interplay between radiation and the tumor microenvironment, particularly through TGF-β signaling, has been shown to influence immune evasion and fibrosis [20]. Understanding these molecular pathways offers opportunities for targeted therapies, such as PARP inhibitors in HR-deficient cancers [15]. Combining radiation with molecularly targeted agents holds potential to enhance therapeutic outcomes and overcome resistance, paving the way for personalized cancer treatment strategies.

18.3.1 Role of Oxidative Stress and DNA Damage Response

Radiation therapy induces cytotoxic effects primarily by generating ROS, which triggers oxidative stress and DNA damage. Recent research has elucidated the intricate molecular pathways linking oxidative stress, DNA damage response (DDR), and radiation sensitivity. Ionizing radiation ionizes cellular water and DNA, producing ROS like hydroxyl radicals, which cause base modifications, single-strand breaks (SSBs), and DSBs [21]. These lesions activate the DDR, orchestrated by ATM and ATR kinases. ATM primarily responds to DSBs, while ATR is activated by SSBs and replication stress, leading to phosphorylation of p53, CHK1, and CHK2, which regulate cell cycle arrest, DNA repair, and apoptosis [22]. Oxidative stress also influences DDR by modulating repair enzymes like OGG1 and APE1, essential for base excision repair (BER) [23]. Persistent ROS accumulation can compromise repair fidelity, driving genomic instability and radioresistance. However, targeting oxidative stress and DDR pathways—such as using PARP inhibitors in homologous recombination-deficient cancers—enhances radiosensitivity [15]. These insights highlight the potential of combining radiation therapy with molecularly targeted agents to optimize therapeutic efficacy.

18.3.2 Apoptosis, Autophagy, and Cell Cycle Regulation

Radiation therapy triggers a network of cellular responses, including apoptosis, autophagy, and cell cycle regulation, which collectively influence treatment outcomes. Apoptosis, a key programmed cell death pathway, is primarily regulated by p53, which responds to radiation-induced DNA damage by activating pro-apoptotic genes such as BAX and PUMA while suppressing anti-apoptotic proteins like BCL-2 [18]. Anti-apoptotic signals are also upregulated by hypoxia-activated pathways such as AKT/mTOR and NF-κB [24, 25]. Autophagy, a cellular stress response, is often upregulated following radiation exposure. Through PI3K/AKT/mTOR signaling, radiation inhibits mTOR, leading to autophagosome formation and the clearance of damaged organelles [26]. While autophagy enhances survival in moderate stress conditions, excessive activation can drive autophagic cell death, highlighting its dual role in cancer therapy. Cell cycle regulation is essential for DNA damage repair post-radiation. The ATM/CHK2 and ATR/CHK1 pathways mediate cell cycle arrest at G1/S and G2/M checkpoints, allowing time for repair [22]. Defects in these checkpoints contribute to genomic instability and radioresistance. Recent advances suggest that combining radiation with CDK4/6 inhibitors enhances tumor radiosensitivity, providing new therapeutic avenues [27]. Understanding these interconnected pathways is critical for optimizing radiotherapy and overcoming treatment resistance.

18.3.3 Interaction Between Hypoxia-Related Genes and Radiation Sensitivity

Deciphering the molecular interactions between hypoxia, radiation response, and tumor adaptation is essential for designing targeted therapeutic strategies to overcome radioresistance and improve radiotherapy outcomes. Hypoxia, a defining feature of many solid tumors, plays an important role in modulating radiation sensitivity through the regulation of hypoxia-related genes. These genes orchestrate key cellular processes, including DNA repair, metabolism, angiogenesis, and survival signaling, ultimately contributing to radioresistance. Radioresistance is driven by hypoxia-induced genes such as *SERPINE1* (Plasminogen Activator Inhibitor-1) and *CXCR4*, which respectively inhibit fibrinolysis and promote tumor cell survival [28], as well as enhance tumor cell migration, survival, and resistance to radiation [29]. Hypoxia-inducible factors (HIFs), particularly HIF-1α and HIF-2α, drive adaptive responses that enable tumor cells to survive in oxygen-deprived conditions. HIF-1α induces *VEGF* to promote angiogenesis and *GLUT1* to enhance glucose uptake, thereby reducing radiation-induced oxidative stress and increasing resistance [30]. Also, hypoxia induces MDR1, enhancing drug and radiation resistance by increasing efflux of cytotoxic agents [31]. HIF-2α further promotes tumor stemness, exacerbating treatment resistance [32]. Also, OCT4 promotes tumor cell stemness and resistance to radiation [33].

Hypoxia also enhances DNA repair capacity, as seen with *RAD51*, which facilitates HRR of radiation-induced DNA double-strand breaks, reinforcing resistance [34]. Additionally, *CA9* regulates pH balance, allowing tumor cells to thrive in acidic environments, reducing radiation efficacy [35]. Metabolic reprogramming under hypoxia further amplifies resistance. Genes such as *LDHA* and *GLUT1* drive glycolysis, ensuring ATP production despite oxygen deprivation [16, 36]. The TME is also significantly altered by hypoxia. MMP2 and MMP9 degrade the extracellular matrix, facilitating invasion, metastasis, and radiation resistance [37]. Additionally, hypoxia-activated pathways such as AKT/mTOR and NF-κB upregulate anti-apoptotic signals, further promoting cell survival [24, 25]. To overcome radioresistance, strategies targeting hypoxia-related pathways are under investigation. HIF inhibitors, hypoxia-activated prodrugs, and metabolic inhibitors such as LDHA blockers show promise in enhancing radiation sensitivity [38]. Combining radiation therapy with these targeted treatments may significantly improve cancer therapy outcomes.

18.4 Radiogenomics in Tumor Response Prediction

Radiogenomic insights are driving the development of targeted therapies to overcome resistance and enhance radiotherapy outcomes. By customizing treatments based on individual tumor biology, radiogenomics offers significant potential for optimizing radiation therapy and advancing precision oncology. The integration of genomic data with radiation response is revolutionizing the prediction of tumor sensitivity to radiotherapy, with recent studies highlighting the promise of radiogenomic biomarkers in personalizing treatment strategies.

18.4.1 Identifying Gene Expression Profiles Linked to Treatment Outcomes

Advancements in genomics have revolutionized personalized radiotherapy by identifying gene expression signatures that predict tumor radiosensitivity and resistance. Recent studies highlight the predictive value of transcriptional profiles in tailoring radiation therapy. Key gene signatures linked to radiation response include DNA repair genes (*ATM, RAD51*), hypoxia-related markers (*HIF-1α, CA9*), and immune modulation genes (*PD-L1, CTLA4*) [39]. Hypoxia signatures, in particular, correlate with poor prognosis due to their role in enhancing radioresistance and tumor aggressiveness [40]. The tumor microenvironment also plays an essential role, as immune-related gene signatures influence radiotherapy outcomes. Elevated expression of immune checkpoint genes is often associated with resistance, while inflammatory gene signatures may predict improved responses [41]. Additionally, machine learning and multi-omics integration are advancing the identification of robust predictive biomarkers, refining patient-specific treatment approaches [42]. These findings underscore the transformative potential of gene expression profiling in precision radiotherapy, enabling tailored treatment strategies that improve clinical outcomes.

18.4.2 Tumor Heterogeneity and Its Impact on Radiotherapy Efficacy

Tumor heterogeneity, characterized by genetic, phenotypic, and microenvironmental diversity within tumors, significantly impacts radiotherapy efficacy. The importance of intratumoral heterogeneity in contributing to variable radiation responses has recently gained significant attention. Genetic heterogeneity, driven by mutations in key pathways such as p53, ATM, and PI3K/AKT, can lead to subpopulations of radioresistant cells [43]. For instance, hypoxic regions within tumors, often associated with HIF-1α overexpression, exhibit reduced radiation sensitivity due to impaired DNA repair and increased survival signaling [44]. Spatial and temporal heterogeneity further complicates radiotherapy outcomes. Clonal evolution and the tumor microenvironment, including stromal and immune cell interactions, create dynamic resistance mechanisms [45]. For example, immune-excluded or immunosuppressive microenvironments can diminish the abscopal effect, limiting radiation‘s systemic efficacy [46]. Advances in single-cell sequencing and spatial transcriptomics are uncovering novel biomarkers of radioresistance, enabling more precise targeting of heterogeneous tumors [47]. Combining radiotherapy with targeted therapies, such as PARP inhibitors or immune checkpoint blockers, shows promise in overcoming resistance driven by heterogeneity [15]. Understanding and addressing tumor heterogeneity is pivotal for improving radiotherapy outcomes.

18.5 Application of Radiogenomics in Personalized Radiotherapy

18.5.1 Genomics-Driven Treatment Planning and Dose Adaptation

Genomics-driven treatment planning and dose adaptation represent a transformative approach in radiation oncology, tailoring therapies based on individual genetic profiles to enhance efficacy and minimize toxicity. The Genomic Adjusted Radiation Dose (GARD) model exemplifies this strategy by integrating tumor-specific genomic information to personalize radiation doses. Studies have validated GARD's utility in adjusting radiation doses for various cancers, including locally advanced rectal cancer, thereby improving treatment outcomes [48].

18.5.2 Potential for AI and Machine Learning Integration

It has been recently proposed that advanced computational approaches, including AI-driven genomic analysis, further refine predictive models for radiation response, improving personalized treatment outcomes [49]. Machine learning (ML) further augments genomics-driven radiotherapy by analyzing complex biological data to predict optimal treatment parameters. ML models can process high-dimensional genomic datasets, facilitating the development of precision biomarkers and guiding dose adaptation strategies. This integration of ML with genomics enables a more nuanced understanding of tumor biology, leading to personalized radiotherapy plans that are both effective and adaptive to individual patient needs [50]. In summary, the convergence of genomics and advanced computational models in treatment planning signifies a significant advancement in precision oncology, offering the potential for more individualized and effective cancer therapies.

18.5.3 Challenges in Clinical Implementation

Despite its potential, the clinical implementation of radiogenomics in personalized radiotherapy faces significant challenges. One major hurdle is the complexity of integrating multi-omics data, including genomic, transcriptomic, and radiomic features, into actionable clinical insights [42]. The lack of standardized protocols for data collection, analysis, and interpretation further complicates its adoption [51]. Another challenge is tumor heterogeneity, which leads to diverse molecular profiles even within the same tumor, making it difficult to predict radiation response accurately [43]. Additionally, the high cost and limited accessibility of advanced genomic technologies restrict widespread clinical application [39]. Ethical and privacy concerns related to genomic data usage also pose barriers, requiring robust regulatory frameworks [52]. Furthermore, the need for large-scale, prospective clinical trials to validate radiogenomic biomarkers and their utility in real-world settings remains unmet [41]. Addressing these challenges through interdisciplinary collaboration, technological advancements, and policy development is decisive for realizing the full potential of radiogenomics in personalized radiotherapy.

18.6 Radiogenomics and Combination Therapies

18.6.1 Synergy with Chemotherapy, Immunotherapy, and Targeted Therapies

Genome-based personalized radiotherapy, which tailors radiation treatment based on an individual's genetic profile, has shown enhanced efficacy when combined with chemotherapy, immunotherapy, and targeted therapies. This integrative approach allows for the development of treatment plans that are specifically designed to address the unique molecular characteristics of a patient's tumor. Combining personalized radiotherapy with chemotherapy has demonstrated superior outcomes compared to radiotherapy alone, particularly in locally advanced cancers. For instance, studies have shown that concomitant radiotherapy and chemotherapy improve treatment efficacy in anal cancer [53]. Incorporating immunotherapy with genome-based radiotherapy has also yielded promising results. This combination can enhance the immune system's ability to recognize and attack cancer cells, leading to improved patient outcomes. Research indicates that such integrative treatments can provide synergistic benefits, improving patient responses [54]. Furthermore, the integration of targeted therapies with personalized radiotherapy allows for the precise inhibition of specific molecular pathways driving tumor growth. In recent years, the critical role of cancer biology, particularly targeting organelle-specific genes and/or organelles like mitochondria, in improving oncological treatment outcomes has been extensively explored and elucidated [55]. This strategy has been effective in various cancers, as it directly addresses the genetic alterations present in the tumor; advancements in genomic technologies have facilitated the identification of these alterations, enabling the development of tailored treatment. In summary, the synergy of genome-based personalized radiotherapy with chemotherapy, immunotherapy, and targeted therapies represents a significant advancement in precision oncology, offering more effective and individualized cancer treatment strategies.

18.6.2 Role in Adaptive Radiotherapy Approaches

As discussed, radiogenomics enhances adaptive radiotherapy by linking genomic biomarkers to real-time tumor response, enabling dynamic treatment adjustments. Key biomarkers, such as DNA repair defects (e.g., *BRCA1/2* mutations) and hypoxia-related genes (e.g., *HIF-1α*, *CA9*), predict radiosensitivity and guide dose optimization [15, 30]. Immune-related markers like *PD-L1* further refine strategies by identifying tumors likely to respond to combined radiotherapy and immunotherapy [41]. Advanced machine learning models integrate multi-omics data (genomic, radiomic, and clinical) to map spatial and temporal tumor heterogeneity, allowing real-time adaptation to evolving resistance mechanisms [42]. Moreover, advances in artificial intelligence have made Adaptive Radiation Therapy (ART) more feasible and efficient. ART with deep learning-based algorithms offers more precise cancer treatment by adapting to changes in the patient's body during the course of therapy, leading to better outcomes with fewer side effects [1, 56]. By aligning treatment with molecular and microenvironmental dynamics, radiogenomics transforms adaptive radiotherapy into a precision tool, potentially improving survival rates and reducing the risk of relapse.

18.7 Ethical, Clinical, and Translational Challenges

Radiogenomics faces ethical, clinical, and translational challenges that hinder its integration into personalized radiotherapy. Ethical concerns include privacy risks in genetic data handling and disparities in access to advanced genomic tools. Clinically, tumor heterogeneity, dynamic molecular changes, and the lack of validated biomarkers limit predictive accuracy. Translationally, regulatory gaps and the need for large-scale validation impede clinical adoption. Addressing these challenges requires interdisciplinary collaboration, robust frameworks, and equitable implementation strategies to ensure radiogenomics fulfills its potential in improving cancer care.

18.7.1 Limitations in Current Radiogenomic Studies

Current radiogenomic studies are constrained by small sample sizes, retrospective designs, and inconsistent methodologies, limiting generalizability. Tumor heterogeneity and evolving resistance mechanisms further reduce reproducibility of biomarkers like *BRCA1/2* and *HIF-1α* [51].

18.7.2 Need for Large-Scale Validation and Regulatory Considerations

Large-scale, prospective trials are critical to validate findings across diverse populations and treatment settings. Regulatory frameworks lag behind technological advancements, lacking standardized guidelines for biomarker approval and clinical use [52]. For instance, the FDA's evolving criteria for companion diagnostics highlight the need for adaptive policies to accommodate radiogenomic innovations. Collaborative consortia, such as the Radiogenomics Consortium, are pioneering efforts to harmonize data collection and accelerate regulatory acceptance, bridging the gap between research and clinical practice.

18.7.3 Ethical Considerations in Genetic Screening for Radiotherapy

Genetic screening in radiogenomics raises ethical dilemmas, including informed consent, data privacy, and potential misuse of genomic information. Patients may face psychological distress or insurance discrimination if genetic predispositions to radioresistance are disclosed [57]. Additionally, biases in genomic databases, which underrepresent minority populations, risk exacerbating health disparities [58].

Transparent consent processes and secure data-sharing protocols are essential to maintain trust. Ethical guidelines, such as the ACMG's recommendations on incidental findings, must be adapted to address radiotherapy-specific scenarios [59]. Equitable access to radiogenomic technologies is equally vital to prevent stratification of care based on socioeconomic status, ensuring benefits are universally accessible (Tables 18.1, 18.2 and Fig. 18.1).

Table 18.1 Summarizes key genes, SNPs, and genetic variations that play a role in radiogenomics and influence radiosensitivity or radioresistance. These genes are involved in DNA repair, hypoxia response, cell cycle regulation, and immune modulation, which are critical determinants of radiation response

Gene	SNP/genetic variation	Function	Impact on radiation response.
ATM	rs1801516	DNA damage sensing and repair	Variants associated with impaired DSB repair, increasing radiosensitivity.
BRCA1/2	Multiple variants	Homologous recombination repair (HRR)	Mutations impair HRR, increasing sensitivity to radiation and PARP inhibitors.
RAD51	rs1801321, rs12593359	Homologous recombination repair	Variants linked to altered DNA repair capacity, affecting radiosensitivity.
CHEK1	rs558351	Cell cycle checkpoint regulation	Variants associated with poor prognosis and therapy resistance in HNSCC.
TP53	rs1042522 (Arg72Pro)	Apoptosis and cell cycle regulation	Variants impair p53 function, leading to radioresistance.
HIF-1α	Multiple variants	Hypoxia response and angiogenesis	Overexpression promotes radioresistance by enhancing survival under hypoxia.
CA9	Multiple variants	pH regulation in hypoxic tumors	Overexpression linked to poor radiation response due to acidification of the tumor microenvironment.
XRCC5	rs828907	Non-homologous end joining (NHEJ) repair	Variants associated with altered DSB repair efficiency, affecting radiosensitivity.
XRCC6	rs2267437	NHEJ repair	Variants linked to impaired DSB repair and increased radiosensitivity.
LIG4	rs1805388	DNA ligase involved in NHEJ repair	Variants associated with reduced repair capacity and increased radiosensitivity.
MRE11	rs2155209	DNA damage sensing and repair	Variants linked to altered DSB repair and radiation response.
NBN	rs2735383	DNA damage response and repair	Variants associated with impaired repair and increased radiosensitivity.
PD-L1	Multiple variants	Immune checkpoint regulation	Overexpression linked to immune evasion and radioresistance.
CTLA4	Multiple variants	Immune checkpoint regulation	Variants influence immune response, affecting radiation efficacy.
GLUT1	Multiple variants	Glucose transport and metabolism	Overexpression supports tumor survival under hypoxia, contributing to radioresistance.

(continued)

Table 18.1 (continued)

Gene	SNP/genetic variation	Function	Impact on radiation response.
LDHA	Multiple variants	Glycolysis and lactate production	Overexpression promotes acidification and radioresistance.
OCT4	Multiple variants	Stem cell maintenance and tumor stemness	Overexpression linked to tumor stemness and radioresistance.
MMP2/ MMP9	Multiple variants	Extracellular matrix remodeling and invasion	Overexpression promotes metastasis and radioresistance.
SERPINE1	Multiple variants	Inhibition of fibrinolysis and promotion of tumor survival	Overexpression linked to poor radiation response.
NF-κB	Multiple variants	Pro-survival signaling and anti-apoptotic pathways	Activation promotes radioresistance by enhancing cell survival.

DSB Double-strand break, *HRR* Homologous recombination repair, *NHEJ* Non-homologous end joining, *HNSCC* Head and neck squamous cell carcinoma

Table 18.2 Radiogenomic Genes and Pathways: This network illustrates the interconnected molecular pathways and genes involved in radiogenomics. Key pathways include DNA damage repair, hypoxia response, cell cycle regulation, immune modulation, and tumor microenvironment remodeling. Variants in these genes influence radiosensitivity or radioresistance, highlighting their potential as biomarkers for personalized radiotherapy

Category	Key genes	Pathways	Impact
DNA damage and repair pathways	ATM, BRCA1/2, RAD51, CHEK1, MRE11, NBN, XRCC5, XRCC6, LIG4	Homologous recombination repair (HRR): BRCA1/2, RAD51 Non-homologous end joining (NHEJ): XRCC5, XRCC6, LIG4 DNA damage sensing: ATM, CHEK1, MRE11, NBN	Variants in these genes impair DNA repair, increasing radiosensitivity or contributing to resistance if repair is overly efficient.
Hypoxia and metabolic adaptation	HIF-1α, CA9, GLUT1, LDHA	Hypoxia response: HIF-1α activates CA9, GLUT1, and LDHA Glycolysis and pH regulation: GLUT1 (glucose uptake), LDHA (lactate production), CA9	Hypoxia-induced genes promote survival under low oxygen, leading to radioresistance.
Cell cycle regulation and apoptosis	TP53, CHEK1	Cell cycle checkpoints: CHEK1 regulates G2/M arrest Apoptosis: TP53 induces cell death in response to DNA damage	Loss of function (e.g., TP53 mutations) leads to radioresistance by preventing apoptosis.

(continued)

Table 18.2 (continued)

Category	Key genes	Pathways	Impact
Immune modulation	PD-L1, CTLA4	Immune checkpoint regulation: PD-L1 and CTLA4 suppress T-cell activity	Overexpression promotes immune evasion and radioresistance, while inhibition enhances the abscopal effect.
Tumor Microenvironment and invasion	MMP2, MMP9, SERPINE1	Extracellular matrix (ECM) remodeling: MMP2/MMP9 facilitate invasion Fibrinolysis inhibition: SERPINE1 promotes tumor survival	Enhances metastasis and resistance to radiation.
Stemness and survival signaling	OCT4, NF-κB	Tumor stemness: OCT4 maintains stem cell properties Pro-survival signaling: NF-κB activates anti-apoptotic genes	Promotes radioresistance by enhancing tumor cell survival and regeneration.

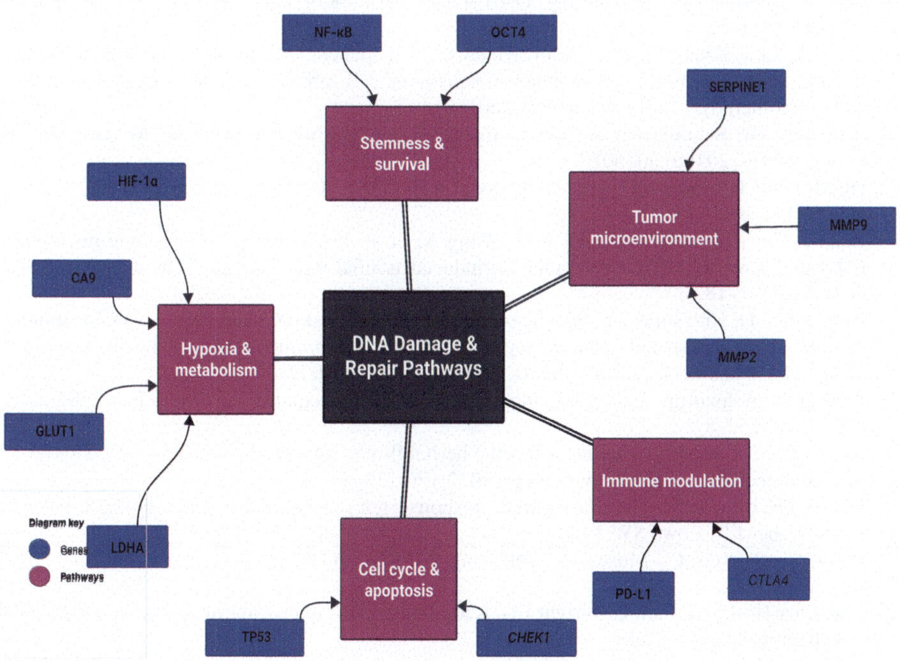

Fig. 18.1 Radiogenomic network of genes and pathways: This network illustrates the interconnected molecular pathways and genes involved in radiogenomics. Key pathways include DNA damage repair, hypoxia response, cell cycle regulation, immune modulation, and tumor microenvironment remodeling. Variants in these genes influence radiosensitivity or radioresistance, highlighting their potential as biomarkers for personalized radiotherapy

References

1. Wang C, Zhu X, Hong JC, Zheng D. Artificial intelligence in radiotherapy treatment planning: present and future. Technol Cancer Res Treat. 2019;18:1533033819873922.
2. Barnett GC, Elliott RM, Alsner J, Andreassen CN, Abdelhay O, Burnet NG, et al. Individual patient data meta-analysis shows no association between the SNP rs1800469 in TGFB and late radiotherapy toxicity. Radiother Oncol. 2012;105:289–95.
3. Scott JG, Sedor G, Scarborough JA, Kattan MW, Torres-Roca JF. Radiotherapy with genomic-adjusted radiation dose – authors' reply. Lancet Oncol. 2021;22:e470–1.
4. Burack WR, Li H, Adlowitz D, Spence JM, Rimsza LM, Shadman M, et al. Subclonal TP53 mutations are frequent and predict resistance to radioimmunotherapy in follicular lymphoma. Blood Adv. 2023;7:5082–90.
5. Barnett GC, Coles CE, Elliott RM, Baynes C, Luccarini C, Conroy D, et al. Independent validation of genes and polymorphisms reported to be associated with radiation toxicity: a prospective analysis study. Lancet Oncol. 2012;13:65–77.
6. Uddin MH, Zhang D, Muqbil I, El-Rayes BF, Chen H, Philip PA, et al. Deciphering cellular plasticity in pancreatic cancer for effective treatments. Cancer Metastasis Rev. 2024;43:393–408.
7. Chen JY, Lin PY, Hong WZ, Yang PC, Chiang SF, Chang HY, et al. Activation of STING by the novel liposomal TLC388 enhances the therapeutic response to anti-PD-1 antibodies in combination with radiotherapy. Cancer Immunol Immunother. 2024;73
8. Shiloh Y. The cerebellar degeneration in ataxia-telangiectasia: a case for genome instability. DNA Repair. 2020;95:102950.
9. Lord CJ, Ashworth A. Bringing DNA repair in tumors into focus. Clin Cancer Res. 2009;15:3241–3.
10. Zhang Y, Li Z, Zhang J, Li H, Qiao Y, Huang C, et al. Genetic variants in MTHFR gene predict ≥ 2 radiation pneumonitis in esophageal squamous cell carcinoma patients treated with thoracic radiotherapy. PLoS One. 2017;12:e0169147.
11. Williams AB, Schumacher B. p53 in the DNA-damage-repair process. Cold Spring Harb Perspect Med. 2016;6:a026070.
12. Hu H, Gatti RA. MicroRNAs: new players in the DNA damage response. J Mol Cell Biol. 2010;3:151–8.
13. Zhao J, Zhi Z, Zhang M, Li Q, Li J, Wang X, et al. Predictive value of single nucleotide polymorphisms in *XRCC1* for radiation-induced normal tissue toxicity. Onco Targets Ther. 2018;11:3901–18.
14. Butkiewicz D, Krześniak M, Gdowicz-Kłosok A, Składowski K, Rutkowski T. DNA double-strand break response and repair gene polymorphisms may influence therapy results and prognosis in head and neck cancer patients. Cancers. 2023;15:4972.
15. Lord CJ, Ashworth A. PARP inhibitors: synthetic lethality in the clinic. Science. 2017;355:1152–8.
16. Naderi E, et al. Meta-GWAS identifies the heritability of acute radiation-induced toxicities in head and neck cancer. Radiother Oncol. 2022;176:138–48.
17. Harper JW, Elledge SJ. The DNA damage response: ten years after. Mol Cell. 2007;28:739–45.
18. Kastenhuber ER, Lowe SW. Putting p53 in context. Cell. 2017;170:1062–78.
19. Porta C, Paglino C, Mosca A. Targeting PI3K/Akt/mTOR signaling in cancer. Front Oncol. 2014;4
20. Barcellos-Hoff MH, Park C, Wright EG. Radiation and the microenvironment – tumorigenesis and therapy. Nat Rev Cancer. 2005;5:867–75.
21. Schieber M, Chandel N. ROS function in redox signaling and oxidative stress. Curr Biol. 2014;24:R453–62.
22. Ciccia A, Elledge SJ. The DNA damage response: making it safe to play with knives. Mol Cell. 2010;40:179–204.
23. Tell G, Fantini D, Quadrifoglio F. Understanding different functions of mammalian AP endonuclease (APE1) as a promising tool for cancer treatment. Cell Mol Life Sci. 2010;67:3589–608.

24. Zhong H, Chiles K, Feldser D, Laughner E, Hanrahan C, Georgescu MM, et al. Modulation of hypoxia-inducible factor 1alpha expression by the epidermal growth factor/phosphatidylinositol 3-kinase/PTEN/AKT/FRAP pathway in human prostate cancer cells: implications for tumor angiogenesis and therapeutics. Cancer Res. 2000;60:1541–5.
25. Koong AC, Chen EY, Giaccia AJ. Hypoxia causes the activation of nuclear factor kappa B through the phosphorylation of I kappa B alpha on tyrosine residues. Cancer Res. 1994;54:1425–30.
26. Levy JMM, Towers CG, Thorburn A. Targeting autophagy in cancer. Nat Rev Cancer. 2017;17:528–42.
27. Goel S, DeCristo MJ, McAllister SS, Zhao JJ. CDK4/6 inhibition in cancer: beyond cell cycle arrest. Trends Cell Biol. 2018;28:911–25.
28. Bajou K, Peng H, Laug WE, Maillard C, Noel A, Foidart JM, et al. Plasminogen activator inhibitor-1 protects endothelial cells from FasL-mediated apoptosis. Cancer Cell. 2008;14:324–34.
29. Staller P, Sulitkova J, Lisztwan J, Moch H, Oakeley EJ, Krek W. Chemokine receptor CXCR4 downregulated by von Hippel–Lindau tumour suppressor pVHL. Nature. 2003;425:307–11.
30. Semenza G. Hypoxia-inducible factors in physiology and medicine. Cell. 2012;148:399–408.
31. Comerford KM, Wallace TJ, Karhausen J, Louis NA, Montalto MC, Colgan SP. Hypoxia-inducible factor-1-dependent regulation of the multidrug resistance (MDR1) gene. Cancer Res. 2002;62:3387–94.
32. Keith B, Johnson RS, Simon MC. HIF1α and HIF2α: sibling rivalry in hypoxic tumour growth and progression. Nat Rev Cancer. 2011;12:9–22.
33. Mathieu J, Zhang Z, Zhou W, Wang AJ, Heddleston JM, Pinna CMA, et al. HIF induces human embryonic stem cell markers in cancer cells. Cancer Res. 2011;71:4640–52.
34. Bindra RS, Schaffer PJ, Meng A, Woo J, Måseide K, Roth ME, et al. Down-regulation of Rad51 and decreased homologous recombination in hypoxic cancer cells. Mol Cell Biol. 2004;24:8504–18.
35. Chia SK, Wykoff CC, Watson PH, Han C, Leek RD, Pastorek J, et al. Prognostic significance of a novel hypoxia-regulated marker, carbonic anhydrase IX, in invasive breast carcinoma. J Clin Oncol. 2001;19:3660–8.
36. Fantin VR, St-Pierre J, Leder P. Attenuation of LDH-A expression uncovers a link between glycolysis, mitochondrial physiology, and tumor maintenance. Cancer Cell. 2006;9:425–34.
37. Egeblad M, Werb Z. New functions for the matrix metalloproteinases in cancer progression. Nat Rev Cancer. 2002;2:161–74.
38. Wilson WR, Hay MP. Targeting hypoxia in cancer therapy. Nat Rev Cancer. 2011;11:393–410.
39. Scott JG, Berglund A, Schell MJ, Mihaylov I, Fulp WJ, Yue B, et al. A genome-based model for adjusting radiotherapy dose (GARD): a retrospective, cohort-based study. Lancet Oncol. 2017;18:202–11.
40. Toustrup K, Sørensen BS, Lassen P, Wiuf C, Alsner J, Overgaard J. Gene expression classifier predicts for hypoxic modification of radiotherapy with nimorazole in squamous cell carcinomas of the head and neck. Radiother Oncol. 2012;102:122–9.
41. Weichselbaum RR, Liang H, Deng L, Fu YX. Radiotherapy and immunotherapy: a beneficial liaison? Nat Rev Clin Oncol. 2017;14:365–79.
42. Trebeschi S, Drago SG, Birkbak NJ, Kurilova I, Călin AM, Delli Pizzi A, et al. Predicting response to cancer immunotherapy using noninvasive radiomic biomarkers. Ann Oncol. 2019;30:998–1004.
43. McGranahan N, Swanton C. Clonal heterogeneity and tumor evolution: past, present, and the future. Cell. 2017;168:613–28.
44. Bristow RG, Hill RP. Hypoxia, DNA repair and genetic instability. Nat Rev Cancer. 2008;8:180–92.
45. Junttila MR, de Sauvage FJ. Influence of tumour micro-environment heterogeneity on therapeutic response. Nature. 2013;501:346–54.
46. Demaria S, Golden EB, Formenti SC. Role of local radiation therapy in cancer immunotherapy. JAMA Oncol. 2015;1:1325.

47. Navin N, Kendall J, Troge J, Andrews P, Rodgers L, McIndoo J, et al. Tumour evolution inferred by single-cell sequencing. Nature. 2011;472:90–4.
48. Xia H, Li Z, Lin Y, Lin Y, Zeng L, Xu B, et al. Validation of a genome-based model for adjusting radiotherapy dose (GARD) in patients with locally advanced rectal cancer. Sci Rep. 2024;14
49. Zhou H, Zhao Q, Xie Q, Peng Y, Chen M, Huang Z, et al. Preoperative prediction model of lymph node metastasis in the inguinal and femoral region based on radiomics and artificial intelligence. Int J Gynecol Cancer. 2024;34:1437–44.
50. Kang J, Coates JT, Strawderman RL, Rosenstein BS, Kerns SL. Genomics models in radiotherapy: from mechanistic to machine learning. Med Phys. 2020;47:e203.
51. Lambin P, Leijenaar RTH, Deist TM, Peerlings J, de Jong EEC, van Timmeren J, et al. Radiomics: the bridge between medical imaging and personalized medicine. Nat Rev Clin Oncol. 2017;14:749–62.
52. Phillips KA, Deverka PA, Hooker GW, Douglas MP. Genetic test availability and spending: where are we now? Where are we going? Health Aff. 2018;37:710–6.
53. Jabbour SK, Kumar R, Anderson B, Chino JP, Jethwa KR, McDowell L, et al. Combinatorial approaches for chemotherapies and targeted therapies with radiation: united efforts to innovate in patient care. Int J Radiat Oncol Biol Phys. 2024;118:1240–61.
54. Garg P, Pareek S, Kulkarni P, Horne D, Salgia R, Singhal SS. Next-generation immunotherapy: advancing clinical applications in cancer treatment. J Clin Med. 2024;13:6537.
55. Taghizadeh-Hesary F, Akbari H, Bahadori M, Behnam B. Targeted anti-mitochondrial therapy: the future of oncology. Genes. 2022;13:1728.
56. Lemus OMD, Cao M, Cai B, Cummings M, Zheng D. Adaptive radiotherapy: next-generation radiotherapy. Cancers. 2024;16(6):1206.
57. Smit AK, Gokoolparsadh A, McWhirter R, Newett L, Milch V, Hermes A, et al. Ethical, legal, and social issues related to genetics and genomics in cancer: a scoping review and narrative synthesis. Genet Med. 2024;26:101270.
58. Popejoy AB, Fullerton SM. Genomics is failing on diversity. Nature. 2016;538:161–4.
59. ACMG Board of Directors. Points to consider for the return of genomic results. Genet Med. 2021;23:803–6.

Chapter 19
The Emerging Role of Microbiome in Radiation Research

Pegah Alizadeh, Ali Mojtahed, and Farzad Taghizadeh-Hesary

Abstract Radiotherapy is a cornerstone of cancer treatment. Initially, research concentrated on the intrinsic characteristics of cancer cells in determining radiosensitivity. However, recent findings underscore the pivotal role of both the tumor microenvironment (TME) and the host microbiota in modulating the response to radiation. The human microbiota, comprising microorganisms that reside within the body, impacts the efficacy of radiotherapy both locally through tumor-infiltrating microbes (TiMs) and systemically, primarily via the gut microbiota. The microbiota modulates radiation response through diverse mechanisms, including immune modulation, regulation of hypoxia, reactivation of oncovirus, induction of immune exhaustion, and activation of TME. Dysbiosis may contribute to increased radioresistance by hindering anti-tumor immune reactivation, exacerbating tumor hypoxia, and fostering an immunosuppressive TME. Conversely, radiotherapy can directly impact microbial composition, which might affect therapeutic efficacy. This chapter offers a comprehensive overview of both experimental and clinical evidence regarding microbiota's influence on radiation response, elucidating potential mechanisms and exploring microbiome-targeted strategies to enhance radiotherapy efficacy. Particular emphasis is placed on the role of probiotics as potential radiosensitizers. Elucidating these interactions presents a promising opportunity to develop Personalized Radiotherapy approaches, incorporating microbiota profiling to optimize cancer treatment outcomes while reducing adverse effects.

Keywords Microbiome · Radiotherapy · Cancer · Tumor microenvironment · Radiosensitivity

---

P. Alizadeh · F. Taghizadeh-Hesary (✉)
ENT and Head and Neck Research Center and Department, The Five Senses Health Institute, School of Medicine, Iran University of Medical Sciences, Tehran, Iran

A. Mojtahed
Department of Microbiology, School of Medicine, Iran University of Medical Sciences, Tehran, Iran

F. Taghizadeh-Hesary et al. (eds.), *From Radiobiology to Radiation Oncology*,
https://doi.org/10.1007/978-3-031-95724-6_19

19.1 Introduction

Radiotherapy is a promising therapeutic approach in cancer treatment, utilized in the care of more than half of cancer patients at some point during their treatment. In conjunction with other disciplines, *Personalized Radiotherapy* has emerged as a promising strategy to enhance the efficacy of radiotherapy on tumor tissue while minimizing treatment-related toxicities [1].

To date, experimental studies have primarily focused on intrinsic factors within cancer cells to elucidate the characteristics that determine the extent of radiotherapy's impact on these cells [2]. However, additional research has identified extrinsic factors outside the cancer cells that play crucial roles in their response to radiation. One such element is the tumor microenvironment (TME). It has been shown that components of TME can influence cancer cells' response to radiation through a variety of mechanisms, including enhancing the mitochondrial function of cancer cells and impeding the infiltration of immune cells into the TME [3].

A growing body of evidence has highlighted the host microbiota as another pivotal factor influencing radiosensitivity, operating both within and beyond the TME. The microbiota consists of a diverse community of microorganisms inhabiting various regions of the human body, including the skin, digestive tract, respiratory system, and urogenital tract [4]. The impact of microbiota on tumors can occur both remotely, primarily through the gut microbiota, and locally, via tumor-infiltrating microbes (TiMs) [5].

It has been established that host microbiota can influence cancer cells' response to radiation [6]. Additionally, they may play a protective role against toxicities induced by radiation [7]. Conversely, radiotherapy itself can alter the host microbial composition [8]. This chapter delves into both foundational and recent studies examining the role of the microbiome in radiation response. The subsequent section will provide an in-depth discussion on the potential mechanisms by which host microbiota may impact radiation response, the experimental and clinical evidence supporting these effects, and the role of probiotics in modulating cancer cells' response to radiation. The intended audience for this chapter includes Radiation Oncologists, Microbiologists, Cancer Scientists, and all those interested in Oncology.

19.2 Mechanistic Overview

This section delineates the various mechanisms through which the composition of the host microbiota may influence tissue response to radiation. The subsequent section will explore the reciprocal effects of radiation on the host microbial composition.

19.2.1 Host Microbial Impacts on Radiation Response

Modulating Anti-tumor Immune Response

Advances in immunology have opened a new frontier in radiobiology. In 2019, Boustani et al. published a comprehensive review highlighting the pivotal role of the immune system in cancer cells' response to radiation, introducing the concept of "immune reactivation" as the 6th "R" of radiobiology, denoted as "Reactivation" [9]. It has been shown that microbial dysbiosis can disrupt the cancer cells' response to radiation by modulating the anti-tumor immune response. For instance, Li et al. (2022) demonstrated in a hepatocellular carcinoma mouse model that gut microbial dysbiosis can enhance the cancer cells' resistance to radiation by impairing immune reactivation. This study utilized 16S rRNA sequencing to analyze the gut microbial composition, a widely employed method for assessing microbiota diversity in specimens. The findings revealed that microbial-derived cyclic diadenylate (c-di-AMP) acts as a second messenger, inhibiting effector T cells by inducing the release of interferon 1 (IFN-1). The study concluded that enhancing gut microbial composition can improve the response to radiation by activating the anti-tumor immune response [6].

Enhancing Tumor Hypoxia

Hypoxia is a characteristic *intrinsic to the TME.* TiMs can enhance the radioresistance of cancer cells by decreasing oxygen availability within the TME. This process occurs through two main mechanisms: the direct consumption of oxygen by the microbiota and the recruitment of innate immune cells to the TME, which further deplete oxygen as a result of their metabolic activities [5].

The resulting hypoxic conditions impede the efficacy of radiation therapy by restricting the production of reactive oxygen species (ROS), thereby diminishing the indirect cytotoxic effects of radiation. Furthermore, under hypoxic conditions, hypoxia-inducible factor 1 (HIF-1) is not degraded and instead orchestrates the adaptation of cancer cells to these conditions. As a master regulator, HIF-1 shifts cancer cells' metabolism from the high-oxygen-demanding oxidative phosphorylation (OXPHOS) to the more adaptable glycolytic pathway. This metabolic adaptation, mediated by HIF-1 overexpression, also confers several additional benefits to cancer cells, including enhanced radioresistance. For example, a metabolic shift toward glycolysis and the pentose phosphate pathway (PPP) enhances radioresistance by elevating the levels of nicotinamide adenine dinucleotide phosphate (NADPH) and glutathione, two crucial cellular antioxidants [10]. Additionally, it is well established that hypoxia can augment cancer cells' radioresistance by promoting autophagy—a process through which cancer cells degrade and recycle damaged, overused organelles and proteins, enhancing survival under stressful conditions [11]. Furthermore, hypoxia can exacerbate radioresistance by driving cancer cells toward the epithelial-to-mesenchymal transition (EMT) process. In an experimental study on head and neck squamous cell carcinoma cells, Wiechec et al. demonstrated that hypoxic conditions enhance cancer cells' radioresistance by upregulating the expression of mesenchymal proteins, such as vimentin and fibronectin, as well as stem cell-associated mediators [12]. Consistent with these findings, numerous experimental studies have shown that malignancies with a larger hypoxic fraction exhibit greater radioresistance [13].

Epstein–Barr Virus Reactivation

A growing body of evidence has established a connection between microbiota disruption and the activation of the Epstein–Barr virus (EBV) in the pathogenesis of nasopharyngeal carcinoma (NPC). The role of microbiota disruption in EBV activation and NPC pathogenesis is a developing area of research, suggesting a complex interplay between microbial dysbiosis and viral oncogenesis. EBV, an oncogenic gamma herpesvirus, is strongly linked to NPC, persisting in and being transmitted primarily through the oral cavity, where it can lead to malignant transformation of nasopharyngeal epithelial cells [5]. A recent study by Liao et al. (2023) revealed that an increased abundance of *Streptococcus sanguinis* in the oral microbiota exacerbates EBV lytic reactivation. This effect is primarily mediated by the release of H_2O_2 from *S. sanguinis*, which leads to the demethylation of the EBV genome in a TNFα/NF-κB dependent manner [14].

Emerging evidence indicates that EBV-associated NPC exhibits enhanced radioresistance. Xu et al. (2021) showed that EBV-encoded latent membrane protein 1 (LMP1) enhances radioresistance in NPC by promoting autophagy in a BNIP3-dependent manner. BNIP3 facilitates autophagy by binding to Beclin-1, thereby competitively inhibiting Bcl-2 from interacting with Beclin-1 [15]. Consistent with these findings, Zhou et al. (2021) demonstrated that EBV-derived miR-BART8-3p contributes to NPC radioresistance by targeting the *PAG1* gene, ultimately leading to EMT [16]. Yang et al. (2015) previously demonstrated that EBV-encoded LMP1 enhances NPC radioresistance by increasing vascular endothelial growth factor (VEGF) secretion via the JNKs/c-Jun/HIF-1 signaling pathway [17]. These findings highlight the complex role of the oral microbiota in the radioresistance of EBV-associated NPC. Given the established connection between microbiota composition and EBV reactivation, the TME shaped by microbial dysbiosis may contribute to treatment resistance. This interaction suggests that targeting the oral microbiota could provide a novel therapeutic approach to modulate EBV reactivation and enhance the radiosensitivity of NPC. Further research is needed to elucidate the mechanistic interplay between microbiota-derived metabolites, EBV-driven oncogenesis, and NPC radioresistance, potentially paving the way for microbiota-targeted adjuvant therapies.

Inflammation-Induced Immune Exhaustion

TiMs play a pivotal role in modulating the TME, impacting both the efficacy of radiotherapy and the host immune response. Recent findings indicate that certain microbial communities within tumors can induce chronic intra-tumoral inflammation, which paradoxically hinders the effectiveness of radiation therapy. This microbial-driven inflammation leads to sustained immune activation, ultimately resulting in immune exhaustion—a condition marked by dysfunction and diminished cytotoxic activity of tumor-infiltrating lymphocytes (TILs) [18, 19].

Radiotherapy is recognized for inducing immunogenic cell death, a process that can enhance anti-tumor immunity by releasing tumor-associated antigens and activating antigen-presenting cells [9]. However, in the context of sustained microbial-induced inflammation, this immunogenic effect is significantly compromised. The persistent production of inflammatory cytokines within the TME creates an immunosuppressive environment, thereby hindering the 6th "R" of Radiobiology: immune reactivation.

A hallmark of immune exhaustion is mitochondrial dysfunction within TILs, which compromises their metabolic fitness and ability to sustain anti-tumor responses [20]. This dysfunction results in decreased ATP production, excessive ROS accumulation, and impaired oxidative phosphorylation, further exacerbating T cell exhaustion [21]. Notably, the expression of programmed cell death protein 1 (PD-1) on immune cells is closely linked to this metabolic dysregulation. PD-1 engagement not only inhibits T cell receptor signaling but also disrupts mitochondrial function, leading to energy depletion and reduced effector activity. Akbari et al. have proposed a mechanism linking mitochondrial dysfunction to elevated PD-1 expression, which further impairs the anti-tumor function of immune cells [22]. This suggests that targeting mitochondrial metabolism could offer a potential strategy to reinvigorate exhausted T cells and enhance radiotherapy outcomes.

Understanding the role of tumor-infiltrating microbes in inducing inflammation-driven immune exhaustion is essential for optimizing radiotherapy outcomes. Targeting microbial dysbiosis or modulating inflammatory pathways within the tumor microenvironment could provide novel therapeutic strategies to overcome radioresistance and enhance the efficacy of radiation therapy in cancer treatment.

Microbial-Induced Tumor microenvironment Activation

Emerging body of evidence underscores the pivotal role of various TME components in promoting radioresistance, recently conceptualized as the 7th "R" of Radiobiology: Reinforcement by tumor microenvironment [3] (see Chap. 2). It has been shown that microbiota can stimulate different TME components, thereby further impairing the effectiveness of radiotherapy on cancer cells. For instance, Eskandari-Malayeri et al. (2024) demonstrated that *Fusobacterium nucleatum* (*F. nucleatum*) can activate cancer-associated fibroblasts (CAFs), as evidenced by increased expression of fibroblast activation protein (FAP) and interleukin-6 (IL-6) [23].

Furthermore, several studies have highlighted the critical role of intestinal microbiota and the recruitment of myeloid-derived suppressor cells (MDSCs) in modulating the immune environment [24, 25]. Consistent with these findings, Fidelle et al. demonstrated that antibiotic-induced gut dysbiosis can facilitate the migration of immunosuppressive regulatory T cells (Tregs) from the intestine to tumors [26]. These immunosuppressive populations suppress the ability of effector T cells to mount a robust response, thereby neutralizing the immunostimulatory effects of radiation therapy. The cumulative impact of these processes shapes an immune landscape that favors tumor persistence and progression, ultimately contributing to radioresistance.

19.2.2 Radiation Impacts on Host Microbiota

Notably, radiotherapy itself can alter the host's microbial composition. A study by Nam et al. (2013) on patients with gynecological malignancies demonstrated that radiotherapy modifies the gut microbiota composition. Specifically, the study found a 10% decrease in the phylum *Firmicutes* and a 3% increase in *Fusobacterium* following radiation. In addition, alpha diversity, measured by the number of species-level taxa, was significantly reduced post-radiotherapy [8]. A systematic review further highlighted the association between ionizing radiation exposure and dysbiosis, characterized by decreased microbiota diversity and richness, an increased abundance of pathogenic bacteria (*Proteobacteria* and *Fusobacteria*), and a decline in beneficial bacteria (*Faecalibacterium* and *Bifidobacterium*) [27].

A recent systematic review of animal models (2023) reinforced the association between radiotherapy and dysbiosis, characterized by reduced microbiota diversity and richness, as well as changes in taxonomic composition. Although specific taxonomic shifts varied between studies, a consistent pattern emerged: increased relative abundances of *Proteobacteria*, *Verrucomicrobia*, *Alistipes*, and *Akkermansia*, while *Bacteroidetes*, *Firmicutes*, and *Lactobacillus* exhibited a relative decline [28].

These findings emphasize the profound impact of radiotherapy on host microbiota, highlighting a shift toward dysbiosis marked by reduced microbial diversity and an imbalance in taxonomic composition. The consistent decline in beneficial bacteria such as *Faecalibacterium, Bifidobacterium, Firmicutes,* and *Lactobacillus*—which are crucial for maintaining gut homeostasis and immune function, suggests that radiation-induced dysbiosis may contribute to treatment-related complications, including gastrointestinal toxicity, systemic inflammation, and impaired immune responses. Conversely, the enrichment of *Proteobacteria* and *Fusobacteria*, known for their pro-inflammatory potential, raises concerns about their role in exacerbating radiation-induced tissue damage and immune dysregulation. Given these findings, microbiome-targeted interventions, such as probiotic and prebiotic supplementation, fecal microbiota transplantation (FMT), and dietary modifications, merit further investigation as potential strategies to mitigate the adverse effects of radiotherapy and restore microbial equilibrium. Future studies should also explore individual variability in microbiota response to radiation, as personalized microbiome-based approaches could enhance treatment tolerability and improve patient outcomes.

19.3 Experimental and Clinical Evidence

This section provides a comprehensive overview of the indexed experimental and clinical evidence, highlighting the significance of host microbiota in influencing the efficacy of radiotherapy (Table 19.1).

Table 19.1 Summary of studies evaluating the effects of microbiota on radiation response

Study	Year	Objectives	Results	References
Dong et al.	2021	Investigate the impact of oral microbiota dysbiosis on rectal tumor response to radiation	*F. nucleatum* impairs radiation efficacy in rectal tumors; metronidazole enhances response and reduces radiation enteritis	[14]
Uribe-Herranz et al.	2020	Examine the role of gut microbiota in radiation-induced immune reactivation	Vancomycin enhances radiation efficacy by promoting tumor antigen presentation to CD8+ T cells	[15]
Yang et al.	2021	Assess the effect of butyrate on radiation-induced immune activation	Butyrate inhibits immune reactivation by blocking STING-induced IFN-1 expression in dendritic cells	[16]
Park et al.	2020	Investigate the radiosensitizing effects of butyrate in colorectal cancer	Butyrate impedes S-phase entry, enhancing radiosensitivity in cancer cells without affecting normal cells	[17]
Shiao et al.	2021	Study the impact of gut bacteria and fungi on radiation-induced immune activation	Bacteria enhance immune reactivation; fungi suppress it by recruiting myeloid-derived suppressor cells	[18]
Ahmadi et al.	2024	Explore the effect of urolithin on gastric cancer cell radiosensitivity	Urolithin enhances radiosensitivity by inhibiting CHK1, disrupting DNA repair	[19]
Cui et al.	2017	Assess the potential of fecal microbiota transplantation (FMT) in reducing radiation-induced toxicity	FMT reduces radiation-induced mortality and hematological toxicity, improves gut integrity	[20]

19.3.1 Impacts of Microbiome on Radiotherapy Efficacy

Oral microbiota can significantly influence the response of rectal tumors to radiation therapy. A novel study by Dong et al. (2021) demonstrated that dysbiosis of the oral microbiota, characterized by an increased presence of *F. nucleatum*, can compromise the effectiveness of radiation therapy in rectal tumors. Their findings revealed that *F. nucleatum* can migrate to the gut and infiltrate colorectal tumors, thereby diminishing the efficacy of radiation treatment. Notably, the administration of metronidazole, an agent effective against *F. nucleatum*, acted as a radiosensitizer by depleting *F. nucleatum* within colorectal tumors [29]. These results underscore the potential impact of oral microbiota dysbiosis on the radiation response of distant tumors within the gastrointestinal tract. Further investigation is warranted to determine whether maintaining oral health during radiotherapy could enhance treatment efficacy.

Gut microbiota may also play a significant role in modulating radiation-induced immune reactivation. Uribe-Herranz et al. (2020) demonstrated that vancomycin, an antibiotic targeting Gram-positive bacteria, can enhance the anti-tumor effects of radiation by acting as a radiosensitizer. This occurs through the promotion of tumor-specific antigen presentation to $CD8^+$ cytotoxic T cells [30]. In support of these findings, Yang et al. (2021) found that gut microbiota can impede radiation-induced immune reactivation by releasing butyrate, which inhibits STING-induced IFN-1 expression in dendritic cells—cells responsible for presenting cancer antigens to effector immune cells [31]. These findings underscore the critical role of gut microbiota in regulating cancer antigen presentation to anti-tumor immune cells. Therefore, modulating gut microbiota composition could potentially enhance radiotherapy efficacy by activating the sixth "R" of radiobiology: Reactivation.

A study on colorectal cancer organoids demonstrated that butyrate itself can act as a radiosensitizer. Park et al. (2020) found that adding butyrate to the organoid enhanced radiation effects by preventing cancer cells from entering the S-phase of the cell cycle [32], thereby impeding cancer cell repopulation. This contrasts with findings from Yang et al. (2021).s study [31], where microbiota-derived butyrate impaired radiation effects by blocking the anti-tumor immune response. These contrasting results highlight the complexity and context-dependent nature of butyrate's role in radiotherapy outcomes. The differential impact of butyrate on immune modulation versus cancer cell repopulation underscores the need for a more nuanced approach when considering microbiota-targeted interventions in radiotherapy. Future studies should focus on defining the conditions under which butyrate exerts predominantly radiosensitizing versus radioresistant effects. Investigating the timing, dosage, and microbiome composition that influence these outcomes could pave the way for personalized microbiota-based strategies to optimize radiotherapy efficacy.

A recent study by Shiao et al. (2021) has revealed an additional facet of the interplay between gut microbiota and radiotherapy. Their findings indicate that gut bacteria and fungi exert contrasting effects on radiation-induced immune activation.

Specifically, antibiotic administration was found to facilitate fungal overgrowth in the gut, leading to immune suppression. In this context, fungi recruit MDSCs, which inhibit the activation of anti-tumor T cells, thereby dampening the immune response to radiation. Conversely, the administration of amphotericin, an antifungal agent, resulted in a shift from fungal to bacterial dominance in the gut microbiota. This shift enhanced the radiation response by promoting the activation of anti-tumor effector T cells [33].

These findings underscore the intricate and dynamic interactions between diverse microbial communities and the host immune system during radiotherapy. While bacterial populations seem to support immune activation, fungal overgrowth may counteract these effects, leading to a compromised therapeutic response. This highlights the necessity for a more nuanced approach to microbiota modulation in cancer treatment. Future research should focus on developing personalized microbiome-targeted strategies, including selective antimicrobial interventions, probiotics, or microbiota transplantation, to optimize radiotherapy efficacy while minimizing immune suppression.

Recently, Ahmadi et al. (2024) demonstrated that gut microbiota-derived urolithin can effectively sensitize gastric cancer cells to irradiation. This in vitro study revealed that urolithin enhances the radiosensitivity of cancer cells by inhibiting checkpoint kinase 1 (CHK1), a crucial enzyme involved in DNA repair. By targeting CHK1, urolithin disrupts the cellular DNA damage response, thereby improving the efficacy of radiation therapy [34]. Consequently, microbial-derived urolithin influences the first "R" of radiobiology: DNA Repair. These findings highlight the potential of microbiota-derived metabolites as natural radiosensitizers, offering a novel approach to enhancing radiotherapy outcomes. Future research should explore the translational potential of urolithin in preclinical and clinical settings, as well as investigate its synergistic effects with existing DNA damage response inhibitors.

19.3.2 Impacts of Microbiome on Radiotherapy Toxicity

Oral microbiota dysbiosis may play a role in exacerbating radiation-induced toxicity in the gastrointestinal tract. Research by Dong et al. (2021) indicates that an imbalance in oral microbiota, leading to an increased presence of *F. nucleatum*, can worsen radiation-induced enteritis. Their study revealed that *F. nucleatum* migrates to the gut, where it infiltrates colorectal tumors and potentially intensifies radiation-related toxicity. Notably, the use of metronidazole, an agent effective against *F. nucleatum*, not only enhanced tumor response to radiation but also effectively reduced radiation-induced enteritis [29]. These findings suggest that modulating oral microbiota could be a strategy to mitigate radiotherapy-associated toxicity, underscoring the importance of further research into the role of oral health in managing radiation side effects.

As previously noted, the study by Park et al. [32] demonstrated that butyrate exhibits radiosensitizing effects on colorectal cancer organoids. In addition to its effects on cancerous tissues, the researchers also investigated whether butyrate has a similar impact on normal organoids, suggesting that it selectively targets cancerous tissues. Consequently, butyrate appears to be a specific agent with radiosensitizing effects that are preferentially directed towards malignant tissues. Future studies can further elucidate this concept.

Fecal microbiota transplantation involves the transfer of gut microbiota from a healthy donor to a recipient, aiming to restore microbial balance. While primarily used to treat recurrent gastrointestinal infections, FMT is also being explored for conditions such as inflammatory bowel disease and metabolic disorders. Additionally, FMT has shown potential in reducing the toxicities associated with radiotherapy. In an experiment on both male and female mice, Cui et al. (2017) demonstrated that FMT could effectively reduce the radiation-induced mortality and hematological toxicities. Furthermore, this study highlighted FMT's ability to improve gastrointestinal tract function and epithelial integrity following radiation exposure [35]. These findings suggest that FMT may offer a novel approach to mitigating radiotherapy-induced toxicities by restoring gut homeostasis. However, further research is necessary to optimize donor selection, standardize transplantation protocols, and assess its long-term safety and efficacy in cancer patients undergoing radiation therapy. Future clinical trials will be pivotal in determining FMT's therapeutic potential in radiotherapy settings.

19.4 Clinical Implications and Future Prospects

19.4.1 Probiotics: A Viable Choice to Reduce Radiotherapy Toxicities

Recent studies underscore the potential of dietary interventions in mitigating radiotherapy-induced toxicities, particularly through gut microbiota modulation [36]. A fiber-rich diet has been shown to enhance gut barrier function, reduce inflammatory processes, and minimize oxidative stress [37–39]. Studies conducted on mice indicate that consuming high-fiber diets both before and after radiation exposure can decrease intestinal damage and inflammation, suggesting a protective effect against radiotherapy-induced gut toxicity [39].

Traditional Chinese Medicine formulations, such as Guiqi Baizhu Decoction (GQBZD), have demonstrated efficacy in alleviating colitis symptoms and inflammation in radiation models. In rat models of radiation-induced proctitis, GQBZD was found to reverse microbiota imbalances characterized by decreased *Firmicutes* and increased *Bacteroidetes*, a dysbiosis pattern commonly observed in multiple animal models and humans. Mechanistically, GQBZD suppressed the phosphorylation of P65, a crucial regulator in the NF-κB inflammatory pathway, thereby reducing intestinal inflammation and permeability [40]. The regulation of gut microbiota and inflammatory pathways by GQBZD highlights its potential in managing radiation-induced gut toxicities. However, further studies are necessary to elucidate the precise mechanisms underlying GQBZD's effects and to identify the specific populations that may benefit from this intervention.

Spirulina platensis (SP), a microalga rich in bioactive compounds, has been explored for its antioxidative, anti-inflammatory, and gut microbiota-modulating properties [41]. Due to its capacity to serve as a natural micro-carrier, SP has been investigated as an oral delivery system for radioprotective agents. When loaded with amifostine (AMF), SP enhanced radioprotection in the small intestine, particularly in the ileum, by facilitating efficient drug distribution throughout the gut. Compared to conventional enteric-coated AMF capsules, SP-loaded AMF (SP@AMF) provided superior radioprotection, suggesting its potential as a targeted drug delivery system for intestinal protection in cancer patients undergoing radiotherapy. Additionally, SP@AMF preserved the gut microbiota composition in irradiated mice, likely due to the prebiotic properties of SP, which include cellulose and polyunsaturated fatty acids [42]. However, the precise microbial changes induced by SP@AMF remain unclear, necessitating further research that integrates metagenomics, metabolomics, and transcriptomics to elucidate these effects. Given its demonstrated efficacy and favorable safety profile, SP@AMF represents a promising strategy for protecting the intestines from radiotherapy-induced damage.

Beyond the gastrointestinal tract, dietary interventions have also been explored for their role in mitigating radiation-induced brain damage. Chang et al. demonstrated that a curcumin derivative, bisdemethoxycurcumin (BDMC), can effectively improve learning and memory in rats with radiotherapy-induced brain injury [43]. Studies have shown that radiation exposure to the central nervous system triggers microglial activation, neuroinflammation, and behavioral changes, including anxiety and cognitive decline. The gut microbiota plays a critical role in these effects through the microbiota-gut-brain axis, which involves pathways such as the hypothalamic-pituitary-adrenal axis, immune signaling, and vagus nerve stimulation. Alterations in gut microbiota composition following radiation exposure are associated with increased intestinal permeability and systemic inflammation, which can further exacerbate brain damage.

Quercetin inclusion complex gels (QICG) have been investigated as a potential therapeutic strategy for mitigating radiation-induced neurotoxicity. In a study involving irradiated mice, QICG treatment was found to improve behavioral outcomes, maintain intestinal barrier integrity, and reduce neuroinflammation. Notably, QICG modulated gut microbiota composition by increasing the abundance of *Bacteroidota* while decreasing Firmicutes, a shift associated with enhanced gut and brain health. However, the therapeutic effects of QICG were absent in antibiotic-treated mice, highlighting the dependence of its efficacy on an intact gut microbiota [44]. By modulating the microbiota-gut-brain axis, QICG offers a novel approach for mitigating central nervous system damage following radiotherapy. Future research should focus on elucidating the precise mechanisms underlying its neuroprotective effects.

Lycium barbarum extract (LBE) represents another microbiota-targeting intervention with potential radioprotective benefits. Studies suggest that LBE supplementation enhances the presence of beneficial gut bacteria, promotes long-term microbial stability, and increases the production of indole derivatives—microbial metabolites derived from dietary tryptophan with immunomodulatory properties. These Indole derivatives play a crucial role in maintaining gut barrier integrity, enhancing antimicrobial defense, and regulating inflammation, making them relevant in mitigating radiation-induced toxicities [45]. Furthermore, a study by Jiang et al. (2023) highlighted the potential of *Lycium barbarum* polysaccharide-glycoprotein in preventing radiation-induced oral mucositis. This study demonstrated that LBE can effectively prevent cellular and mitochondrial damage following radiation exposure [46].

Collectively, these findings emphasize the potential of dietary and microbiota-based strategies in reducing radiotherapy-induced toxicities. While preclinical studies provide compelling evidence supporting the role of fiber-rich diets, herbal formulations, probiotics, and bioactive compounds in mitigating radiation damage, further clinical research is needed to establish optimal dosing regimens, treatment duration, and long-term safety profiles. Given the complexity of host–microbiota interactions, personalized approaches may be required to maximize the benefits of these interventions. Nonetheless, integrating microbiota-modulating strategies into cancer care holds considerable promise for improving patient outcomes and quality of life during radiotherapy.

19.4.2 Probiotics: A Candidate Choice for Improving Radiotherapy Efficacy

As discussed in Sect. 19.2, the microbiota can influence the tumor cell resistance to radiotherapy through several mechanisms, including modulating the anti-tumor immune response, enhancing tumor hypoxia, reactivating EBV (in the case of NPC), inducing immune exhaustion, and reinforcing the TME. Given the significant role of microbiota in shaping radiation response, researchers have begun exploring probiotics as a potential adjuvant strategy to enhance radiotherapy efficacy. *Probiotics*, defined as live microorganisms that confer health benefits when administered in sufficient quantities, have shown promise in modulating microbiota composition and counteracting tumor-associated dysbiosis [47]. As such, future clinical trials are invited to evaluate the potential of probiotic supplementation in patients with cancer undergoing radiation therapy, with a focus on assessing their impact on treatment response, immune modulation, and overall patient outcomes. However, careful consideration is needed when selecting appropriate probiotic strains, as the effects of microbiota modulation can vary depending on tumor type, treatment regimen, and patient-specific factors.

19.5 Conclusions

Radiotherapy remains a fundamental component of cancer treatment, yet its efficacy varies significantly among individuals, underscoring the need for Personalized Radiotherapy approaches to optimize treatment outcomes while minimizing toxicities. Initially, research focused on intrinsic tumor factors; however, the host microbiota has increasingly been recognized as a crucial determinant of radiosensitivity.

The host microbiota, both systemically (gut microbiota) and locally (TiMs), plays a dual role in radiation response. Certain microbial communities can enhance tumor resistance to radiation by modulating immune responses, increasing tumor hypoxia, promoting inflammation-induced immune exhaustion, reactivating oncoviruses, and interacting with various TME components. Given the emerging evidence, integrating microbiome profiling into radiotherapy decision-making could be a transformative step toward precision oncology. Future research should focus on identifying microbial biomarkers, elucidating the mechanisms underlying microbiota-driven radiation response, and developing microbiota-based therapeutics. Moreover, clinical trials evaluating microbiome interventions in radiotherapy settings will be essential to translating these findings into practice. Consequently, the intersection of radiotherapy, immunology, and microbiome research presents a promising frontier in oncology. Harnessing the microbiome‘s role in radiation response has the potential to redefine radiotherapy strategies, ultimately improving patient survival and quality of life.

References

1. VSK M, Taghizadeh-Hesary F. Advances in personalized radiotherapy. BMC Cancer. 2024;24(1):556. s12885-024-12317–3
2. Liberal FDCG, McMahon SJ. Characterization of intrinsic radiation sensitivity in a diverse panel of normal, cancerous and CRISPR-modified cell lines. Int J Mol Sci. 2023;24(9):7861.
3. Taghizadeh-Hesary F. Reinforcement by tumor microenvironment: the seventh “R” of radiobiology. Int J Radiat Oncol Biol Phys. 2023;
4. Chunxi L, Haiyue L, Yanxia L, Jianbing P, Jin S. The gut microbiota and respiratory diseases: new evidence. J Immunol Res. 2020;2020:1–12.
5. Ashique S, Houshyari M, Islam A, Pal R, Ghazanfar S, Taghizadeh-Hesary F. The role of microbiota in nasopharyngeal cancer: where do we stand? Oral Oncol. 2024;158:106982.
6. Li Z, Zhang Y, Hong W, Wang B, Chen Y, Yang P, et al. Gut microbiota modulate radiotherapy-associated antitumor immune responses against hepatocellular carcinoma Via STING signaling. Gut Microbes. 2022;14(1):2119055.
7. Manichanh C, Varela E, Martinez C, Antolin M, Llopis M, Dor J, et al. The gut microbiota predispose to the pathophysiology of acute postradiotherapy diarrhea. Am J Gastroenterol. 2008;103(7):1754–61.
8. Nam YD, Kim HJ, Seo JG, Kang SW, Bae JW. Impact of pelvic tadiotherapy on gut microbiota of gynecological cancer patients revealed by massive pyrosequencing. Heimesaat MM, editor. PLoS ONE. 2013;8(12):e82659.
9. Boustani J, Grapin M, Laurent PA, Apetoh L, Mirjolet C. The 6th R of radiobiology: reactivation of anti-tumor immune response. Cancers (Basel). 2019;11(6)

10. Huang R, Zhou PK. HIF-1 signaling: a key orchestrator of cancer radioresistance. Radiat Med Prot. 2020;1(1):7–14.
11. Hill RM, Fok M, Grundy G, Parsons JL, Rocha S. The role of autophagy in hypoxia-induced radioresistance. Radiother Oncol. 2023;189:109951.
12. Wiechec E, Matic N, Ali A, Roberg K. Hypoxia induces radioresistance, epithelial-mesenchymal transition, cancer stem cell-like phenotype and changes in genes possessing multiple biological functions in head and neck squamous cell carcinoma. Oncol Rep. 2022;47(3):58.
13. Beckers C, Pruschy M, Vetrugno I. Tumor hypoxia and radiotherapy: a major driver of resistance even for novel radiotherapy modalities. Semin Cancer Biol. 2024;98:19–30.
14. Liao Y, Zhang JB, Lu LX, Jia YJ, Zheng MQ, Debelius JW, et al. Oral microbiota alteration and roles in Epstein-Barr virus reactivation in nasopharyngeal carcinoma. Microbiol Spectr. 2023;11(1):03448–22.
15. Xu S, Zhou Z, Peng X, Tao X, Zhou P, Zhang K, et al. EBV-LMP1 promotes radioresistance by inducing protective autophagy through BNIP3 in nasopharyngeal carcinoma. Cell Death Dis. 2021;12(4):344.
16. Zhou X, Lin Y, Chen Y, Wang L, Peng X, Liao J, et al. Epstein–Barr virus (EBV) encoded microRNA BART8-3p drives radioresistance-associated metastasis in nasopharyngeal carcinoma. J Cell Physiol. 2021;236(9):6457–71.
17. Yang L, Liu L, Xu Z, Liao W, Feng D, Dong X, et al. EBV-LMP1 targeted DNAzyme enhances radiosensitivity by inhibiting tumor angiogenesis via the JNKs/HIF-1 pathway in nasopharyngeal carcinoma. Oncotarget. 2015;6(8):5804–17.
18. Chen J, Pitmon E, Wang K. Microbiome, inflammation and colorectal cancer. Semin Immunol. 2017;32:43–53.
19. Fang L, Liu K, Liu C, Wang X, Ma W, Xu W, et al. Tumor accomplice: T cell exhaustion induced by chronic inflammation. Front Immunol. 2022;13:979116.
20. Houshyari M, Taghizadeh-Hesary F. Is mitochondrial metabolism a new predictive biomarker for antiprogrammed cell death Protein-1 immunotherapy? JCO Oncol Pract. 2023;19(3):123–4.
21. Wu H, Zhao X, Hochrein SM, Eckstein M, Gubert GF, Knöpper K, et al. Mitochondrial dysfunction promotes the transition of precursor to terminally exhausted T cells through HIF-1α-mediated glycolytic reprogramming. Nat Commun. 2023;14(1):6858.
22. Akbari H, Taghizadeh-Hesary F, Bahadori M. Mitochondria determine response to anti-programmed cell death protein-1 (anti-PD-1) immunotherapy: an evidence-based hypothesis. Mitochondrion. 2022;62:151–8.
23. Eskandari-Malayeri F, Rezeai M, Narimani T, Esmaeil N, Azizi M. Investigating the effect of Fusobacterium nucleatum on the aggressive behavior of cancer-associated fibroblasts in colorectal cancer. Discov Oncol. 2024;15(1):292.
24. Chen J, Liu X, Zou Y, Gong J, Ge Z, Lin X, et al. A high-fat diet promotes cancer progression by inducing gut microbiota–mediated leucine production and PMN-MDSC differentiation. Proc Natl Acad Sci USA. 2024;121(20):e2306776121.
25. Zhang Q, Ma C, Duan Y, Heinrich B, Rosato U, Diggs LP, et al. Gut microbiome directs hepatocytes to recruit MDSCs and promote cholangiocarcinoma. Cancer Discov. 2021;11(5):1248–67.
26. Fidelle M, Rauber C, Alves Costa Silva C, Tian AL, Lahmar I, De La Varende ALM, et al. A microbiota-modulated checkpoint directs immunosuppressive intestinal T cells into cancers. Science. 2023;380(6649):eabo2296.
27. Fernandes A, Oliveira A, Soares R, Barata P. The effects of ionizing radiation on gut microbiota, a systematic review. Nutrients. 2021;13(9):3025.
28. Fernandes A, Oliveira A, Soares R, Barata P. The effects of ionizing radiation on gut microbiota: what can animal models tell us?—a systematic review. CIMB. 2023;45(5):3877–910.
29. Dong J, Li Y, Xiao H, Zhang S, Wang B, Wang H, et al. Oral microbiota affects the efficacy and prognosis of radiotherapy for colorectal cancer in mouse models. Cell Rep. 2021;37(4):109886.
30. Uribe-Herranz M, Rafail S, Beghi S, Gil-de-Gómez L, Verginadis I, Bittinger K, et al. Gut microbiota modulate dendritic cell antigen presentation and radiotherapy-induced antitumor immune response. J Clin Invest. 2019;130(1):466–79.

31. Yang K, Hou Y, Zhang Y, Liang H, Sharma A, Zheng W, et al. Suppression of local type I interferon by gut microbiota–derived butyrate impairs antitumor effects of ionizing radiation. J Exp Med. 2021;218(3):e20201915.
32. Park M, Kwon J, Shin HJ, Moon SM, Kim SB, Shin US, et al. Butyrate enhances the efficacy of radiotherapy via FOXO3A in colorectal cancer patient-derived organoids. Int J Oncol. 2020;57(6):1307–18.
33. Shiao SL, Kershaw KM, Limon JJ, You S, Yoon J, Ko EY, et al. Commensal bacteria and fungi differentially regulate tumor responses to radiation therapy. Cancer Cell. 2021;39(9):1202–1213.e6.
34. Ahmadi A, Hosseini F, Ostadi H, Gholamhosseinian H, Iranshahy M, Rassouli FB. Enhancing the efficacy of radiotherapy in gastric adenocarcinoma cells by gut microbiota metabolites: implications for urolithin combination therapy. Nat Prod Commun. 2024;19(12):1934578X241310020.
35. Cui M, Xiao H, Li Y, Zhou L, Zhao S, Luo D, et al. Faecal microbiota transplantation protects against radiation-induced toxicity. EMBO Mol Med. 2017;9(4):448–61.
36. Wang L, Mo YY, Huang Y, Echeveste CE, Wang H, Chen J, et al. Effects of dietary interventions on gut microbiota in humans and the possible impacts of foods on patients' responses to cancer immunotherapy. eFood. 2020;1(4):279–87.
37. Guan ZW, Yu EZ, Feng Q. Soluble dietary fiber, one of the most important nutrients for the gut microbiota. Molecules. 2021;26(22):6802.
38. Zhang F, Fan D, Huang J, lin, Zuo T. The gut microbiome: linking dietary fiber to inflammatory diseases. Med Microecol. 2022;14:100070.
39. Patel P, Malipatlolla DK, Devarakonda S, Bull C, Rascón A, Nyman M, et al. Dietary oat bran reduces systemic inflammation in mice subjected to pelvic irradiation. Nutrients. 2020;12(8):2172.
40. Zhang LY, Zhou T, Zhang YM, Xu XM, Li YY, Wei KX, et al. Guiqi Baizhu decoction alleviates radiation inflammation in rats by modulating the composition of the gut microbiota. Evid Based Complement Alternat Med. 2020;2020:9017854.
41. Lu L, Li F, Gao Y, Kang S, Li J, Guo J. Microbiome in radiotherapy: an emerging approach to enhance treatment efficacy and reduce tissue injury. Mol Med. 2024;30(1):105.
42. Zhang D, Zhong D, Ouyang J, He J, Qi Y, Chen W, et al. Microalgae-based oral microcarriers for gut microbiota homeostasis and intestinal protection in cancer radiotherapy. Nat Commun. 2022;13(1):1413.
43. Chang YQ, Zhou GJ, Wen HM, He DQ, Xu CL, Chen YR, et al. Treatment of radiation-induced brain injury with bisdemethoxycurcumin. Neural Regen Res. 2023;18(2):416–21.
44. Hu J, Jiao W, Tang Z, Wang C, Li Q, Wei M, et al. Quercetin inclusion complex gels ameliorate radiation-induced brain injury by regulating gut microbiota. Biomed Pharmacother. 2023;158:114142.
45. Alexeev EE, Lanis JM, Kao DJ, Campbell EL, Kelly CJ, Battista KD, et al. Microbiota-derived indole metabolites promote human and murine intestinal homeostasis through regulation of interleukin-10 receptor. Am J Pathol. 2018;188(5):1183–94.
46. Jiang SJ, Xiao X, Li J, Mu Y. Lycium barbarum polysaccharide-glycoprotein ameliorates ionizing radiation-induced epithelial injury by regulating oxidative stress and ferroptosis via the Nrf2 pathway. Free Radic Biol Med. 2023;204:84–94.
47. Azad MAK, Sarker M, Li T, Yin J. Probiotic species in the modulation of gut microbiota: an overview. Biomed Res Int. 2018;2018:9478630.

Chapter 20
Future Directions in Radiobiology and Clinical Radiation Oncology

Debasmita Pal, Utpal Bhui, Satadeep Bandyapadhyay, Sumel Ashique, and Biplab Debnath

Abstract Radiobiology, often known as clinical radiation oncology, is at the forefront of revolutionary developments that could increase therapeutic efficacy and change patient outcomes. Expanding our knowledge of DNA damage, repair pathways, and tumor microenvironment modulation, central to these advancements is the study of cellular and molecular mechanisms in radiobiology. Emerging as the pillar of individualized medicine, precision radiation treatment uses cutting-edge imaging and delivery methods to maximize tumor control and minimize damage to healthy tissues. Combining radiomics with artificial intelligence (AI) into radiation oncology marks a new chapter that will allow predictive analytics, treatment optimization, and real-time adaptation. Advancements in proton treatment and other innovative radiation techniques provide good routes for accurately treating difficult and radioresistant tumors. Combining treatments and molecular interventions helps to overcome radiation resistance, although this is still a challenging task. Examined through the perspective of radiobiology as connected with radiation oncology, the advancements in radiation oncology research and the field of radiation sciences have been highlighted. Emphasizing their potential to redefine standards of treatment in radiation oncology and their part in reaching equal, effective cancer treatment globally, this chapter explores future directions.

D. Pal · S. Bandyapadhyay · S. Ashique · B. Debnath (✉)
Department of Pharmaceutical Technology, Bharat Technology,
Uluberia, West Bengal, India

U. Bhui
NSHM Institute of Pharmaceutical Technology, NSHM Knowledge Campus,
Durgapur, West Bengal, India

F. Taghizadeh-Hesary et al. (eds.), *From Radiobiology to Radiation Oncology*,
https://doi.org/10.1007/978-3-031-95724-6_20

Keywords Radiobiology · Cellular Mechanisms · Molecular Mechanisms · Radiomics · Artificial Intelligence (AI) · Proton Therapy · Radiation Modalities · Radiation Resistance · Oncology Advancements · Technological Integration · Future Oncology Trends

20.1 Introduction

The chapter "Future Directions in Radiobiology and Clinical Radiation Oncology" focuses on the major developments influencing radiation-based cancer treatments going forward. Highlighting cellular and molecular mechanisms, technological innovations, and global accessibility, this chapter provides a framework for increasing therapeutic efficacy and patient outcomes [1]. Historically, the management and treatment of many cancers have benefited significantly from radiation oncology. Ongoing advancements in technology and scientific insights to meet the rising needs for accuracy, efficacy, and worldwide availability are rapidly transforming this field. Advanced research and technological developments will drive major changes in radiobiology and clinical radiation oncology that define their prospects for the future. This chapter explores critical domains influencing the future of this field, encompassing advancements in radiobiology, precision radiation therapy, the amalgamation of radiomics and artificial intelligence (AI), proton therapy and other innovative radiation modalities, strategies to surmount radiation resistance, and initiatives to enhance global access and equity in radiation oncology [2].

New radiobiology research has tremendously improved our knowledge of the interactions between ionizing radiation and cellular as well as molecular structures. The current ideas clarify how chromatin remodeling and repair mechanisms for double-stranded breaks affect cellular sensitivity to radiation by means of the DNA damage response (DDR). Recent tumor microenvironment, hypoxia, and immune modulation studies have found new radiosensitization targets. Developing biomarkers like gamma-H2AX, ATM, and ATR kinases to predict therapeutic responses and improve patient selection. Radiation-induced immunogenic cell death with radiobiology and immunotherapy may provide long-term tumor control. Real-time imaging, adaptive planning, and precision radiation therapy are changing the field of oncologic treatment. Stereotactic body radiation treatment (SBRT) and intensity-modulated radiation treatment (IMRT) help to accurately administer doses, which helps reduce the damage to surrounding healthy tissues [3]. Adaptive radiation therapy, or ART, is a sometimes-used method that alters treatment plans based on anatomical changes. Four-dimensional imaging and advances in image-guided radiation therapy (IGRT) provide opportunities to treat organ motion and tumor dynamics. Currently under investigation are tailored approaches for treatment mostly based on radiogenomics studies, which link DNA variations to radiation sensitivity [4].

These advances reduce side effects and help to maximize tumor control. Radiomics and artificial intelligence (AI) will together fundamentally change radiation oncology. Radiomics is the high-dimensional data extraction from medical images that helps to characterize tumor phenotypes outside conventional imaging criteria. Artificial intelligence-powered algorithms can examine data to predict treatment responses, spot prognostic biomarkers, and guide decisions. Trained to automatically create treatment plans, machine learning models improve accuracy and efficiency. Deep learning methods help to segment organs-at-risk and tumors, so reducing inter-observer variability. In radiation oncology, proton treatment and other cutting-edge radiation modalities, like heavy ion chemotherapy, show encouraging effects. These modalities allow ideal tumor irradiation by better dose distribution and lowering of surrounding healthy tissue exposure [5, 6]. Integrating artificial intelligence into radiotherapy treatments presents great possibilities to improve accuracy and support a data-driven approach in oncology. Proton therapy has shown excellent results in curing pediatric cancers, skull base tumors, and other anatomically complex malignancies. To improve clinical results, current studies look at combining proton therapy with systemic treatments, including immunotherapy and targeted agents. It is expected that developments in small proton therapy systems will enhance accessibility, so increasing the availability of this advanced treatment [7, 8].

Radiation resistance is a major barrier frequently resulting in therapy failure and disease recurrence in oncology. Research on molecular mechanics of resistance has revealed vital pathways, including p53 mutations, cancer stem cells, and epithelial-mesenchymal transition (EMT). Targeted treatments meant to modulate these pathways show promise in preclinical and clinical settings [9, 10]. Overcoming resistance has proved rather successful by combining radiation with radiosensitizers, including PARP inhibitors, DNA damage repair agents, and drugs aimed at hypoxia. Durable therapeutic effects call for knowledge of and capability to cope with the resistance pathways. Disparities in radiation oncology access still cause a major worldwide problem even with progress. Among the several challenges developing countries have to face are poor infrastructure, underdeveloped personnel, and limited resources.

International projects like the Global Task Force on Radiotherapy for Cancer Control (GTFRCC) make internationally fair access to radiation treatment possible. If we are to solve inequalities, support remote treatment planning, and forward education, telemedicine and digital platforms are becoming ever more essential [11, 12].

Affordable technologies, including linear accelerators made for low-resource environments, will help to lower global cancer treatment inequalities. Radiobiology and clinical radiation oncology present many exciting future directions. Therapeutic potential will be redefined by new radiation modalities, cell and molecular mechanisms, artificial intelligence integration, and precision therapy developments. Together, academics, doctors, and legislators must solve issues including radiation resistance and global inequalities. Adoption of these innovations will assist the oncology community to expect more fair access to life-saving treatments and better patient outcomes. This cooperation is necessary for effective implementation and accessibility of these innovations to every patient, irrespective of location or socio-economic level. Working together among the oncology community will help to transform the field and raise outcomes for cancer patients all around. Clinical radiation oncology and radiobiology seem to be about to see major changes [13]. Development in precision therapy, artificial intelligence integration, new radiation modalities, and a better knowledge of cellular and molecular mechanisms will redefine therapeutic possibilities. Scholars, doctors, and legislators have to cooperate to address problems including global inequality and radiation resistance. Adoption of these innovations will assist the oncology community to expect more fair access to life-saving treatments and better patient outcomes [14, 15].

20.2 Advancement in Radiobiology: Cellular and Molecular Mechanisms

The concept of cellular response to radiation exposure is well-established. Radiobiology plays an essential part in improving cancer patient outcomes by reducing radiation-related morbidities. This allows for a thorough evaluation and quantification of the risk of carcinogenesis when radiation is present. These advances, fueled by improved experimental models, novel molecular biology methods, and multidisciplinary research, have shed new light on the cellular and molecular processes that underpin radiation impacts [15]. These developments, particularly in the field of radiation therapy for cancer treatment, have far-reaching implications for both fundamental science and practical applications. Recent advances in radiobiology will be highlighted, with a focus on their significance to the direction of clinical radiation oncology.

20.2.1 Ionizing Radiation Causes DNA Damage

Radiobiology is fundamentally the study of how ionizing radiation interacts with DNA, which is the primary target of radiation damage. Ionizing radiation produces a variety of DNA lesions, including single-stranded breaks (SSBs), double-stranded breaks (DSBs), base damage, and DNA crosslinks. Among these, DSBs are the most damaging and difficult to fix since, if improperly corrected, they cause chromosomal abnormalities and cell death. Comprising a sophisticated network of channels, the DNA damage response (DDR) detects, signals, and repairs DNA damage. New regulators and mechanisms within this network have been revealed by recent research including: Chromatin Dynamics, which is today understood as a fundamental component of the DDR [15, 16]. By improving access to damaged sites, elements including histone modifications—e.g., phosphorylation of H2AX to generate γ-H2AX—and chromatin remodelers—e.g., SWI/SNF complexes—help in repair. MicroRNAs (miRNAs) and long non-coding RNAs (lncRNAs) have become clear regulators of DDR pathways. Like, particular lncRNAs have been shown to stabilize repair proteins at damage sites, so affecting the results of repairs. DDR signaling crosses cellular metabolic paths. Linking cellular energy status to repair efficiency, NAD + -dependent enzymes such as PARP serve both functions in energy metabolism and DNA repair [17, 18].

20.2.2 Cellular Responses to Radiation

Cellular responses to ionizing radiation encompass apoptosis, autophagy, cell cycle arrest, and senescence. Comprehending these responses is crucial for enhancing radiation therapy and minimizing adverse effects. Radiation-induced DNA damage activates cell cycle checkpoints, primarily via the ATM and ATR kinases. DDR signaling tightly controls cyclin-dependent kinases (CDKs), so regulating the change between cell cycle phases. Particularly relevant in cancer cells, prolonged checkpoint activation can cause mitotic catastrophe, a type of cell death linked with aberrant mitosis.

A main result of sub-lethal radiation exposure is senescence, a permanent state of cell cycle arrest. Senescence-associated secretory phenotype (SASP) cells secrete proteases, growth factors, and pro-inflammatory cytokines [19, 20]. This phenotype has two functions: It helps to suppress tumors and causes tissue damage under radiation. Senescence is established and maintained in part by epigenetic modifications including histone changes and DNA methylation. Autophagy, an aspect of cellular self-digestion, is becoming more recognized as a survival mechanism subsequent to radiation exposure. Radiation-induced autophagy can alleviate damage by eliminating impaired organelles and reutilizing cellular constituents. Preventing autophagy enhances the sensitivity of cancer cells to radiation, indicating a possible therapeutic strategy.

20.2.3 Molecular Mechanisms of Radiation-Induced Cell Death

Radiation predominantly causes cell death via apoptosis, necrosis, and ferroptosis. Comprehending the molecular foundations of these pathways has resulted in innovative therapeutic approaches. Apoptosis, a type of programmed cell death, is governed by intrinsic (mitochondrial) and extrinsic (death receptor-mediated) pathways. After radiation exposure, the cellular outcome is determined by pro-apoptotic proteins (such as BAX and BAK) and anti-apoptotic proteins (such as BCL-2 and BCL-XL) [21, 22]. The execution of apoptosis relies on caspases, particularly caspase-3 and caspase-9. Research on small molecules that can enhance radiation sensitivity by targeting these proteins is currently in progress. Ferroptosis, an iron-dependent mode of cellular demise, is receiving increased focus as a mechanism induced by radiation. The buildup of lipid peroxides induces ferroptosis. Reactive oxygen species (ROS) induced by radiation aggravate this process. Targeting ferroptosis regulators such as GPX4 has demonstrated potential in improving radiation efficacy [23, 24].

20.2.4 Radiation-Induced Immune Modulation

Radiation can target tumor cells and also modulate the immune system, opening up new avenues for synergistic treatment. Immunogenic cell death (ICD) is a specific kind of cell death that activates the immune system by releasing damage-associated molecular patterns (DAMPs) [25, 26]. Molecules such as HMGB1, ATP, and calreticulin are essential for the recruitment and activation of immune cells following radiation exposure. Enhancing antitumor immunity can be accomplished by combining radiation therapy with immune checkpoint inhibitors, including anti-PD-1/PD-L1 agents. The activation of the immune system facilitates the abscopal effect, wherein localized radiation induces tumor regression in distant sites within the body. The abscopal effect depends on CD8+ T cells, and radiation enhances their recruitment and cytotoxicity. Investigations into the identification of predictive biomarkers, including T cell receptor clonality, linked to the abscopal effect are currently underway [27, 28].

20.2.5 Radiation-Induced Epigenetic Changes

Radiation-induced epigenetic alterations have the ability to influence cellular behavior and possibly contribute to late effects like carcinogenesis, even after exposure has ended. The cellular reaction to radiation is greatly influenced by epigenetic alterations, such as DNA methylation, histone modifications, and non-coding RNAs [29, 30]. In exposure to radiation, histone acetylation and methylation modulate gene expression. One mechanism by which radiation can cause malignancy is through the aberrant methylation of tumor suppressor genes. Scientists are looking into epigenetic drugs that might improve radiation sensitivity. These drugs include those that inhibit histone deacetylase (HDAC) and DNA methyltransferase [31, 32].

20.2.6 Advances in Radiation Biology Techniques

The evolution of technology has significantly altered the landscape of radiobiology research, facilitating a comprehensive investigation into the cellular and molecular reactions to radiation. Methods like live-cell imaging and super-resolution microscopy facilitate the real-time observation of DNA repair mechanisms and chromatin dynamics. Advanced methods of sequencing and genetic research have revealed the mutational patterns prompted by radiation exposure. Implementing genomics and next-generation sequencing (NGS), investigators essentially discovered mutational signatures induced by radiation [33–35]. Mass spectrometry proteomics clarified post-translational modifications and deviations in protein expression. Metabolic alterations induced by radiation offer a profound understanding of the dynamics of cellular energy. Models that more accurately reflect physiological conditions for the study of radiation responses, including organoids and three-dimensional culture systems, effectively emulate the tissue microenvironment [36–38]. The progress in radiation oncology is markedly improved through the confluence of radiobiology and the burgeoning domains of systems biology. Mass spectrometry-based proteomics elucidates post-translational modifications and variations in protein expression. Radiation-induced metabolic alterations provide insight into cellular energy dynamics. More physiologically relevant models for investigating radiation responses, such as organoids and 3D culture systems, replicate the tissue microenvironment. The advancement of radiation oncology is significantly enhanced by the integration of radiobiology with emerging fields such as systems biology, artificial intelligence, and nanotechnology. Radiobiology's developments are opening the path for tailored radiotherapy depending on specific tumor and patient traits. Molecular profiling of tumors, for example, can direct the use of radiosensitizers or dose escalation [39–41]. An interesting frontier is combining immune-based treatments with radiotherapy. Maximizing these combinations will depend on an awareness of the interactions between radiation and the immune system. Priority still remains the development of agents to safeguard normal tissues or reduce side effects brought on by radiation [42]. Under active study are molecules aiming at ROS, inflammation, and DDR pathways. Large-scale radiobiological data is being analyzed using artificial intelligence and machine learning to find predictive biomarkers and refine therapy plans. Thanks to developments in cellular and molecular knowledge of radiobiology, the discipline has been revolutionized, and a basis for creative treatment ideas is now available. Leveraging these insights will help radiation oncology to provide less toxic treatments, more efficient, customized treatments, so enhancing patient outcomes all around [22, 43, 44].

20.3 Precision Radiation Therapy as Future Advancement

Precision radiation therapy (PRT) represents a major advance in the dynamic domain of cancer treatment and has the potential of improving outcomes for millions of patients across the world [45]. Molecular imaging, artificial intelligence (AI), genomics, and immunotherapy are employed in this novel methodology to optimize therapeutic efficacy of radiation while minimizing injury to healthy cells. Integrating recent advances in science with available therapeutic procedures, PRT has evolved to be a key move in the field of contemporary cancer treatment [46]. We thus use conventional radiation treatment, a fundamental part of oncology, to target tumor sites with high-energy radiation to destroy malignant cells. However, the constraints that have frequently been applied to it, such as unintentional injury of neighboring healthy tissues, have frequently been the subject of criticism due to its absence of adaptability to tumor evolution [47]. In doing so, PRT overcomes these problems by applying the most sophisticated procedures and tools to tailor treatment programs to the needs and specific biological characteristics of each individual. One of the key aspects of PRT is its reliance on the modern imaging technologies that provide unprecedented resolution for dependence in visualizing tumors [48]. Most of this has translated into innovations such as MRI-guided linear accelerators (MRI-linacs), which have changed the accuracy of radiation delivery. Real-time imaging during therapy is thus assured, as these instruments guarantee precise targeting of the radiation to the tumor, even in anatomically complex or mobile regions such as the lungs or the liver. Together with sophisticated MRI protocols, functional imaging modalities such as positron emission tomography (PET) are used to uncover vital information on tumor biology, i.e., hypoxia, metabolic activity, and proliferation rates. With this knowledge, clinicians can adjust radiation dosages on the fly to increase efficacy while limiting the amount of radiation harming healthy tissues by driving it into saturation [49].

AI greatly improves the field of radiation oncology. AI systems do this to help improve treatment planning, forecast patient response, and spot problems, such as. AI-based contouring tools provide directly defined, stable tumor boundaries from practitioners and workflows. By predicting tumor response to different radiation dosages, machine learning (ML) methods help increase adaptability and allow treatment regimens to quickly adjust to the behavior of the tumor in real time [50]. Researchers have enhanced the speed and quality of imaging by incorporating AI into MRI-Linacs, facilitating seamless modifications during therapy sessions and assuring consistent optimization of radiation delivery. Genomics and molecular biology have initiated a novel phase of personalized cancer treatment, with PRT utilizing these advancements to elucidate the genetic and epigenetic composition of tumors. Because resistance cannot be overcome and areas of resistance must be found, this is an important way to find areas of radiation-resistant tumors that clinicians can then adjust dosage on [51]. Liquid biopsies to vivo monitor real-time, real-world tumor dynamics and concepts for the care of circulating tumor DNA (ctDNA). Since tumors progress, monitoring fluctuations in ctDNA levels guarantees that treatment remains effective, and, in turn, this also allows clinicians to evaluate treatment efficacy and adjust treatment as needed [52]. The further domain of PRT is immuno-radio-oncology, that is, combining radiation with immunotherapy to enhance overall therapeutic efficacy. It appears to have opened cancer cells to immune-mediated eradication by modifying the tumor microenvironment to become immunogenic. Radiation, when combined with immune checkpoint inhibitors or other immunotherapies, can elicit a systemic antitumor response, including a noteworthy abscopal effect, in which local radiation induces a systemic antitumor response against metastases. Clinical trials are presently investigating appropriate dose and sequencing strategies to enhance the synergy between different therapy modalities. The clinical effectiveness of PRT is proven by multiple investigations [53].

MRI-guided radiation has demonstrated enhanced tumor control and fewer side effects in lung cancer patients relative to conventional techniques. Similarly, prostate cancer adaptive radiation therapy, directed through sophisticated imaging, has achieved elevated cure rates with decreased treatment-associated harm. Genomic profiling has revolutionized diagnosis and treatment in head and neck tumors by allowing patients to receive tailored radiation dosages based on genomic predictors of radioresistance with improved survival rates and fewer comorbidities [54]. Notwithstanding its transformative potential, PRT has numerous hurdles that impede its extensive implementation. The substantial expenses linked to advanced equipment, such as MRI-Linacs and AI-driven platforms, provide a considerable obstacle, especially in resource-constrained environments. The integration of molecular and genomic data into clinical workflows necessitates a strong infrastructure and specialized training for healthcare personnel [55]. However, standardization is a major challenge because the protocols for acute radiation therapy, genomic-guided treatment, and AI integration must all be fully developed. These problems must be addressed by coherent efforts of academics, physicians, and politicians alike to satisfy PRT access and quality in treatment uniformity among healthcare systems. So, too, therefore, is the efficacy of PRT largely dependent on interdisciplinary teamwork. Realizing the complete potential of these new technologies requires the collaboration of radiation oncologists, medical physicists, radiologists, molecular biologists, and AI experts [56]. Delivering comprehensive, personalized cancer care is through the integrated oncology centers that promote collaboration. Continuous research advances PRT's capabilities, and PRT's future is replete with potential. Proton treatment and carbon ion therapy are innovative technologies that provide highly precise delivery, delivering high radiation doses only to tumors while sparing healthy tissues. Gains in nanotechnology are enabling the design of radiosensitizers that augment tumor sensitivity to radiation yet protect healthy cells [57]. The increasing sharing of worldwide data and collaboration, combined with genomic insights and AI integration into radiation oncology, will begin to mature towards ever more sophisticated integration of real-time tumor biological and patient response-informed modifications of treatment plans. This advancement will change the direction of oncology care, revolutionizing cancer treatment and the hope it now brings to patients worldwide [58].

20.4 Integration of Radiomics and AI in Radiation Oncology in Near Future

Radiomics integrated with AI in radiation oncology stands together as a game changer in cancer treatment, opening up exciting opportunities to enhance the precision, personalization, and efficacy of therapeutic interventions [59]. Radiomics is the extraction of massive quantitative data from the conventional medical imaging modalities of CT, MRI, and PET scans. This data encapsulates intricate details regarding the tumor's biology, microenvironment, and behavior, frequently surpassing the insights attainable by human visual interpretation. The integration of this extensive dataset with the computational capabilities of AI, specifically ML and deep learning (DL) algorithms, facilitates enhanced comprehension of tumor biology, forecasting treatment results, and customizing individualized therapeutic regimens. This integration is believed to create a new radiation oncology paradigm of data-driven decision-making that will revolutionize clinical workflows and result in radiation oncology outcomes in a precision medicine sense [60]. The key element of integrating AI into the radiomics paradigm is AI's ability to handle and analyze the large volumes of data generated from radiomics. Conventional tumor assessment approaches typically depend on restricted descriptors like size and location, which inadequately represent the complex heterogeneity of cancer. Radiomics enhances this comprehension by extracting characteristics such as texture, shape, intensity, and spatial correlations within the tumor and adjacent tissues [61]. AI systems can look at these traits and find intricate patterns and relations almost lost to the human eye. These discoveries are of such tremendous importance to all phases of the radiation oncology workflow, from diagnosis to treatment planning to follow-up care. The ability of AI-driven radiomics models to predict the tumor response to radiation therapy to allow customization of radiation techniques for optimal efficacy has been demonstrated. Predictive models have been especially useful in cancers characterized by significant variability in treatment response, including non-small cell lung cancer, glioblastoma, and head and neck cancers [62]. AI-enhanced radiomics significantly contributes to tumor segmentation, an essential phase in radiation therapy planning. Precise identification of the tumor and adjacent organs at risk (OARs) is crucial to guarantee that therapeutic radiation doses are accurately administered to malignant tissues while reducing unintended harm to healthy structures. Manual segmentation is labor-intensive and prone to interobserver variability, which can impair uniformity in treatment and outcomes. The success of automating this process with DL-based AI algorithms has been exceptionally precise and efficient [63]. We have trained convolutional neural networks (CNNs) to identify tumor boundaries very accurately, which surpasses conventional methods and dramatically reduces the time required for planning. In addition to optimizing clinical workflow uses, these innovations ensure a safer working environment at the point of care by reducing mistakes and variability. Aside from segmentation, AI and radiomics are helping to deliver adaptive radiotherapy, which is an adaptive method to change the treatment plan in real time based on changes in tumor morphology, patient anatomy, as well as the therapeutic response [64].

However, this feature is essential when tumors have significant shrinking or expansion during treatment, as fixed plans can dose the tumor to an inadequate level or can over-irradiate healthy tissues with unnecessary radiation. Sequential imaging can be used by AI models to evaluate the associated radiomics of if changes are occurring early, thereby enabling doctors to proactively change treatment strategies. The work showed that AI-guided adaptive radiotherapy would reduce toxicities while reducing the target control rate and thus would become a routine treatment in radiation oncology. Radiomics and AI are primary applications for predicting or mitigating the adverse outcomes of treatment-related toxicities, a major clinical crisis in radiation oncology [65]. Although radiotherapy is effective at treating cancer, the side effects are often severe and can reduce a patient's quality of life. AI models can use the analysis of the radiomic characteristics of the tumor as well as surrounding normal tissue to determine the likelihood of specific toxicities, such as radiation-induced pneumonitis, mucositis, or fibrosis. They are able to use this information to determine how to customize radiation dosages and field configuration to try to reduce hazards while trying to maintain treatments. Radiomics provides the basis for the development of prediction models for the prediction of esophageal toxicity in lung cancer patients undergoing radiotherapy, which permit alleviation of these effects through preventative measures [66].

Numerous research studies have demonstrated the transformational potential of merging radiomics and AI in radiation oncology. An exemplary case is the application of radiomic signatures to forecast survival outcomes and recurrence in lung cancer patients undergoing stereotactic body radiotherapy (SBRT). Researchers have built models that categorize individuals into risk groups based on examination of characteristics such as tumor morphology and textural variability to help inform adjuvant treatment choices or enhanced monitoring [67]. Radiomics has been used in concert with AI to identify biomarkers related to treatment resistance in glioblastoma, in order to both address this challenge and improve outcomes, yet these encouraging outcomes are tempered by numerous barriers to fully realize the potential of radiomics and their AI in the clinical setting. Reproducibility of radiomic data is particularly influenced by differences in imaging methodology, scanner configurations, and radiomic feature extraction techniques. There are initiatives underway to set up uniform rules and frameworks for radiomics research, which include the Image Biomarker Standardization Initiative (IBSI), an effort intended to standardize feature definitions and analytical processes across multiple institutions [68].

Moreover, the opaque nature of numerous AI models poses an obstacle to wider implementation, as clinicians frequently necessitate interpretability and transparency to trust and utilize these systems in decision-making. Researchers are investigating methods like explainable AI (XAI) to mitigate this issue and provide clarity on the rationale behind model predictions. Moreover, the additional problems in the use of radiomics and AI in existing clinical workflows require strong infrastructure and trained health care practitioners. Data Sciences, radiology, medical physics, radiation oncology: critical interdisciplinary cooperation to get these technologies working correctly. Such cooperation is better promoted by institutions that have the capacity to exploit radiomics and AI with the use of the whole capabilities of those sciences, commonly via integrated cancer centers or consortia [69]. Now when it comes to applying healthcare AI use specifically in radiation oncology, there will be much deep thought on the ethical and legal side of where AI can go in the realm of radiation oncology and patient privacy, any data security, any algorithmic bias, etc. There's definitely a need for deep regulation and deep control.

Radiomics and AI will have a much brighter future in radiology and have potential for further advancements in the technology and in the research. With the improvements of imaging techniques, like high-resolution MRI and dual-energy CT, we expect more complete datasets for radiomic analysis [70]. Three omics (genomes, proteomics, and metabolomics) are combined with radiomics to expect a revolution in our understanding of tumor biology and therapeutic response. Commonly referred to as "radiogenomics" and tested on this comprehensive method, this has shown the potential to discern genetic subgroups of cancer that respond differently to radiation therapy. With the global expansion of data sharing and collaboration, the creation of extensive, diversified datasets will facilitate the training of more resilient and generalizable AI models [71]. Federated learning, a decentralized methodology enabling institutions to cooperatively train models without the exchange of raw data, is emerging as a potent way to address data privacy issues. These activities are anticipated to expedite the transition of radiomics and AI from research to clinical application, hence enhancing global access to precision radiation oncology [72].

20.5 Proton Therapy and Radiation Modalities: Future Window

Proton therapy is a highly complex form of radiation oncology, with a major advantage over conventional photon-based therapies. Its defining characteristic is that proton treatment delivers a dose of radiation to the tumor with minimal damage to surrounding healthy tissue, and it does so because the dose distribution is unique—called the Bragg peak. The accuracy of this location is especially useful for addressing tumors near important structures like the brain, spinal cord, and radiation-sensitive organs. Encouraging results have been shown for proton therapy, mostly in pediatric oncology, where minimizing radiation exposure to growing tissues is vital [73]. Proton treatment has been especially effective in treating pediatric brain tumors, as traditional photon radiation may result in cognitive deficits and subsequent malignancies. Clinical investigations have shown that proton therapy yields superior long-term cognitive function and a reduced incidence of subsequent malignancies in these patients. Proton therapy is different from traditional radiation, as it is unique in delivering therapy to tumors while protecting surrounding healthy tissue, which may make it a desirable treatment option for certain cancers that are particularly hard to treat with conventional radiation because of how close they are to the eye, brain, or spinal cord [74].

However, because conventional radiation therapy can produce serious developmental side effects in pediatric patients, it is particularly important for this accuracy. In addition, proton therapy is also associated with favorable outcomes in lung malignancies, with clinical research showing that proton therapy reduces radiation-related side effects compared to conventional techniques. Proton therapy also has an advantage in the management of head and neck malignancies, as preserving vital tissues like the salivary glands, the brain stem, and the spinal cord, all of which frequently receive sizeable radiation doses in traditional treatments, is possible [75]. Proton treatment enhances the patient's quality of life by minimizing harm to sensitive regions, hence alleviating side effects such as xerostomia, dysphagia, and neurological impairment.

Advanced research is investigating the merger of proton treatment with ultra-high dose rate radiation delivery, referred to as FLASH therapy. FLASH therapy administers radiation at exceptionally high dose rates, demonstrating in preclinical trials the ability to improve tumor control while markedly decreasing normal tissue harm. This advancement represents an advancement that could substantially improve the therapeutic ratio (gaining greater tumor control with fewer adverse effects) by making radiation therapy administration significantly more efficient. While clinical research is underway examining the integration of FLASH therapy with proton beams, clinical validation is in progress, showing promising results, particularly in preclinical models [76]. In addition, the precision of proton therapy makes it an ideal component of other sophisticated treatments, such as MRI-guided radiation therapy, which involves real-time imaging of the tumor to alter radiation administration [77]. The integration permits later changes to treatment protocols in recognition of the tumor displacement, to deal with tumors whose location changes, like those found within the pulmonary or abdominal areas.

Proton therapy is advancing through its integration with other treatment modalities, including immunotherapy and targeted medicines, to increase tumor sensitivity to radiation and concurrently activate the immune system to combat the tumor [78]. The integration of proton therapy with immunotherapy may leverage the immune response elicited by radiation, thus enhancing results in malignancies that are typically resistant to treatment. Proton radiation is specifically promising as a way of administering proton radiation directly at the tumor while at the same time strengthening the immune system's capacity to get rid of cancer [79]. For example, traditional treatments fail in non-small cell lung cancer, head and neck cancers, and melanoma, and this combination could be especially useful. Currently, researchers are investigating the use of proton therapy-aided immunotherapies through clinical studies to augment the effectiveness of cancer treatment with the hope that this technique will soon become part of the standard of care in cancer. Proton therapy demonstrates considerable potential in the management of thoracic malignancies. To treat lung tumors with elevated radiation doses while avoiding damage to surrounding critical structures such as the heart and spinal cord, proton therapy is being integrated within stereotactic body radiation therapy (SBRT) [80].

Despite the clear advantages offered by proton therapy in this setting, difficulties remain in managing mobility, because lung tumors can shift with respiration. Advances in proton therapy include making more sophisticated motion tracking devices to accommodate the variations, thus allowing more accurate radiation administration for each treatment session [81]. Proton therapy also has promise for treating oligometastatic disease (defined as limited metastases) that can be addressed, and preliminary studies have shown its efficacy for treating people with localized metastatic disease, e.g., brain or liver metastases [82]. The ability of proton therapy to deliver enhanced precision in the treatment of patients and to spare the normal tissues may make it an important modality for the management of metastatic malignancies where standard therapies are difficult to gain access to. Radiobiology is needed to provide the science of proton therapy by which proton irradiation can add extra biological effects to the treatment technique, which can markedly refine the treatment method. Since proton treatment uses molecular mechanisms to effectively treat cancer cells, researchers are exploring ways to boost the efficacy of proton treatment [83].

The radiobiological effects of proton treatment on the tumor microenvironment are the focus of this research, with interest directed to the effects of proton energy deposition patterns on tumor vasculature, immunological response, and tumor growth. Research from the cellular level is being carried out to better understand how proton therapy affects radiobiologies and tumor resistance mechanisms with a view to creating strategies to overcome them and help improve patient results [84]. The forthcoming generation of proton therapy apparatus, featuring tiny proton accelerators, is anticipated to enhance the accessibility of this technology to hospitals globally. The substantial expense and considerable dimensions of proton therapy facilities hinder numerous institutions, especially in low-resource environments, from implementing this treatment approach. Nonetheless, recent advances seek to diminish the cost and dimensions of proton treatment devices, thereby enhancing their viability for wider application [85]. The proliferation of proton therapy centers, especially in developing markets, suggests a promising future for proton therapy, enhancing global patient access to this advanced treatment modality. However, despite these advances, the application of proton therapy continues to struggle to optimize for all tumor types, to ensure access for all patients, and to improve precision and efficacy of proton beam delivery. Beyond the use of proton therapy historically, much of the future involves studying its integration with AI and ML algorithms [86].

AI has the capacity to significantly improve the planning and administration of proton therapy by facilitating personalized treatment strategies, optimizing beam delivery according to patient-specific variables, and forecasting tumor response. AI-driven systems can evaluate extensive clinical data to enhance therapy precision, automate treatment planning, and forecast potential treatment results. Alongside AI, progress in radiomics, which involves the extraction of quantitative features from medical pictures, may further improve the accuracy of proton therapy [87]. Radiomics can enhance treatment plans for individual patients by analyzing pre-treatment imaging and anticipating tumor behavior, thus enhancing outcomes and minimizing unnecessary harm. The integration of these technologies is anticipated to yield more tailored, efficacious, and efficient therapeutic methodologies in the forthcoming years. Proton treatment signifies a notable progression in radiation oncology, providing enhanced tumor targeting while minimizing harm to surrounding healthy tissues. Proton therapy is an evolving technology and holds great promise for transformation in cancer treatment through continuous advancements leading to FLASH therapy, motion tracking, integration with immunotherapy, and AI-assisted treatment planning [88]. However, to realize its advantage fully and make it available to patients around the world, sustained investment in research, advances in technology, and development of the infrastructure will be required. Proton therapy will take a step forward and become a core element of personalized cancer treatment and deliver better outcomes with fewer side effects and improved quality of life for patients [89] (Table 20.1).

20.6 Overcoming Radiation Resistance—A Success to Clinical Oncologist

One excellent and incredibly affordable single-modality treatment for cancer is radiation therapy. High radiation doses are used in radiation therapy to either kill or halt the growth of tumors. X-rays and gamma (γ)-rays are examples of ionizing radiation (IR), which is typically used for the treatment of cancer because it may infiltrate tissue, disrupt chemical bonds, and ionize atoms by eliminating their electrons [100]. Despite the fact that radiotherapy is a successful cancer treatment, many individuals later develop resistance to radiation and their malignancies return [100]. Thus, radiation enhancement techniques aim to either increase the impact on cancer or decrease the effect on healthy tissues. In order to accomplish these goals, the normal tissues must not be sensitized, and the tumor must not be protected [101].

Various kinds of cell signaling mechanisms are involved to develop resistance against ionizing radiation (IR). Like adaptive pathway, DNA damages repair pathway, adhesion pathway, inflammation pathway, developmental pathway, hypoxic pathway, etc. [100]. Therefore, in order to overcome tumor radioresistance and enhance the results of radiation, new therapeutic radiosensitizers are crucial. For clinical application, an increasing number of inhibitors that target particular elements of radioresistance pathway are now being explored. Furthermore, an attempt has been made to make up for the shortcomings of synthetic inhibitors by developing natural radiosensitizers. To overcome IR resistance in cancer treatment, a number of approaches have been proposed [100].

20.6.1 Changes in radiation Fractionation

Patient will be benefited by taking fractionated radiation therapy over the single administration of radiation [100]. The best therapeutic ratio is generally achieved in the curative scenario of solid tumors by delivering doses of 50–70 Gy utilizing repeated fractions of 1.8–2 Gy per fraction [102]. High-dose radiation is administered in 1–5 fractions during SBRT and stereotactic radiosurgery (SRS). In addition to other tumors, SBRT and SRS have been successfully implemented to treat lung, liver, brain, prostate, and recurring head and neck cancers [103–107]. In conventional radiation therapy per day 2.6 Gy radiation is given to the patient; unluckily,

Table 20.1 Advancements in radiation oncology

Sl. No.	Study	Focus area	Intervention/technology	Outcome measures	Key findings	References
1	Grover et al. (2024)	Precision Radiation Therapy	AI-enhanced MRI-guided linear accelerators	Tumor targeting accuracy, radiation exposure	AI integration improved real-time tumor tracking, reducing radiation to healthy tissues	[90]
2	Avenda Health. (2024)	Integration of AI in Radiation Oncology	AI tool "Unfold AI" for prostate cancer detection	Detection accuracy, treatment precision	AI detected prostate cancer with 84% accuracy, surpassing doctors' 67% accuracy	[91]
3	Pietsch et al. (2022)	Proton Therapy	AI for detecting treatment deviations in proton therapy	Detection accuracy, treatment safety	AI improved detection and classification of treatment deviations, enhancing safety	[92]
4	Chamseddine et al. (2023)	Precision Radiation Therapy	AI model for predicting liver toxicity in proton therapy	Prediction accuracy, personalized treatment	AI model aided in personalized selection of proton vs. photon therapy, reducing toxicity	[93]
5	Jiao et al. (2022)	Integration of Radiomics and AI	Deep learning radiomics with circulating tumor cell counts	Recurrence risk prediction, patient stratification	Integration improved accuracy in predicting recurrence risks, aiding in treatment decisions	[94]
6	Lin et al. (2020)	Proton Beam Therapy	Intensity-Modulated Radiation Therapy	Cost-effectiveness, sustainability	Study highlighted challenges in cost analyses; suggested ML could improve accuracy	[95]
7	Hurkmans et al. (2024)	Integration of AI in Radiation Oncology	Guidelines for AI model development in radiation therapy	Model development standards, clinical validation	Provided guidelines to ensure transparency and efficacy in AI model development	[96]
8	Ono et al. (2024)	Proton Therapy	AI applications in quality assurance for proton therapy	QA efficiency, treatment accuracy	Identified potential AI applications to enhance QA procedures in proton therapy	[97]
9	Maniaci et al. (2024)	Integration of Radiomics and AI	Quantitative radiomics approach for tumor phenotype decoding	Imaging analysis, tumor characterization	Demonstrated potential of radiomics in non-invasive tumor characterization	[98]
10	Chang et al. (2023)	Proton Therapy	Early in vivo radiation damage quantification using MRI	Damage quantification, treatment monitoring	MRI-based approach aided in early detection of radiation damage, improving patient management	[99]

because of its limited effectiveness and the development of adverse effects, this technique is unable to adequately control locally advanced malignancies and does not satisfy patients. To overcome this condition, hypofractionated radiation therapy has come into the picture [100]. Hypofractionated regimen can destroy tumor cell DNA [108]. This therapy includes delivering daily doses more than 2.0 Gy, shorter treatment times, and decreased total doses [100].

Compared to conventional radiation, this hypofractionated method delivers a lower overall dose even though the single fraction dose is larger. These strategies counteract fast repopulation of tumor tissues by utilizing increased cell death per fraction of the cell survival curve. With hyperfractionation strategy a lower radiation dose like 1.2 Gy is being administered twice daily at around 6-h intervals, which leads to increase the overall radiation dose and reducing adverse effects. A comparatively high radiation dose that is 1.6 Gy is administered twice day during accelerated fractionation, which is beneficial, particularly for tumors that are growing extremely quickly [100].

20.6.2 Multimodality Therapy

Significant clinical benefits have been shown in a variety of tumor types when chemotherapy and radiation therapy combinations are given [109]. The development of tumor-specific radiosensitizers has been the subject of extensive research in an attempt to get around tumor radioresistance. The use of combined anticancer therapy in clinical settings is becoming more common. This is predicated on the idea that whereas a single chemotherapeutic or radiotherapeutic treatment has little therapeutic effect on a variety of tumor types, combining therapeutic agents with different modes of action may have additive or synergistic anti-neoplastic effects [100].

Synthetic Derivatives with a Specific Target

It has been demonstrated that certain Akt, mTOR, and Chk1/2 inhibitors function as possible radiosensitizers [110–117]. Hyperactive homologous recombination (HR) pathway which is a DNA repair pathway plays a major role in radioresistance. Some radiosensitizers have been discovered to inhibit HR pathway like Gemcitabine, TAS-106, gimeracil, pentoxifylline, and caffeine are a few examples of nucleoside and base analogues [118].

Naturally Occurring Radiosensitizers

Naturally derived radiosensitizers are more advantages over synthetic derivatives due to their easy availability in foods. These have antioxidant and immune boosting properties [100]. Curcumin (a polyphenol derivative) reduces NF-kB expression to increase the radiosensitivity [119, 120]. It has been demonstrated that resveratrol, a polyphenol derivative enhances the apoptotic effects of IR, cytokines, and chemotherapeutic agents [120–122]. Additionally, by encouraging its de novo production, resveratrol enhances the IR-induced buildup of ceramide, a possible anticancer agent [123]. A tyrosine protein kinase inhibitor genistein which is an isoflavone glycoside that is naturally present in soybeans. For instance, by inhibiting radiation-induced activation of survival signals like Akt and Erk, genistein significantly increases the radiosensitivity of human esophageal cancer cells [124]. Quercetin, a flavonoid in combination with radiation therapy, has been shown to improve tumor radiosensitivity both in vitro and in vivo, due to its ability to target the ATM-mediated pathway in response to radiation [125].

20.6.3 Changes in Inflammatory Pathway

Tumor-associated macrophage suppression may increase IR's anticancer effects. For instance, by suppressing IR-induced VEGF in macrophages, VEGF-neutralizing antibodies can improve the antitumor response to IR [126]. Some innovative methods have tried radiotherapy in conjunction with cancer vaccines or adoptive T-cell transfer because it induces in situ vaccination. It is also possible to effectively treat antibodies that target immunological checkpoint receptors and/or inducible T-cell costimulatory receptors with radiation therapy [127].

20.6.4 Strategies for Preventing Tumor Hypoxia

Since hypoxic tumor cells have a higher chance of surviving IR, tumor hypoxia is regarded as one of the main issues with radiation therapy. Numerous treatment approaches have been created to combat tumor hypoxia. Radiation fractionation, the use of high linear energy transfer (LET) particles in the treatment of hypoxic tumors, and the application of bioreductive medications, such as adjuvant treatment, are the most widely used techniques to overcome tumor hypoxia-related resistance. Apart from this, it has been found that RBC transfusion, hyperbaric oxygenation, erythropoietin administration and other some methods deliver oxygen to the hypoxic tumor cells [100].

Misonidazole and other nitroimidazole derivatives were discovered in the 1970s to imitate the action of oxygen in the radiochemical process, which opened the door for their application to increase the cytotoxic effect of IR in hypoxic environments [128] [129]. Since HIF-1 actively contributes to the radioresistance of hypoxic cancer cells, it has been highlighted as a key molecular target for sensitization to the therapeutic impact of radiation [130]. Additionally, it has been shown to suppress the growth and spread of tumors by lowering the expression of the HIF-1 target gene in hypoxic environments [131, 132].

Radioresistance in lung cancer cells is greatly reduced by combined IR and deguelin treatment [133]. To increase radiosensitivity, a gene therapy approach that targets hypoxic cancer cells might also be a viable choice [128]. Radiation therapy and angiogenesis inhibition together have a bright future in the treatment of cancer [134]. Recent years have seen the development of novel medications like VEGF antibody bevacizumab and avastin and direct antiangiogenic peptide anginex can temporarily restore normalcy to the vasculature inside tumors in order to enhance oxygen delivery [135].

These are the methods that have been successfully implemented for eliminating the radioresistance—a great success to clinical oncologist.

20.7 Global Access and Equity in radiation Oncology

Radiation oncology is the most resource-intensive area of comprehensive cancer care due to the high initial costs of equipment, the need for specialized construction, the need for a diverse workforce, and ongoing maintenance expenses [136]. Around the world, access to radiation therapy (RT) facilities is still extremely unequal. Many low- and middle-income countries (LMICs) still suffer from a serious lack of RT facilities, despite the fact that high-income countries share them in plenty [137] [138, 139]. An essential component of all-encompassing cancer care, radiation treatment is very successful in increasing cancer patients' chances of survival and easing their pain. However, efforts to bridge the cancer divide have mostly overlooked access, which is unequally divided between rich and poor nations [140]. Misconceptions regarding the expense and intricacy of radiotherapy have restricted its accessibility in nations with conflicting agendas, inadequate infrastructure, and weak regulatory frameworks. Nonetheless, the information that is now available suggests that it is cost-effective, particularly when value is determined by characteristics other than survival [141].

When chemotherapy and/or radiation are postponed and resources like routine blood products and [18F]-fluorodeoxyglucose positron emission tomography imaging are unavailable, particular instructions for LMICs would offer direction on how to treat patients. Since high-income and low-income nations may have different resources for cancer care, some LMICs have already taken steps to create their own recommendations for specific malignancies [142, 143]. Sustainable access to cancer control resources, with an emphasis on granting access to radiation therapy, is crucial, in addition to research aimed at enhancing cancer outcomes in LMIC settings. Building radiation treatment capacity in LMICs has long been a focus of the International Atomic Energy Association, which also played a key role in Zambia's radiation therapy capacity expansion [144]. Pharmaceutical and device companies' philanthropic investments, in consultation with health ministries and academic training institutions, offer a roadmap for partnerships to increase the capacity of health service delivery, clinical training, and research for health care professionals to satisfy the unique needs of women with cervical cancer in LMICs. In fact, a similar partnership was recently established to improve the outcomes of pediatric cancer care [145]. In LMICs, about 50% of patients in need of radiation therapy lack access to care. In LIC (Low income countries), where the percentage exceeds 90%, the situation is dramatic. There are 4221 teletherapy machines in LMICs, but between 4300 and 7000 more are required. Even while things can get better, many patients who have radiation indications won't be able to receive treatment anytime soon. To build radiotherapy communities that could serve as the foundation for future growth, one potential tactic in the interim is to coordinate efforts to open at least one radiotherapy center in nations [137].

20.8 Conclusions

The aging population will lead to an increase in the incidence of cancer, the leading cause of disease and disability worldwide. The number of patients undergoing radiation therapy has been gradually rising. By drastically reducing or eliminating tumors, radiotherapy can help individuals for those patients whom surgery is not an option. Tumors exposed to infrared radiation undergo a range of immediate or long-term changes, from minor alterations in metabolism to numerous forms of death. The extent and/or kind of cellular damage can be ascertained by the total dose and number of IR fractionations. By causing several DNA lesions, including DSBs, SSBs, DNA crosslinks, and base alterations, IR has anti-cancer effects. Thus, there is a pressing clinical need to create radiosensitizers in order to reduce tumor radioresistance and thereby increase the efficacy of radiation therapy. Many tumor types were successfully treated in clinical trials using a combination of chemotherapy and radiation. A lot of the radiosensitization techniques used today focus on damaging of the DNA. As a result, new radiosensitizing medications should be the subject of more thorough and meticulous preclinical research in order to better understand their mechanisms of action before beginning clinical trials. In order to improve the effectiveness and economics of radiotherapy, the patient should also receive more precisely calibrated radiation treatment. Improvements in tumor targeting and delineation brought about by biological and technological advancements have improved tumor coverage and decreased exposure to normal tissue. Proton therapy which is a great approach may improve local control, increase survival, and reduce toxicity in cancer patients when used in a therapeutic context. The technique's development may help patients with tumors near vital organs, such as hepatic, esophageal, and lung cancer. Biomarkers are being used more and more in the precision medicine age to forecast how different medicines will work. It is clear that depending on the genetic makeup of each cancer, the effectiveness and toxicity of radiation therapy will probably differ. However, attempts to combine targeted medication therapies with radiation have not been very successful, and the -omic variables underpinning radiotherapy response are still poorly understood.

The application of artificial intelligence in radiation oncology has numerous chances to improve treatment results, accuracy, and efficiency. AI-driven advancements in patient monitoring, real-time adaptive therapy, diagnostic accuracy, and customized treatment plans have the potential to completely transform the industry and improve both patient care and operational effectiveness. The field may overcome these obstacles and achieve revolutionary breakthroughs in radiation therapy by prioritizing cooperative research, ongoing education, and patient-centered strategies.

References

1. Hall EJ, Giaccia AJ. Radiobiology for the radiologist. Lippincott Williams & Wilkins; 2006.
2. Till JE, Mc CE. A direct measurement of the radiation sensitivity of normal mouse bone marrow cells. Radiat Res. 1961;14:213–22.
3. Kastan MB, Zhan Q, el-Deiry WS, et al. A mammalian cell cycle checkpoint pathway utilizing p53 and GADD45 is defective in ataxia-telangiectasia. Cell. 1992;71(4):587–97.
4. Durante M, Cucinotta FA. Heavy ion carcinogenesis and human space exploration. Nat Rev Cancer. 2008;8(6):465–72.
5. Williams JP, Brown SL, Georges GE, et al. Animal models for medical countermeasures to radiation exposure. Radiat Res. 2010;173(4):557–78.
6. Tiwana GS, Prevo R, Buffa FM, et al. Identification of vitamin B1 metabolism as a tumor-specific radiosensitizing pathway using a high-throughput colony formation screen. Oncotarget. 2015;6(8):5978–89.
7. Higgins GS, Prevo R, Lee YF, et al. A small interfering RNA screen of genes involved in DNA repair identifies tumor-specific radiosensitization by POLQ knockdown. Cancer Res. 2010;70(7):2984–93.
8. Roberson JD 2nd, Burnett OL 3rd, Robin N. Radiogenomics: towards a personalized radiation oncology. Curr Opin Pediatr. 2016;28(6):713–7.
9. Labay E, Mauceri HJ, Efimova EV, et al. Repurposing cephalosporin antibiotics as pro-senescent radiosensitizers. Oncotarget. 2016;7(23):33919–33. 22 Molecular Imaging and Radiobiology 280
10. Wei D, Parsels LA, Karnak D, et al. Inhibition of protein phosphatase 2A radiosensitizes pancre atic cancers by modulating CDC25C/CDK1 and homologous recombination repair. Clin Cancer Res. 2013;19(16):4422–32.
11. Hatano K, Kumar B, Zhang Y, et al. A functional screen identifies miRNAs that inhibit DNA repair and sensitize prostate cancer cells to ionizing radiation. Nucleic Acids Res. 2015;43(8):4075–86.
12. Durante M, Loeffler JS. Charged particles in radiation oncology. Nat Rev Clin Oncol. 2010;7(1):37–43.
13. Zhou BB, Bartek J. Targeting the checkpoint kinases: chemo sensitization versus chemoprotection. Nat Rev Cancer. 2004;4(3):216–25.
14. Fokas E, Prevo R, Pollard JR, et al. Targeting ATR in vivo using the novel inhibitor VE-822 results in selective sensitization of pancreatic tumours to radiation. Cell Death Dis. 2012;3:e441.
15. Hall AB, Newsome D, Wang Y, et al. Potentiation of tumor responses to DNA damaging therapy by the selective ATR inhibitor VX-970. Oncotarget. 2014;5(14):5674–85.

16. Lambin P, Leijenaar RTH, Deist TM, Peerlings J, de Jong EEC, van Timmeren J, Sanduleanu S, Larue RTHM, Even AJG, Jochems A, van Wijk Y, Woodruff H, van Soest J, Lustberg T, Roelofs E, van Elmpt W, Dekker A, Mottaghy FM, Wildberger JE, Walsh S. Radiomics: the bridge between medical imaging and personalized medicine. Nat Rev Clin Oncol. 2017;14(12):749–62.
17. De Craene B, Berx G. Regulatory networks defining EMT during cancer initiation and progression. Nat Rev Cancer. 2013;13(2):97–110.
18. Feron O. Pyruvate into lactate and back: from the Warburg effect to symbiotic energy fuel exchange in cancer cells. Radiother Oncol. 2009;92(3):329–33.
19. Li Y, Liu J, Gao L, Liu Y, Meng F, Li X, Qin FX. Targeting the tumor microenvironment to overcome immune checkpoint blockade therapy resistance. Immunol Lett. 2020;220:88–96.
20. Cairns RA, Harris IS, Mak TW. Regulation of cancer cell metabolism. Nat Rev Cancer. 2011;11(2):85–95.
21. Menon SS, Uppal M, Randhawa S, et al. Radiation metabolomics: current status and future directions. Front Oncol. 2016;6:20.
22. Baumann M, Krause M, Overgaard J, et al. Radiation oncology in the era of precision medicine. Nat Rev Cancer. 2016;16(4):234–49.
23. Bao S, Wu Q, McLendon RE, et al. Glioma stem cells promote radioresistance by preferential activation of the DNA damage response. Nature. 2006;444(7120):756–60.
24. Kroonen J, Nassen J, Boulanger YG, et al. Human glioblastoma-initiating cells invade specifically the subventricular zones and olfactory bulbs of mice after striatal injection. Int J Cancer. 2011;129(3):574–85.
25. Evers P, Lee PP, DeMarco J, et al. Irradiation of the potential cancer stem cell niches in the adult brain improves progression-free survival of patients with malignant glioma. BMC Cancer. 2010;10:384.
26. Joiner MC, van der KogeJ A, editors. Basic clinical radiobiology. CRC Press; 2025.
27. Vissink A, Mitchell JB, Baum BJ, et al. Clinical management of salivary gland hypofunction and xerostomia in head-and-neck cancer patients: successes and barriers. Int J Radiat Oncol Biol Phys. 2010;78(4):983–91.
28. Lombaert IM, Brunsting JF, Wierenga PK, et al. Cytokine treatment improves parenchymal and vascular damage of salivary glands after irradiation. Clin Cancer Res. 2008;14(23):7741–50.
29. Knox SM, Lombaert IM, Haddox CL, et al. Parasympathetic stimulation improves epithelial organ regeneration. Nat Commun. 2013;4:1494.
30. Barker HE, Paget JT, Khan AA, Harrington KJ. The tumour microenvironment after radiotherapy: mechanisms of resistance and recurrence [published correction appears in Nat Rev Cancer. 2015;15(8):509]. Nat Rev Cancer. 2015;15(7):409–25.
31. Lombaert IM, Brunsting JF, Wierenga PK, et al. Keratinocyte growth factor prevents radiation damage to salivary glands by expansion of the stem/progenitor pool. Stem Cells. 2008;26(10):2595–601.
32. Barcellos-Hoff MH, Park C, Wright EG. Radiation and the microenvironment – tumorigenesis and therapy. Nat Rev Cancer. 2005;5(11):867–75.
33. Lombaert IM, Abrams SR, Li L, et al. Combined KIT and FGFR2b signaling regulates epithelial progenitor expansion during organogenesis. Stem Cell Rep. 2013;1(6):604–19.
34. Xiao N, Lin Y, Cao H, et al. Neurotrophic factor GDNF promotes survival of salivary stem cells. J Clin Invest. 2014;124(8):3364–77.
35. Zeidan YH, Xiao N, Cao H, et al. Botulinum toxin confers radioprotection in murine salivary glands. Int J Radiat Oncol Biol Phys. 2016;94(5):1190–7.
36. van Luijk P, Pringle S, Deasy JO, et al. Sparing the region of the salivary gland containing stem cells preserves saliva production after radiotherapy for head and neck cancer. Sci Transl Med. 2015;7(305):305ra147.
37. Formenti SC, Demaria S. Combining radiotherapy and cancer immunotherapy: a paradigm shift. J Natl Cancer Inst. 2013;105(4):256–65.

38. Demaria S, Ng B, Devitt ML, et al. Ionizing radiation inhibition of distant untreated tumours (abscopal effect) is immune mediated. Int J Radiat Oncol Biol Phys. 2004;58(3):862–70.
39. Dewan MZ, Galloway AE, Kawashima N, et al. Fractionated but no single dose radiotherapy induces an immune-mediated abscopal effect when combined with anti-CTLA-4 antibody. Clin Cancer Res. 2009;15(17):5379–88.
40. Prise KM, O'Sullivan JM. Radiation-induced bystander signalling in cancer therapy. Nat Rev Cancer. 2009;9(5):351–60.
41. Levental KR, Yu H, Kass L, et al. Matrix crosslinking forces tumor progression by enhancing integrin signalling. Cell. 2009;139(5):891–906.
42. La Thangue NB, Kerr DJ. Predictive biomarkers: a paradigm shift towards personalized cancer medicine. Nat Rev Clin Oncol. 2011;8(10):587–96.
43. Yard B, Chie EK, Adams DJ, et al. Radiotherapy in the era of precision medicine. Semin Radiat Oncol. 2015;25(4):227–36.
44. Scott JG, Berglund A, Schell MJ, et al. A genome-based model for adjusting radiotherapy dose (GARD): a retrospective, cohort-based study. Lancet Oncol. 2017;18(2):202–11.
45. Jaffray DA, Das S, Jacobs PM, Jeraj R, Lambin P. How advances in imaging will affect precision radiation oncology. Int J Radiat Oncol Biol Phys. 2018;101(2):292–8.
46. Huynh E, Hosny A, Guthier C, Bitterman DS, Petit SF, Haas-Kogan DA, et al. Artificial intelligence in radiation oncology. Nat Rev Clin Oncol. 2020;17(12):771–81.
47. Allison RR. Radiobiological modifiers in clinical radiation oncology: current reality and future potential. Future Oncol. 2014;10(15):2359–79.
48. Lawrence YR, Werner-Wasik M, Dicker AP. Biologically conformal treatment: biomarkers and functional imaging in radiation oncology. Future Oncol. 2008;4(5):689–704.
49. Otazo R, Lambin P, Pignol JP, Ladd ME, Schlemmer HP, Baumann M, et al. MRI-guided radiation therapy: an emerging paradigm in adaptive radiation oncology. Radiology. 2021;298(2):248–60.
50. Yakar M, Etiz D. Artificial intelligence in radiation oncology. Artif Intell Med Imaging. 2021;2(2):13–31.
51. García-Figueiras R, Baleato-González S, Luna A, Padhani AR, Vilanova JC, Carballo-Castro AM, et al. How imaging advances are defining the future of precision radiation therapy. Radiographics. 2024;44(2):e230152.
52. Turabi K, Klute K, Radhakrishnan P. Decoding the dynamics of circulating tumor DNA in liquid biopsies. Cancers (Basel). 2024;16(13):2432.
53. Derer A, Deloch L, Rubner Y, Fietkau R, Frey B, Gaipl US. Radio-immunotherapy-induced immunogenic cancer cells as basis for induction of systemic anti-tumor immune responses—pre-clinical evidence and ongoing clinical applications. Front Immunol. 2015;6:505.
54. Nath SK, Simpson DR, Rose BS, Sandhu AP. Recent advances in image-guided radiotherapy for head and neck carcinoma. J Oncol. 2009;2009:1–10.
55. Cobanaj M, Corti C, Dee EC, McCullum L, Boldrini L, Schlam I, et al. Advancing equitable and personalized cancer care: novel applications and priorities of artificial intelligence for fairness and inclusivity in the patient care workflow. Eur J Cancer. 2024;198:113504.
56. Jeong C, Goh Y, Kwak J. Challenges and opportunities to integrate artificial intelligence in radiation oncology: a narrative review. Ewha Med J. 2024;47(4)
57. Perez CA, Mutic S. Advances and future of radiation oncology. Rep Pract Oncol Radiother. 2013;18(6):329–32.
58. Bitterman DS, Gensheimer MF, Jaffray D, Pryma DA, Jiang SB, Morin O, et al. Cancer informatics for cancer centers: sharing ideas on how to build an artificial intelligence–ready informatics ecosystem for radiation oncology. JCO Clin Cancer Inform. 2023:7.
59. Maniaci A, Lavalle S, Gagliano C, Lentini M, Masiello E, Parisi F, et al. The integration of radiomics and artificial intelligence in modern medicine. Life. 2024;14(10):1248.
60. Arimura H, Soufi M, Kamezawa H, Ninomiya K, Yamada M. Radiomics with artificial intelligence for precision medicine in radiation therapy. J Radiat Res. 2019;60(1):150–7.

61. Chen M, Copley SJ, Viola P, Lu H, Aboagye EO. Radiomics and artificial intelligence for precision medicine in lung cancer treatment. Semin Cancer Biol. 2023;93:97–113.
62. Mireştean CC, Iancu RI, Iancu DPT. Radiomics breakthrough could spark the head and neck cancer radiotherapy revolution. J Med Radiat Oncol. 2023;3(1):11–9.
63. Goel A. From manual to AI - based tumor contouring: advancements in radiation therapy planning. Int J Sci Res (IJSR). 2023;12(7):697–703.
64. Kiser KJ, Fuller CD, Reed VK. Artificial intelligence in radiation oncology treatment planning: a brief overview. J Med Artif Intell. 2019;2:9–9.
65. Kawamura M, Kamomae T, Yanagawa M, Kamagata K, Fujita S, Ueda D, et al. Revolutionizing radiation therapy: the role of AI in clinical practice. J Radiat Res. 2024;65(1):1–9.
66. Bourbonne V, Lucia F, Dissaux G, Bert J, Visvikis D, Pradier O, et al. Pulmonary and esophageal toxicity in lung cancer treated by (chemo)-radiotherapy: a radiomics-based prediction model. Int J Radiat Oncol Biol Phys. 2020;108(3):S31.
67. Ramasamy G, Muanza T. Radiomics as biomarkers for the treatment of non-small cell lung cancer with stereotactic body radiation therapy: a review of concepts. Cureus. 2024;
68. Chaddad A, Kucharczyk MJ, Daniel P, Sabri S, Jean-Claude BJ, Niazi T, et al. Radiomics in glioblastoma: current status and challenges facing clinical implementation. Front Oncol. 2019;9:374.
69. Rane N, Choudhary S, Rane J. Explainable artificial intelligence (XAI) in healthcare: interpretable models for clinical decision support. SSRN Electron J. 2023;
70. Lahmi L, Mamzer MF, Burgun A, Durdux C, Bibault JE. Ethical aspects of artificial intelligence in radiation oncology. Semin Radiat Oncol. 2022;32(4):442–8.
71. Saxena S, Jena B, Gupta N, Das S, Sarmah D, Bhattacharya P, et al. Role of artificial intelligence in radiogenomics for cancers in the era of precision medicine. Cancers (Basel). 2022;14(12):2860.
72. Truhn D, Tayebi Arasteh S, Saldanha OL, Müller-Franzes G, Khader F, Quirke P, et al. Encrypted federated learning for secure decentralized collaboration in cancer image analysis. Med Image Anal. 2024;92:103059.
73. Wilson VC, McDonough J, Tochner Z. Proton beam irradiation in pediatric oncology. J Pediatr Hematol Oncol. 2005;27(8):444–8.
74. Kahalley LS, Ris MD, Grosshans DR, Okcu MF, Paulino AC, Chintagumpala M, et al. Comparing intelligence quotient change after treatment with proton versus photon radiation therapy for pediatric brain tumors. J Clin Oncol. 2016;34(10):1043–9.
75. Thomas H, Timmermann B. Paediatric proton therapy. Br J Radiol. 2020;93:1107.
76. Fenwick JD, Mayhew C, Jolly S, Amos RA, Hawkins MA. Navigating the straits: realizing the potential of proton FLASH through physics advances and further pre-clinical characterization. Front Oncol. 2024;14:1420337.
77. Pham TT, Whelan B, Oborn BM, Delaney GP, Vinod S, Brighi C, et al. Magnetic resonance imaging (MRI) guided proton therapy: a review of the clinical challenges, potential benefits and pathway to implementation. Radiother Oncol. 2022;170:37–47.
78. Lee HJ Jr, Zeng J, Rengan R. Proton beam therapy and immunotherapy: an emerging partnership for immune activation in non-small cell lung cancer. Transl Lung Cancer Res. 2018;7(2):180–8.
79. Bockel S, Durand B, Deutsch E. Combining radiation therapy and cancer immune therapies: from preclinical findings to clinical applications. Cancer/Radiothérapie. 2018;22(6–7):567–80.
80. Badiyan SN, Roach MC, Chuong MD, Rice SR, Onyeuku NE, Remick J, et al. Combining immunotherapy with radiation therapy in thoracic oncology. J Thorac Dis. 2018;10(S21):S2492–507.
81. Zhang M, Zou W, Teo BKK. Image guidance in proton therapy for lung cancer. Transl Lung Cancer Res. 2018;7(2):160–70.
82. De Ruysscher D, Sterpin E, Haustermans K, Depuydt T. Tumour movement in proton therapy: solutions and remaining questions: a review. Cancers (Basel). 2015;7(3):1143–53.

83. Lühr A, von Neubeck C, Pawelke J, Seidlitz A, Peitzsch C, Bentzen SM, et al. "Radiobiology of proton therapy": results of an international expert workshop. Radiother Oncol. 2018;128(1):56–67.
84. Jumaniyazova E, Smyk D, Vishnyakova P, Fatkhudinov T, Gordon K. Photon- and proton-mediated biological effects: what has been learned? Life. 2022;13(1):30.
85. Schippers JM, Lomax A, Garonna A, Parodi K. Can technological improvements reduce the cost of proton radiation therapy? Semin Radiat Oncol. 2018;28(2):150–9.
86. Yan S, Ngoma TA, Ngwa W, Bortfeld TR. Global democratisation of proton radiotherapy. Lancet Oncol. 2023;24(6):e245–54.
87. Al Hagawi WSA, Jubran EAH, Almohammedsaleh FYA, Ayed AS, ANA H, Alnasser AM, et al. Integrating artificial intelligence in radiotherapy: challenges and opportunities in clinical workflows. J Ecohumanism. 2024;3(8)
88. Ger RB, Wei L, El NI, Wang J. The promise and future of radiomics for personalized radiotherapy dosing and adaptation. Semin Radiat Oncol. 2023;33(3):252–61.
89. Mohan R, Bortfeld T. Proton therapy: clinical gains through current and future treatment programs. Front Radiat Ther Oncol. 2011;43:440–64.
90. Grover J, Liu P, Dong B, Shan S, Whelan B, Keall P, et al. Super-resolution neural networks improve the spatiotemporal resolution of adaptive MRI-guided radiation therapy. Commun Med. 2024;4(1):1–12.
91. Brisbane WG, Priester AM, Nguyen AV, Topoozian M, Mota S, Delfin MK, et al. Focal therapy of prostate cancer: use of artificial intelligence to define tumour volume and predict treatment outcomes. BJUI Compass. 2024;6(1):e456.
92. Pietsch J, Khamfongkhruea C, Berthold J, Janssens G, Stützer K, Löck S, et al. Automatic detection and classification of treatment deviations in proton therapy from realistically simulated prompt gamma imaging data. Med Phys. 2023;50(1):506–17.
93. Chamseddine I, Kim Y, De B, Naqa IE, Duda DG, Wolfgang JA, Pursley J, Wo JY, Hong TS, Paganetti H, Koay EJGC. Predictive model of liver toxicity to aid the personalized selection of proton versus photon therapy in hepatocellular carcinoma. Int J Radiat Oncol Biol Phys. 2023;5(116):1234–43.
94. Jiao Z, Li H, Xiao Y, Dorsey J, Simone CB, Feigenberg S, et al. Integration of deep learning radiomics and counts of circulating tumor cells improves prediction of outcomes of early stage NSCLC patients treated with stereotactic body radiation therapy. Int J Radiat Oncol Biol Phys. 2022;112(4):1045–54.
95. Lin SH, Hobbs BP, Verma V, Tidwell RS, Smith GL, Lei X, et al. Randomized phase IIB trial of proton beam therapy versus intensity-modulated radiation therapy for locally advanced esophageal cancer. J Clin Oncol. 2020;38(14):1569–78.
96. Hurkmans C, Bibault JE, Brock KK, van Elmpt W, Feng M, David Fuller C, et al. A joint ESTRO and AAPM guideline for development, clinical validation and reporting of artificial intelligence models in radiation therapy. Radiother Oncol. 2024;197(May):110345.
97. Ono T, Iramina H, Hirashima H, Adachi T, Nakamura M, Mizowaki T. Applications of artificial intelligence for machine- and patient-specific quality assurance in radiation therapy: current status and future directions. J Radiat Res. 2024;65(4):421–32.
98. Maniaci A, Lavalle S, Gagliano C, Lentini M, Masiello E, Parisi F, et al. The integration of radiomics and artificial intelligence in modern medicine. Life. 2024;14(10):1–14.
99. Chang CW, Goette M, Kadom N, Wang Y, Wynne J, Wang T, et al. Early in vivo radiation damage quantification for pediatric craniospinal irradiation using longitudinal MRI for intensity modulated proton therapy. Adv Radiat Oncol. 2023;8(5):101267.
100. Kim BM, Hong Y, Lee S, Liu P, Lim JH, Lee YH, Lee TH, Chang KT, Hong Y. Therapeutic implications for overcoming radiation resistance in cancer therapy. Int J Mol Sci. 2015;16(11):26880–913.
101. Begg AC, Stewart FA, Vens C. Strategies to improve radiotherapy with targeted drugs. Nat Rev Cancer. 2011;11(4):239–53.

102. Arnold KM, Flynn NJ, Raben A, Romak L, Yu Y, Dicker AP, Mourtada F, Sims-Mourtada J. The impact of radiation on the tumor microenvironment: effect of dose and fractionation schedules. Cancer Growth Metastasis. 2018;11:1179064418761639.
103. Mahajan A, Ahmed S, McAleer MF, Weinberg JS, Li J, Brown P, Settle S, Prabhu SS, Lang FF, Levine N, McGovern S. Post-operative stereotactic radiosurgery versus observation for completely resected brain metastases: a single-centre, randomised, controlled, phase 3 trial. Lancet Oncol. 2017;18(8):1040–8.
104. Aridgides P, Bogart J. Stereotactic body radiation therapy for stage I non–small cell lung cancer. Thorac Surg Clin. 2016;26(3):261–9.
105. Qiu H, Moravan MJ, Milano MT, Usuki KY, Katz AW. SBRT for hepatocellular carcinoma: 8-year experience from a regional transplant center. J Gastrointest Cancer. 2018;49:463–9.
106. Kishan AU, King CR. Stereotactic body radiotherapy for low-and intermediate-risk prostate cancer. Semin Radiat Oncol. 2017;27(3):268–78. WB Saunders
107. Baliga S, Kabarriti R, Ohri N, Haynes-Lewis H, Yaparpalvi R, Kalnicki S, Garg MK. Stereotactic body radiotherapy for recurrent head and neck cancer: a critical review. Head Neck. 2017;39(3):595–601.
108. Kim MS, Kim W, Park IH, Kim HJ, Lee E, Jung JH, Cho LC, Song CW. Radiobiological mechanisms of stereotactic body radiation therapy and stereotactic radiation surgery. Radiat Oncol J. 2015;33(4):265.
109. Higgins GS, O'Cathail SM, Muschel RJ, McKenna WG. Drug radiotherapy combinations: review of previous failures and reasons for future optimism. Cancer Treat Rev. 2015;41(2):105–13.
110. Fujiwara K, Iwado E, Mills GB, Sawaya R, Kondo S, Kondo Y. Akt inhibitor shows anticancer and radiosensitizing effects in malignant glioma cells by inducing autophagy. Int J Oncol. 2007;31(4):753–60.
111. Albert JM, Kim KW, Cao C, Lu B. Targeting the Akt/mammalian target of rapamycin pathway for radiosensitization of breast cancer. Mol Cancer Ther. 2006;5(5):1183–9.
112. Chautard E, Loubeau G, Tchirkov A, Chassagne J, Vermot-Desroches C, Morel L, Verrelle P. Akt signaling pathway: a target for radiosensitizing human malignant glioma. Neuro-Oncology. 2010;12(5):434–43.
113. Liu WL, Gao M, Tzen KY, Tsai CL, Hsu FM, Cheng AL, Cheng JC. Targeting phosphatidylinositide3-Kinase/Akt pathway by BKM120 for radiosensitization in hepatocellular carcinoma. Oncotarget. 2014;5(11):3662.
114. Ekshyyan O, Rong Y, Rong X, Pattani KM, Abreo F, Caldito G, Chang JK, Ampil F, Glass J, Nathan CA. Comparison of radiosensitizing effects of the mammalian target of rapamycin inhibitor CCI-779 to cisplatin in experimental models of head and neck squamous cell carcinoma. Mol Cancer Ther. 2009;8(8):2255–65.
115. Kim KW, Myers CJ, Jung DK, Lu B. NVP-BEZ-235 enhances radiosensitization via blockade of the PI3K/mTOR pathway in cisplatin-resistant non-small cell lung carcinoma. Genes Cancer. 2014;5(7–8):293.
116. Ma Z, Yao G, Zhou B, Fan Y, Gao S, Feng X. The Chk1 inhibitor AZD7762 sensitises p53 mutant breast cancer cells to radiation in vitro and in vivo. Mol Med Rep. 2012;6(4):897–903.
117. Borst GR, McLaughlin M, Kyula JN, Neijenhuis S, Khan A, Good J, Zaidi S, Powell NG, Meier P, Collins I, Garrett MD. Targeted radiosensitization by the Chk1 inhibitor SAR-020106. Int J Radiat Oncol Biol Phys. 2013;85(4):1110–8.
118. Mladenov E, Magin S, Soni A, Iliakis G. DNA double-strand break repair as determinant of cellular radiosensitivity to killing and target in radiation therapy. Front Oncol. 2013;3:113.
119. Chendil D, Ranga RS, Meigooni D, Sathishkumar S, Ahmed MM. Curcumin confers radiosensitizing effect in prostate cancer cell line PC-3. Oncogene. 2004;23(8):1599–607.
120. Garg AK, Buchholz TA, Aggarwal BB. Chemosensitization and radiosensitization of tumors by plant polyphenols. Antioxid Redox Signal. 2005;7(11–12):1630–47.

121. Rashid A, Liu C, Sanli T, Tsiani E, Singh G, Bristow RG, Dayes I, Lukka H, Wright J, Tsakiridis T. Resveratrol enhances prostate cancer cell response to ionizing radiation. Modulation of the AMPK, Akt and mTOR pathways. Radiat Oncol. 2011;6:1–2.
122. Kma L. Synergistic effect of resveratrol and radiotherapy in control of cancers. Asian Pac J Cancer Prev. 2013;14(11):6197–208.
123. Scarlatti F, Sala G, Ricci C, Maioli C, Milani F, Minella M, Botturi M, Ghidoni R. Resveratrol sensitization of DU145 prostate cancer cells to ionizing radiation is associated to ceramide increase. Cancer Lett. 2007;253(1):124–30.
124. Akimoto T, Nonaka T, Ishikawa H, Sakurai H, Saitoh JI, Takahashi T, Mitsuhashi N. Genistein, a tyrosine kinase inhibitor, enhanced radiosensitivity in human esophageal cancer cell lines in vitro: possible involvement of inhibition of survival signal transduction pathways. Int J Radiat Oncol Biol Phys. 2001;50(1):195–201.
125. Lin C, Yu Y, Zhao HG, Yang A, Yan H, Cui Y. Combination of quercetin with radiotherapy enhances tumor radiosensitivity in vitro and in vivo. Radiother Oncol. 2012;104(3):395–400.
126. Meng Y, Beckett MA, Liang H, Mauceri HJ, van Rooijen N, Cohen KS, Weichselbaum RR. Blockade of tumor necrosis factor α signaling in tumor-associated macrophages as a radiosensitizing strategy. Cancer Res. 2010;70(4):1534–43.
127. Formenti SC, Demaria S. Combining radiotherapy and cancer immunotherapy: a paradigm shift. JNCI J Natl Cancer Inst. 2013;105(4):256–65.
128. Harada H. How can we overcome tumor hypoxia in radiation therapy? J Radiat Res. 2011;52(5):545–56.
129. Overgaard J. Hypoxic radiosensitization: adored and ignored. J Clin Oncol. 2007;25(26):4066–74.
130. Meijer TW, Kaanders JH, Span PN, Bussink J. Targeting hypoxia, HIF-1, and tumor glucose metabolism to improve radiotherapy efficacy. Clin Cancer Res. 2012;18(20):5585–94.
131. Yeo EJ, Chun YS, Cho YS, Kim J, Lee JC, Kim MS, Park JW. YC-1: a potential anticancer drug targeting hypoxia-inducible factor 1. J Natl Cancer Inst. 2003;95(7):516–25.
132. Shin DH, Kim JH, Jung YJ, Kim KE, Jeong JM, Chun YS, Park JW. Preclinical evaluation of YC-1, a HIF inhibitor, for the prevention of tumor spreading. Cancer Lett. 2007;255(1):107–16.
133. Kim WY, Oh SH, Woo JK, Hong WK, Lee HY. Targeting heat shock protein 90 overrides the resistance of lung cancer cells by blocking radiation-induced stabilization of hypoxia-inducible factor-1α. Cancer Res. 2009;69(4):1624–32.
134. Kleibeuker EA, Griffioen AW, Verheul HM, Slotman BJ, Thijssen VL. Combining angiogenesis inhibition and radiotherapy: a double-edged sword. Drug Resist Updat. 2012;15(3):173–82.
135. Dings RP, Loren M, Heun H, McNiel E, Griffioen AW, Mayo KH, Griffin RJ. Scheduling of radiation with angiogenesis inhibitors anginex and Avastin improves therapeutic outcome via vessel normalization. Clin Cancer Res. 2007;13(11):3395–402.
136. Laskar SG, Sinha S, Krishnatry R, Grau C, Mehta M, Agarwal JP. Access to radiation therapy: from local to global and equality to equity. JCO Glob Oncol. 2022;(8):e2100358.
137. Zubizarreta EH, Fidarova E, Healy B, Rosenblatt E. Need for radiotherapy in low and middle income countries–the silent crisis continues. Clin Oncol. 2015;27(2):107–14.
138. Yap ML, Zubizarreta E, Bray F, Ferlay J, Barton M. Global access to radiotherapy services: have we made progress during the past decade? J Glob Oncol. 2016;2(4):207–15.
139. Elmore SN, Polo A, Bourque JM, Pynda Y, van der Merwe D, Grover S, Hopkins K, Zubizarreta E, Abdel-Wahab M. Radiotherapy resources in Africa: an International Atomic Energy Agency update and analysis of projected needs. Lancet Oncol. 2021;22(9):e391–9.
140. Knaul FM, editor. Closing the cancer divide: an equity imperative. Harvard University Press; 2012.
141. Sullivan R, Peppercorn J, Sikora K, Zalcberg J, Meropol NJ, Amir E, Khayat D, Boyle P, Autier P, Tannock IF, Fojo T. Delivering affordable cancer care in high-income countries. Lancet Oncol. 2011;12(10):933–80.

142. Chopra SJ, Mathew A, Maheshwari A, Bhatla N, Singh S, Rai B, Surappa ST, Ghosh J, Sharma D, Bhaumik J, Biswas M. National cancer grid of India consensus guidelines on the management of cervical cancer. J Glob Oncol. 2018;4:1–5.
143. Pramesh CS, Badwe RA, Sinha RK. The national cancer grid of India. Indian J Med Paediatr Oncol. 2014;35(03):226–7.
144. Lin LL, Msadabwe SC, Chiao E. Improving access to radiation therapy globally to meet World Health Organization goals to eliminate cervical cancer. Int J Radiat Oncol Biol Phys. 2023;116(2):459–62.
145. Colwell J. $100 M Africa initiative targets childhood cancer. Cancer Discov. 2017;7(5):447–8.

Index

F. Taghizadeh-Hesary et al. (eds.), *From Radiobiology to Radiation Oncology*, https://doi.org/10.1007/978-3-031-95724-6

The manufacturer's authorised representative in the EU is Springer Nature Customer Service Centre GmbH, Europaplatz 3, 69115 Heidelberg, Germany. If you have any concerns regarding our products, please contact ProductSafety@springernature.com

Printed and bound by CPI Group (UK) Ltd, Croydon, CR0 4YY
21/09/2026
02220015-0013